# ADVANCED
# MECHANICS OF MATERIALS

**Madhukar Vable**

Michigan Technological University

EXPANDING EDUCATIONAL HORIZONS, LLC

Houghton, Michigan
madhuvable.org
2015

Paper copy: ISBN: 978-0-9912446-4-5

Print edition: November 19, 2015

To my guru
Professor David L. Sikarskie
1937-1999

# CONTENTS


| | | | |
|---|---|---|---|
| PREFACE | | | X |
| ACKNOWLEDGMENTS | | | X |

**CHAPTER 1 STRESS AND STRAIN 1**

| | | | |
|---|---|---|---|
| 1.1 | | Stress on a surface | 1 |
| 1.2 | | Stress at a point | 2 |
| | 1.2.1 | Sign convention for stress | 3 |
| | 1.2.2 | Stress elements | 4 |
| | 1.2.3 | Construction of a stress element | 4 |
| | 1.2.4 | Plane stress | 5 |
| 1.3 | | Stress transformation in two dimensions | 5 |
| 1.3.1 | | Matrix method in two dimensions | 6 |
| 1.3.2 | | Principal stresses | 7 |
| 1.4.4 | | Maximum shear stress | 9 |
| 1.4 | | Stress transformation in three dimensions | 8 |
| | 1.4.1 | Principal stresses | 8 |
| | 1.4.2 | Principal stress convention | 9 |
| | 1.4.3 | Characteristic equation and stress invariants | 9 |
| | 1.4.4 | Octahedral stresses | 10 |
| | 1.4.5 | Deviatoric stress | 11 |
| 1.5 | | Average strains | 13 |
| | 1.5.1 | Units of average strains | 14 |
| 1.6 | | Strain at a point | 14 |
| | 1.6.1 | Plane strain | 15 |
| | 1.6.2 | Polar coordinates | 16 |
| 1.7 | | Strain transformation | 16 |
| 1.7.1 | | Principal strains | 17 |
| 1.8 | | Material model | 18 |
| | 1.8.1 | Linear material model | 18 |
| | 1.8.2 | Isotropic material | 19 |
| | 1.8.3 | Plane stress and plane strain | 20 |
| | 1.8.4 | Nonlinear material models | 23 |
| | 1.8.5 | Elastic–perfectly plastic material model | 23 |
| | 1.8.6 | Linear strain-hardening material model | 23 |
| | 1.8.7 | Power law material model | 24 |
| 1.9 | | Effects of temperature | 25 |
| 1.10 | | Failure theories | 26 |
| | 1.10.1 | Maximum shear stress theory | 27 |
| | 1.10.2 | Maximum octahedral shear stress theory | 27 |
| | 1.10.3 | Maximum normal stress theory | 28 |
| | 1.10.4 | Modified Mohr's theory | 28 |
| 1.11 | | Closure | 30 |
| 1.12 | | Synopsis of equations | 31 |
| | | Problems | 32 |

**CHAPTER 2 ONE-DIMENSIONAL STRUCTURAL MEMBERS 35**

| | | | |
|---|---|---|---|
| 2.1 | | Internal forces and moments | 35 |

| | | | |
|---|---|---|---|
| | 2.1.1 | Decoupling axial, bending, and torsion analysis | 36 |
| 2.2 | | Logic in mechanics of materials | 40 |
| 2.3 | | Classical theories of 1-D structural members | 40 |
| | 2.3.1 | Limitations | 41 |
| | 2.3.2 | Definition of variables and sign conventions | 41 |
| | 2.3.4 | Shear stress in thin symmetric beams | 48 |
| 2.4 | | Boundary value problems | 51 |
| | 2.4.1 | Axial displacement | 51 |
| | 2.4.2 | Torsional rotation | 52 |
| | 2.4.3 | Beam deflection | 52 |
| 2.5 | | Discontinuity functions | 56 |
| | 2.5.1 | Definitions | 56 |
| | 2.5.2 | Axial displacement | 58 |
| | 2.5.3 | Torsional rotation | 58 |
| | 2.5.4 | Beam deflection | 59 |
| 2.6 | | Closure | 63 |
| 2.7 | | Synopsis of equations | 64 |
| | | Problems | 65 |

| **CHAPTER 3** | | **INFLUENCE FUNCTIONS** | **71** |
|---|---|---|---|
| 3.1 | | Basic concepts | 71 |
| | 3.1.1 | Definitions | 71 |
| | 3.1.2 | Force (source) singularity influence functions in beams | 71 |
| | 3.1.3 | Moment (doublet) singularity influence functions in beams | 72 |
| | 3.1.4 | Numerical integration | 73 |
| | 3.1.5 | Non-dimensional variables | 73 |
| 3.2 | | Influence functions for classical beams | 74 |
| | 3.2.1 | Influence functions for simply supported beam | 74 |
| 3.3 | | Beams on elastic foundations | 80 |
| 3.4 | | Fundamental solutions | 80 |
| | 3.4.1 | Some properties of fundamental solutions associated with force singularity | 82 |
| 3.5 | | Finite beams | 86 |
| | 3.5.1 | Symmetric loading | 88 |
| | 3.5.2 | Asymmetric loading and boundary conditions | 88 |
| | 3.5.3 | General loading and boundary conditions | 88 |
| 3.6 | | Closure | 91 |
| 3.7 | | Synopsis of equations | 92 |
| 3.8 | | Mathematical details of the derivation of fundamental solutions | 93 |
| | | Problems | 95 |

| **CHAPTER 4** | | **STABILITY** | **97** |
|---|---|---|---|
| 4.1 | | Euler buckling | 97 |
| | 4.1.1 | Buckling modes | 98 |
| | 4.1.2 | Effects of end conditions | 99 |
| | 4.1.3 | Classification of columns | 99 |
| 4.2 | | Imperfect columns | 103 |
| 4.2 | | Imperfect columns | 103 |
| 4.3 | | Closure | 107 |
| 4.4 | | Synopsis of equations | 107 |
| | | Problems | 108 |

| | | | |
|---|---|---|---|
| **CHAPTER 5** | | **PLATES** | **111** |
| 5.1 | | Limitations on plate. | 111 |
| 5.2 | | Classical plate theory | 112 |
| | 5.2.1 | Kinematics | 112 |
| | 5.2.2 | Material model | 114 |
| | 5.2.3 | Static equivalency | 114 |
| | 5.2.4 | Location of neutral surface | 115 |
| | 5.2.5 | Stress formulas | 115 |
| | 5.2.6 | Equilibrium | 116 |
| | 5.2.7 | Differential equation | 116 |
| | 5.2.8 | Boundary conditions | 117 |
| | 5.2.9 | Complexity map | 118 |
| 5.3 | | Navier's solution for rectangular plates | 124 |
| 5.4 | | Nadai-Levy solution for rectangular plates | 128 |
| | 5.4.1 | Homogeneous solution | 129 |
| | 5.4.2 | Particular solution | 129 |
| 5.5 | | Circular plates | 132 |
| 5.6 | | Axisymmetric plates | 134 |
| | 5.6.1 | Uniform load | 135 |
| | 5.6.2 | Solid circular plates | 135 |
| | 5.6.3 | Annular plates | 136 |
| | 5.6.4 | Non-dimensional equations in annular plates | 136 |
| 5.7 | | Closure | 139 |
| 5.8 | | Synopsis of equations | 140 |
| | | Problems | 141 |

| | | | |
|---|---|---|---|
| **CHAPTER 6** | | **INTRODUCTION TO ELASTICITY** | **143** |
| 6.1 | | Elasticity equations | 143 |
| | 6.1.1 | Strain-displacement equations | 143 |
| | 6.1.2 | Compatibility equations | 143 |
| | 6.1.3 | Plane stress and plane strain | 144 |
| | 6.1.4 | Equilibrium equations | 144 |
| | 6.1.5 | Boundary conditions | 146 |
| 6.2 | | Axisymmetric problems | 146 |
| | 6.2.1 | Axisymmetric plane strain | 146 |
| | 6.2.2 | Axisymmetric plane stress | 147 |
| 6.3 | | Rotating disks | 147 |
| 6.4 | | Airy stress function | 148 |
| 6.5 | | Solution by polynomials | 149 |
| | 6.5.1 | Quadratic polynomials | 149 |
| | 6.5.2 | Cubic polynomials | 150 |
| | 6.5.3 | Fourth-order polynomials | 150 |
| 6.6 | | Displacements from strains in 2-D | 151 |
| | 6.6.1 | Rigid body motion | 152 |
| 6.7 | | Torsion of non-circular shafts | 153 |
| | 6.7.1 | Saint-Venant's method | 153 |
| | 6.7.2 | Prandtl's method | 154 |
| | 6.7.3 | Procedure for solving problems of torsion of non-circular shafts | 156 |
| 6.8 | | Membrane analogy | 158 |
| 6.8.1 | | Membrane analogy for cross section with holes | 159 |
| 6.9 | | Torsion of thin-walled open section | 160 |

| | | | |
|---|---|---|---|
| 6.9.1 | | End effects on rectangular cross sections | 161 |
| 6.10 | | Closure | 162 |
| 6.11 | | Synopsis of equations | 163 |
| | | Problems | 164 |

**CHAPTER 7 VARIATIONAL AND ENERGY METHODS 167**

| | | | |
|---|---|---|---|
| 7.1 | | Basic concepts in variational calculus | 167 |
| | 7.1.1 | Extremum and stationary Values | 168 |
| | 7.1.2 | Functionals | 169 |
| 7.2 | | Work | 169 |
| 7.3 | | Strain energy | 170 |
| | 7.3.1 | Strain energy in symmetric bending of beams | 171 |
| | 7.3.2 | Strain energy in bending of thin plates | 171 |
| | 7.3.3 | Strain energy in plane stress elasticity | 172 |
| | 7.3.4 | Strain energy in form of bilinear functional | 173 |
| 7.4 | | Virtual work | 173 |
| 7.5 | | Minimum potential energy | 173 |
| 7.6 | | Mathematical preliminAries | 174 |
| 7.7 | | Stationary value of a definite line integral | 175 |
| | 7.7.1 | Stationary value of a functional with first order derivatives | 176 |
| | 7.7.2 | Boundary value problem | 176 |
| | 7.7.3 | Stationary value of a functional with second order derivatives | 177 |
| | 7.7.4 | Generalization | 178 |
| 7.8 | | Stationary value of a definite area integral | 180 |
| | 7.8.1 | Stationary value of a functional with first order derivatives | 180 |
| 7.9 | | Functionals with second order derivatives | 182 |
| | 7.9.1 | Boundary value problem for rectangular geometries | 182 |
| | 7.9.2 | Boundary value problem for geometries with curvilinear boundaries | 185 |
| 7.10 | | Stationary value of a definite volume integral | 189 |
| 7.11 | | Nonlinearities | 189 |
| | 7.11.2 | Material nonlinearity | 191 |
| | 7.11.1 | Geometric nonlinearity | 189 |
| 7.12 | | Rayleigh-Ritz Method | 193 |
| 7.13 | | Finite element method | 200 |
| | 7.13.1 | Lagrange polynomials in one dimension | 201 |
| | 7.13.2 | Natural coordinates | 202 |
| | 7.13.3 | Vector arithmetic | 203 |
| | 7.13.4 | Lagrange polynomials in two dimensions | 203 |
| | 7.13.5 | Element stiffness matrix and right hand side vector | 205 |
| | 7.13.6 | Lagrange polynomials in three dimensions | 205 |
| 7.14 | | Closure | 209 |
| 7.15 | | Synopsis of equations | 210 |
| | | Problems | 212 |

**CHAPTER 8 INDICIAL NOTATION 215**

| | | | |
|---|---|---|---|
| 8.1 | | Basic definitions | 215 |
| | 8.1.1 | Summation convention | 215 |
| | 8.1.2 | Kronecker $\delta$ function | 216 |
| | 8.1.3 | Permutation function | 216 |
| | 8.1.4 | Derivative notation | 217 |

| | | | |
|---|---|---|---|
| 8.2 | | Equations of elasticity and thin plate in indicial notation | 217 |
| 8.4 | | Variational calculus | 220 |
| | 8.4.1 | Stationary value of a functional with first order derivatives | 220 |
| | 8.4.2 | Stationary value of a functional with second order derivatives | 221 |
| 8.5 | | Nonlinear strains | 222 |
| | 8.5.1 | A perspective of reference frame | 222 |
| | 8.5.2 | Lagrangian strain in two dimension | 223 |
| | 8.5.3 | Eulerian strain in two dimension | 224 |
| | 8.5.4 | Indicial notation and non-linear strain in three dimension | 224 |
| 8.6 | | Closure | 227 |
| 8.7 | | Synopsis of equations | 227 |
| | | Problems | 228 |

**APPENDIX A** **BIBLIOGRAPHY** **229**

**APPENDIX B** **ANSWERS TO SELECTED PROBLEMS** **230**

**INDEX** **233**

# PREFACE

In 2010, I taught *Advanced Mechanics of Materials* for the first time. The books available for teaching the course were similar to the ones in use when I was a graduate student more than forty years ago. The question I needed to answer was: What are the needs of the students today that should be addressed by a graduate level course in mechanics of materials? This book is my answer as elaborated next.

The early developments of structural theories were very intuitive. These developments were driven by a desire to get an analytical solution for use in design of structural members. In the past forty years we have pushed the design envelope to create more efficient structures. This efficiency has been obtained by modifying our classical structural theories with complexities such as anisotropy, inelastic strains, material non-homogeneity, material non-linearity, geometric non-linearity, shear in beams and plates, etc. Though getting an analytical solution is still a desirable goal, it is no longer a limitation it was in the past. Now we can find numerical solutions after incorporating complexities of interest into our theory. However, this need-driven incorporation of complexities has created a very large and baroque knowledge base that is becoming a challenge to master for most students of mechanics. *Synthesizing this tremendous growth of knowledge into a logical framework is the primary objective of this book.* The logic shown symbolically on the cover of the book provides the framework of adding complexities. Examples and post-text problems highlight the modularity of the logic and demonstrate the addition of complexities to the classical theories of axial members, torsion of circular shafts, and bending of beams and plates. *Complexity Maps* of Sections 2.3.3 and 5.2.9 lists complexities and associated examples and problems in the chapters of one dimensional structural members and bending of plates. Understanding the methodology of incorporating complexity into the classical theories opens the door for incredible number of possibilities that arise from combinations of complexities.

The other topics which make my book more contemporary are described below in order of greatest differences from existing books. The learning objectives are described at the beginning of each chapter. This book has more material then can be covered in a three credit graduate course. Instructor's choice dictates which topics to include in their course.

Newtonian mechanics and differential calculus have been the main stay of development of structural theories. Strands of Eulerian perspective and variational calculus have always been present in structural mechanics in form of energy methods. But energy methods are a small subset of variational calculus. Variational calculus is a very elegant and general way of obtaining boundary value problems by minimizing the potential energy of a structural member. Potential energy expressions are easy to obtain even with aforementioned complexities, particularly with large strains. Thus, students learn an alternative way of obtaining boundary value problems that may be difficult to obtain with differential calculus. An added bonus is that the principles of variational calculus are the bedrocks of optimization methods and approximate methods. Approximate methods of Rayleigh-Ritz and finite element method are discussed as another demonstration of variational calculus in Chapter 7.

Indicial notation is a compact way of deriving equations and formulas. It uses are demonstrated for plate bending, elasticity, and variational calculus. The combination of indicial notation with variational calculus reduce derivations of boundary value problems to less than a page when the same derivation in expanded notation by differential calculus may take many pages. Indicial notation is discussed in Chapter 8.

All *Advanced Mechanics of Materials* books have a chapter on beams on elastic foundations. In Chapter 3, influence functions are used to formulate and solve problems of classical beam theory, infinite beam on elastic foundations and finite beams on elastic foundations. When loading is simple then analytical solutions are obtained. When loading is complex then numerical integration is used for solving the problem. Numerical integration scheme requires use of spreadsheets. All students are familiar with spreadsheets and would have no difficulty in following the solution procedure explained in the text.

Chapters 1, 4, and 6 are relatively traditional. Chapter 1 discusses stress and strain and material models—it serves both as a review and to introduce concepts used in subsequent chapters. Chapter 4 discusses Euler buckling theory of column buckling with various complexities. Chapter 6 introduces elasticity and Airy stress function and their usage to obtain solutions to mechanics of materials problem.

All advanced mechanics of materials books contain lots of equations. To help students, a synopsis of equations is presented at the end of each chapter before the posttext problems.

On my website madhuvable.org there are several brief notebooks reviewing topics needed in this book. Topics such as statics, mechanics of materials, matrix methods, and numerical methods—all an engineering student has seen in an undergraduate curriculum. Although consistent in its design and notation with my introductory and intermediate mechanics of materials books, this book does not depend upon any book used in the undergraduate curriculum.

I welcome any comments, suggestions, concerns, or corrections that will help me improve the book. You can reach me through my website madhuvable.org.

# ACKNOWLEDGMENTS

Professor Ibrahim Miskioglu reviewed the entire book and made many valuable suggestions for which I am very grateful. My thanks also to Professor William Bulliet for reviewing Chapter 3 and his suggestions are also gratefully acknowledged.

# 1 Stress and Strain

## LEARNING OBJECTIVES

1. To understand the concepts of stress and strain.
2. To understand stress and strain transformations in three dimensions.
3. To understand the material models relating stresses and strains.

A building collapses because it was not strong enough to support the loads acting on it. The lid of a sauce bottle cannot be opened because the bond between the bottle and the lid is too strong. A crankshaft vibrates and causes noise because it is not stiff enough. A diver cannot jump high enough to execute her dives because the diving board is too stiff. Strength and stiffness are both important in the design of structural elements. Stress, a measure of the intensity of internal forces, and strain, a measure of the intensity of deformation, are two fundamental variables in the mechanics of materials that are used in assessing strength and stiffness.

In this chapter we start by briefly reviewing the concepts of stress and strain at a point. The notation of double subscripts on stresses and strain is described. Of particular importance is the use of the subscripts in determining the direction of a stress component on a surface.

Analysis of forces and deformation, hence of stresses and strains, is conducted in a coordinate system that is chosen to simplify analysis. But the maximum stresses and strains that may cause failure in a material may exist on planes and in directions that are different from the chosen coordinate system. Stress and strain transformation equations provide a means of moving from one coordinate system to another. In this chapter we start with the equations of stress and strain transformation in two dimensions that were studied in the introductory course on the mechanics of materials. We cast these familiar equations in matrix form and use the matrix method to derive familiar conclusions in two dimensions. We then generalize the matrix method approach to solve problems in stress and strain transformation in three dimensions.

Stress and strains are two definitions related by equations obtained experimentally. The experimentally determined equations depend upon the material model constructed to approximate the material behavior. In this chapter we briefly review linear and non-linear material models, the effect of temperature, and failure theories. We will use these material models in subsequent chapters to develop theories of structural members.

## 1.1 STRESS ON A SURFACE

The forces of attraction and repulsion between two particles (atoms or molecules) in a body are assumed to act along the line that joins the two particles. Forces that act along the line joining two particles are called **central forces**. The concept of central forces started with Newton's universal gravitation law. Central forces vary inversely as an exponent of the radial distance separating the two particles. At atomic levels the exponent is a power of 8 or 10.Thus, every particle exerts a force on every other particle, as shown symbolically on an imaginary surface of a body in Figure 1.1*a*. These forces between the particles hold the body together and are referred to as the internal forces. The body changes shape when we apply external forces on the body. The change in shape implies that the distance between the particles must change, which further implies that the forces between the particles (internal forces) must change. When the change in the internal forces exceeds some characteristic material value, the body will break. Thus, the strength of the material can be characterized by the measure of *change in the intensity* of internal forces. This measure of change in the intensity of internal forces is what we call stress.

We can replace all the forces that are exerted on any single particle in Figure 1.1*a* by the resultant of these forces on that particle as shown in Figure 1.1*b*. The magnitude and direction of these resultant forces will vary with the location of the particle (point), as shown in Figure 1.1*b*. In other words, when external forces are applied, an internal distributed force system is generated in the material. The internal distributed force on an imaginary cut surface of a body is called the **stress on a surface**. Furthermore, we note that force is a vector; hence the internal distributed forces (stress on a surface) can be resolved into normal

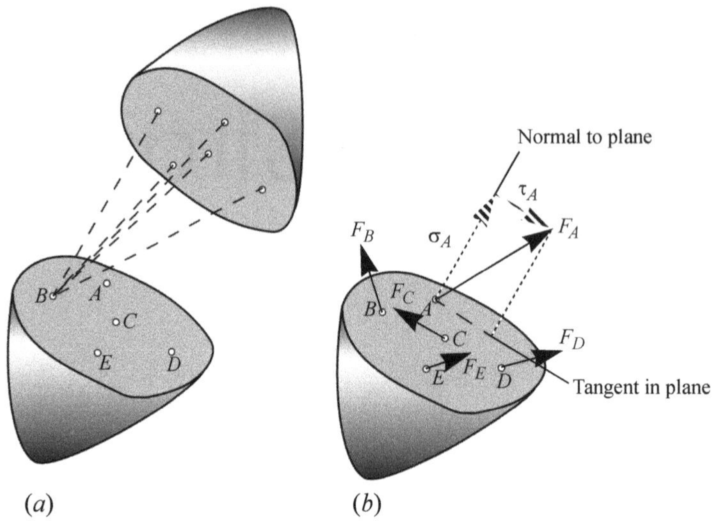


**Figure 1.1** Internal forces between particles on the two sides of an imaginary cut. (a) Forces between particles in a body shown on particle *A*. (b) The resultant force on each particle.

(perpendicular to the surface) and tangential (parallel to the surface) distributed forces as shown in Figure 1.1*b*. The internal distributed force that is normal to the surface of an imaginary cut is called the **normal stress** on a surface. The internal distributed force that is parallel to the surface of an imaginary cut surface is called the **shear stress** on the surface.

Table 1.1 shows the various units of stress used in this book. It should be noted that, *one psi* is equal to 6.95 kPa or *approximately 7 kPa*. Alternatively, 1 kPa is equal to 0.145 psi or *approximately 0.15 psi*.

**Table 1.1** Units of Stress

| Units | Description | Basic Units |
|---|---|---|
| psi | Pounds per square inch | $lb/in^2$ |
| ksi | Kilopounds (kips) per square inch | $10^3\ lb/in^2$ |
| Pa | Pascals | $N/m^2$ |
| kPa | Kilopascals | $10^3\ N/m^2$ |
| MPa | Megapascals | $10^6\ N/m^2$ |
| GPa | Gigapascals | $10^9\ N/m^2$ |

Normal stress on a surface may be viewed as the internal forces that are developed owing to the material resistance to the *pulling apart or pushing together of two adjoining planes* of an imaginary cut. Like pressure, normal stress is always perpendicular to the surface of the imaginary cut. But unlike pressure, which can only be compressive, normal stress can be tensile. A normal stress that pulls the surface away from a body is called a **tensile stress**. A normal stress that pushes the surface into a body is called a **compressive stress.**

In other words, tensile stress acts in the direction of the outward normal to the surface, while compressive stress is opposite to the direction of the outward normal to the surface. Normal stress is usually reported as tensile (T) or compressive (C), not as positive or negative. Thus, $\sigma = 100$ MPa (T) and $\sigma = 10$ ksi (C) are the conventional ways of reporting tensile and compressive normal stresses, respectively.

Shear stress on a surface may be viewed as the internal forces that are developed owing to the resistance of the material to the *sliding of two adjoining planes* along the imaginary cut. Like friction, shear stresses act tangentially to the plane in the direction opposite the impending motion of the surface. But unlike friction, shear stress is not related to the normal stress.

## 1.2 STRESS AT A POINT

The breaking of a structure starts at a point where the internal force intensity (i.e., stress) exceeds some material characteristic value. This implies that we need to refine our definition of "stress on a surface" to "stress at a point." But an infinite number of planes (surfaces) can pass through a point. Which imaginary surface do we shrink to zero? When we shrink the surface area to zero, which internal force component should we use? Both difficulties can be addressed by assigning directions to the orientation of the imaginary surface and to the direction of the internal force on this surface and then carrying the description of the directions as the subscripts of the stress components just as we carried *x*, *y*, and *z* as subscripts to describe the components of a vector.

Figure 1.2 shows a body cut by an imaginary plane that has an outward normal in the *i* direction. On this surface we have a differential area $\Delta A_i$ on which a resultant force[1] acts. The component of the force in the *j* direction is $\Delta F_j$. A component

1. If a resultant moment is included, then the stress is referred to as **couple stress**. Couple stress is important if stress analysis is conducted at such a small scale that the moment transmitted by the bonds between molecules must be incorporated. See Fung [1965] for additional details.

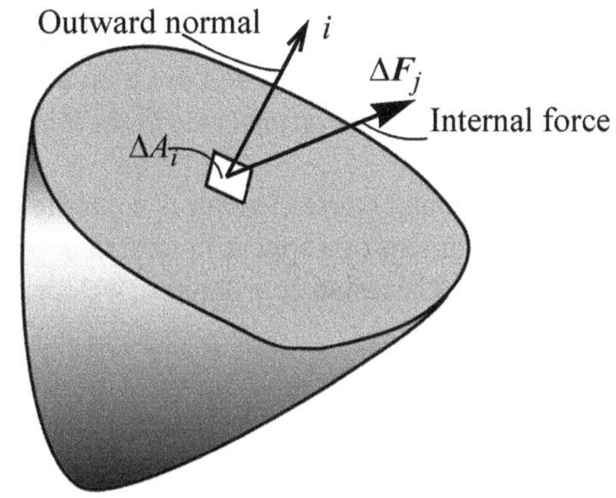


**Figure 1.2** Stress at a point.

of average stress is $(\Delta F_j / \Delta A_i)$. If we shrink $\Delta A_i$ to zero, we get the definition of a stress component at a point as shown by Equation (1.1).

$$\sigma_{ij} = \lim_{\Delta A_i \to 0} \left( \frac{\Delta F_j}{\Delta A_i} \right) \tag{1.1}$$

Direction of outward normal to the imaginary cut surface (first subscript $i$); Direction of the internal force component (second subscript $j$)

Now when we look at a stress component, the first subscript tells us the orientation of the imaginary surface and the second subscript tells us the direction of the internal force.

In three dimensions, each subscript *i* and *j* can refer to the *x*, *y*, or *z* direction. In other words, there are nine possible combinations of the two subscripts, as shown in the stress matrix in Figure 1.3. The diagonal elements in the stress matrix are normal stresses, and all off-diagonal elements in the stress matrix represent shear stresses.

$$[\sigma] = \begin{bmatrix} \sigma_{xx} & \tau_{xy} & \tau_{xz} \\ \tau_{yx} & \sigma_{yy} & \tau_{yz} \\ \tau_{zx} & \tau_{zy} & \sigma_{zz} \end{bmatrix}$$

**Figure 1.3** Stress matrix in three dimensions.

Note that to specify the stress at a point, we must have a magnitude and two directions. Table 1.2 shows the number of components needed to specify a scalar, a vector, and a stress. Now force, moment, velocity, and acceleration are all different quantities, but all are called vectors. In a similar manner, stress belongs to a category called *tensors* (See Synge and Schild [1978]) or more specifically, *stress is a second-order tensor*, where "second order" refers to the exponent in the last row. In this terminology, a vector is a tensor of order one, and a scalar is a tensor of order zero.

**Table 1.2** Comparison of Number of Components

| Quantity | Dimensions | | |
|---|---|---|---|
| | One | Two | Three |
| Scalar | $1 = 1^0$ | $1 = 2^0$ | $1 = 3^0$ |
| Vector | $1 = 1^1$ | $2 = 2^1$ | $3 = 3^1$ |
| Stress | $1 = 1^2$ | $4 = 2^2$ | $9 = 3^2$ |

In the introductory course on the mechanics of materials, it was explained that shear stress is symmetric, as shown by Equation (1.2).

$$\tau_{xy} = \tau_{yx} \qquad \tau_{yz} = \tau_{zy} \qquad \tau_{zx} = \tau_{xz} \tag{1.2}$$

The symmetry of shear stress given by Equations (1.2) implies that *stress at a point has nine stress components, but only six are independent*. Similarly, in two dimensions stress at a point has four nonzero components in general, but only three are independent.

### 1.2.1 Sign convention for stress

We will consider $\Delta A_i$ *to be positive if the outward normal to the surface is in the positive i direction.*

If the outward normal is in the negative *i* direction, then $\Delta A_i$ will be considered to be negative

With this convention in mind, we can immediately deduce the sign for stress. A stress component can be positive in two ways. Both the numerator and denominator are positive in Equation (1.1), or both the numerator and denominator are negative in Equation (1.1). A stress component is positive if the numerator and denominator have the same sign in Equation (1.1).

### 1.2.2 Stress elements

The discussion in the preceding section shows that stress at a point is an abstract quantity, and developing an intuitive feel for it is difficult. Stress on a surface, however, is easier to visualize as a distributed force on a surface. A **stress element** is an imaginary object that helps us visualize stress at a point by allowing us to *construct surfaces that have outward normal in the directions of the coordinates.*

It shall be seen in the discussion that follows that a stress element is a parallelepiped in Cartesian coordinates, a fragment of a cylinder in cylindrical coordinates, and a fragment of a sphere in spherical coordinates. We start our discussion in the next section with the construction of a stress element in Cartesian coordinates. We can use a similar process to draw stress elements in cylindrical and spherical coordinate systems.

### 1.2.3 Construction of a stress element

Consider the point at which we want to describe stress. Around this point, imagine an object that has sides whose outward normal are in directions of the coordinates. A parallelepiped has six surfaces with outward normal that are either in the positive or negative coordinate direction. Thus, we have accounted for the first subscript in our stress definition. We know that force is in the positive or negative direction of the second subscript. We use our sign convention to show the stress in the direction of the force on each of the six surfaces.

To demonstrate the construction just described, we will assume that all nine stress components shown in the stress matrix in Figure 1.3 are positive.

Let us consider the first row in the stress matrix in Figure 1.3. The first subscript gives us the direction of the outward normal, which is the $x$ direction. Surfaces $A$ and $B$ in Figure 1.4 have the outward normal in the $x$ direction, and it is on these surfaces that the stress component of the first row will be shown.

The outward normal on surface $A$ is in the positive $x$ direction [denominator is positive in Equation (1.1)]. For the stress component to be positive on surface $A$, the force must be in the positive direction [numerator must be positive], as shown in Figure 1.4.

The outward normal on surface $B$ is in the negative $x$ direction [denominator is negative in Equation (1.1)]. For the stress component to be positive on surface $B$, the force must be in the negative direction [numerator must be negative], as shown in Figure 1.4.

Let us now consider the second row in the stress matrix in Figure 1.3. From the first subscript, we know that the normal to the surface is in the $y$ direction. Surface $C$ has an outward normal in the positive $y$ direction; therefore all forces on surface $C$ are in the positive direction of the second subscript, as shown in Figure 1.4. Surface $D$ has an outward normal in the negative $y$ direction. Therefore, all forces on surface $D$ are in the negative direction of the second subscript, as shown in Figure 1.4.

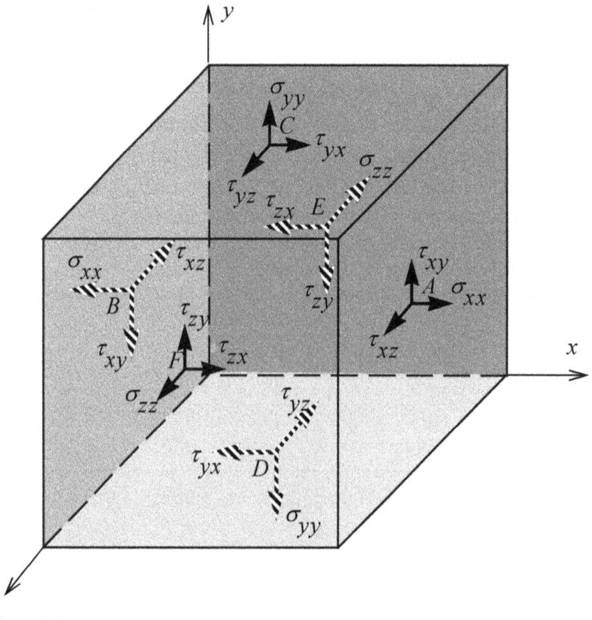


**Figure 1.4** stress element showing all positive stress components.

The components of the third row in the stress matrix are shown on surfaces $E$ and $F$ in Figure 1.4, in accordance with the foregoing logic.

The positive normal stress components (e.g., $\sigma_{xx}$) are pulling the parallelepiped in opposite directions; that is, the parallelepiped is in tension owing to a positive normal stress component. As mentioned earlier, normal stresses are reported as tension or compression, not as positive or negative.

Figure 1.4 shows that the symmetric pair of shear stress components either points toward an edge or away from it—this observation can be used in drawing shear stresses on the three surfaces of the stress element once the shear stress on one of its surfaces has been drawn.

### 1.2.4 Plane stress

Plane stress is one of the two types of two-dimensional simplification in the mechanics of materials. In Section 1.6.1, we will study plane strain, the other type of two-dimensional simplification, and in Section 1.8.3, we will see the difference between the two types of simplification.

**Plane stress** implies all stresses on a plane are zero. We choose $z$ to be the coordinate and set all stresses with subscript $z$ to be zero. Figure 1.5*a* shows the stress matrix in plane stress, Figure 1.5*b* the associated stress element for the point, and Figure 1.5*c* the simplified two-dimensional representation of the stress element as viewed from the $z$ axis. We note that the plane with outward normal in the $z$ direction is stress free. *Stress-free* surfaces are also called **free surfaces**.

$$\begin{bmatrix} \sigma_{xx} & \tau_{xy} & 0 \\ \tau_{yx} & \sigma_{yy} & 0 \\ 0 & 0 & 0 \end{bmatrix}$$

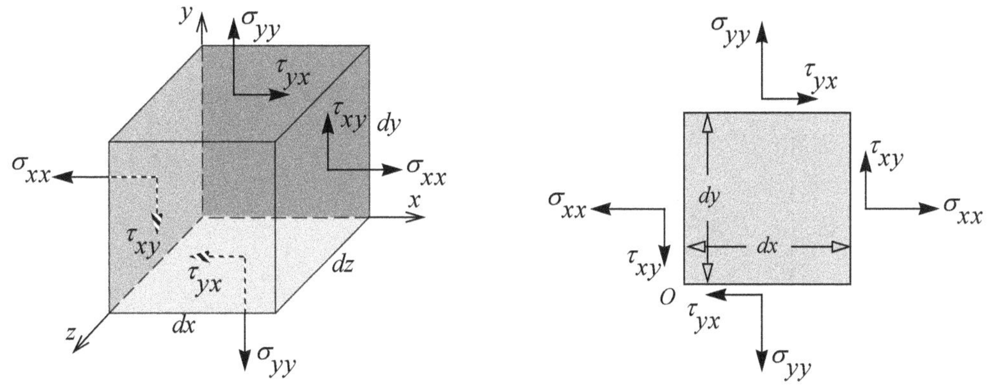


**Figure 1.5** stress element in plane stress.

## 1.3 STRESS TRANSFORMATION IN TWO DIMENSIONS

The stress transformation equations from the introductory course on the mechanics of materials are briefly reviewed in this section and cast in a matrix form. This gives another perspective on familiar concepts. In Section 1.4, this new perspective will be extended to stress transformation in three dimensions.

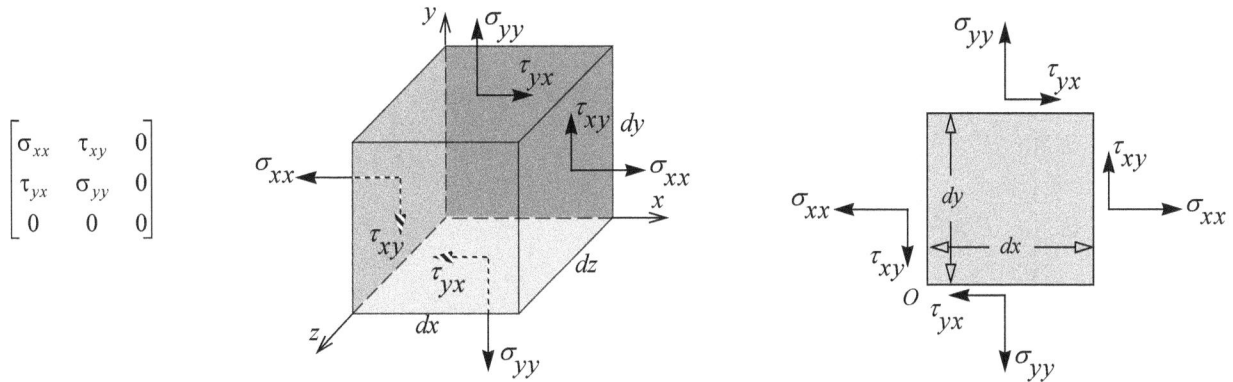

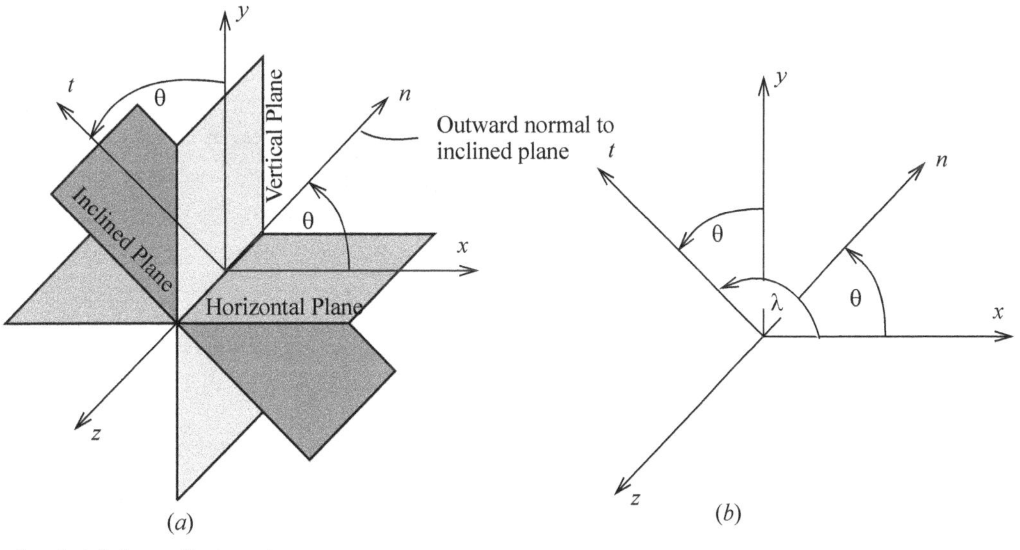


**Figure 1.6** Local and global coordinate systems.

The relationship of stresses at *a point* in different coordinate systems, that is, transformation of stress components with coordinate systems, is called **stress transformation**. Stress transformation can also be viewed as relating stresses on different planes that pass through *a point*. The normal of the plane define the axis of a coordinate system to which we are transforming the stress components. The fixed reference coordinate system in which the entire problem is described is called the **global coordinate system**. A coordinate system that can be fixed at any point on the body and has an orientation that is defined with respect to the global coordinate system is called the **local coordinate system**.

In this book the global coordinate system will most often be the $x$-$y$-$z$ Cartesian coordinate system. Relating internal forces and moments to external forces and moments is usually done in a global coordinate system. The internal quantities are then used to obtain stresses in the global coordinate system.

We assume that the point is in plane stress, and we know the stresses in the $x$-$y$-$z$ coordinate system, that is, on the horizontal and vertical planes shown in Figure 1.6*a*. We seek to find stresses on inclined planes, which can be obtained by rotation about the $z$ axis. Alternatively, we seek to find stresses in the $n$-$t$-$z$ coordinate system, where $n$ is the normal direction to a plane through that point as shown in Figure 1.6*a* and *b*, and $t$ is the tangent direction to the plane such that $n$-$t$-$z$ is a right-handed coordinate system as shown in Figure 1.6*b*.

In the introductory course on the mechanics of materials, the stresses in the *n-t-z* coordinate system were related to stresses in the Cartesian coordinate system by means of the wedge method, which is briefly elaborated here. Figure 1.7*a* shows the stresses acting on the vertical, horizontal, and inclined planes that compose the stress wedge. The stresses are converted to forces by multiplying by the areas of each plane on which the stresses act, as shown on the force wedge in Figure 1.7*b*. By equilibrium of forces in the *n* and *t* directions, we obtain

$$\sigma_{nn} = \sigma_{xx}\cos^2\theta + \sigma_{yy}\sin^2\theta + 2\tau_{xy}\sin\theta\cos\theta \qquad \tau_{nt} = -\sigma_{xx}\cos\theta\sin\theta + \sigma_{yy}\sin\theta\cos\theta + \tau_{xy}(\cos^2\theta - \sin^2\theta) \quad \textbf{(1.2a)}$$

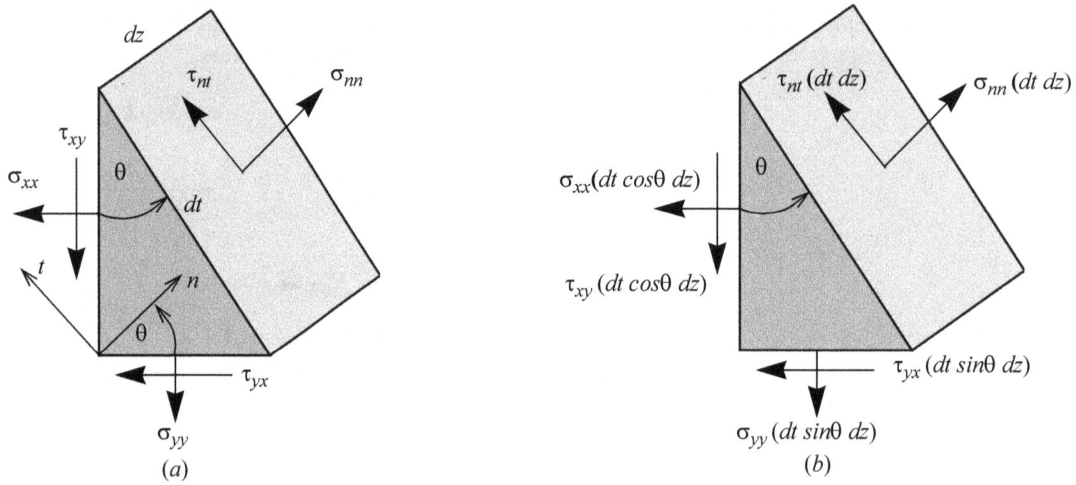


**Figure 1.7** (*a*) Stress wedge. (*b*) Force wedge.

By substituting $90° + \theta$ for $\theta$ in Equation (1.2a), we obtain the normal stress in the *t* direction as

$$\sigma_{tt} = \sigma_{xx}\sin^2\theta + \sigma_{yy}\cos^2\theta - 2\tau_{xy}\cos\theta\sin\theta \quad \textbf{(1.2b)}$$

On adding Equation (1.2a) and (1.2b), we obtain

$$\sigma_{nn} + \sigma_{tt} = \sigma_{xx} + \sigma_{yy} \quad \textbf{(1.3)}$$

Equation (1.3) implies that the sum of normal stresses in an orthogonal system is independent of coordinate transformation.

### 1.3.1 Matrix method in two dimensions

Figure 1.6*c* shows the normal (*n*) and tangent (*t*) directions of a plane. The direction cosines of the unit vectors in the *n* and *t* directions can be written as in Equation (1.4a).

$$n_x = \cos\theta \qquad n_y = \sin\theta \qquad t_x = \cos\lambda \qquad t_y = \sin\lambda \quad \textbf{(1.4a)}$$

Noting that $\lambda = 90° + \theta$, we obtain Equation (1.4b).

$$t_x = -n_y \qquad t_y = n_x \quad \textbf{(1.4b)}$$

To develop stress transformation equations in matrix form, we introduce the matrix notation given in Equation (1.4c).

$$\{n\} = \begin{Bmatrix} n_x \\ n_y \end{Bmatrix} \qquad \{t\} = \begin{Bmatrix} t_x \\ t_y \end{Bmatrix} \qquad [\sigma] = \begin{bmatrix} \sigma_{xx} & \tau_{xy} \\ \tau_{yx} & \sigma_{yy} \end{bmatrix} \quad \textbf{(1.4c)}$$

The symmetry of shear stresses implies that the stress matrix is symmetric. In matrix notation, this symmetry is expressed as

$$[\sigma]^T = [\sigma] \quad \textbf{(1.5)}$$

where $[\ ]^T$ implies the transpose of the matrix. In matrix notation, Equations (1.2a) and (1.2b) can be written as

$$\boxed{\sigma_{nn} = \{n\}^T[\sigma]\{n\} \qquad \tau_{nt} = \{t\}^T[\sigma]\{n\} \qquad \sigma_{tt} = \{t\}^T[\sigma]\{t\}} \quad \textbf{(1.6)}$$

where $\{\ \}^T$ implies the transpose of the column. Equation (1.6) show that we can obtain a stress component by pre- and post-multiplying the stress matrix by the unit vectors in the direction of the subscripts. Because the shear stress is symmetric, we could have premultiplied by the unit normal and postmultiplied by the unit tangent. In other words, the order of multiplication is immaterial in the calculation of shear stress.

Stress on a surface is called **traction** or a **stress vector**. Note that stress at a point is a second-order tensor. When we specify a surface, the orientation of the surface is defined; hence the stress on the surface needs only one direction to specify, which implies that stress on a surface is a vector quantity. Mathematically the stress vector $\{S\}$ is defined in Equation (1.7).

$$\boxed{\{S\} = [\sigma]\{n\}} \quad \textbf{(1.7)}$$

Equation (1.7) in expanded notation can be written as

$$S_x = \sigma_{xx}n_x + \tau_{xy}n_y \qquad S_y = \tau_{yx}n_x + \sigma_{yy}n_y \quad \textbf{(1.8)}$$

Since direction cosines do not have units, the stress vector has the units of stress. Thus, we now know three different quantities that have units of force per unit area:

- Pressure, which is a scalar quantity.
- Traction, which is a vector quantity.
- Stress, which is a second-order tensor.

To better appreciate the concept of traction, we once more consider the force wedge shown in Figure 1.7*b*. We represent the incline area $dt\,dz = dA$ and use the direction cosine notation of Equation (1.4a) to show the two-dimensional representation of the force wedge, as in Figure 1.8*a*. The forces shown on the inclined plane could be resolved into forces in the $x$ and $y$ coordinates, as shown in Figure 1.8*b*. By equilibrium of forces in Figure 1.8*b*, we obtain Equation (1.8).

Alternatively, the stress vector $\{S\}$ could be written as

$$\{S\} = \sigma_{nn}\{n\} + \tau_{nt}\{t\} \tag{1.9}$$

Equations (1.7) and (1.9) are representations of the vector $\{S\}$ in two different coordinate systems, as shown in Figure 1.8*c*.

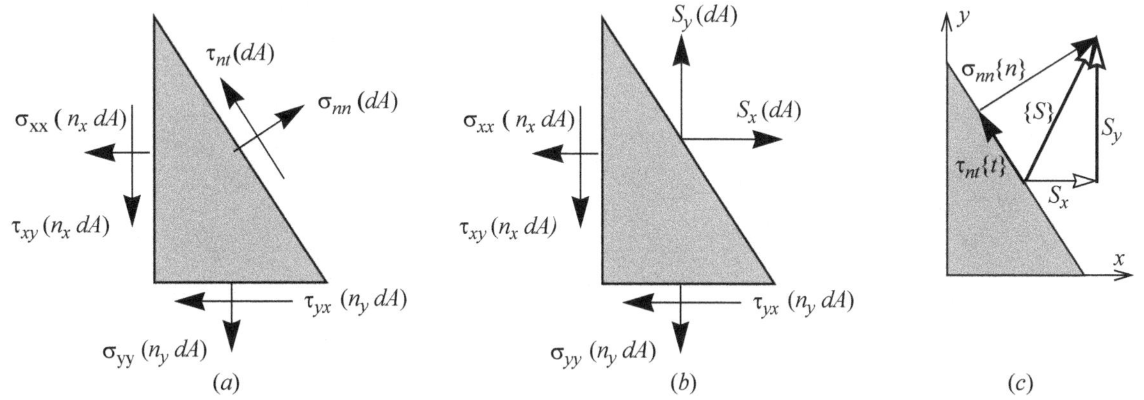


**Figure 1.8** Statically equivalent force wedges.

## 1.3.2 Principal stresses

Recall the following definitions from the introductory course on the mechanics of materials. Planes on which the shear stresses are zero are called the **principal planes**. The normal direction to the principal planes is referred to as the **principal direction,** or the **principal axis**. The angles the principal axis makes with the global coordinate system are called **principal angles**. Normal stress on a principal plane is called **principal stress**. The greatest principal stress is called **principal stress 1**.

In plane stress, the plane with outward normal in the $z$ direction has zero shear stress. Therefore by definition, the $z$ direction is a principal direction. Since there are three principal planes, there are three principal stresses at a point. The three principal stresses are labeled $\sigma_1$, $\sigma_2$, and $\sigma_3$.

Let the unit normal vector of a principal plane be given by $\{p\}$ and the corresponding principal stress by $\sigma_p$. By definition, the plane with normal $\{p\}$ will have zero shear stress and the normal stress will the principal stress and Equations (1.7) and (1.9) can be written as

$$\{S\} = [\sigma]\{p\} = \sigma_p\{p\} \tag{1.10}$$

Equation (1.10) in expanded notation can be written as follows:

$$\{S\} = \begin{bmatrix} \sigma_{xx} & \tau_{xy} \\ \tau_{yx} & \sigma_{yy} \end{bmatrix} \begin{Bmatrix} p_x \\ p_y \end{Bmatrix} = \begin{bmatrix} \sigma_p & 0 \\ 0 & \sigma_p \end{bmatrix} \begin{Bmatrix} p_x \\ p_y \end{Bmatrix} \tag{1.11a}$$

Alternatively, Equation (1.11a) can be written as Equation (1.11b).

$$\begin{bmatrix} (\sigma_{xx} - \sigma_p) & \tau_{xy} \\ \tau_{yx} & (\sigma_{yy} - \sigma_p) \end{bmatrix} \begin{Bmatrix} p_x \\ p_y \end{Bmatrix} = 0 \tag{1.11b}$$

Equation (1.11b) represents two equations in the two unknowns $p_x$ and $p_y$. For a *nontrivial* (*nonzero*) solution to exist, the determinant of the matrix must be zero. This is a classic statement of an eigenvalue problem. To show that the eigenvalues of the stress matrix are the principal stresses, we set the determinant of the matrix in Equation (1.11b) to zero and obtain

$$\sigma_p^2 - \sigma_p(\sigma_{xx} + \sigma_{yy}) + (\sigma_{xx}\sigma_{yy} - \tau_{xy}^2) = 0 \tag{1.12}$$

Equation (1.12) is called the **characteristic equation** in an eigenvalue problem. The above equation is a quadratic whose roots are

$$\sigma_{1,2} = [(\sigma_{xx} + \sigma_{yy}) \pm \sqrt{(\sigma_{xx} + \sigma_{yy})^2 - 4(\sigma_{xx}\sigma_{yy} - \tau_{xy}^2)}]/2 \tag{1.13a}$$

The terms in Equation (1.13a) can be rearranged to obtain

$$\sigma_{1,2} = \left[ \left( \frac{\sigma_{xx} + \sigma_{yy}}{2} \right) \pm \sqrt{\left( \frac{\sigma_{xx} - \sigma_{yy}}{2} \right)^2 + \tau_{xy}^2} \right] \tag{1.13b}$$

Equation (1.13b) gives the formulas for principal stresses we saw in the introductory course on the mechanics of materials. We conclude, *the eigenvalues of the stress matrix are the principal stresses and the eigenvectors of the stress matrix are the principal directions.* To determine the eigenvectors, we can use *either* of the two equations represented in (1.11b), along with the fact that the sum of the square of direction cosines is 1.

## 1.4 STRESS TRANSFORMATION IN THREE DIMENSIONS

The concepts and formulas developed in two dimensions can be extended to three dimensions by noting that the matrix relationships described in Section 1.3 do not depend upon the size of the matrix. Figure 1.9(a) shows the direction cosines of the unit normal in three dimensions.

$$n_x = \cos \theta_x$$
$$n_y = \cos \theta_y$$
$$n_z = \cos \theta_z$$

(a)  (b)

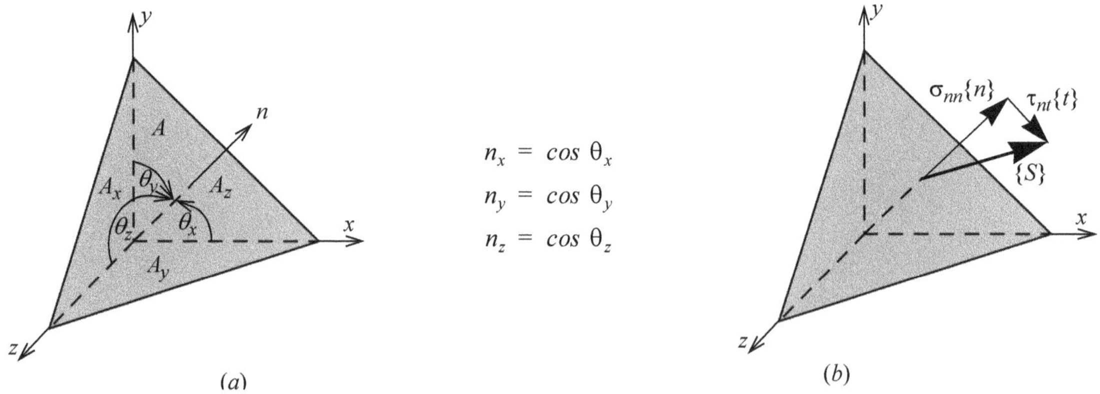

**Figure 1.9** (*a*) Direction cosines of a unit normal. (*b*) Equilibrating shear stress.

We define the following quantities in matrix notation

$$\{n\} = \begin{Bmatrix} n_x \\ n_y \\ n_z \end{Bmatrix} \qquad \{S\} = \begin{Bmatrix} S_x \\ S_y \\ S_z \end{Bmatrix} \qquad [\sigma] = \begin{bmatrix} \sigma_{xx} & \tau_{xy} & \tau_{xz} \\ \tau_{yx} & \sigma_{yy} & \tau_{yz} \\ \tau_{zx} & \tau_{zy} & \sigma_{zz} \end{bmatrix} \tag{1.14}$$

The normal stress can be found from Equation (1.6). In two dimensions, the normal and tangent directions were related. In three dimensions, however, there are an infinite number of tangents in the inclined plane shown in Figure 1.9a. The calculation of shear stress in three dimensions entails problems of two types.

1. The tangent direction is known. The shear stress in that direction may be found from Equation (1.6). Shear stress found from Equation (1.6) represents the component of the stress vector in the given tangent direction; this may not be the shear stress necessary for equilibrium of the wedge shown in Figure 1.9a.

2. The shear stress that is necessary for equilibrium may be found from Equation (1.9), as shown in Figure 1.9b. We can find the stress vector $\{S\}$ from Equation (1.7) and $\sigma_{nn}$ from Equation (1.6). From Equation (1.9) we obtain $\tau_{nt}\{t\} = \{S\} - \sigma_{nn}\{n\}$. The magnitude and direction of the shear stress can be determined from the vector $\tau_{nt}\{t\}$.

### 1.4.1 Principal stresses

The equivalent form of Equation (1.11a) and (1.11b) in three dimensions can be written as follows:

$$\begin{bmatrix} \sigma_{xx} & \tau_{xy} & \tau_{xz} \\ \tau_{yx} & \sigma_{yy} & \tau_{yz} \\ \tau_{zx} & \tau_{zy} & \sigma_{zz} \end{bmatrix} \begin{Bmatrix} p_x \\ p_y \\ p_z \end{Bmatrix} = \begin{bmatrix} \sigma_p & 0 & 0 \\ 0 & \sigma_p & 0 \\ 0 & 0 & \sigma_p \end{bmatrix} \begin{Bmatrix} p_x \\ p_y \\ p_z \end{Bmatrix} \quad \text{or} \quad \begin{bmatrix} (\sigma_{xx} - \sigma_p) & \tau_{xy} & \tau_{xz} \\ \tau_{yx} & (\sigma_{yy} - \sigma_p) & \tau_{yz} \\ \tau_{zx} & \tau_{zy} & (\sigma_{zz} - \sigma_p) \end{bmatrix} \begin{Bmatrix} p_x \\ p_y \\ p_z \end{Bmatrix} = 0 \tag{1.15}$$

The matrix Equation (1.15) represents three equations in the three unknown direction cosines. For a nontrivial solution, the determinant of the matrix must be zero. Once more we observe that the eigenvalues of the stress matrix are the principal stresses, and the eigenvectors are the principal directions. To determine the eigenvectors corresponding to an eigenvalue, we can use any two of the three equations in Equation (1.15), along with the knowledge that the square of the direction cosines is 1, as shown in Equation (1.16).

$$p_x^2 + p_y^2 + p_z^2 = 1 \tag{1.16}$$

### 1.4.2 Principal stress convention

Principal stresses may be ordered or unordered. Generally speaking, the following convention is observed for ordering of principal stresses. In three dimensions the principal stresses are reported such that $\sigma_1 > \sigma_2 > \sigma_3$. In two dimensions the principal stresses are reported such that $\sigma_1 > \sigma_2$, but $\sigma_3$ is not governed by any order. This is because $\sigma_3$ is equal to $\sigma_{zz}$, which depends upon whether the state of stress is plane stress or plane strain. By convention, calculated principal stresses are reported in ordered form.

The principal angles are calculated from the direction cosines of the principal directions. By convention, the angles reported are between zero and 180°, that is,

$$0° \le \theta_x, \theta_y, \theta_z \le 180° \tag{1.17}$$

### 1.4.3 Characteristic equation and stress invariants

We set the determinant of the matrix in Equation (1.15) to zero to obtain the characteristic equation below.

$$\boxed{\sigma_p^3 - I_1\sigma_p^2 + I_2\sigma_p - I_3 = 0} \tag{1.18}$$

$$I_1 = \sigma_{xx} + \sigma_{yy} + \sigma_{zz} \qquad I_2 = \begin{vmatrix} \sigma_{xx} & \tau_{xy} \\ \tau_{yx} & \sigma_{yy} \end{vmatrix} + \begin{vmatrix} \sigma_{yy} & \tau_{yz} \\ \tau_{zy} & \sigma_{zz} \end{vmatrix} + \begin{vmatrix} \sigma_{xx} & \tau_{xz} \\ \tau_{zx} & \sigma_{zz} \end{vmatrix} \qquad I_3 = \begin{vmatrix} \sigma_{xx} & \tau_{xy} & \tau_{xz} \\ \tau_{yx} & \sigma_{yy} & \tau_{yz} \\ \tau_{zx} & \tau_{zy} & \sigma_{zz} \end{vmatrix} \tag{1.19}$$

where, | | is the determinant of the quantity enclosed in the verticals.

The three roots[2] of the characteristic equation yield the principal stresses, as observed in Section 1.3.2. *Principal stresses at a point are unique* and do not depend upon the coordinate system in which the stress matrix was specified. If roots of the characteristic equation are independent of the coordinate system then the coefficients in the characteristic equation should be independent of the coordinate system. Hence, the coefficients $I_1$, $I_2$, and $I_3$ are called **stress invariants** because the values of these quantities are independent of coordinate transformation. Notice that the sign in the terms containing odd stress invariants is negative.

- The first stress invariant $I_1$ is the sum of diagonal terms of the stress matrix.
- The second stress invariant $I_2$ is the sum of minor determinants of the stress matrix.
- The third stress invariant $I_3$ is the determinant of the entire stress matrix.

Since stress invariants are independent of the coordinate system, the value of these quantities will be the same in a principal coordinate system. The stress matrix in a principal coordinate system can be written as

$$\begin{bmatrix} \sigma_1 & 0 & 0 \\ 0 & \sigma_2 & 0 \\ 0 & 0 & \sigma_3 \end{bmatrix} \tag{1.20}$$

The stress invariants can be written in terms of principal stress as

$$I_1 = \sigma_1 + \sigma_2 + \sigma_3 \qquad I_2 = \begin{vmatrix} \sigma_1 & 0 \\ 0 & \sigma_2 \end{vmatrix} + \begin{bmatrix} \sigma_2 & 0 \\ 0 & \sigma_3 \end{bmatrix} + \begin{bmatrix} \sigma_1 & 0 \\ 0 & \sigma_3 \end{bmatrix} = \sigma_1\sigma_2 + \sigma_2\sigma_3 + \sigma_3\sigma_1 \qquad I_3 = \begin{vmatrix} \sigma_1 & 0 & 0 \\ 0 & \sigma_2 & 0 \\ 0 & 0 & \sigma_3 \end{vmatrix} = \sigma_1\sigma_2\sigma_3 \tag{1.21}$$

The stress invariants, particularly the first and the third, provide a quick and easy check on the laboriously calculated principal stresses.

### 1.4.4 Maximum shear stress

In the introductory course on the mechanics of materials, we saw that the maximum shear stress exists on a plane that is 45° to the principal planes. The magnitude of the maximum shear stress is given by

$$\boxed{\tau_{max} = \max\left(\left|\frac{\sigma_1 - \sigma_2}{2}\right|, \left|\frac{\sigma_2 - \sigma_3}{2}\right|, \left|\frac{\sigma_3 - \sigma_1}{2}\right|\right)} \tag{1.22}$$

where the term on the right-hand side is the magnitude of maximum difference between two principal stresses. Figures 1.10 through 1.12 show the various planes on which the maximum shear stress will exist, depending upon the relative values of the principal stresses.

2. The roots of equation $x^3 - I_1x^2 + I_2x - I_3 = 0$ are $x_1 = 2A\cos\alpha + I_1/3$, and $x_{2,3} = -2A\cos(\alpha \pm 60°) + I_1/3$, where $A = \sqrt{(I_1/3)^2 - I_2/3}$ and $\cos 3\alpha = [2(I_1/3)^3 - (I_1/3)I_2 + I_3]/(2A^3)$.

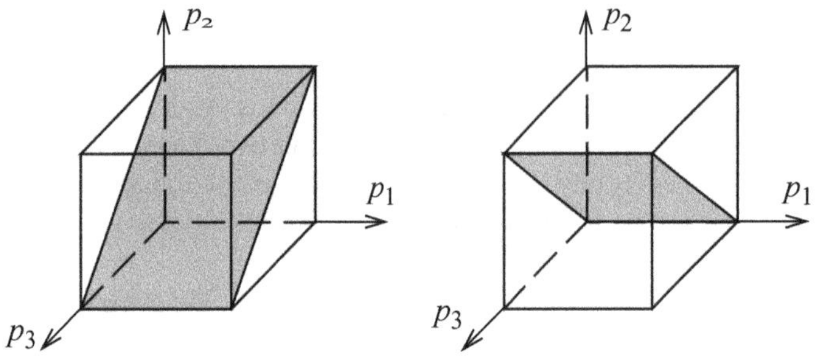


$$\tau_{23} = -\tau_{32} = \frac{\sigma_2 - \sigma_3}{2}$$

**Figure 1.10** Planes of maximum shear stress that are 45° to principal planes 2 and 3.

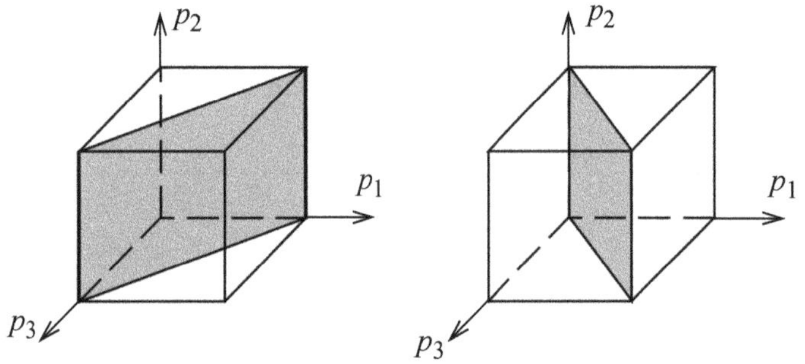


$$\tau_{31} = -\tau_{13} = \frac{\sigma_3 - \sigma_1}{2}$$

**Figure 1.11** Planes of maximum shear stress that are 45° to principal planes 1 and 3.

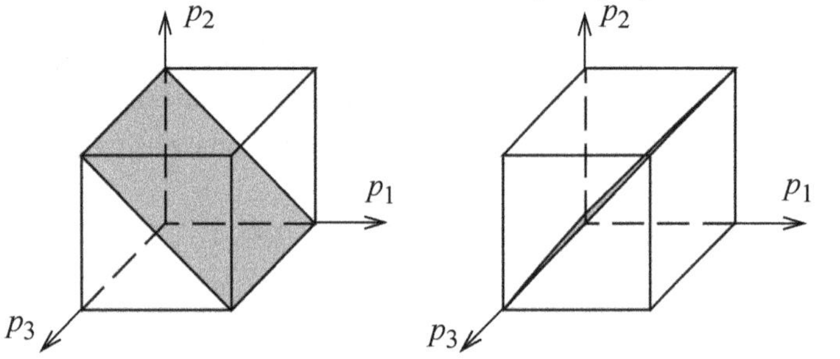


$$\tau_{12} = -\tau_{21} = \frac{\sigma_1 - \sigma_2}{2}$$

**Figure 1.12** Planes of maximum shear stress that are 45° to principal planes 1 and 2.

## 1.4.4 Octahedral stresses

A plane that makes equal angles with the principal planes is called an **octahedral plane**. The stresses on the octahedral plane are called **octahedral stresses**. In Section 1.10 we will use the octahedral stresses to formulate a failure theory.

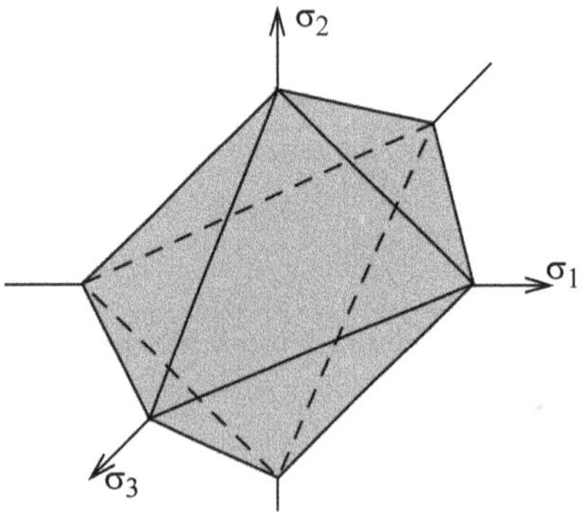


**Figure 1.13** The eight octahedral planes.

Figure 1.13 shows eight (octal) planes that make equal angles with the principal planes. We use $n_1$, $n_2$, and $n_3$ to represent the direction cosines of the outward normal of a plane measured from the principal axis. The normal stress and shear stress on a plane with direction cosines $n_1$, $n_2$, and $n_3$ can be shown to be given by Equation (1.23).

$$\sigma_{nn} = \sigma_1 n_1^2 + \sigma_2 n_2^2 + \sigma_3 n_3^2 \qquad \tau_{nt} = \sqrt{(\sigma_1^2 n_1^2 + \sigma_2^2 n_2^2 + \sigma_3^2 n_3^2) - \sigma_{nn}^2} \tag{1.23}$$

Though the sign of the direction cosines changes with each of the eight octahedral planes, their magnitude is the same for all eight planes, that is, $|n_1| = |n_2| = |n_3| = 1/\sqrt{3}$. Substituting the direction cosines into Equation (1.23) we obtain the octahedral normal stress as

$$\boxed{\sigma_{oct} = (\sigma_1 + \sigma_2 + \sigma_3)/3} \tag{1.24a}$$

Equation (1.24a) shows $\sigma_{oct} = I_1/3$, where $I_1$ is the first stress invariant. A more interesting interpretation of octahedral normal stress is obtained by considering a point in fluid. For a point in fluid, the shear stresses are zero and all normal stresses (principal stresses) are compressive and equal to the pressure at that point, that is, $\sigma_1 = \sigma_2 = \sigma_3 = -p$, where $p$ is the hydrostatic pressure. From Equation (1.24a) we obtain $\sigma_{oct} = -p$; in other words, the octahedral normal stress is similar to the hydrostatic pressure.

Substituting the direction cosines for the octahedral planes in Equation (1.23) we obtain the magnitude of shear stress, which can be simplified as shown in Equation (1.24b).

$$\tau_{oct} = \sqrt{\frac{(\sigma_1^2+\sigma_2^2+\sigma_3^2)}{3} - \frac{(\sigma_1+\sigma_2+\sigma_3)^2}{9}} = \frac{1}{3}\sqrt{(2\sigma_1^2+2\sigma_2^2+2\sigma_3^2) - 2(\sigma_1\sigma_2+\sigma_2\sigma_3+\sigma_3\sigma_1)}$$

$$\boxed{\tau_{oct} = \frac{1}{3}\sqrt{(\sigma_1-\sigma_2)^2+(\sigma_2-\sigma_3)^2+(\sigma_3-\sigma_1)^2}} \qquad \textbf{(1.24b)}$$

### 1.4.5 Deviatoric stress

A **deviatoric stress matrix** is a stress matrix from which the effect of hydrostatic state of stress ($I_1/3$) is removed. It plays an important role in theory of plasticity (See Mendelson [1968]). The deviatoric stress matrix can be written in the Cartesian coordinates or in principal coordinates as shown in Figure 1.14.

$$[D] = \begin{bmatrix} \sigma_{xx}-I_1/3 & \tau_{xy} & \tau_{xz} \\ \tau_{yx} & \sigma_{yy}-I_1/3 & \tau_{yz} \\ \tau_{zx} & \tau_{zy} & \sigma_{zz}-I_1/3 \end{bmatrix} \quad \text{or} \quad [D] = \begin{bmatrix} \sigma_1-I_1/3 & 0 & 0 \\ 0 & \sigma_2-I_1/3 & 0 \\ 0 & 0 & \sigma_3-I_1/3 \end{bmatrix}$$

(*a*) (*b*)

**Figure 1.14** Stress deviatoric matrix. in (*a*) Cartesian coordinates. (*b*) Principal coordinates.

We can find the invariants of deviatoric stress matrix in the same manner as we found for the ordinary stress matrix. If we don't have the stress matrix in principal coordinates, then it is algebraically less tedious to find the deviatoric stress invariants directly for the matrix in Cartesian (or cylindrical / spherical) coordinates. However, for derivation purposes it is easier to work with the matrix in principal coordinates. It can be shown (see problems 1.10 and 1.11) that the three stress deviatoric stress invariants are as shown below.

$$J_1 = 0 \qquad J_2 = -\left(\frac{1}{6}\right)[(\sigma_1-\sigma_2)^2 + +(\sigma_2-\sigma_3)^2+(\sigma_3-\sigma_1)^2] \qquad J_3 = \frac{1}{27}(2\sigma_1-\sigma_2-\sigma_3)(2\sigma_2-\sigma_3-\sigma_1)(2\sigma_3-\sigma_1-\sigma_2) \quad \textbf{(1.25)}$$

Substituting Equation (1.24b) into Equation (1.25) we obtain

$$J_2 = -(3/2)\tau_{oct}^2 \qquad \textbf{(1.26)}$$

## EXAMPLE 1.1

Show that the principal directions are orthogonal if the principal stresses are not equal.

**SOLUTION**

Let $\{p_n\}$ and $\{p_m\}$ be the principal directions corresponding to two principal stresses $\sigma_n$ and $\sigma_m$. Equation (1.10) can be written as

$$[\sigma]\{p_n\} = \sigma_n\{p_n\} \qquad \text{(E1)}$$

$$[\sigma]\{p_m\} = \sigma_m\{p_m\} \qquad \text{(E2)}$$

We can premultiply Equation (E1) by $\{p_m\}^T$ and Equation (E2) by $\{p_n\}^T$ to obtain

$$\{p_m\}^T[\sigma]\{p_n\} = \sigma_n\{p_m\}^T\{p_n\} \qquad \text{(E3)}$$

$$\{p_n\}^T[\sigma]\{p_m\} = \sigma_m\{p_n\}^T\{p_m\} \qquad \text{(E4)}$$

The left- and right-hand sides of Equation (E3) are scalar quantities, and hence we can take the transpose of both sides and write Equation (E3) and (E4) by using matrix identities:

$$[\{p_m\}^T[\sigma]\{p_n\}]^T = \sigma_n[\{p_m\}^T\{p_n\}]^T \quad \text{or} \quad \{p_n\}^T[\sigma]^T\{p_m\} = \sigma_n\{p_n\}^T\{p_m\} \qquad \text{(E5)}$$

By noting that the stress matrix is symmetric as given by Equation (1.5), we can rewrite the above equation as

$$\{p_n\}^T[\sigma]\{p_m\} = \sigma_n\{p_n\}^T\{p_m\} \qquad \text{(E6)}$$

Subtracting Equation (E6) from Equation (E4), we obtain

$$[\sigma_m-\sigma_n]\{p_n\}^T\{p_m\} = 0 \qquad \text{(E7)}$$

Since $\sigma_m \neq \sigma_n$, the above equation implies that

$$\{p_n\}^T\{p_m\} = 0 \qquad \text{(E8)}$$

Equation (E8) is the condition of orthogonality of two vectors. Thus we have shown that any two principal directions are orthogonal if the principal stresses are not equal.

**COMMENTS**

1. The orthogonality of eigenvectors is a property of all symmetric matrices regardless of the matrix size. Because the stress matrix is a symmetric matrix and the principal directions are the eigenvectors of the stress matrix, the result that principal directions are orthogonal is not surprising.
2. If $\sigma_m = \sigma_n$, then all directions in the plane containing the vectors $\{p_n\}$ and $\{p_m\}$ are principal directions.

## EXAMPLE 1.2

Determine the second and third deviatoric stress invariants for the stress matrix given below.

$$\begin{bmatrix} 30 & 0 & 20 \\ 0 & 30 & -10 \\ 20 & -10 & 0 \end{bmatrix} \text{MPa}$$

**PLAN: Method I**

We first find the three principal stresses, then find the deviatoric stress matrix in principal coordinate and find its invariants.

**SOLUTION**

We can find the three stress invariants from Equation (1.19) as shown below.

$$I_1 = 30 + 30 + 0 = 60 \tag{E1}$$

$$I_2 = \begin{vmatrix} 30 & 0 \\ 0 & 30 \end{vmatrix} + \begin{vmatrix} 30 & -10 \\ -10 & 0 \end{vmatrix} + \begin{vmatrix} 30 & 20 \\ 20 & 0 \end{vmatrix} = 900 - 100 - 400 = 400 \tag{E2}$$

$$I_3 = 30\begin{vmatrix} 30 & -10 \\ -10 & 0 \end{vmatrix} + 20\begin{vmatrix} 0 & 30 \\ 20 & -10 \end{vmatrix} + 0 = -3000 - 12{,}000 = -15{,}000 \tag{E3}$$

The characteristic equation is $\sigma_p^3 - 60\sigma_p^2 + 400\sigma_p + 15{,}000 = 0$. The roots of the equation can be found from the formulas given in footnote 2 on page 9 as shown below.

$$A = \sqrt{(60/3)^2 - 400/3} = 16.390 \tag{E4}$$

$$cos\ 3\alpha = \frac{2(60/3)^3 - (60/3)400 + (-15{,}000)}{2(16.39)^3} = -0.8037 \quad \text{or} \quad \alpha = 143.49/3 = 47.83 \tag{E5}$$

$$x_1 = 2(16.39)cos\ 47.83 + 60/3 = 41.9258 \tag{E6}$$

$$x_3 = (-2(16.39)cos(-12.17) + (60)/3) = -11.9258 \tag{E7}$$

$$x_2 = -2(16.39)cos(107.83) + (60)/3 = 30 \tag{E8}$$

The principal stresses are

$$\sigma_1 = 41.9 \text{ MPa (T)} \qquad \sigma_2 = 30 \text{ MPa (T)} \qquad \sigma_3 = 11.9 \text{ MPa (C)} \tag{E9}$$

*Checking Results*: We can check our results by finding stress invariants 1 and 3 from Equation (1.19).

$$I_1 = 41.927 + 29.998 - 11.925 = 60 \qquad \text{Checks.}$$

$$I_3 = (41.927)(29.998)(-11.925) = -14{,}998 \qquad \text{Checks.}$$

Subtracting $I_1/3 = 20$ from the diagonal terms of principal stress matrix we obtain the deviatoric stress matrix.

$$[D] = \begin{bmatrix} 41.9258 - 20 & 0 & 0 \\ 0 & 30 - 20 & 0 \\ 0 & 0 & -11.9258 - 20 \end{bmatrix} = \begin{bmatrix} 21.9258 & 0 & 0 \\ 0 & 10 & 0 \\ 0 & 0 & -31.9258no \end{bmatrix} \text{MPa} \tag{E10}$$

The second and third deviatoric stress invariant can be written as

$$J_2 = \begin{vmatrix} 21.9258 & 0 \\ 0 & 10 \end{vmatrix} + \begin{vmatrix} 10 & 0 \\ 0 & -31.9258 \end{vmatrix} + \begin{vmatrix} 21.9258 & 0 \\ 0 & -31.9258 \end{vmatrix} = -800 \tag{E11}$$

$$J_3 = \begin{vmatrix} 21.9258 & 0 & 0 \\ 0 & 10 & 0 \\ 0 & 0 & -31.9258 \end{vmatrix} = -7000 \tag{E12}$$

**ANS.** $J_2 = -800\text{MPa}^2 \qquad J_3 = -7000\text{MPa}^3$

**PLAN: Method II**

We first find the deviatoric stress matrix from the one that is given and find the second and third invariant of the matrix.

**SOLUTION**

We subtract $I_1/3 = 20$ from the diagonal terms of the given stress matrix to obtain the deviatoric stress matrix.

$$[D] = \begin{bmatrix} 30-20 & 0 & 20 \\ 0 & 30-20 & -10 \\ 20 & -10 & 0-20 \end{bmatrix} = \begin{bmatrix} 10 & 0 & 20 \\ 0 & 10 & -10 \\ 20 & -10 & -20 \end{bmatrix} \quad \text{(E13)}$$

The second and third deviatoric stress invariant can be written as

$$J_2 = \begin{vmatrix} 10 & 0 \\ 0 & 10 \end{vmatrix} + \begin{vmatrix} 10 & -10 \\ -10 & -20 \end{vmatrix} + \begin{vmatrix} 10 & 20 \\ 20 & -20 \end{vmatrix} = -800 \quad \text{(E14)}$$

$$J_3 = \begin{vmatrix} 10 & 0 & 20 \\ 0 & 10 & -10 \\ 20 & -10 & -20 \end{vmatrix} = -7000 \quad \text{(E15)}$$

**ANS.** $J_2 = -800\text{MPa}^2 \qquad J_3 = -7000\text{MPa}^3$

**COMMENTS**

1. The formulas for deviatoric stress invariants in terms of principal stress are simpler than in Cartesian coordinates. However, calculations of the deviatoric stress invariants in Cartesian coordinates are simpler as principal stresses need not be calculated.
2. The calculation of principal stresses permits calculation of maximum shear stress and octahedral stresses as shown below.

As per Equation (1.22), the maximum shear stress is the magnitude of the largest difference between the principal stresses, which in this example is between principal stresses 1 and 3. Its value is

$$\tau_{max} = |\sigma_3 - \sigma_1|/2 = |-11.925 - 41.927|/2 = 26.9 \text{ MPa}$$

Substituting principal stress values in Equation (1.24a) *and* (1.24b) yields the octahedral stresses.

$$\sigma_{oct} = I_1/3 = 20 \text{ MPa (T)} \qquad \tau_{oct} = \sqrt{(41.927-29.998)^2 + [29.998+11.925]^2 + [-11.925-41.927]^2}/3 = 23.1 \text{ MPa}$$

3. The maximum shear stress is different from the octahedral shear stress. One failure theory uses maximum shear stress as a failure criterion and another uses octahedral shear stress or von-Mises stress. The results in this example highlight that two theories predict different failure values depending upon which failure criterion we use.

## 1.5 AVERAGE STRAINS

Strain is a measure of the intensity of change in the shape of a body. A change in shape can be described by displacements of points on the body and is thus a problem in geometry (kinematics) as depicted in Figure 1.15.

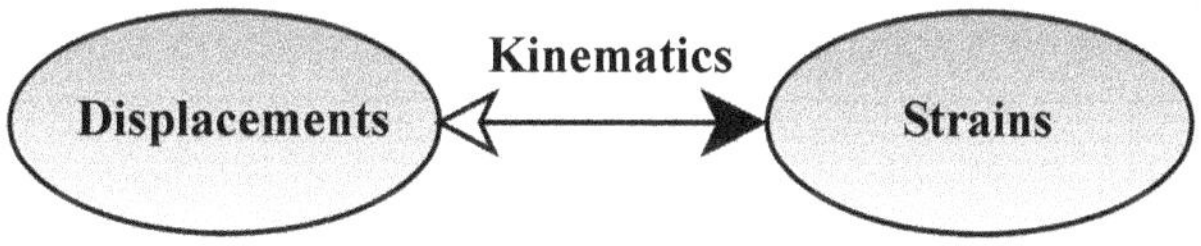


**Figure 1.15** Strains and displacements.

The *total movement* of a point with respect to a fixed reference coordinates is called **displacement**. The *relative movement* of a point with respect to another point on the body is called **deformation**. Thus displacement of a point represents changes due to deformation plus rigid body motion.

In describing changes, we must specify the reference value from which we are measuring change. **Lagrangian strain** is computed from deformation by using the original *undeformed geometry* as the reference geometry. **Eulerian strain** is computed from deformation by using the *final deformed* geometry as the reference geometry. The Lagrangian description is usually used in solid mechanics. The Eulerian description is usually used in fluid mechanics. When a material undergoes very large deformations, such as in a soft rubber or in projectile penetration of metals, either description may be used, depending upon the analysis. Non-linear strain definitions are discussed in Chapter 8. We will use Lagrangian strain unless it is stated otherwise.

Recall that there are two kinds of strain—normal strains and shear strains. Normal strains are measures of change of length and shear strains are measures of change of angles.

Normal strain is usually designated by the Greek letter *epsilon* ($\varepsilon$). The change of length, also called deformation, is designated by the lowercase Greek letter delta ($\delta$). The average normal Lagrangian strain of a line is the ratio of deformation $\delta$ of the line to the original length $L_0$ as follows:

$$\boxed{\varepsilon = \delta / L_0} \quad \textbf{(1.27)}$$

**Elongations** ($\delta > 0$) result in *positive* normal strains. **Contractions** ($\delta < 0$) result in *negative* normal strains.

In *small-strain* approximations ($\varepsilon < 0.01$), the deformation $\delta$ can be approximated by the deformation component in the direction of the undeformed line element. Figure 1.16 shows a bar $AP$ that deforms to the position $AP_1$. For small-strain calculations, the deformed length of $AP_1$ is approximated by length $AP_2$, which is obtained by drawing a perpendicular from $P_1$ to the original direction $AP$. Another perspective on small-strain approximation is that the deformation $\delta$ is the component of deformation vector $PP_1$ in the original direction. The small-strain approximation results in linear analysis and greatly simplifies calculations.

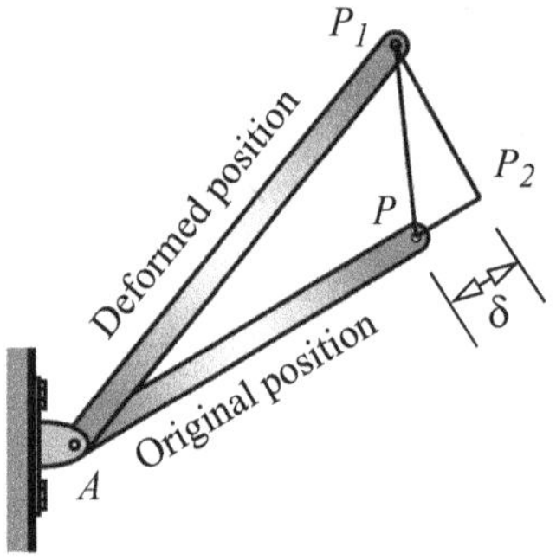


**Figure 1.16** Approximation of a small normal strain.

Shear strain is usually designated by the Greek letter *gamma* ($\gamma$). The average Lagrangian shear strain is defined as a change of angle from the right angle as

$$\boxed{\gamma = \frac{\pi}{2} - \alpha} \tag{1.28}$$

where, the Greek letter *alpha* ($\alpha$) is the final angle measured in radians (rad), and the Greek letter *pi* ($\pi$) = 3.14159 rad. *Decreases* in the angle ($\alpha < \pi/2$) result in *positive* shear strain. *Increases* in the angle ($\alpha > \pi/2$) result in *negative* shear strain.

For small shear strain ($\gamma < 0.01$), the following approximation of the trigonometric functions may be used.

$$\tan\gamma \approx \gamma \qquad \sin\gamma \approx \gamma \qquad \cos\gamma \approx 1$$

### 1.5.1 Units of average strains

Equation (1.27) shows that normal strain is dimensionless, hence should have no units. However, to differentiate average strain and strain at a point (discussed in Section 1.6), units of length per unit length are used for average normal strains. Thus, average normal strains are reported with the units of inch/inch (or cm/cm, m/m, etc.). Average shear strains are reported in radians.

In reporting experimental results and for describing very large deformations, strains are given by means of a percentage change, that is, the right-hand side of Equation (1.27) and (1.28) is multiplied by 100 before the results are reported. Thus, a normal strain of 0.5% is equal to a strain of 0.005. The Greek letter mu ($\mu$) is often used in reporting small strains because it stands for *micro* ($\mu = 10^{-6}$). Thus, a strain of 1000 $\mu$in/in represents a normal strain of 0.001 inch per inch.

## 1.6 STRAIN AT A POINT

Let $u$, v, and $w$ be displacements in the $x$, $y$, and $z$ directions, respectively. The **engineering strains** at a point can be visualized by considering the deformation of a parallelepiped of infinitesimal dimensions $\Delta x$, $\Delta y$, and $\Delta z$, shown in Figure 1.17.

As the dimensions of the parallelepiped tend to zero, we get the mathematical definition of engineering strain at a point. Because the limiting operation is in a given direction, we obtain partial derivatives, not the ordinary derivatives. The double subscript in the case of shear strain indicates the change of angle between two coordinate lines defined by the subscripts. We also record double subscripts for normal strain for sake of consistency, as well as for our subsequent matrix definition of strain. The mathematical definitions of engineering strains at a point are as given in Equation (1.29a) through (1.29d).

$$\varepsilon_{xx} = \lim_{\Delta x \to 0}\left(\frac{\Delta u}{\Delta x}\right) = \frac{\partial u}{\partial x} \qquad \varepsilon_{yy} = \lim_{\Delta y \to 0}\left(\frac{\Delta v}{\Delta y}\right) = \frac{\partial v}{\partial y} \qquad \varepsilon_{zz} = \lim_{\Delta z \to 0}\left(\frac{\Delta w}{\Delta z}\right) = \frac{\partial w}{\partial z} \tag{1.29a}$$

$$\gamma_{xy} = \gamma_{yx} = \lim_{\substack{\Delta x \to 0 \\ \Delta y \to 0}}\left(\frac{\Delta u}{\Delta y} + \frac{\Delta v}{\Delta x}\right) = \frac{\partial u}{\partial y} + \frac{\partial v}{\partial x} \tag{1.29b}$$

$$\gamma_{yz} = \gamma_{zy} = \lim_{\substack{\Delta y \to 0 \\ \Delta z \to 0}}\left(\frac{\Delta v}{\Delta z} + \frac{\Delta w}{\Delta y}\right) = \frac{\partial v}{\partial z} + \frac{\partial w}{\partial y} \tag{1.29c}$$

$$\gamma_{zx} = \gamma_{xz} = \lim_{\substack{\Delta x \to 0 \\ \Delta z \to 0}}\left(\frac{\Delta w}{\Delta x} + \frac{\Delta u}{\Delta z}\right) = \frac{\partial w}{\partial x} + \frac{\partial u}{\partial z} \tag{1.29d}$$

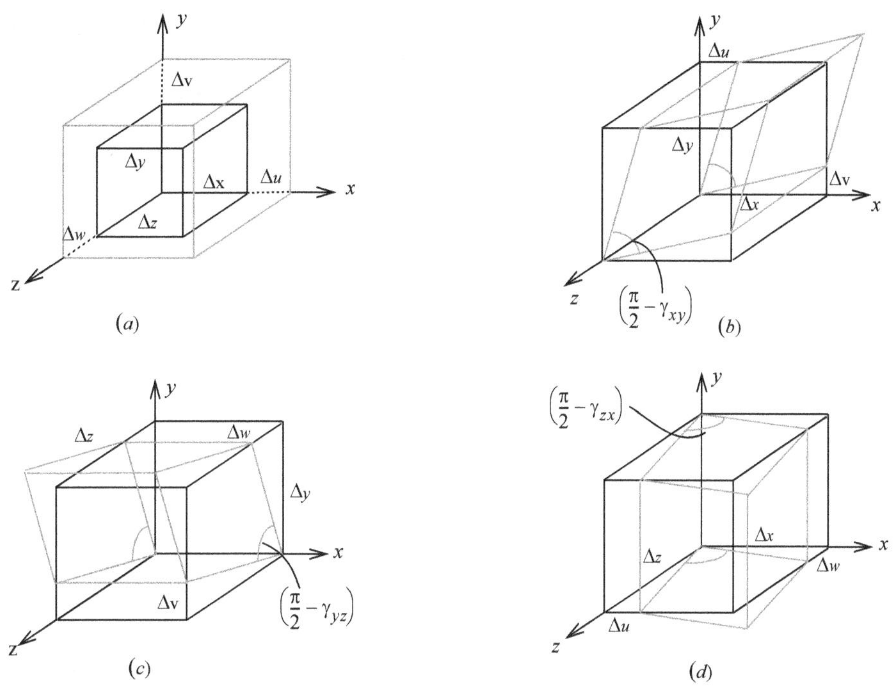


**Figure 1.17** Strain components: (*a*) normal strains, (*b*) shear strain $\gamma_{xy}$, (c) shear strain $\gamma_{yz}$, and (d) shear strain $\gamma_{zx}$.

Equation (1.29b) through (1.29d) show that shear strain is symmetric and that the order of subscripts is immaterial. The symmetry of shear strain makes intuitive sense. The change of angle between the *x* and *y* directions is obviously the same as between the *y* and *x* directions. Note that the sign of shear strain depends only upon whether the angle increases or decreases, not upon the order in which we consider *x* and *y*.

Figure 1.18 shows the engineering strain components as an engineering strain matrix. The matrix is symmetric because of the symmetry of shear strain. Thus, in three-dimensional problems strain has nine components, but only six of the nine components are independent because of symmetry of shear strain.

$$\begin{bmatrix} \varepsilon_{xx} & \gamma_{xy} & \gamma_{xz} \\ \gamma_{yx} & \varepsilon_{yy} & \gamma_{yz} \\ \gamma_{zx} & \gamma_{zy} & \varepsilon_{zz} \end{bmatrix}$$

**Figure 1.18** Engineering strain matrix.

We see that like stress, engineering strain has two subscripts, which would seem to suggest that engineering strain is also a second-order tensor like stress. However unlike stress, engineering strain does not satisfy certain coordinate transformation laws, which we will study in Section 1.7. Hence it is not a second-order tensor but is related to it, as shown in Equation (1.30).

$$\boxed{\text{tensor normal strain} = \text{engineering normal strain} \qquad \text{tensor shear strain} = \text{engineering shear strain}/2} \qquad \textbf{(1.30)}$$

In Section 1.7, we shall see that the factor of 1/2 that changes the engineering shear strain to tensor shear strain permits the extension of stress transformation results to strain transformation.

### 1.6.1 Plane strain

Plane strain is the other type of two-dimensional problems in the study of the mechanics of materials. The differences between plane stress and plane strain are discussed in Section 1.8.3. By "two-dimensional," we imply that one of the coordinates does not play a role in the description of the problem, thus we set all strains with subscript *z* to be zero as shown in the strain matrix in Figure 1.19. Although four components of strain are needed in plane strain, because of the symmetry of shear strain, only three of the four components are independent.

$$\begin{bmatrix} \varepsilon_{xx} & \gamma_{xy} & 0 \\ \gamma_{yx} & \varepsilon_{yy} & 0 \\ 0 & 0 & 0 \end{bmatrix}$$

**Figure 1.19** Plane strain matrix.

The assumption of plane strain is often made in analyzing very thick bodies such as points around tunnels or mine shafts, or a point in the middle of a thick cylinder such as a submarine hull. In thick bodies, it is argued that to deform, a point must push a lot of material in the thickness direction; hence the strains can be expected to be small in the thickness direction. Thus, the strain in the thickness direction is not exactly zero but is small enough to neglect. Plane strain is a mathematical approximation made to simplify the analysis.

### 1.6.2 Polar coordinates

*ABCD* is the differential element in polar coordinate in undeformed state shown Figure 1.20. The differential element *ABCD* deforms to $A_1B_1C_1D_1$ due to radial displacement $u_r$ in Figure 1.20*a*. Point *A* moves to point $A_1$ due to the radial displacement $u_r$. Point *B* move to points $B_1$ by an increased amount as $u_r$ increases in the *r* direction. Movement of point *D* to $D_1$ is caused by the radial displacement $u_r$ plus the increase of $u_r$ in the θ direction which also changes the angle by α between the radial and tangential direction. This change of angle will contribute towards the shear strain. The radial normal strain is given by

$$\varepsilon_{rr} = \frac{A_1B_1 - AB}{AB} = \lim_{\Delta r \to 0}\left[\frac{u_r + (\partial u_r/\partial r)\Delta r - u_r}{\Delta r}\right] = \frac{\partial u_r}{\partial r}$$

$$\boxed{\varepsilon_{rr} = \frac{\partial u_r}{\partial r}} \qquad \textbf{(1.31a)}$$

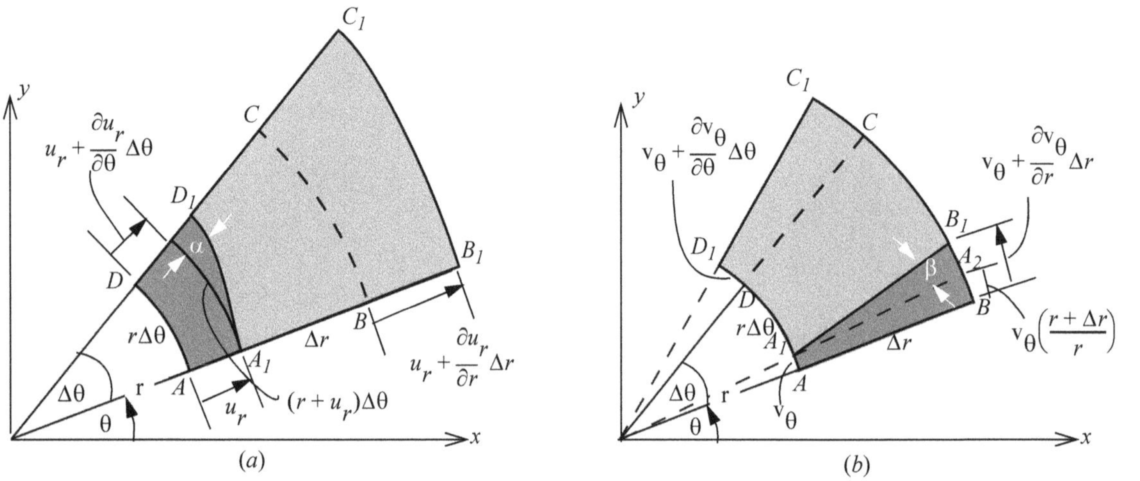


**Figure 1.20** Strains in polar coordinates from (*a*) just radial displacement $u_r$. (*b*) just tangential displacement $v_\theta$.

The differential element *ABCD* deforms to $A_1B_1C_1D_1$ due to tangential displacement $v_\theta$ in Figure 1.20*b*. Points *A* and *D* move to points $A_1$ and $D_1$ due to the tangential displacement $v_\theta$. Movement of point *B* to $B_1$ is due to the increase in $v_\theta$ in the θ direction. The change of angle β between the radial and tangential direction requires us to subtract the distance $A_2B$ which is due to increase of radius from *r* to $r + \Delta r$. The distance $A_2B$ can be found using similar triangle and is as shown.

The length *AD* changes due to both displacements as seen in both figures, hence the tangential normal strain is

$$\varepsilon_{\theta\theta} = \frac{A_1D_1 - AD}{AD} = \lim_{\Delta\theta \to 0}\left[\frac{(r+u_r)\Delta\theta - r\Delta\theta}{r\Delta\theta} + \frac{v_\theta + (\partial v_\theta/\partial\theta)\Delta\theta - v_\theta}{r\Delta\theta}\right]$$

$$\boxed{\varepsilon_{\theta\theta} = \frac{u_r}{r} + \frac{1}{r}\frac{\partial v_\theta}{\partial\theta}} \qquad \textbf{(1.31b)}$$

We can find each of the angle as follows

$$tan\alpha \approx \alpha = \lim_{\Delta\theta \to 0}\left[\frac{u_r + (\partial u_r/\partial\theta)\Delta\theta - u_r}{(r+u_r)\Delta\theta}\right] \approx \lim_{\Delta\theta \to 0}\left[\frac{(\partial u_r/\partial\theta)\Delta\theta}{r\Delta\theta}\right] \approx \frac{1}{r}\frac{\partial u_r}{\partial\theta}$$

$$tan\beta \approx \beta = \lim_{\Delta r \to 0}\left[\frac{v_\theta + (\partial v_\theta/\partial r)\Delta r - v_\theta(r+\Delta r)/r}{\Delta r}\right] = \frac{\partial v_\theta}{\partial r} - \frac{v_\theta}{r}$$

The shear strain is the change of angle *DAB*, hence is the sum of the two angles, that is, $\gamma_{r\theta} = \alpha + \beta$.

$$\boxed{\gamma_{r\theta} = \frac{1}{r}\frac{\partial u_r}{\partial\theta} + \frac{\partial v_\theta}{\partial r} - \frac{v_\theta}{r}} \qquad \textbf{(1.31c)}$$

## 1.7 STRAIN TRANSFORMATION

Strain transformation is relating strains in two coordinate systems. We once more start with the two-dimensional problem of plane strain. Obtain equations that were obtained in the introductory course on the mechanics of materials and show how the stress transformation equations can be used for strain transformation.

Figure 1.21 shows two coordinate systems. The *n*, *t*, *z* coordinate system is obtained from *x*, *y*, *z* coordinate system by rotating counterclockwise about the *z* axis by an angle of θ. The coordinates can be related as shown below.

$$x = ncos\theta - tsin\theta \qquad y = nsin\theta + tcos\theta \qquad \textbf{(1.32a)}$$

The displacements in Cartesian coordinates ($u$, v) can be related to displacements ($u_n$, $v_t$) as shown below.

$$u_n = u\cos\theta + v\sin\theta \qquad v_t = -u\sin\theta + v\cos\theta \tag{1.32b}$$

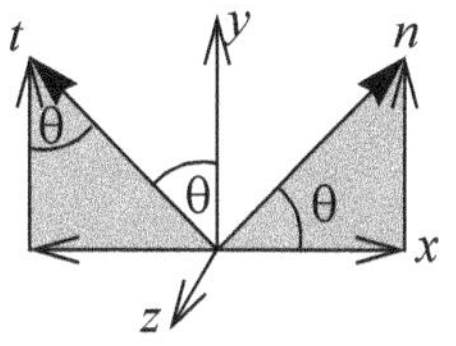


**Figure 1.21** Coordinate transformation.

Using chain rule we can write the normal strain in $n$ direction as

$$\varepsilon_{nn} = \frac{\partial u_n}{\partial n} = \frac{\partial u_n}{\partial x}\frac{\partial x}{\partial n} + \frac{\partial u_n}{\partial y}\frac{\partial y}{\partial n} \tag{1.32c}$$

From Equation (1.32a) we have $\partial x/\partial n = \cos\theta$ and $\partial y/\partial n = \sin\theta$. Substituting these results and Equation (1.32b) we obtain the following.

$$\varepsilon_{nn} = \frac{\partial}{\partial x}(u\cos\theta + v\sin\theta)\cos\theta + \frac{\partial}{\partial y}(u\cos\theta + v\sin\theta)\sin\theta$$

$$\varepsilon_{nn} = \left(\frac{\partial u}{\partial x}\right)\cos^2\theta + \left(\frac{\partial v}{\partial x}\right)\sin\theta\cos\theta + \left(\frac{\partial u}{\partial y}\right)\cos\theta\sin\theta + \left(\frac{\partial v}{\partial y}\right)\sin^2\theta$$

$$\varepsilon_{nn} = \varepsilon_{xx}\cos^2\theta + \varepsilon_{yy}\sin^2\theta + \gamma_{xy}\sin\theta\cos\theta \tag{1.33a}$$

In a similar manner we obtain the following equations

$$\varepsilon_{tt} = \varepsilon_{xx}\sin^2\theta + \varepsilon_{yy}\cos^2\theta - \gamma_{xy}\sin\theta\cos\theta \tag{1.33b}$$

$$\gamma_{nt} = -2\varepsilon_{xx}\sin\theta\cos\theta + 2\varepsilon_{yy}\sin\theta\cos\theta + \gamma_{xy}(\cos^2\theta - \sin^2\theta) \tag{1.33c}$$

The above equations are similar to the stress transformation relations [Equations (1.2a) through (1.2b)] with the difference that the coefficient of shear stress term is twice the coefficient of the shear strain term. This difference exists because we are using engineering strain instead of tensor strain. If we convert the engineering strains to tensor strains, by dividing all engineering shear strains by 2, then the results for the matrix method in stress transformation can be adapted for strain transformation. Figure 1.22 shows the tensor strain matrix that is obtained from the engineering strains.

$$[\varepsilon] = \begin{bmatrix} \varepsilon_{xx} & \varepsilon_{xy} = \gamma_{xy}/2 & \varepsilon_{xz} = \gamma_{xz}/2 \\ \varepsilon_{yx} = \gamma_{yx}/2 & \varepsilon_{yy} & \varepsilon_{yz} = \gamma_{yz}/2 \\ \varepsilon_{zx} = \gamma_{zx}/2 & \varepsilon_{zy} = \gamma_{zy}/2 & \varepsilon_{zz} \end{bmatrix}$$

**Figure 1.22** Tensor strain matrix obtained from an engineering strain matrix.

The two perpendicular directions $n$ and $t$ in Figure 1.21 can be represented by unit vectors $\{n\}$ and $\{t\}$. Analogous to Equation (1.6) we obtain the strain transformation equations for the normal and shear strain in the $n$-$t$ directions as shown below.

$$\boxed{\varepsilon_{nn} = \{n\}^T[\varepsilon]\{n\} \qquad \varepsilon_{nt} = \{t\}^T[\varepsilon]\{n\} \qquad \varepsilon_{tt} = \{t\}^T[\varepsilon]\{t\}} \tag{1.34}$$

The shear strain $\varepsilon_{nt}$ in Equation (1.34) is tensor shear strain and must be multiplied by 2 to obtain the engineering shear strain $\gamma_{nt}$. The strain vector in the direction of $n$ is analogous to the stress vector and can be defined as shown in Equation (1.35).

$$\{E\} = [\varepsilon]\{n\} \tag{1.35}$$

### 1.7.1 Principal strains

We recall the following definitions from the introductory course on the mechanics of material. The **principal coordinate directions** for strain are the axes in which the *shear strain is zero.* Normal strains in the principal directions are called **principal strains**. The *greatest* principal strain is called **principal strain 1**. The angles the principal axis makes with the global coordinate system are called *principal angles*.

In plane strain, the shear strains with subscripts $z$ are zero, which implies that one principal direction lies along the $z$ axis. In plane strain problems we will label the $z$ direction as principal direction 3. There are three principal strains at a point, which are labeled as $\varepsilon_1$, $\varepsilon_2$, and $\varepsilon_3$.

The principal strains $\varepsilon_p$ are the eigenvalues of the tensor strain matrix, and the principal directions $\{p\}$ are the eigenvectors of the strain matrix, which are found in exactly the same manner as the principal stresses and the associated principal directions. The characteristic equation analogous to Equation (1.18) is

$$\varepsilon_p^3 - I_1\varepsilon_p^2 + I_2\varepsilon_p - I_3 = 0 \tag{1.36}$$

and its roots of the characteristic yield the principal strains. The coefficients $I_1$, $I_2$, and $I_3$ are called the strain invariants because their value is independent of coordinate transformation. Notice that the sign in terms containing the odd stress invariants is negative.

The strain invariants $I_1$, $I_2$, and $I_3$, analogous to stress invariants in Equation (1.19) are

$$I_1 = \varepsilon_{xx} + \varepsilon_{yy} + \varepsilon_{zz} \qquad I_2 = \begin{vmatrix} \varepsilon_{xx} & \varepsilon_{xy} \\ \varepsilon_{yx} & \varepsilon_{yy} \end{vmatrix} + \begin{vmatrix} \varepsilon_{yy} & \varepsilon_{yz} \\ \varepsilon_{zy} & \varepsilon_{zz} \end{vmatrix} + \begin{vmatrix} \varepsilon_{xx} & \varepsilon_{xz} \\ \varepsilon_{zx} & \varepsilon_{zz} \end{vmatrix} \qquad I_3 = \begin{vmatrix} \varepsilon_{xx} & \varepsilon_{xy} & \varepsilon_{xz} \\ \varepsilon_{yx} & \varepsilon_{yy} & \varepsilon_{yz} \\ \varepsilon_{zx} & \varepsilon_{zy} & \varepsilon_{zz} \end{vmatrix} \tag{1.37}$$

The strain invariants in terms of principal strains are

$$I_1 = \varepsilon_1 + \varepsilon_2 + \varepsilon_3 \qquad I_2 = \begin{vmatrix} \varepsilon_1 & 0 \\ 0 & \varepsilon_2 \end{vmatrix} + \begin{bmatrix} \varepsilon_2 & 0 \\ 0 & \sigma_3 \end{bmatrix} + \begin{bmatrix} \varepsilon_1 & 0 \\ 0 & \varepsilon_3 \end{bmatrix} = \varepsilon_1\varepsilon_2 + \varepsilon_2\varepsilon_3 + \varepsilon_3\varepsilon_1 \qquad I_3 = \begin{vmatrix} \varepsilon_1 & 0 & 0 \\ 0 & \varepsilon_2 & 0 \\ 0 & 0 & \varepsilon_3 \end{vmatrix} = \varepsilon_1\varepsilon_2\varepsilon_3 \tag{1.38}$$

The strain invariants, particularly the first and the third, provide a quick and easy check on the laboriously calculated principal strains.

The maximum engineering shear strain is related to the difference in principal strains and can be expressed as

$$\frac{\gamma_{max}}{2} = \max\left(\left|\frac{\varepsilon_1 - \varepsilon_2}{2}\right|, \left|\frac{\varepsilon_2 - \varepsilon_3}{2}\right|, \left|\frac{\varepsilon_3 - \varepsilon_1}{2}\right|\right) \tag{1.39}$$

## 1.8 MATERIAL MODEL

In mechanics of materials, the experimentally determined material relationship between stresses and strains is incorporated into the logical framework of mechanics to produce formulas for the analysis and design of structural members. Figure 1.23 symbolically shows the relationship between stress and strain. If a material model does not fit the experimental data well, there will be a high degree of error in the theoretical predictions. Yet a material model that fits the experimental data very accurately may be so complex that no analytical model (theory) can be built. Thus, the choice of material model is dictated both by the experimental data and by the accuracy needs of the analysis.

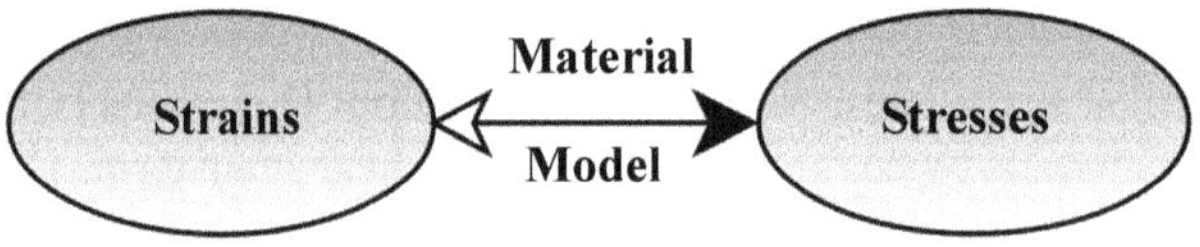


**Figure 1.23** Material model.

### 1.8.1 Linear material model

The simplest material model is a linear relationship between stresses and strains. With no additional assumptions, the linear relationship of the six independent strain components to six independent stress components can be written as shown in Equation (1.40). The matrix $C_{ij}$ in Equation (1.40) is called the compliance matrix. Equation (1.40) implies that we need 36 material constants to describe the most general linear relationship between stresses and strains. However, it can be shown that the matrix formed by the constants $C_{ij}$ is a symmetric matrix; that is, $C_{ij} = C_{ji}$, where $i$ and $j$ can be any number from 1 to 6. This symmetry can be proved by using the requirement that the strain energy—energy stored in a material owing to deformation, must always be positive. The symmetry requirement reduces the number of independent constants to 21 for the most general linear relationship between stress and strain.

$$\begin{aligned}
\varepsilon_{xx} &= C_{11}\sigma_{xx} + C_{12}\sigma_{yy} + C_{13}\sigma_{zz} + C_{14}\tau_{yz} + C_{15}\tau_{zx} + C_{16}\tau_{xy} \\
\varepsilon_{yy} &= C_{21}\sigma_{xx} + C_{22}\sigma_{yy} + C_{23}\sigma_{zz} + C_{24}\tau_{yz} + C_{25}\tau_{zx} + C_{26}\tau_{xy} \\
\varepsilon_{zz} &= C_{31}\sigma_{xx} + C_{32}\sigma_{yy} + C_{33}\sigma_{zz} + C_{34}\tau_{yz} + C_{35}\tau_{zx} + C_{36}\tau_{xy} \\
\gamma_{yz} &= C_{41}\sigma_{xx} + C_{42}\sigma_{yy} + C_{43}\sigma_{zz} + C_{44}\tau_{yz} + C_{45}\tau_{zx} + C_{46}\tau_{xy} \\
\gamma_{zx} &= C_{51}\sigma_{xx} + C_{52}\sigma_{yy} + C_{53}\sigma_{zz} + C_{54}\tau_{yz} + C_{55}\tau_{zx} + C_{56}\tau_{xy} \\
\gamma_{xy} &= C_{61}\sigma_{xx} + C_{62}\sigma_{yy} + C_{63}\sigma_{zz} + C_{64}\tau_{yz} + C_{65}\tau_{zx} + C_{66}\tau_{xy}
\end{aligned}
\quad \text{or} \quad
\begin{Bmatrix} \varepsilon_{xx} \\ \varepsilon_{yy} \\ \varepsilon_{zz} \\ \gamma_{yz} \\ \gamma_{zx} \\ \gamma_{xy} \end{Bmatrix} = \begin{bmatrix} C_{11} C_{12} C_{13} C_{14} C_{15} C_{16} \\ C_{12} C_{22} C_{23} C_{24} C_{25} C_{26} \\ C_{13} C_{23} C_{33} C_{34} C_{35} C_{36} \\ C_{41} C_{42} C_{43} C_{44} C_{45} C_{46} \\ C_{51} C_{52} C_{53} C_{54} C_{55} C_{56} \\ C_{61} C_{62} C_{63} C_{64} C_{65} C_{66} \end{bmatrix} \begin{Bmatrix} \sigma_{xx} \\ \sigma_{yy} \\ \sigma_{zz} \\ \tau_{yz} \\ \tau_{zx} \\ \tau_{xy} \end{Bmatrix} \tag{1.40}$$

Equation (1.40) presupposes that the relation between the stress and the strain in the $x$ direction is different from the relation of stress and strain in the $y$ or $z$ directions. Alternatively stated, Equation (1.40) implies that if we apply a force (stress) in the $x$ direction and observe the deformation (strain), this deformation will be different from the deformation that would be produced if we applied the same force in the $y$ direction. This phenomenon is not observable by the naked eye for most metals, but for metals at the crystal-size level, the number of constants needed to describe the stress–strain relationship does depend upon the crystal structure. Thus, are we conducting the analysis at the eye level or at the crystal-size level? For analysis at the eye level, we average the impact of the crystal structure, to obtain the simplest material description. An **isotropic material** has stress–strain relationships that are independent of the orientation of the coordinate system at a point.

An **anisotropic** material is a material that is not isotropic. The most general anisotropic material requires *21 independent material constants* to describe its linear stress–strain relationships. An isotropic body requires only *two independent* material constants to describe its linear stress–strain relationships. Some of the factors that influence whether we treat a material as isotropic or anisotropic are the degree of difference in material properties with orientation, the scale at which the analysis is being conducted, and the kind of information that is desired from the analysis.

Homogeneity is another approximation that is often used to describe a material behavior. A material is said to be **homogeneous** if the material properties are the same at all points on the body. Alternatively, if the material constants $C_{ij}$ are functions of the coordinates $x$, $y$, or $z$, the material is called **nonhomogeneous.** Most materials at the atomic level, crystalline level, or grain-size level are nonhomogeneous. The treatment of a material as homogeneous or nonhomogeneous depends once more on the type of information that is required from the analysis. Homogenization of material properties is the process of averaging different material properties by an overall material property. Any body can be treated as a homogeneous body if the scale at which analysis is conducted is made sufficiently large.

Isotropic–homogeneous, anisotropic–homogeneous, isotropic–nonhomogeneous, and anisotropic–nonhomogeneous are all possible descriptions of material behavior.

There are 31 types of crystal. Crystalline bodies can be grouped into classes for the purpose of defining the independent material constants needed in the linear stress–strain relationship. In between the isotropic material and the most general anisotropic material, there are material groups of several types, some of which are discussed briefly below.

1. **Monoclinic materials**: The $xy$ plane is the plane of symmetry in monoclinic materials. This implies that the stress–strain relationship is the same in the positive $z$ and negative $z$ directions. A monoclinic material requires *13 independent material constants in three dimensions.*
2. **Orthotropic materials**: Orthotropic materials have two axis of symmetry. That is, if we rotate by $90^o$ about the $x$ or the $y$ axis, we will obtain the same stress–strain relation. An orthotropic material requires *nine independent constants in three dimensions and four independent constants in two dimensions.*
3. **Transversely isotropic materials**: Materials are isotropic in a plane. That is, rotation by an arbitrary angle about the $z$ axis does not change the stress–strain relations, and the material is isotropic in the $x$-$y$ plane. Transversely isotropic material requires *five independent material constants* in three dimensions.

## 1.8.2 Isotropic material

An isotropic material requires only *two independent material constants.* There are many constants used to describe isotropic material, but only two are independent. Rotation about the $x$, $y$, or $z$ axis by any arbitrary angle results in the same stress–strain relationship. In matrix notation, the stress–strain relationship for isotropic material is given by Equation (1.41).

$$\begin{Bmatrix} \varepsilon_{xx} \\ \varepsilon_{yy} \\ \varepsilon_{zz} \\ \gamma_{yz} \\ \gamma_{zx} \\ \gamma_{xy} \end{Bmatrix} = \begin{bmatrix} C_{11} & C_{12} & C_{12} & 0 & 0 & 0 \\ C_{12} & C_{11} & C_{12} & 0 & 0 & 0 \\ C_{12} & C_{12} & C_{11} & 0 & 0 & 0 \\ 0 & 0 & 0 & 2(C_{11}-C_{12}) & 0 & 0 \\ 0 & 0 & 0 & 0 & 2(C_{11}-C_{12}) & 0 \\ 0 & 0 & 0 & 0 & 0 & 2(C_{11}-C_{12}) \end{bmatrix} \begin{Bmatrix} \sigma_{xx} \\ \sigma_{yy} \\ \sigma_{zz} \\ \tau_{yz} \\ \tau_{zx} \\ \tau_{xy} \end{Bmatrix} \quad \textbf{(1.41)}$$

Equation (1.42) along with Equation (1.43), are called the **Generalized Hooke's law** that was studied in the introductory course of mechanics of materials. Equation (1.41) can be written using three constants: the modulus of elasticity $E$, the Poisson ratio $\nu$, and the shear modulus of elasticity $G$ as shown in Equation (1.42).

$$\boxed{\begin{matrix} \varepsilon_{xx} = [\sigma_{xx} - \nu(\sigma_{yy}+\sigma_{zz})]/E & \varepsilon_{yy} = [\sigma_{yy} - \nu(\sigma_{zz}+\sigma_{xx})]/E & \varepsilon_{zz} = [\sigma_{zz} - \nu(\sigma_{xx}+\sigma_{yy})]/E \\ \gamma_{xy} = \tau_{xy}/G & \gamma_{yz} = \tau_{yz}/G & \gamma_{zx} = \tau_{zx}/G \end{matrix}} \quad \textbf{(1.42)}$$

Comparing Equation (1.42) with Equation (1.41), we obtain: $C_{11} = 1/E \qquad C_{12} = -\nu/E \qquad 2(C_{11} - C_{12}) = 1/G$
Substituting $C_{11}$ and $C_{12}$ in the last relationship, we obtain

$$\boxed{G = \frac{E}{2(1+\nu)}} \quad \textbf{(1.43)}$$

An alternative form for Equation (1.42) that may be easier to remember is the matrix form given below.

$$\begin{Bmatrix} \varepsilon_{xx} \\ \varepsilon_{yy} \\ \varepsilon_{zz} \end{Bmatrix} = \frac{1}{E}\begin{bmatrix} 1 & -\nu & -\nu \\ -\nu & 1 & -\nu \\ -\nu & -\nu & 1 \end{bmatrix} \begin{Bmatrix} \sigma_{xx} \\ \sigma_{yy} \\ \sigma_{zz} \end{Bmatrix} \quad \textbf{(1.44)}$$

The Generalized Hooke's law for *isotropic material* is valid for *any orthogonal coordinate system* at a point, including a point in a non homogeneous material. Thus, we could write equivalent forms for Equation (1.42) for cylindrical (polar) coordinates $(r, \theta, z)$ and for spherical coordinates $(r, \theta, \phi)$. The principal coordinate systems for stresses and strains are also orthogonal. For *only isotropic materials*, the principal directions for stresses and strains are identical; hence we can write the generalized Hooke's law in principal coordinate form as shown below.

$$\boxed{\varepsilon_1 = [\sigma_1 - \nu(\sigma_2 + \sigma_3)]/E \qquad \varepsilon_2 = [\sigma_2 - \nu(\sigma_3 + \sigma_1)]/E \qquad \varepsilon_3 = [\sigma_3 - \nu(\sigma_1 + \sigma_2)]/E} \tag{1.45}$$

*Principal stresses and principal strains are different for orthotropic and other anisotropic materials* and similar equations cannot be written for these materials.

The normal stresses can be solved in terms of normal strains by using the Equation (1.42) to obtain.

$$\sigma_{xx} = \frac{E[(1-\nu)\varepsilon_{xx} + \nu\varepsilon_{yy} + \nu\varepsilon_{zz}]}{(1-2\nu)(1+\nu)} \qquad \sigma_{yy} = \frac{E[\nu\varepsilon_{xx} + (1-\nu)\varepsilon_{yy} + \nu\varepsilon_{zz}]}{(1-2\nu)(1+\nu)} \qquad \sigma_{zz} = \frac{E[\nu\varepsilon_{xx} + \nu\varepsilon_{yy} + (1-\nu)\varepsilon_{zz}]}{(1-2\nu)(1+\nu)} \tag{1.46}$$

## 1.8.3 Plane stress and plane strain

If we take the definitions of plane stress and plane strain and apply them to Equation (1.42), we will obtain the right-most matrices in Figure 1.24. The differences between the pair of two-dimensional idealizations of material behavior are in the zero and non-zero values of the normal strain and normal stress in the $z$ direction. In plane stress, $\sigma_{zz} = 0$, which implies that the normal strain in the $z$ direction is $\varepsilon_{zz} = -\nu(\sigma_{xx} + \sigma_{yy})/E$. In plane strain, $\varepsilon_{zz} = 0$, which implies that the normal stress in the $z$ direction is $\sigma_{zz} = \nu(\sigma_{xx} + \sigma_{yy})$.

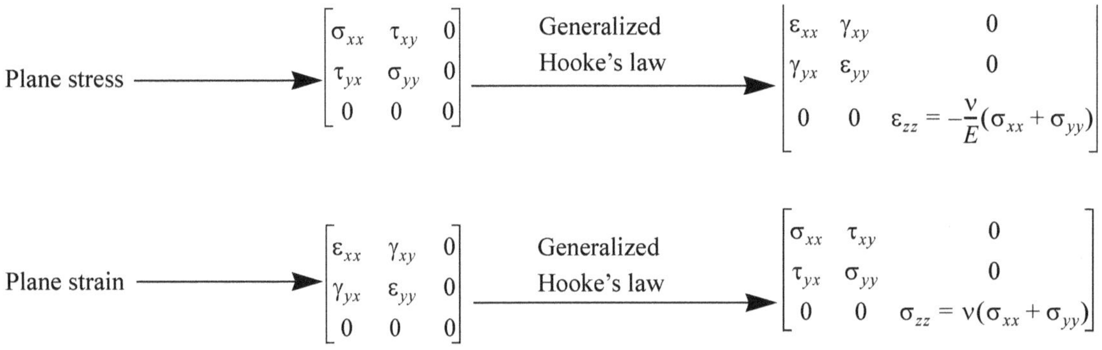


$$\text{Plane stress} \rightarrow \begin{bmatrix} \sigma_{xx} & \tau_{xy} & 0 \\ \tau_{yx} & \sigma_{yy} & 0 \\ 0 & 0 & 0 \end{bmatrix} \xrightarrow{\text{Generalized Hooke's law}} \begin{bmatrix} \varepsilon_{xx} & \gamma_{xy} & 0 \\ \gamma_{yx} & \varepsilon_{yy} & 0 \\ 0 & 0 & \varepsilon_{zz} = -\frac{\nu}{E}(\sigma_{xx} + \sigma_{yy}) \end{bmatrix}$$

$$\text{Plane strain} \rightarrow \begin{bmatrix} \varepsilon_{xx} & \gamma_{xy} & 0 \\ \gamma_{yx} & \varepsilon_{yy} & 0 \\ 0 & 0 & 0 \end{bmatrix} \xrightarrow{\text{Generalized Hooke's law}} \begin{bmatrix} \sigma_{xx} & \tau_{xy} & 0 \\ \tau_{yx} & \sigma_{yy} & 0 \\ 0 & 0 & \sigma_{zz} = \nu(\sigma_{xx} + \sigma_{yy}) \end{bmatrix}$$

**Figure 1.24** Stress and strain matrices in plane stress and plane strain.

To better appreciate the difference between plane stress and plane strain, consider the two plates shown in Figure 1.25, on which only compressive normal stresses in the $x$ and $y$ direction are applied. The top and bottom surfaces on the plate in Figure 1.25*a* are free surfaces (plane stress); but because the plate is free to expand, the deformation (strain) in the $z$ direction is not zero. The plate in Figure 1.25*b* is constrained from expanding in the $z$ direction by the rigid plates. As the material pushes on the plate, a reaction force develops, and this reaction force results in a nonzero value of normal stress in the $z$ direction.

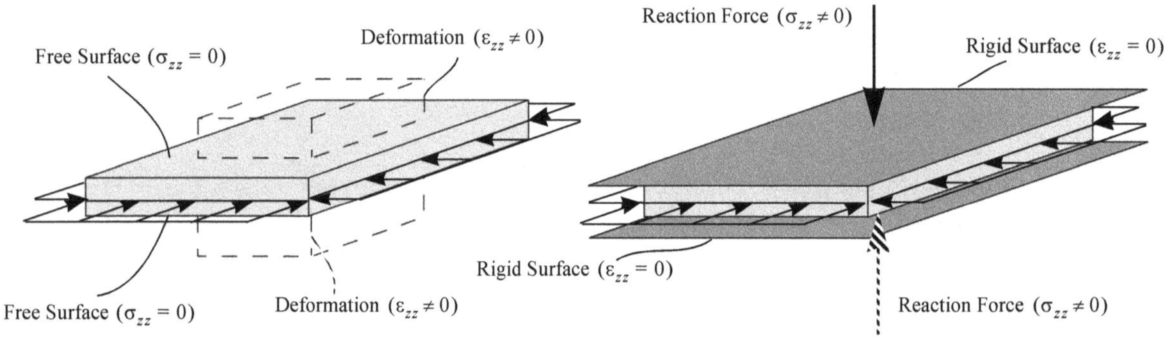


**Figure 1.25** (*a*) Plane stress. (*b*) Plane strain.

Though the example in Figure 1.25 helps explain the difference, it should be emphasized that plane stress or plane strain often is an approximation made to simplify analysis. The approximation of plane stress is often made in thin plates and shells, such as the skin of an aircraft or the walls of pressure vessels. The approximation of plane strain is often made in thick plates and shells, such as submarine hulls or points beneath the earth's surface. A difference of an order of magnitude or more between the thickness and other dimensions is generally used to classify a body as "*thin.*"

It should be recognized that in plane strain and plane stress there are only three independent quantities—even though the number of nonzero quantities is more than three. For example, if we know $\sigma_{xx}$, $\sigma_{yy}$, and $\tau_{xy}$, we can calculate $\varepsilon_{xx}$, $\varepsilon_{yy}$, $\gamma_{xy}$, $\varepsilon_{zz}$, and $\sigma_{zz}$ for plane stress and plane strain. Similarly, if we know $\varepsilon_{xx}$, $\varepsilon_{yy}$, and $\gamma_{xy}$, we can calculate $\sigma_{xx}$, $\sigma_{yy}$, $\tau_{xy}$, $\sigma_{zz}$, and $\varepsilon_{zz}$ for plane stress and plane strain. Thus, in both plane stress and plane strain, the number of independent stress or strain components is three, although the number of nonzero components is greater than three.

## EXAMPLE 1.3

The engineering strains at a point on a building frame were determined as $\varepsilon_{xx} = 600\ \mu$, $\varepsilon_{yy} = 400\ \mu$, $\varepsilon_{zz} = 200\ \mu$, $\gamma_{xy} = 800\ \mu$, $\gamma_{yz} = 400\ \mu$, $\gamma_{zx} = 600\ \mu$. The modulus of elasticity of the frame material is 200 GPa and Poisson's ratio is 1/3. Determine the principal stresses, maximum shear stress and principal stress direction 1.

**PLAN: Method I**

We construct the tensor strain matrix and determine the principal strains and principal direction 1. For isotropic material the principal directions for strain and stress are the same. Using Generalized Hooke's law in principal coordinates we determine the principal stresses and maximum shear stress from maximum shear strain.

**SOLUTION**

The tensor strain matrix can be written as

$$[\varepsilon] = \begin{bmatrix} 600 & 400 & 300 \\ 400 & 400 & 200 \\ 300 & 200 & 200 \end{bmatrix} \mu \tag{E1}$$

We can find the three strain invariants from Equation (1.37) as shown below.

$$I_1 = 600 + 400 + 200 = 1200\ \mu \tag{E2}$$

$$I_2 = \begin{vmatrix} 600 & 400 \\ 400 & 400 \end{vmatrix} + \begin{vmatrix} 400 & 200 \\ 200 & 200 \end{vmatrix} + \begin{vmatrix} 600 & 300 \\ 300 & 200 \end{vmatrix} = 150(10^3)\ \mu^2 \tag{E3}$$

$$I_3 = 600\begin{vmatrix} 400 & 200 \\ 200 & 200 \end{vmatrix} - 400\begin{vmatrix} 400 & 200 \\ 300 & 200 \end{vmatrix} + 300\begin{vmatrix} 400 & 400 \\ 300 & 200 \end{vmatrix} = 4000(10^3)\ \mu^3 \tag{E4}$$

The characteristic equation thus is given by

$$\varepsilon_p^3 - 1,200\varepsilon_p^2 + 150(10^3)\varepsilon_p - 4000(10^3) = 0 \tag{E5}$$

The principal strains are calculated using the formulas given in footnote 2 on page 9 as shown below.

$$A = \sqrt{(1200/3)^2 - 150(10^3)/3} = 331.67 \tag{E6}$$

$$\cos 3\alpha = \left[2\left(\frac{1200}{3}\right)^3 - \left(\frac{1200}{3}\right)(150)(10^3) + 4000(10^3)\right] / 2(331.67)^3 = 0.9867 \quad \text{or} \quad \alpha = 9.356/3 = 3.111 \tag{E7}$$

$$x_1 = 2(331.67)\cos 3.111 + 1200/3 = 1062.34 \tag{E8}$$

$$x_2 = -2(331.67)\cos(63.111) + 1200/3 = 100.0 \qquad x_3 = -2(331.67)\cos(-56.889) + 1200/3 = 37.65 \tag{E9}$$

$$\textbf{ANS.}\quad \varepsilon_1 = 1062\ \mu\ ;\ \varepsilon_2 = 100\ \mu\ ;\ \varepsilon_3 = 37.6\ \mu \tag{E10}$$

*Checking Results*: The first and third strain invariants are

$I_1 = 1062.36 + 100.07 + 37.57 = 1200 \rightarrow$ Checks $\qquad I_3 = (1062.36)(100.07)(37.57) = 1200 \rightarrow$ Checks

The maximum engineering shear strain is

$$\gamma_{max}/2 = |(\varepsilon_1 - \varepsilon_3)|/2 = 1024.7\ \mu \tag{E11}$$

On substituting $\varepsilon_1$ in the equations corresponding to the first two rows of the matrix, we obtain

$$(600 - 1062.36)p_x + 400p_y + 300p_z = 0 \quad \text{or} \quad 400p_y + 300p_z = 462.4p_x \tag{E12}$$

$$400p_x + (400 - 1062.36)p_y + 200p_z = 0 \quad \text{or} \quad -662.4p_y + 200p_z = -400p_x \tag{E13}$$

Solving Equation (E12) and (E13) in terms of $p_x$ we obtain

$$p_y = 0.7624p_x \qquad p_z = 0.5247p_x \tag{E14}$$

Noting that the sum of square of direction cosine is one, we obtain

$$p_x^2 + (0.7624p_x)^2 + (0.5247p_x)^2 = 1 \quad \text{or} \quad p_x = \pm 0.7339 \tag{E15}$$

Substituting $p_x$ into Equation (E14), we obtain $p_y = \pm 0.5595$ and $p_z = \pm 0.3851$. The two solutions for principal direction are:

$$\textbf{ANS.}\quad \theta_x = 42.8°;\theta_y = 56°;\theta_z = 67.4° \quad \text{or} \quad \theta_x = 137.2°;\theta_y = \theta_y = 124°;\theta_z = 112.6°$$

*Checking Results*: The normal strain in principal direction 1 should be the principal strain 1.

$$\varepsilon_1 = \{p_1\}^T \begin{bmatrix} 600 & 400 & 300 \\ 400 & 400 & 200 \\ 300 & 200 & 200 \end{bmatrix} \begin{Bmatrix} 0.7339 \\ 0.5595 \\ 0.3851 \end{Bmatrix} = \begin{Bmatrix} 0.7339 \\ 0.5595 \\ 0.3851 \end{Bmatrix}^T \begin{Bmatrix} 779.7 \\ 594.4 \\ 409.1 \end{Bmatrix} = 1062.3 \rightarrow \text{Checks}$$

The shear modulus of elasticity and the constant in Equation (1.46) can be found as shown below.

$$G = \frac{E}{2(1+\nu)} = \frac{200}{2(1+1/3)} = 75\text{GPa} \qquad \frac{E}{(1-2\nu)(1+\nu)} = \frac{(200)}{(1-2/3)(1+1/3)} = 450\text{GPa} \tag{E16}$$

The maximum shear stress is

$$\tau_{max} = G\gamma_{max} = 75(10^9)(1024.7)(10^{-6}) = 76.85(10^6)\ \text{N/m}^2 \quad \text{(E17)}$$

Equation (1.46) can be written in principal coordinates and principal stresses calculated as shown below.

$$\sigma_1 = \frac{E[(1-\nu)\varepsilon_1 + \nu\varepsilon_2 + \nu\varepsilon_3]}{(1-2\nu)(1+\nu)} = (450)(10^9)\left[\frac{2}{3}(1062.36) + \frac{1}{3}(100) + \frac{1}{3}(37.65)\right](10^{-6}) = 339.4(10^6)\ \text{N/m}^2 \quad \text{(E18)}$$

$$\sigma_2 = \frac{E[\nu\varepsilon_1 + (1-\nu)\varepsilon_2 + \nu\varepsilon_3]}{(1-2\nu)(1+\nu)} = (450)(10^9)\left[\frac{2}{3}(100) + \frac{1}{3}(37.65) + \frac{1}{3}(1062.36)\right](10^{-6}) = 195.0(10^6)\ \text{N/m}^2 \quad \text{(E19)}$$

$$\sigma_3 = \frac{E[\nu\varepsilon_1 + \nu\varepsilon_2 + (1-\nu)\varepsilon_3]}{(1-2\nu)(1+\nu)} = (450)(10^9)\left[\frac{2}{3}(37.65) + \frac{1}{3}(1062.36) + \frac{1}{3}100\right](10^{-6}) = 185.6(10^6)\ \text{N/m}^2 \quad \text{(E20)}$$

**ANS.** $\sigma_1 = 339.4\ \text{MPa(T)}; \sigma_2 = 195.0\ \text{MPa(T)}; \sigma_3 = 185.6\ \text{MPa(T)}; \tau_{max} = 76.85\ \text{MPa}$

**PLAN: METHOD II**

We find the stresses in the Cartesian coordinates and find the principal stresses and principal direction 1.

**SOLUTION**

From Equation (1.46) we have the normal stress in Cartesian coordinates as shown below.

$$\sigma_{xx} = \frac{E[(1-\nu)\varepsilon_{xx} + \nu\varepsilon_{yy} + \nu\varepsilon_{zz}]}{(1-2\nu)(1+\nu)} = (450)(10^9)\left[\frac{2}{3}(600) + \frac{1}{3}(400) + \frac{1}{3}(200)\right](10^{-6}) = 270(10^6)\ \text{N/m}^2 \quad \text{(E21)}$$

$$\sigma_{yy} = \frac{E[\nu\varepsilon_{xx} + (1-\nu)\varepsilon_{yy} + \nu\varepsilon_{zz}]}{(1-2\nu)(1+\nu)} = (450)(10^9)\left[\frac{1}{3}(600) + \frac{2}{3}(400) + \frac{1}{3}(200)\right](10^{-6}) = 240(10^6)\ \text{N/m}^2 \quad \text{(E22)}$$

$$\sigma_{zz} = \frac{E[\nu\varepsilon_{xx} + \nu\varepsilon_{yy} + (1-\nu)\varepsilon_{zz}]}{(1-2\nu)(1+\nu)} = (450)(10^9)\left[\frac{1}{3}(600) + \frac{1}{3}(400) + \frac{2}{3}(200)\right](10^{-6}) = 210(10^6)\ \text{N/m}^2 \quad \text{(E23)}$$

The shear stresses can be found and the stress matrix written as shown below.

$$\tau_{xy} = G\gamma_{xy} = 60\ \text{MPa} \qquad \tau_{xz} = G\gamma_{xz} = 45\ \text{MPa} \qquad \tau_{yz} = G\gamma_{yz} = 30\ \text{MPa} \quad \text{(E24)}$$

$$\begin{bmatrix} 270 & 60 & 45 \\ 60 & 240 & 30 \\ 45 & 30 & 210 \end{bmatrix} \text{MPa} \quad \text{(E25)}$$

The stress invariants can be found as

$$I_1 = 720 \qquad I_2 = 165375 \qquad I_3 = 12285000 \quad \text{(E26)}$$

The variables in the footnote 2 on page 9 can be found as

$$A = 49.7494 \qquad \alpha = 3.111 \qquad \sigma_1 = x_1 = 339.35 \qquad \sigma_2 = x_2 = 195 \qquad \sigma_3 = x_3 = 185.65 \quad \text{(E27)}$$

*Checking Results*: The first and third stress invariants from principal stresses are

$$I_1 = 339.35 + 195 + 185.65 = 720 \rightarrow \text{Checks} \qquad I_3 = (339.35)(195)(185.65) = 12285063 \rightarrow \text{Checks}$$

The maximum shear stress is

$$\tau_{max} = |339.35 - 185.65|/2 = 76.85 \quad \text{(E28)}$$

**ANS.** $\sigma_1 = 339.4\ \text{MPa(T)}; \sigma_2 = 195.0\ \text{MPa(T)}; \sigma_3 = 185.6\ \text{MPa(T)}; \tau_{max} = 76.85\ \text{MPa}$

On substituting $\sigma_1$ in the equations corresponding to the first two rows of the matrix, we obtain

$$(270 - 339.4)p_x + 60p_y + 45p_z = 0 \quad \text{or} \quad 60p_y + 45p_z = 69.35p_x \quad \text{(E29)}$$

$$60p_x + (240 - 339.4)p_y + 30p_z = 0 \quad \text{or} \quad -99.4p_y + 30p_z = -60p_x \quad \text{(E30)}$$

Solving Equation (E29) and (E30) in terms of $p_x$ we obtain

$$p_y = 0.7624p_x \qquad p_z = 0.5247p_x \quad \text{(E31)}$$

Noting that the sum of square of direction cosine is one, we obtain

$$p_x^2 + (0.7624p_x)^2 + (0.5247p_x)^2 = 1 \quad \text{or} \quad p_x = \pm 0.7339 \quad \text{(E32)}$$

Substituting $p_x$ into Equation (E32), we obtain $p_y = \pm 0.5595$ and $p_z = \pm 0.3851$. The two solutions for principal direction are:

**ANS.** $\theta_x = 42.8°; \theta_y = 56°; \theta_z = 67.4°$ or $\theta_x = 137.2°; \theta_y = \theta_y = 124°; \theta_z = 112.6°$

## COMMENTS

1. This example shows that once an engineering strain matrix has been converted to a tensor strain matrix, the calculations in strain transformation are similar to those in stress transformation.
2. Method I will not work for orthotropic and other anisotropic materials, because the compliance matrix relating stresses and strains will change dramatically with coordinate transformation. Only for isotropic material is the compliance matrix independent of coordinate transformation and principal direction for stresses and strains the same.

### 1.8.4 Nonlinear material models

Rubber, plastics, and muscles and other organic tissues exhibit nonlinearity in the stress–strain relationship, even at small strains. Metals also the exhibit nonlinearity after yield stress. In this section, we consider various nonlinear material models: that is, various forms of equations that are used for representing the stress–strain nonlinear relationship. The constants in the equations relating stress and strain are material constants. We shall consider three material models:

1. Elastic–perfectly plastic, in which the nonlinearity is approximated by a constant.
2. Linear strain hardening, in which the nonlinearity is approximated by a linear function.
3. Power law, in which the nonlinearity is approximated by one term, a nonlinear function.

The choice of material model to use depends not only upon the material stress–strain curve, but also upon the degree of accuracy needed and the resulting complexity of the analysis.

### 1.8.5 Elastic–perfectly plastic material model

Figure 1.26 shows the stress–strain curves describing an elastic–perfectly plastic behavior of a material. It is assumed that the material has the same behavior in tension and in compression. Similarly for shear stress–strain, the material behavior is same for positive and negative stresses and strains.

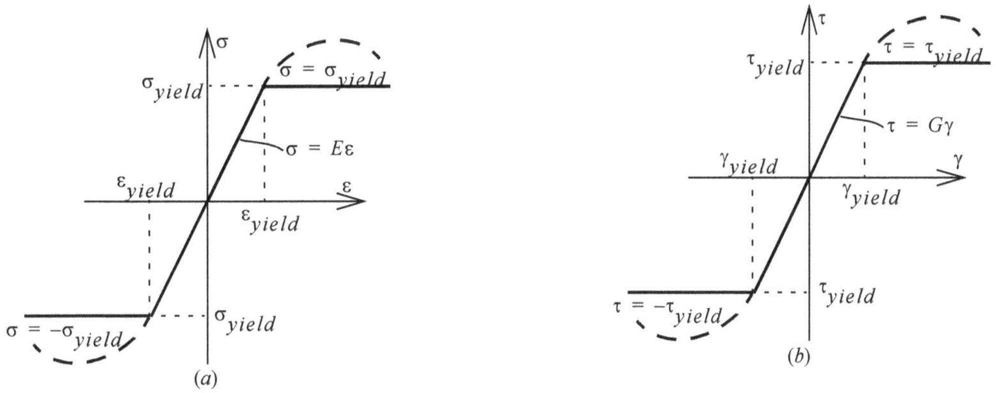


**Figure 1.26** Elastic–perfectly plastic material behavior: (*a*) normal and (*b*) shear.

Before yield stress, the stress–strain relationship is given by Hooke's law, after yield stress, the stress is a constant. Elastic–perfectly plastic material behavior is a simplifying approximation used to conduct an elastic–plastic analysis. The approximation is a conservative approximation, since it ignores the capacity of the material to carry stresses higher than the yield stress.

The equations describing the stress–strain curve are as follows:

$$\sigma = \begin{cases} \sigma_{\text{yield}} & \varepsilon \geq \varepsilon_{\text{yield}} \\ E\varepsilon & -\varepsilon_{\text{yield}} \leq \varepsilon \leq \varepsilon_{\text{yield}} \\ -\sigma_{\text{yield}} & \varepsilon \leq -\varepsilon_{\text{yield}} \end{cases} \quad \text{and} \quad \tau = \begin{cases} \tau_{\text{yield}} & \gamma \geq \gamma_{\text{yield}} \\ G\gamma & -\gamma_{\text{yield}} \leq \gamma \leq \gamma_{\text{yield}} \\ -\tau_{\text{yield}} & \gamma \leq -\gamma_{\text{yield}} \end{cases} \tag{1.47}$$

The set of points forming the boundary between the elastic and plastic regions on a body is called the **elastic–plastic boundary**. Determining the location of the elastic–plastic boundary is one of the critical issues in elastic–plastic analysis. The location of the elastic–plastic boundary is determined using the following observations:

1. On the elastic–plastic boundary, strain must be equal to yield strain, and stress equal to yield stress.
2. Deformations and strains are continuous at all points, including points at the elastic– plastic boundary.

### 1.8.6 Linear strain-hardening material model

Figure 1.27, shows stress–strain curves for the linear strain-hardening model, also referred to as the **bilinear material model**. It is assumed that the material has the same behavior in tension and in compression. Similarly for shear stress–strain, the material behavior is same for positive and negative stresses and strains.

This is another conservative, simplifying approximation of material behavior in which we once more ignore the ability of the material to carry stresses higher than those shown by the straight lines. Determining the location of the elastic–plastic boundary is once more a critical issue in the analysis and is determined as described in Section 1.8.5.

The equations describing the stress–strain curve are as follows:

$$\sigma = \begin{cases} \sigma_{\text{yield}} + E_2(\varepsilon - \varepsilon_{\text{yield}}) & \varepsilon \geq \varepsilon_{\text{yield}} \\ E_1\varepsilon & -\varepsilon_{\text{yield}} \leq \varepsilon \leq \varepsilon_{\text{yield}} \\ -\sigma_{\text{yield}} + E_2(\varepsilon + \varepsilon_{\text{yield}}) & \varepsilon \geq \varepsilon_{\text{yield}} \end{cases} \quad \text{and} \quad \tau = \begin{cases} \tau_{\text{yield}} + G_2(\gamma - \gamma_{\text{yield}}) & \gamma \geq \gamma_{\text{yield}} \\ G_1\gamma & -\gamma_{\text{yield}} \leq \gamma \leq \gamma_{\text{yield}} \\ -\tau_{\text{yield}} + G_2(\gamma + \gamma_{\text{yield}}) & \gamma \leq \gamma_{\text{yield}} \end{cases} \tag{1.48}$$

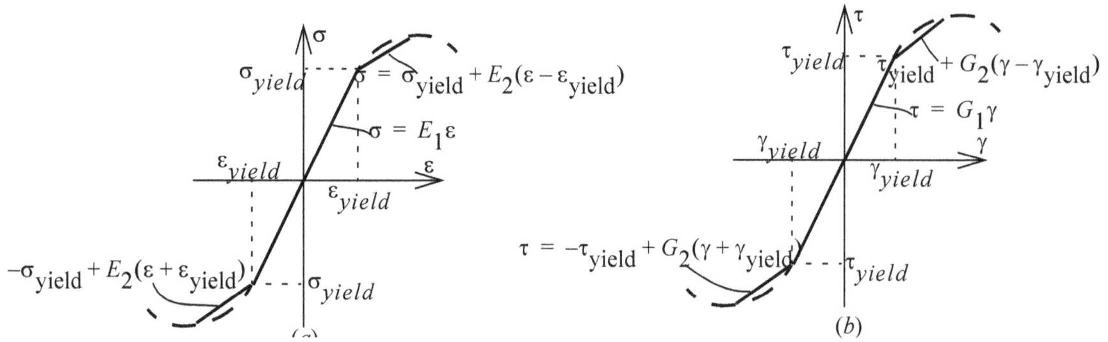


**Figure 1.27** Linear strain hardening model: (*a*) normal and (*b*) shear.

## 1.8.7 Power law material model

Figure 1.28 shows a power law representation of the nonlinear stress–strain curve. It is assumed that the material has the same behavior in tension and in compression. Similarly, for shear stress–strain, the material behavior is same for positive and negative stresses and strains. The equations describing the stress–strain relationship are as follows:

$$\sigma = \begin{cases} E\varepsilon^n & \varepsilon \ge 0 \\ -E(-\varepsilon)^n & \varepsilon < 0 \end{cases} \quad \text{and} \quad \tau = \begin{cases} G\gamma^n & \gamma \ge 0 \\ -G(-\gamma)^n & \gamma < 0 \end{cases} \tag{1.49}$$

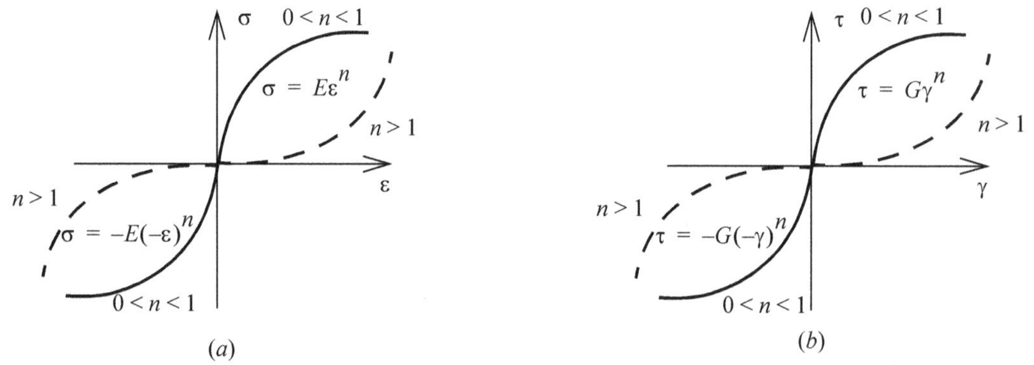


**Figure 1.28** Nonlinear stress–strain curves: (*a*) normal and (*b*) shear.

The constants *E* and *n*, called the strength coefficient and the strain-hardening coefficient. Most metals in the plastic region and most plastics are represented by the solid curve, with a strain-hardening coefficient less than 1. Soft rubber, as well as muscles and other organic materials, represented by the dashed line in Figure 1.28, have a strain-hardening coefficient greater than 1.

From Equation (1.49) we note that when strain is negative, the term on the right becomes positive, permitting evaluation of the number to fractional powers. Furthermore, with negative strain we obtain negative stress, as we should.

We could combine a linear equation for the linear part and nonlinear equation for the nonlinear part as we did in Section 1.8.6; or we could combine two nonlinear equations, thus creating additional material models. Which material model to use? The answer depends upon the material stress–strain curve, the degree of accuracy needed, and the resulting complexity of the analysis.

### EXAMPLE 1.4

Aluminum has a yield stress of $\sigma_{yield} = 40$ ksi in tension, a yield strain of $\varepsilon_{yield} = 0.004$, an ultimate stress of $\sigma_{ult} = 45$ ksi, and the corresponding ultimate strain of $\varepsilon_{ult} = 0.017$. Determine the material constants for (a) the elastic–perfectly plastic model, (b) the linear strain-hardening model, and (c) the nonlinear power law model.

**PLAN**

We have coordinates of three points on the curve: $P_0$ ($\sigma_0 = 0.00$, $\varepsilon_0 = 0.000$), $P_1$ ($\sigma_1 = 40.0$, $\varepsilon_1 = 0.004$), and $P_2$ ($\sigma_2 = 45.0$, $\varepsilon_2 = 0.017$). We can use these data to find the various constants in the material models.

**SOLUTION**

(a) The modulus of elasticity $E$ is the slope between points $P_0$ and $P_1$ and can be found as

$$E_1 = \frac{\sigma_1 - \sigma_0}{\varepsilon_1 - \varepsilon_0} = \frac{40}{0.004} = 10{,}000 \text{ ksi} \tag{E1}$$

After yield stress, the stress is a constant. The stress–strain behavior can be written as

$$\textbf{ANS.} \quad \sigma = \begin{cases} 10{,}000\varepsilon \text{ ksi} & |\varepsilon| \le 0.004 \\ 40 \text{ ksi} & |\varepsilon| \ge 0.004 \end{cases} \tag{E2}$$

(b) In the linear strain-hardening model, the slope of the straight line before the yield stress is as calculated in Equation (E1). After the yield stress, the slope of the line can be found from the coordinates of points $P_1$ and $P_2$ as shown below.

$$E_2 = \frac{\sigma_2 - \sigma_1}{\varepsilon_2 - \varepsilon_1} = \frac{5}{0.013} = 384.6 \text{ ksi} \qquad \text{(E3)}$$

The stress–strain behavior can be written as

$$\textbf{ANS.} \quad \sigma = \begin{cases} 10{,}000\varepsilon \text{ ksi} & |\varepsilon| \le 0.004 \\ 40 + 384.6(\varepsilon - 0.004) \text{ ksi} & |\varepsilon| \ge 0.004 \end{cases} \qquad \text{(E4)}$$

(c) The two constants $E$ and $n$ in $\sigma = E\varepsilon^n$ can be found by substituting the coordinates of the two points $P_1$ and $P_2$, to obtain

$$40 = E(0.004)^n \qquad 45 = E(0.017)^n \qquad \text{(E5)}$$

Dividing one equation by the other and taking the logarithm of both sides, we can solve for $n$ as shown below.

$$ln\left(\frac{0.017}{0.004}\right)^n = ln\left(\frac{45}{40}\right) \quad \text{or} \quad n\, ln(4.25) = ln(1.125) \quad \text{or} \quad n = 0.0814 \qquad \text{(E6)}$$

Substituting Equation (E6) into Equation (E5), we can obtain the value of $E$ as follows:

$$E = 40/(0.004)^{0.0814} = 62.7 \text{ ksi} \qquad \text{(E7)}$$

We can now write the stress–strain equations for the power law model as

$$\textbf{ANS.} \; \sigma = \begin{cases} 62.7\varepsilon^{0.0814} \text{ ksi} & \varepsilon \ge 0 \\ -62.7(-\varepsilon)^{0.0814} \text{ ksi} & \varepsilon < 0 \end{cases} \qquad \text{(E8)}$$

**COMMENT**

**1.** The example demonstrates how data from tension test can be used to obtain constants of material model for purpose of analysis. In next chapter we will incorporate material models to develop theories of one-dimensional structural members.

## 1.9 EFFECTS OF TEMPERATURE

A material expands with an increase in temperature and contracts with a decrease in temperature. If the change in temperature is *uniform*, and if the material is *isotropic* and *homogeneous*, then all lines on the material will change dimensions by equal amounts. The expansion will result in a normal strain, but no shear strain will be produced because there will be no change in the angles between any two lines. Experimental observations confirm this deduction. Experiments also show that the change in temperature ($\Delta T$) is related to the thermal normal strain ($\varepsilon_T$) as

$$\varepsilon_T = \alpha\, \Delta T \qquad \textbf{(1.50)}$$

where the Greek letter alpha ($\alpha$) is called the linear coefficient of thermal expansion. The linear relationship given by Equation (1.50) is valid for metals in the region far from the melting point. In this linear region, the strains for most metals are small and the usual units for $\alpha$ are $\mu$/°F or $\mu$/°C. Throughout the discussion in this section we shall assume that we are in the linear region.

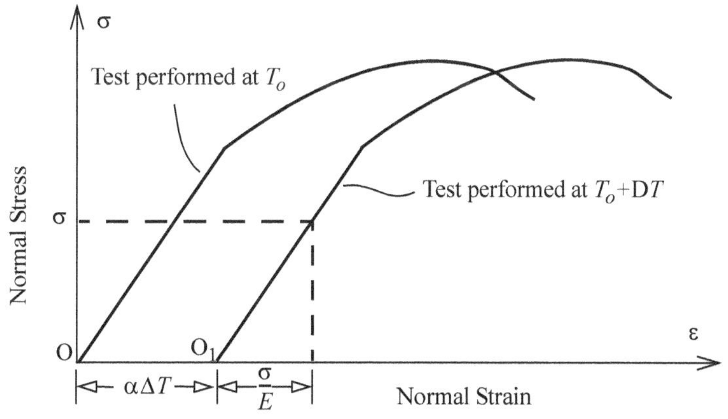


**Figure 1.29** Effect of temperature on stress–strain curve.

The tension test to determine material constants is conducted at an ambient temperature. We expect the stress–strain curve to have the same character at two different ambient temperatures. If we raise the temperature before we apply the force $P$ on the specimen, then the specimen will expand and result in a thermal strain as given by Equation (1.50). But because there is no external force, there will be no resulting internal forces, hence no stresses. Thus, the increase in temperature before the application of the force causes the starting point of the stress–strain curve to move from $O$ to $O_1$ as shown in Figure 1.29. The total strain at any point is the sum of mechanical strain and thermal strains, which can be written as

$$\boxed{\varepsilon = \frac{\sigma}{E} + \alpha\, \Delta T} \qquad \textbf{(1.51)}$$

If there are no internal forces generated in a body owing to changes in temperature, then no stresses will be produced. Material non homogeneity, material anisotropy, nonuniform temperature distribution, or reaction forces from body constraints are the reasons for the generation of stresses from temperature changes. We record the following observation.

- No thermal stresses are produced in a homogeneous, isotropic, unconstrained body as a result of uniform temperature changes.

The generalized version of Hooke's law relates mechanical strains to stresses. The total normal strain, as seen from Equation (1.51), is the sum of mechanical and thermal strains. For isotropic materials with temperature changes, the generalized law can be written as shown below. We do not have to comment on the homogeneity of the material or the uniformity of temperature change because Hooke's law is written for a point, not the whole body.

$$\varepsilon_{xx} = [\sigma_{xx} - \nu(\sigma_{yy} + \sigma_{zz})]/E + \alpha\,\Delta T \qquad \textbf{(1.52a)}$$

$$\varepsilon_{yy} = [\sigma_{yy} - \nu(\sigma_{zz} + \sigma_{xx})]/E + \alpha\,\Delta T \qquad \textbf{(1.52b)}$$

$$\varepsilon_{zz} = [\sigma_{zz} - \nu(\sigma_{xx} + \sigma_{yy})]/E + \alpha\,\Delta T \qquad \textbf{(1.52c)}$$

$$\gamma_{xy} = \tau_{xy}/G \qquad \textbf{(1.52d)}$$

$$\gamma_{yz} = \tau_{yz}/G \qquad \textbf{(1.52e)}$$

$$\gamma_{zx} = \tau_{zx}/G \qquad \textbf{(1.52f)}$$

Mechanical strain Thermal strain

### EXAMPLE 1.5

The strains at a point on aluminum ($E = 70$ GPa, $G = 28$ GPa, $\nu = 0.25$, and $\alpha = 23\ \mu/°C$) were found to be

$$\varepsilon_{xx} = 650\ \mu \qquad \varepsilon_{yy} = 300\ \mu \qquad \gamma_{xy} = 750\ \mu$$

If the temperature increases by 20°C and the point is in plane stress, determine the stresses $\sigma_{xx}$, $\sigma_{yy}$, and $\tau_{xy}$ and the strain $\varepsilon_{zz}$.

**PLAN**

The shear strain can be calculated by using Equation (1.52d). Substituting the known values of the variables in Equation (1.52a) and (1.52b), we generate two equations in the two unknown stresses $\sigma_{xx}$ and $\sigma_{yy}$, which we solve. Then, from Equation (1.52c), the normal strain $\varepsilon_{zz}$ can be found.

**SOLUTION**

From Equation (1.52d) we can find the shear stress as

$$\tau_{xy} = G\gamma_{xy} = 28(10^9)750(10^{-6}) = 21(10^6)\ \text{N/m}^2 \qquad \text{(E1)}$$

**ANS.** $\tau_{xy} = 21\,\text{MPa}$

Knowing $\Delta T = 20$, we find the thermal strain as $\alpha\,\Delta T = 460(10^{-6})$. Equations (1.52a) and (1.52b) can be rewritten with $\sigma_{zz} = 0$ as shown below.

$$\sigma_{xx} - \nu\sigma_{yy} = E(\varepsilon_{xx} - \alpha\,\Delta T) = 70(10^9)(650 - 460)(10^{-6})\ \text{N/m}^2 \quad \text{or} \quad \sigma_{xx} - 0.25\sigma_{yy} = 13.3\ \text{MPa} \qquad \text{(E2)}$$

$$\sigma_{yy} - \nu\sigma_{xx} = E(\varepsilon_{yy} - \alpha\,\Delta T) = 70(10^9) \times (300 - 460)(10^{-6})\ \text{N/m}^2 \quad \text{or} \quad \sigma_{yy} - 0.25\sigma_{xx} = -11.2\ \text{MPa} \qquad \text{(E3)}$$

By solving Equations (E2) and (E3) we obtain the stresses.

**ANS.** $\sigma_{xx} = 11.2$ MPa (T) $\sigma_{yy} = 8.4$ MPa (C)

From Equation (1.52c) with $\sigma_{zz} = 0$, we obtain

$$\varepsilon_{zz} = \frac{[-\nu(\sigma_{xx} + \sigma_{yy})]}{E} + \alpha\,\Delta T = \frac{[-0.25(11.2 - 8.4)(10^6)]}{70(10^9)} + 460(10^{-6}) = 450(10^{-6})$$

**ANS.** $\varepsilon_{zz} = 450\ \mu$

**COMMENT**

1. In a similar manner temperature can be incorporated in plane strain problems.

## 1.10 FAILURE THEORIES

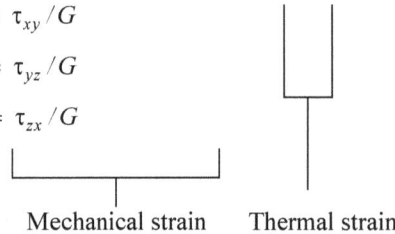

The maximum strength of a material is its atomic strength. In bulk materials, however, the distribution of flaws (impurities, micro holes, microcracks, etc.) creates large stress gradients. These large stress gradients cause the bulk strength of the material to be orders of magnitude lower than the atomic strength of the material. Failure theories assume a homogeneous material in which the effects of flaws have been averaged in some manner. This assumption of homogeneity results in average material strength values that is adequate for most problems in engineering design and analysis.

For homogeneous–isotropic-materials, the characteristic failure stress is either the yield stress or the ultimate stress, usually obtained from a uniaxial tensile test, a test in which there is only one nonzero stress component. How do we relate the one stress component of a uniaxial stress state to the stress components in two- and three-dimensional states of stress? A **failure theory** is a statement about the relationship of stress components to material failure characteristics values. No one theory is applicable to all materials.

We shall consider four theories (see Table 1.3). Maximum shear stress theory and the maximum octahedral shear stress theory are generally used for ductile materials. In ductile materials, failure is characterized by yield stress. The maximum normal stress theory and modified Mohr's theory are generally used for brittle materials. In brittle materials, failure is characterized by ultimate stress.

**Table 1.3** Synopsis of Failure Theories

| | Ductile Material | Brittle Material |
|---|---|---|
| Characteristic failure stress | Yield stress | Ultimate stress |
| Theories | 1. Maximum shear stress<br>2. Maximum octahedral shear stress | 1. Maximum normal stress<br>2. Modified Mohr |

## 1.10.1 Maximum shear stress theory

For ductile materials, the theory states:

*A material will fail when the maximum shear stress exceeds the shear stress at the yield obtained from a uniaxial tensile test.*

The failure criterion described is written as

$$\tau_{max} \le \tau_{yield} \quad \textbf{(1.53)}$$

where $\tau_{yield}$ is the maximum shear stress at the yield point found in a uniaxial tension test. The maximum shear stress is given by the Equation (1.15). Substituting the stresses at the yield point in a uniaxial tension test as $\sigma_1 = \sigma_{yield}$, $\sigma_2 = 0$, and $\sigma_3 = 0$ in Equation (1.15) and equating the result to $\tau_{yield}$, we obtain the maximum shear stress at yield point as follows:

$$\tau_{yield} = \sigma_{yield}/2 \quad \textbf{(1.54)}$$

Substituting Equations (1.53) and (1.15) into Equation (1.54), we obtain the **Tresca's yield criterion** given below.

$$\boxed{\max(|\sigma_1 - \sigma_2|, |\sigma_2 - \sigma_3|, |\sigma_3 - \sigma_1|) \le \sigma_{yield}} \quad \textbf{(1.55)}$$

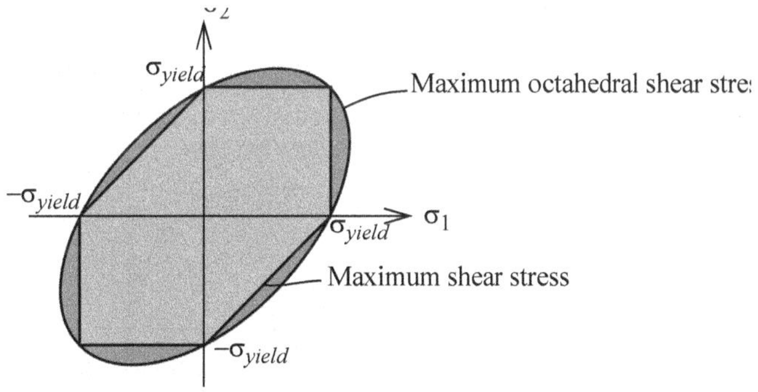


**Figure 1.30** Failure envelopes for ductile materials in plane stress.

## 1.10.2 Maximum octahedral shear stress theory

Also called **von-Mises theory.** Experiments have shown that ductile materials can withstand very large hydrostatic pressures and yielding is governed by $\tau_{oct}$ reaching a critical value. The maximum octahedral shear stress theory for ductile materials is:

*A material will fail when the maximum octahedral shear stress exceeds the octahedral shear stress at the yield obtained from a uniaxial tensile test.*

The failure criterion is

$$\tau_{oct} \le \overline{\tau_{yield}} \quad \textbf{(1.56)}$$

where $\overline{\tau_{yield}}$ is the octahedral shear stress at yield point in a uniaxial tensile test. Substituting $\sigma_1 = \sigma_{yield}$, $\sigma_2 = 0$, and $\sigma_3 = 0$, the stresses at yield point in a uniaxial tension test, into the expression of octahedral shear stress in Equation (1.24d), we obtain $\overline{\tau_{yield}} = \sqrt{2}\sigma_{yield}/3$. Substituting this and Equation (1.24d) into Equation (1.56), we obtain

$$\frac{1}{\sqrt{2}}\sqrt{(\sigma_1 - \sigma_2)^2 + (\sigma_2 - \sigma_3)^2 + (\sigma_3 - \sigma_1)^2} \le \sigma_{yield} \quad \textbf{(1.57)}$$

The left-hand side of Equation (1.57) is referred to as **von-Mises stress** $\sigma_{von}$. Because von Mises stress $\sigma_{von}$ is extensively used in the design of structures and machines, we formally define it as follows:

$$\boxed{\sigma_{von} = \frac{1}{\sqrt{2}}\sqrt{(\sigma_1 - \sigma_2)^2 + (\sigma_2 - \sigma_3)^2 + (\sigma_3 - \sigma_1)^2}} \qquad \textbf{(1.58)}$$

In uniaxial tension test, at yield point we have $\sigma_1 = \sigma_{yield}$, and $\sigma_2 = \sigma_3 = 0$. Thus, at yield point $\tau_{oct} = (\sqrt{2}/3)\sigma_{yield}$, while $\sigma_{von} = \sigma_{yield}$, which is simpler and used in failure criterion for ductile material and referred to as the von Mises yield criterion.

$$\boxed{\sigma_{von} \le \sigma_{yield}} \qquad \textbf{(1.58a)}$$

Equations (1.55) and (1.57) are failure envelopes in a space in which the axes are principal stresses. **Failure envelopes** separate the design space from failure space. In drawing failure envelopes, the convention $\sigma_1 > \sigma_2$ is ignored. If the convention were enforced, there would be no envelope in the second quadrant and only an envelope below the 45° line would be admissible in the third quadrant, resulting in very strange-looking envelopes.

For a plane stress ($\sigma_3 = 0$) problem, we can show these failure envelopes as in Figure 1.30. Notice that the maximum octahedral shear stress envelope encompasses the maximum shear stress envelope. Experiments show that for most ductile materials, the maximum octahedral shear stress theory gives better results than the maximum shear stress theory, but maximum shear stress theory is simpler to use.

## 1.10.3 Maximum normal stress theory

The maximum normal stress theory for brittle materials is:

*A material will fail when the maximum normal stress at a point exceeds the ultimate normal stress ($\sigma_{ult}$) obtained from a uniaxial tension test.*

The theory gives good results for brittle materials provided the principal stress one is tensile. The failure criterion is

$$\boxed{|\max(\sigma_1, \sigma_2, \sigma_3)| \le \sigma_{ult}} \qquad \textbf{(1.59)}$$

For concrete and many other brittle materials, the ultimate stress in tension is far less than the ultimate stress in compression because microcracks tend to grow in tension and to close in compression. But the simplicity of the failure criterion makes the theory attractive, and it can be used *if principal stress 1 is tensile* and is the *dominant* principal stress.

## 1.10.4 Modified Mohr's theory

The Mohr's theory for brittle materials is:

*A material will fail if a stress state is on the envelope that is tangent to the three Mohr's circles corresponding to uniaxial ultimate stress in tension, uniaxial ultimate stress in compression, and pure shear.*

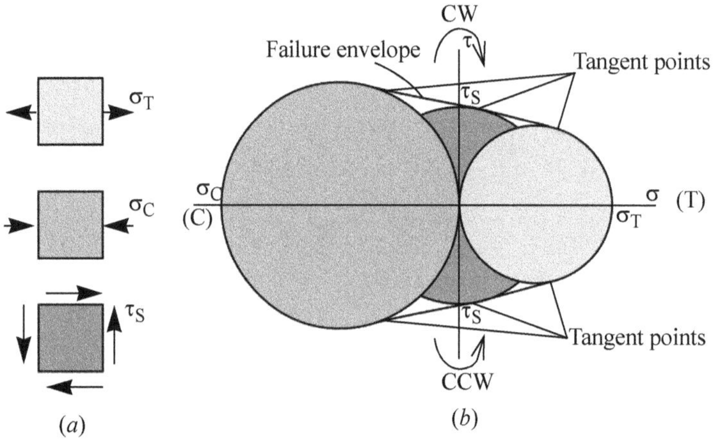


**Figure 1.31** (*a*) stress elements. (*b*) Mohr's failure envelope.

We can conduct three experiments and determine the ultimate stress in tension $\sigma_T$, the ultimate stress in compression $\sigma_C$, and the ultimate shear stress in pure shear $\tau_S$. The three stress states are shown on the stress elements in Figure 1.31*a*. A Mohr's circle for each of the three stress states is then drawn. Finally an envelope that is tangent to the three circles is drawn, which represents the failure envelope (Figure 1.31*b*). If a Mohr's circle corresponding to a stress state just touches the envelope at any point, the material is at incipient failure. If any part of the Mohr's circle for a stress state is outside the envelope, the material has failed at that point.

We can also plot the failure envelope of Figure 1.31 by using the principal stresses as the coordinate axis. In plane stress this envelope is represented by the solid line in Figure 1.32. For most brittle materials, the pure shear test is often ignored. In such a case the line tangent to the circles of uniaxial compression and tension would be a straight line in Figure 1.31. The resulting simplification for plane stress is shown as a dashed line in Figure 1.32 and describes what is called *Modified Mohr's theory.*

Figure 1.32 emphasizes the following points.

1. If both principal stresses are tensile, then the maximum normal stress must be less than the ultimate tensile strength.
2. If both principal stresses are negative, then the maximum normal stress must be less than the ultimate compressive strength.
3. If the principal stresses are of different signs, then for the modified Mohr's theory the failure is governed by Equation (1.60), in which $\sigma_1$ is tensile and $\sigma_2$ is compressive.

$$\boxed{\left|\frac{\sigma_2}{\sigma_C} - \frac{\sigma_1}{\sigma_T}\right| \le 1} \qquad \textbf{(1.60)}$$

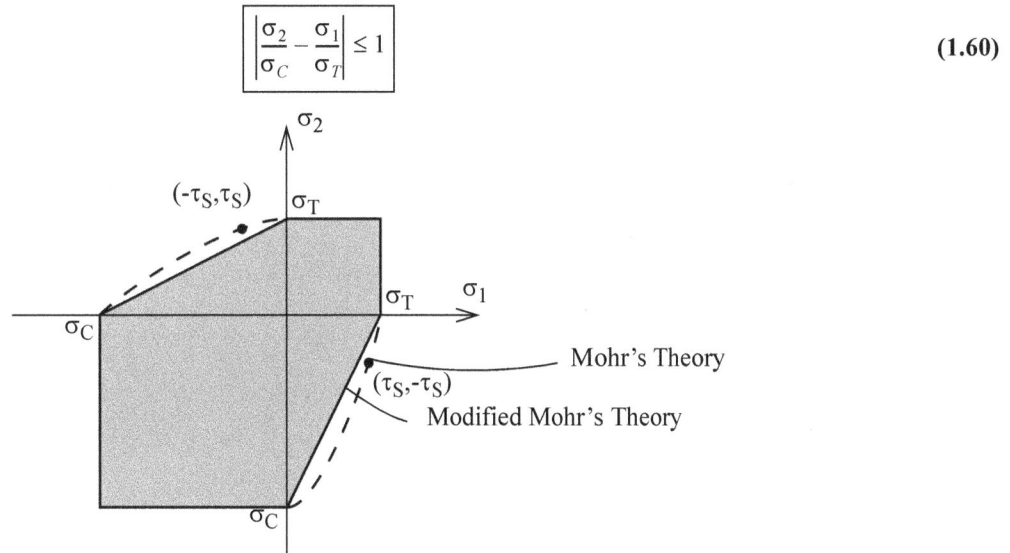

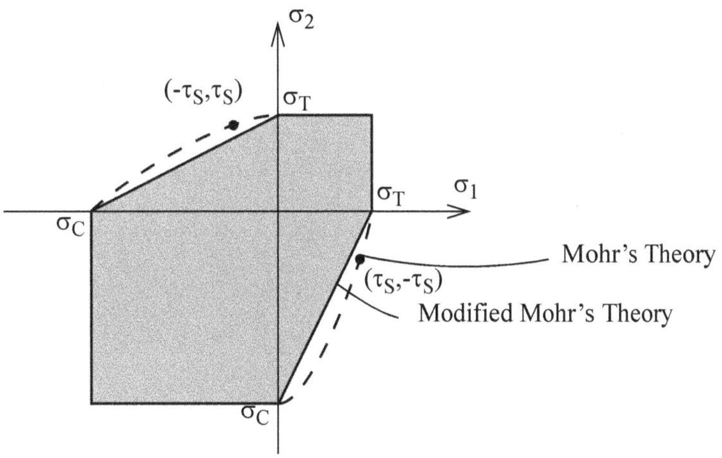


**Figure 1.32** Failure envelope for plane stress according to Mohr's theory. In drawing failure envelopes, the convention that $\sigma_1 > \sigma_2$ is ignored.

## EXAMPLE 1.6

At a critical point on a machine part made of steel, the stress components are $\sigma_{xx}$ = 100 MPa (T), $\sigma_{yy}$ = 50 MPa (C), and $\tau_{xy}$ = 30 MPa. Assuming that the point is in plane stress and the yield stress in tension is $\sigma_{yield}$ = 220 MPa, determine the factor of safety by using (a) the maximum shear stress theory and (b) the maximum octahedral shear stress theory.

**PLAN**

We find the principal stresses and the maximum shear stress from the given stress values. (a) The factor of safety $k_\tau = [\sigma_{yield}/2]/\tau_{max}$ as failure stress in shear is half the yield stress in tension. (b) The factor of safety is $k_\sigma = \sigma_{yield}/\sigma_{von}$.

**SOLUTION**

We can obtain the principal stresses and maximum shear stress as follows:

$$\sigma_{1,2} = \frac{100-50}{2} \pm \sqrt{\left(\frac{100+50}{2}\right)^2 + 30^2} = 25 \pm 80.8$$

$$\sigma_1 = 105.8 \text{ MPa} \qquad \sigma_2 = -55.8 \text{ MPa} \qquad \tau_{max} = \sigma_1 - \sigma_2/2 = 80.8 \text{ MPa} \tag{E1}$$

(a) For the maximum shear stress theory, the factor of safety $k_\tau = [\sigma_{yield}/2]/\tau_{max}$ can be calculated as shown below.

$$k_\tau = 110/80.8 = 1.36 \tag{E2}$$

**ANS.** $k_\tau = 1.36$

(b) The von Mises stress can be found from Equation (1.58) as follows:

$$\sigma_{von} = \frac{1}{\sqrt{2}}\sqrt{[105.8-(-55.8)]^2 + (-55.8)^2 + (105.8)^2} = 142.2 \text{ MPa} \tag{E3}$$

For the maximum octahedral shear stress theory, the factor of safety $k_\sigma = \sigma_{yield}/\sigma_{von}$ can be calculated as shown below.

$$k_\sigma = 220/142.2 = 1.55 \tag{E4}$$

**ANS.** $k_\sigma = 1.55$

**COMMENTS**

1. The failure envelopes corresponding to the yield stress of 220 MPa are shown in Figure 2.33. The coordinates of point S (105.8, –55.8) are the principal stresses. The line joining origin *O* and *S* is the load line. It may be verified that $k_\tau = OT/OS$ = 1.36 and $k_\sigma = OV/OS$ = 1.55. Thus, failure envelopes provide a graphical interpretation of factors of safety. A factor of safety can be interpreted as a measure of how far the current design is from the failure envelope.

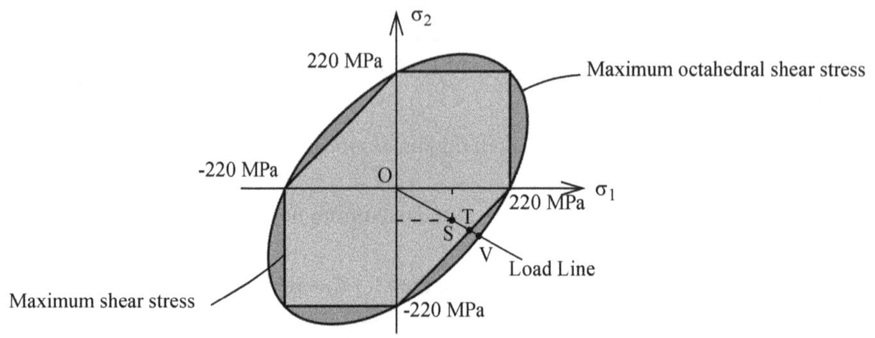


**Figure 2.33** Failure envelope for Example 1.6.

2. Because the failure envelope for the maximum shear stress criterion is always inscribed inside the failure envelope of the maximum octahedral shear stress, the factor of safety based on the maximum octahedral shear stress will always be greater than the factor of safety based on the maximum shear stress.

## EXAMPLE 1.7

Given a load P, the stresses at a point on a free surface were found to be $\sigma_{xx} = 3P$ ksi (C), $\sigma_{yy} = 5P$ ksi (T), and $\tau_{xy} = -2P$ ksi, where $P$ is measured in kips. The brittle material has a tensile strength of 18 ksi and a compressive strength of 36 ksi. Determine the maximum value of the load $P$ that can be applied on the structure using the modified Mohr's theory.

**PLAN**

We can find the principal stresses in terms of $P$. From Equation (1.60) we can determine the maximum value of $P$.

**SOLUTION**

The principal stresses can be found by as shown below

$$\sigma_{1,2} = (-3P+5P)/2 \pm \sqrt{[(-3P-5P)/2]^2+(2P)^2} = P \pm 4.47P \quad \text{or} \quad \sigma_1 = 5.57P \quad \sigma_2 = -3.57P \tag{E1}$$

Substituting the principal stresses in Equation (1.60) and noting that $\sigma_T = 18$ ksi and $\sigma_C = 36$ ksi, we can obtain

$$|(-3.37P)/36-(5.57P)/18| \le 1 \quad \text{or} \quad 0.4031P \le 1 \quad \text{or} \quad P \le 2.481 \tag{E2}$$

**ANS.** $P_{max} = 2.48$ kips

**COMMENTS**

We could not have used the maximum normal stress theory for this material because the tensile and compressive strengths are significantly different, and it is the compressive strength that is the dominant strength, not the tensile strength.

## 1.11 CLOSURE

In this chapter we saw that stress on a surface is an internal distributed force system that is a vector quantity, while stress at a point is a second-order tensor whose sign is determined from the direction of the outward normal of an imaginary surface through the point and the direction of the internal force. We also saw that engineering strain, which is not a second-order tensor, may be transformed into a second-order tensor by dividing all shear strains by a factor of 2. Principal stresses and strains are the eigenvalues, and principal directions are the eigenvectors, of the stress and tensor strain matrices. We saw that the principal stresses and strains at a point are unique, which implies that the coefficients of the characteristic equation are unique. The coefficients of the characteristic equations are the stress and strain invariants that do not depend upon the coordinate system—an observation we used to check our calculations of principal stresses and strains.

We saw that small-strain approximation is a linear relationship between displacements and strains that is valid for strains less than 1%. Small-strain approximations allow significant simplification, including calculation of normal strain, in which we approximate the actual deformation by the component of the deformation in the original direction of the line element, and shear strain, in which the sine and tangent functions of shear strain are approximated by their arguments.

We saw that generalized Hooke's law for isotropic material can be written for any orthogonal systems as the stress strain relationship is same in all direction. Principal directions, which are orthogonal, for stress and strains are same only for isotropic materials.

We saw temperature changes generate only thermal normal strains. If a structural member is unconstraint (statically determinate) then no thermal stresses will be generated.

We saw that failure theories are statements relating the six independent stress components to yield stress and ultimate stress values obtained from tension tests.

We will use the concepts and information introduced in this chapter in subsequent chapters for solving problems and building theories of structural members.

## 1.12 SYNOPSIS OF EQUATIONS

**Stress**

Stress at a point

$$\sigma_{ij} = \lim_{\Delta A_i \to 0} (\Delta F_j / \Delta A_i)\ ;\ \tau_{xy} = \tau_{yx} \qquad \tau_{yz} = \tau_{zy} \qquad \tau_{zx} = \tau_{xz}$$

Transformation in 2-D

$$\sigma_{nn} = \sigma_{xx} cos^2\theta + \sigma_{yy} sin^2\theta + 2\tau_{xy} sin\ \theta\ cos\ \theta \qquad \tau_{nt} = -\sigma_{xx} cos\ \theta\ sin\ \theta + \sigma_{yy} sin\ \theta\ cos\ \theta + \tau_{xy}(cos^2\theta - sin^2\theta)$$

$$\sigma_{1,2} = \left[\left(\frac{\sigma_{xx}+\sigma_{yy}}{2}\right) \pm \sqrt{\left(\frac{\sigma_{xx}-\sigma_{yy}}{2}\right)^2 + \tau_{xy}^2}\right]\ ;\ tan 2\theta_p = \frac{2\tau_{xy}}{(\sigma_{xx}-\sigma_{yy})}$$

$$\sigma_{nn} = \{n\}^T[\sigma]\{n\} \qquad \tau_{nt} = \{t\}^T[\sigma]\{n\} \qquad \sigma_{tt} = \{t\}^T[\sigma]\{t\}\ ;\ \{S\} = [\sigma]\{n\}$$

Transformation in 2-D and 3-D

$$I_1 = \sigma_{xx} + \sigma_{yy} + \sigma_{zz} \qquad I_2 = \begin{vmatrix} \sigma_{xx} & \tau_{xy} \\ \tau_{yx} & \sigma_{yy} \end{vmatrix} + \begin{vmatrix} \sigma_{yy} & \tau_{yz} \\ \tau_{zy} & \sigma_{zz} \end{vmatrix} + \begin{vmatrix} \sigma_{xx} & \tau_{xz} \\ \tau_{zx} & \sigma_{zz} \end{vmatrix} \qquad I_3 = \begin{vmatrix} \sigma_{xx} & \tau_{xy} & \tau_{xz} \\ \tau_{yx} & \sigma_{yy} & \tau_{yz} \\ \tau_{zx} & \tau_{zy} & \sigma_{zz} \end{vmatrix}$$

$$I_1 = \sigma_1 + \sigma_2 + \sigma_3 \qquad I_2 = \begin{vmatrix} \sigma_1 & 0 \\ 0 & \sigma_2 \end{vmatrix} + \begin{bmatrix} \sigma_2 & 0 \\ 0 & \sigma_3 \end{bmatrix} + \begin{bmatrix} \sigma_1 & 0 \\ 0 & \sigma_3 \end{bmatrix} = \sigma_1\sigma_2 + \sigma_2\sigma_3 + \sigma_3\sigma_1 \qquad I_3 = \begin{vmatrix} \sigma_1 & 0 & 0 \\ 0 & \sigma_2 & 0 \\ 0 & 0 & \sigma_3 \end{vmatrix} = \sigma_1\sigma_2\sigma_3$$

$$\tau_{\max} = \max(|(\sigma_1 - \sigma_2)/2|, |(\sigma_2 - \sigma_3)/2|, |(\sigma_3 - \sigma_1)/2|)$$

$$x^3 - I_1 x^2 + I_2 x - I_3 = 0\ ;\ x_1 = 2A\,cos\alpha + I_1/3\ ;\ x_{2,3} = -2A\,cos(\alpha \pm 60^o) + I_1/3\ ;\ A = \sqrt{(I_1/3)^2 - I_2/3}\ ;$$

$$cos 3\alpha = [2(I_1/3)^3 - (I_1/3)I_2 + I_3]/(2A^3)$$

Octahedral stresses

$$\sigma_{\text{oct}} = (\sigma_1 + \sigma_2 + \sigma_3)/3\ ;\ \tau_{\text{oct}} = \sqrt{(\sigma_1-\sigma_2)^2 + (\sigma_2-\sigma_3)^2 + (\sigma_3-\sigma_1)^2}/3$$

Deviatoric stress invariants

$$J_1 = 0 \qquad J_2 = -[(\sigma_1-\sigma_2)^2 + +(\sigma_2-\sigma_3)^2 + (\sigma_3-\sigma_1)^2]/6 \qquad J_3 = (2\sigma_1-\sigma_2-\sigma_3)(2\sigma_2-\sigma_3-\sigma_1)((2\sigma_3-\sigma_1-\sigma_2)/27)$$

**Strain**

Cartesian Coordinates

$$\varepsilon_{xx} = \frac{\partial u}{\partial x} \qquad \varepsilon_{yy} = \frac{\partial v}{\partial y} \qquad \varepsilon_{zz} = \frac{\partial w}{\partial z} \qquad \gamma_{xy} = \gamma_{yx} = \frac{\partial u}{\partial y} + \frac{\partial v}{\partial x} \qquad \gamma_{yz} = \gamma_{zy} = \frac{\partial v}{\partial z} + \frac{\partial w}{\partial y} \qquad \gamma_{zx} = \gamma_{xz} = \frac{\partial w}{\partial x} + \frac{\partial u}{\partial z}$$

Polar Coordinates

$$\varepsilon_{rr} = \frac{\partial u_r}{\partial r} \qquad \varepsilon_{\theta\theta} = \frac{u_r}{r} + \frac{1}{r}\frac{\partial v_\theta}{\partial \theta} \qquad \gamma_{r\theta} = \frac{1}{r}\frac{\partial u_r}{\partial \theta} + \frac{\partial v_\theta}{\partial r} - \frac{v_\theta}{r}$$

Transformation in 2-D

$$\varepsilon_{nn} = \varepsilon_{xx} cos^2\theta + \varepsilon_{yy} sin^2\theta + \gamma_{xy} sin\theta\, cos\theta \qquad \gamma_{nt} = -2\varepsilon_{xx} sin\theta\, cos\theta + 2\varepsilon_{yy} sin\theta\, cos\theta + \gamma_{xy}(cos^2\theta - sin^2\theta)$$

$$\varepsilon_{nn} = \{n\}^T[\varepsilon]\{n\} \qquad \varepsilon_{nt} = \{t\}^T[\varepsilon]\{n\} \qquad \varepsilon_{tt} = \{t\}^T[\varepsilon]\{t\}\ ;$$

Transformation in 2-D and 3-D

$$I_1 = \varepsilon_{xx} + \varepsilon_{yy} + \varepsilon_{zz} \qquad I_2 = \begin{vmatrix} \varepsilon_{xx} & \varepsilon_{xy} \\ \varepsilon_{yx} & \varepsilon_{yy} \end{vmatrix} + \begin{vmatrix} \varepsilon_{yy} & \varepsilon_{yz} \\ \varepsilon_{zy} & \varepsilon_{zz} \end{vmatrix} + \begin{vmatrix} \varepsilon_{xx} & \varepsilon_{xz} \\ \varepsilon_{zx} & \varepsilon_{zz} \end{vmatrix} \qquad I_3 = \begin{vmatrix} \varepsilon_{xx} & \varepsilon_{xy} & \varepsilon_{xz} \\ \varepsilon_{yx} & \varepsilon_{yy} & \varepsilon_{yz} \\ \varepsilon_{zx} & \varepsilon_{zy} & \varepsilon_{zz} \end{vmatrix}$$

$$I_1 = \varepsilon_1 + \varepsilon_2 + \varepsilon_3 \qquad I_2 = \begin{vmatrix} \varepsilon_1 & 0 \\ 0 & \varepsilon_2 \end{vmatrix} + \begin{bmatrix} \varepsilon_2 & 0 \\ 0 & \sigma_3 \end{bmatrix} + \begin{bmatrix} \varepsilon_1 & 0 \\ 0 & \varepsilon_3 \end{bmatrix} = \varepsilon_1\varepsilon_2 + \varepsilon_2\varepsilon_3 + \varepsilon_3\varepsilon_1 \qquad I_3 = \begin{vmatrix} \varepsilon_1 & 0 & 0 \\ 0 & \varepsilon_2 & 0 \\ 0 & 0 & \varepsilon_3 \end{vmatrix} = \varepsilon_1\varepsilon_2\varepsilon_3$$

**Material Models**

Isotropic Material

$$\varepsilon_{xx} = [\sigma_{xx} - \nu(\sigma_{yy} + \sigma_{zz})]/E \qquad \varepsilon_{yy} = [\sigma_{yy} - \nu(\sigma_{zz} + \sigma_{xx})]/E \qquad \varepsilon_{zz} = [\sigma_{zz} - \nu(\sigma_{xx} + \sigma_{yy})]/E \qquad G = \frac{E}{2(1+\nu)}$$

$$\gamma_{xy} = \tau_{xy}/G \qquad \gamma_{yz} = \tau_{yz}/G \qquad \gamma_{zx} = \tau_{zx}/G$$

$$\sigma_{xx} = \frac{E[(1-\nu)\varepsilon_{xx} + \nu\varepsilon_{yy} + \nu\varepsilon_{zz}]}{(1-2\nu)(1+\nu)} \qquad \sigma_{yy} = \frac{E[\nu\varepsilon_{xx} + (1-\nu)\varepsilon_{yy} + \nu\varepsilon_{zz}]}{(1-2\nu)(1+\nu)} \qquad \sigma_{zz} = \frac{E[\nu\varepsilon_{xx} + \nu\varepsilon_{yy} + (1-\nu)\varepsilon_{zz}]}{(1-2\nu)(1+\nu)}$$

Elastic-perfectly plastic

$$\sigma = \begin{cases} \sigma_{\text{yield}} & \varepsilon \ge \varepsilon_{\text{yield}} \\ E\varepsilon & -\varepsilon_{\text{yield}} \le \varepsilon \le \varepsilon_{\text{yield}} \\ -\sigma_{\text{yield}} & \varepsilon \le -\varepsilon_{\text{yield}} \end{cases}\ ;\ \tau = \begin{cases} \tau_{\text{yield}} & \gamma \ge \gamma_{\text{yield}} \\ G\gamma & -\gamma_{\text{yield}} \le \gamma \le \gamma_{\text{yield}} \\ -\tau_{\text{yield}} & \gamma \le -\gamma_{\text{yield}} \end{cases}$$

Linear strain hardening

$$\sigma = \begin{cases} \sigma_{\text{yield}} + E_2(\varepsilon - \varepsilon_{\text{yield}}) & \varepsilon \ge \varepsilon_{\text{yield}} \\ E_1\varepsilon & -\varepsilon_{\text{yield}} \le \varepsilon \le \varepsilon_{\text{yield}} \\ -\sigma_{\text{yield}} + E_2(\varepsilon + \varepsilon_{\text{yield}}) & \varepsilon \ge \varepsilon_{\text{yield}} \end{cases}\ ;\ \tau = \begin{cases} \tau_{\text{yield}} + G_2(\gamma - \gamma_{\text{yield}}) & \gamma \ge \gamma_{\text{yield}} \\ G_1\gamma & -\gamma_{\text{yield}} \le \gamma \le \gamma_{\text{yield}} \\ -\tau_{\text{yield}} + G_2(\gamma + \gamma_{\text{yield}}) & \gamma \le \gamma_{\text{yield}} \end{cases}$$

Power law

$$\sigma = \begin{cases} E\varepsilon^n & \varepsilon \ge 0 \\ -E(-\varepsilon)^n & \varepsilon < 0 \end{cases}\ ;\ \tau = \begin{cases} G\gamma^n & \gamma \ge 0 \\ -G(-\gamma)^n & \gamma < 0 \end{cases}$$

Temperature effect

$$\varepsilon_{xx} = [\sigma_{xx} - \nu(\sigma_{yy} + \sigma_{zz})]/E + \alpha\,\Delta T \qquad \varepsilon_{yy} = [\sigma_{yy} - \nu(\sigma_{zz} + \sigma_{xx})]/E + \alpha\,\Delta T \qquad \varepsilon_{zz} = [\sigma_{zz} - \nu(\sigma_{xx} + \sigma_{yy})]/E + \alpha\,\Delta T$$

**Failure Theories**

Maximum shear stress theory.

$$\max(|\sigma_1 - \sigma_2|, |\sigma_2 - \sigma_3|, |\sigma_3 - \sigma_1|) \le \sigma_{yield}$$

Von-misses theory.

$$\sigma_{von} = \sqrt{(\sigma_1-\sigma_2)^2 + (\sigma_2-\sigma_3)^2 + (\sigma_3-\sigma_1)^2}/\sqrt{2} \le \sigma_{yield}$$

Maximum normal stress theory

$$|\max(\sigma_1, \sigma_2, \sigma_3)| \le \sigma_{ult}$$

Modified Mohr's theory.

$$|(\sigma_2/\sigma_C) - (\sigma_1/\sigma_T)| \le 1$$

# PROBLEMS

## Sections 1.1-1.2

**1.1** Show the nonzero stress components on the *A*, *B*, and *C* faces of the parallelepiped shown in Fig. P1.1.

$$\begin{bmatrix} \sigma_{xx} = 100\text{ MPa (T)} & \tau_{xy} = 200\text{ MPa} & \tau_{xz} = -125\text{ MPa} \\ \tau_{yx} = 200\text{ MPa} & \sigma_{yy} = 175\text{ MPa (C)} & \tau_{yz} = 225\text{ MPa} \\ \tau_{zx} = -125\text{ MPa} & \tau_{zy} = 225\text{ MPa} & \sigma_{zz} = 150\text{ MPa (C)} \end{bmatrix}$$

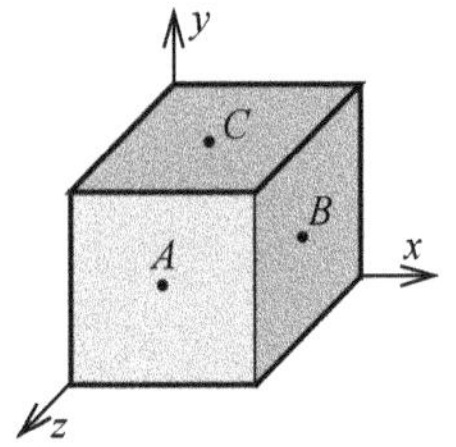


**Fig. P1.1**

**1.2** Show the nonzero stress components in the *r*-θ-*x* cylindrical coordinate system on the *A*, *B*, and *C* faces of the stress element shown in Fig. P1.2.

$$\begin{bmatrix} \sigma_{rr} = 10\text{ ksi (C)} & \tau_{r\theta} = 22\text{ ksi} & \tau_{rx} = 32\text{ ksi} \\ \tau_{\theta r} = 22\text{ ksi} & \sigma_{\theta\theta} = 0 & \tau_{\theta x} = 25\text{ ksi} \\ \tau_{xr} = 32\text{ ksi} & \tau_{x\theta} = 25\text{ ksi} & \sigma_{xx} = 20\text{ ksi (T)} \end{bmatrix}$$

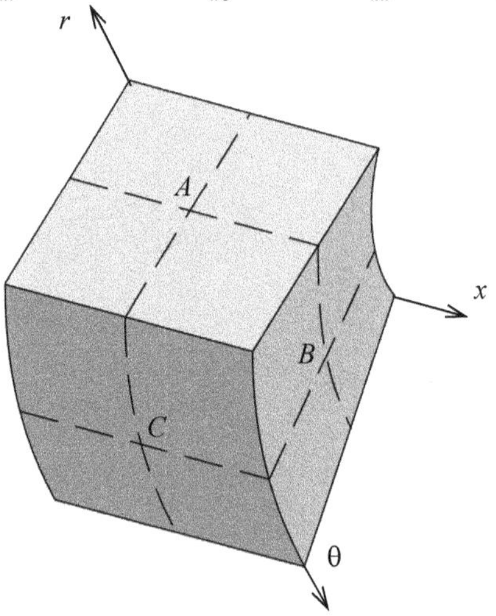


**Fig. P1.2**

**1.3** Show the stress components of a point in plane stress on the stress element shown in Fig. P1.3 in polar coordinates.

$$\begin{bmatrix} \sigma_{rr} = 125\text{ MPa (T)} & \tau_{r\theta} = -65\text{ MPa} \\ \tau_{\theta r} = -65\text{ MPa} & \sigma_{\theta\theta} = 90\text{ MPa (C)} \end{bmatrix}$$

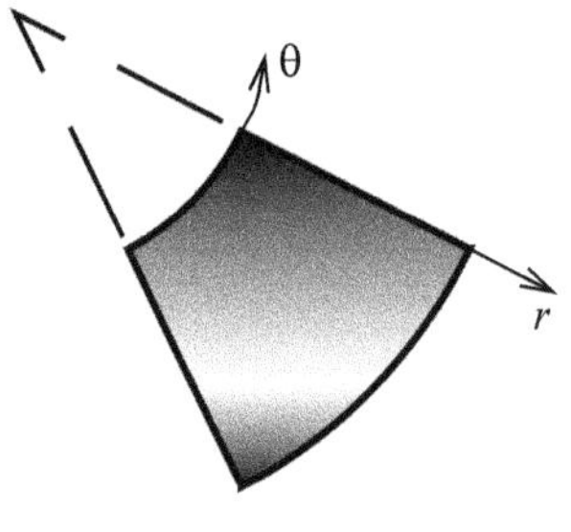


**Fig. P1.3**

**1.4** Show the nonzero stress components in the *r*-θ-ϕ spherical coordinate system on the *A*, *B*, and *C* faces of the stress-element shown in Fig. P1.4.

$$\begin{bmatrix} \sigma_{rr} = 135\text{ MPa (C)} & \tau_{r\theta} = 100\text{ MPa} & \tau_{r\phi} = -125\text{ MPa} \\ \tau_{\theta r} = 100\text{ MPa} & \sigma_{\theta\theta} = 160\text{ MPa (C)} & \tau_{\theta\phi} = 175\text{ MPa} \\ \tau_{\phi r} = -125\text{ MPa} & \tau_{\phi\theta} = 175\text{ MPa} & \sigma_{\phi\phi} = 150\text{ MPa (T)} \end{bmatrix}$$

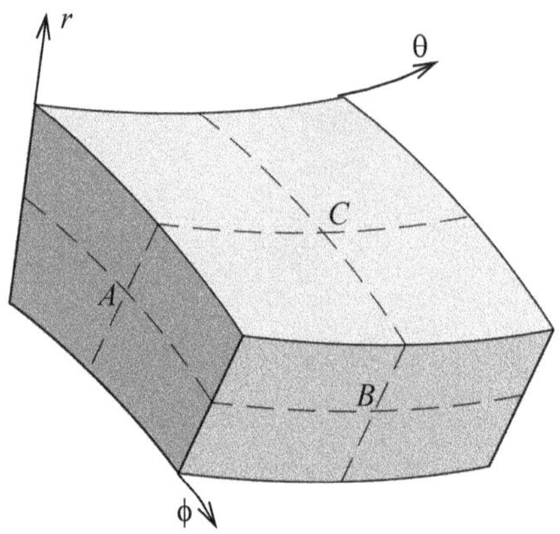


**Fig. P1.4**

**1.5** Starting with Equation (1.6) and taking the transpose of both sides, show that $\tau_{tn} = \tau_{nt}$.

## Sections 1.3-1.4

**1.6** Starting with a stress matrix in a principal coordinate system, show that the normal and shear stresses on a plane with direction cosines $n_1$, $n_2$, and $n_3$ are given by Equation (1.23).

**1.7** Show that the normal stress in three dimensions can be written as follows:

$$\sigma_{nn} = \sigma_{xx}n_x^2 + \sigma_{yy}n_y^2 + \sigma_{zz}n_z^2 + 2\tau_{xy}n_xn_y + 2\tau_{yz}n_yn_z + 2\tau_{zx}n_zn_x$$

**1.8** By expanding the determinant of the matrix in Equation (1.15), show that the invariants are as given by Equation (1.19).

**1.9** The stresses at a point on a building frame were determined and are as shown by the following stress matrix:

$$\begin{bmatrix} 100 & 125 & -150 \\ 125 & 12 & -60 \\ -150 & -60 & 0 \end{bmatrix}\text{MPa}$$

Determine (a) the normal and shear stresses on a plane that has an outward normal at 60°, –60°, and 45° to the *x*, *y*, and *z* directions, respectively, (b) the principal stresses, (c) the third principal direction, (d) the magnitude of the octahedral shear stress and maximum shear stress, and (e) stress deviatoric invariants.

**1.10** Show that the second deviatoric stress invariant can be written as $J_2 = I_2 - I_1^2/3$. Substitute for the stress invariants $I_1$ and $I_2$ and show that $J_2$ can be written as Equation (1.25).

**1.11** Show that the third deviatoric stress invariant can be written as given below.

$$J_3 = I_3 + (I_1I_2)/3 + (2I_1^3)/27$$

Substitute for the stress invariants $I_1$, $I_2$, and $I_3$ and show that $J_3$ can be written as Equation (1.25).

**1.12** Noting that the components of the stress vector *S* in terms of principal stresses can be written as $S_x = n_x\sigma_1$, $S_y = n_y\sigma_2$, and $S_x = n_z\sigma_3$, show that Equation is valid.

$$\frac{S_x^2}{\sigma_1^2} + \frac{S_y^2}{\sigma_2^2} + \frac{S_z^2}{\sigma_3^2} = 1$$

Equation is the equation of a stress ellipsoid, the surface of which is shown in the first quadrant in Fig. P1.12. The stress ellipsoid graphically depicts that the principal stresses are the maximum and minimum stresses at a point. The surface of the ellipsoid describes all possible stress states at a point.

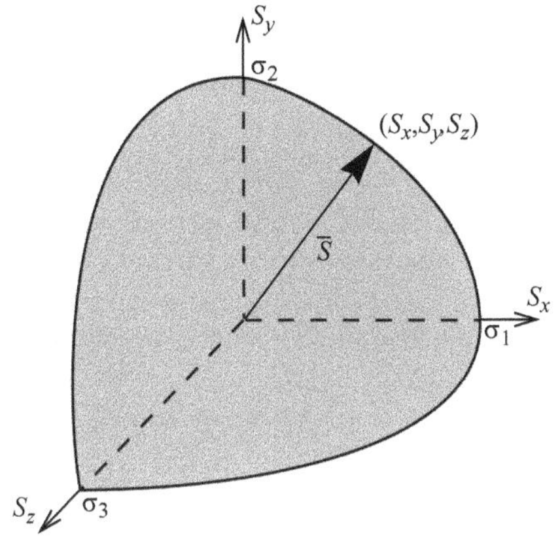


**Fig. P1.12Stress ellipsoid.**

## Sections 1.5-1.6

**1.13** Use a small-strain approximation to determine the deformation in bars *AP* and *BP* in Fig. P1.13.

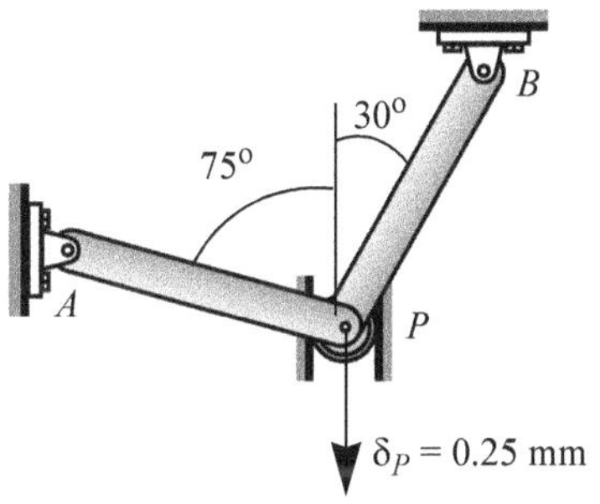


**Fig. P1.13**

**1.14** Use a small-strain approximation to determine the deformation in bars *AP* and *BP* in Fig. P1.14.

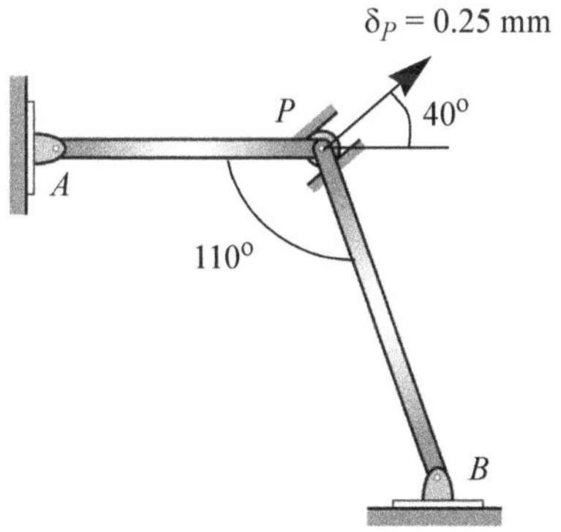


**Fig. P1.14**

**1.15** The axial strain in a bar of length *L* that is due to its own weight was found to be

$$\varepsilon_{xx} = K\left[4L - 2x - \frac{8L^3}{(4L-2x)^2}\right] \qquad 0 \le x \le L$$

where *K* is a constant for a given material and cross-sectional dimensions. Determine the total extension in terms of *K* and *L*.

**1.16** True strain ($\varepsilon_T$) is calculated from $d\varepsilon_T = (du)/(L_0 + u)$, where *u* is the deformation at any given instance and $L_0$ is the original undeformed length. Thus the increment in true strain is the ratio of change in length at any instant to the length at that given instant. If ε represents engineering strain, show that at any instant the relationship between true strain and engineering strain is given by the following:

$$\varepsilon_T = ln(1 + \varepsilon)$$

**1.17** A differential element subjected to only normal strains is shown in Fig. P1.17. The ratio of change in volume ($\Delta V$) to the original volume ($V$) is called the volumetric strain ($\varepsilon_V$) or *dilatation.* Prove

$$\varepsilon_V = \Delta V / V = I_1$$

where $I_1$ is the first strain invariant.

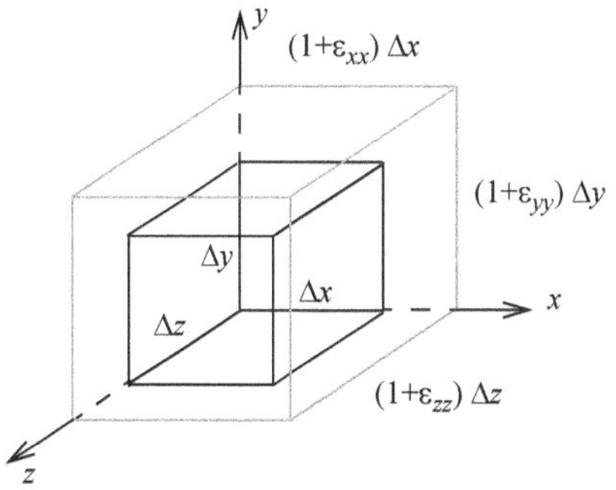


**Fig. P1.17**

**1.18** The displacements in a body are given by

$u = 500x - 300y - 200\ \mu\text{in} \qquad v = -100x + 200y + 250\ \mu\text{in}$

Determine the strains $\varepsilon_{xx}$, $\varepsilon_{yy}$, and $\gamma_{xy}$ at $x = 3$ in and $y = 2.5$ in.

**1.19** The displacements in a body are given by

$$u = [0.5(x^2 - y^2) + 0.5xy](10^{-3})\ \text{mm}$$

$$v = [0.25(x^2 - y^2) - xy](10^{-3})\ \text{mm}$$

Determine the strains $\varepsilon_{xx}$, $\varepsilon_{yy}$, and $\gamma_{xy}$ at $x = 5$ mm and $y = 7$ mm.

**1.20** The displacements at a point in polar coordinates are $u = K\ cos\theta(ln\ r)$, $v = -K\ sin\theta(ln\ r + 1)$, where *K* is a constant. Determine the strains $\varepsilon_{rr}$, $\varepsilon_{\theta\theta}$, and $\gamma_{r\theta}$.

**1.21** The displacements at a point in polar coordinates are

$$u = r^2(-C_1 cos\ 3\theta + C_2 sin\ 3\theta) \qquad v = r^2(-C_1 sin\ 3\theta + C_2 cos\ 3\theta)$$

where $C_1$ and $C_2$ are constants. Determine the strains $\varepsilon_{rr}$, $\varepsilon_{\theta\theta}$, and $\gamma_{r\theta}$.

## Section 1.7

**1.22** Obtain Equation (1.33b) from Equations (1.32a) and (1.32b).

**1.23** Obtain Equation (1.33c) from Equations (1.32a) and (1.32b).

**1.24** At a point on a building frame, the engineering strains were determined as $\varepsilon_{xx} = 720\ \mu$, $\varepsilon_{yy} = 320\ \mu$, $\varepsilon_{zz} = 0$, $\gamma_{xy} = 210\ \mu$, $\gamma_{yz} = 0$, and $\gamma_{zx} = 210\ \mu$. Determine (a) the normal strain along a line oriented at 70.5°, 48.2°, and 48.2° with the *x*, *y*, and *z* axis, respectively, and (b) the principal strains and maximum engineering shear strain, and (c) the principal strain direction 1.

## Section 1.8

**1.25** Use $E = 200$ GPa and $\nu = 0.32$ to calculate $\varepsilon_{xx}$, $\varepsilon_{yy}$, $\gamma_{xy}$, $\varepsilon_{zz}$, and $\sigma_{zz}$ (a) assuming plane stress and (b) assuming plane strain for the following stresses at a point:

$$\sigma_{xx} = 100\ \text{MPa(T)} \qquad \sigma_{yy} = 150\ \text{MPa(T)} \qquad \tau_{xy} = -125\ \text{MPa}$$

**1.26** Use $E = 70$ GPa and $G = 28$ GPa, to calculate $\varepsilon_{rr}$, $\varepsilon_{\theta\theta}$, $\gamma_{x\theta}$, $\varepsilon_{zz}$, and $\sigma_{zz}$ (a) assuming plane stress and (b) assuming plane strain for the following stresses:

$$\sigma_{rr} = 225\text{ MPa(C)} \qquad \sigma_{\theta\theta} = 125\text{ MPa(T)} \qquad \tau_{r\theta} = 150\text{ MPa}$$

**1.27** For a point in plane stress, show

$$\sigma_{xx} = [\varepsilon_{xx} + \nu\varepsilon_{yy}]\frac{E}{(1-\nu^2)} \qquad \sigma_{yy} = [\varepsilon_{yy} + \nu\varepsilon_{xx}]\frac{E}{(1-\nu^2)}$$

**1.28** For a point in plane stress, show

$$\varepsilon_{zz} = -\left(\frac{\nu}{1-\nu}\right)(\varepsilon_{xx} + \varepsilon_{yy})$$

**1.29** For a point in plane strain, show

$$\varepsilon_{xx} = [(1-\nu)\sigma_{xx} - \nu\sigma_{yy}]\frac{(1+\nu)}{E}$$

$$\varepsilon_{yy} = [(1-\nu)\sigma_{yy} - \nu\sigma_{xx}]\frac{(1+\nu)}{E}$$

**1.30** For a point in plane strain, show

$$\sigma_{xx} = \frac{E[(1-\nu)\varepsilon_{xx} + \nu\varepsilon_{yy}]}{(1-2\nu)(1+\nu)} \qquad \sigma_{yy} = \frac{E[(1-\nu)\varepsilon_{yy} + \nu\varepsilon_{xx}]}{(1-2\nu)(1+\nu)}$$

**1.31** At a point on the frame of an aircraft, the engineering strains were determined as $\varepsilon_{xx} = 800\ \mu$, $\varepsilon_{yy} = 500\ \mu$, $\varepsilon_{zz} = -200\ \mu$, $\gamma_{xy} = 300\ \mu$, $\gamma_{yz} = 0$, and $\gamma_{zx} = 400$ The modulus of elasticity and Poisson's ratio of the frame material are 70 GPa and 0.25, respectively. Determine (a) the principal stresses, (b) the maximum shear stress, and (c) the principal stress direction 1.

**1.32** Bronze has a yield stress of $\sigma_{\text{yield}} = 18$ ksi in tension, a yield strain of $\varepsilon_{\text{yield}} = 0.0012$, and ultimate stress of $\sigma_{\text{ult}} = 50$ ksi; the corresponding ultimate strain is $\varepsilon_{\text{ult}} = 0.50$. Determine the material constants for (a) the elastic–perfectly plastic model, (b) the linear strain-hardening model, and (c) the nonlinear power law model.

**1.33** Cast iron has a yield stress of $\sigma_{\text{yield}} = 220$ MPa in tension, a yield strain of $\varepsilon_{\text{yield}} = 0.00125$, and ultimate stress of $\sigma_{\text{ult}} = 340$ MPa; the corresponding ultimate strain is $\varepsilon_{\text{ult}} = 0.20$. Determine the material constants for (a) the elastic–perfectly plastic model, (b) the linear strain-hardening model, (c) and the nonlinear power law model.

## Section 1.9

**1.34** Calculate $\varepsilon_{xx}$, $\varepsilon_{yy}$, $\gamma_{xy}$, $\varepsilon_{zz}$, and $\sigma_{zz}$ (a) assuming plane stress and (b) assuming plane strain for the following stresses at a point, material properties, and change in temperature.

$$\sigma_{xx} = 300\text{ MPa(C)} \qquad \sigma_{yy} = 300\text{ MPa(T)} \qquad \tau_{xy} = 150\text{ MPa}$$

$$G = 15\text{ GPa} \qquad \nu = 0.2 \qquad \alpha = 26.0\mu/^{o}C$$

$$\Delta T = 75^{o}C$$

**1.35** Calculate $\varepsilon_{xx}$, $\varepsilon_{yy}$, $\gamma_{xy}$, $\varepsilon_{zz}$, and $\sigma_{zz}$ (a) assuming plane stress and (b) assuming plane strain for the following stresses at a point, material properties, and change in temperature.

$$\sigma_{xx} = 22\text{ ksi(C)} \qquad \sigma_{yy} = 25\text{ ksi(C)} \qquad \tau_{xy} = -15\text{ ksi}$$

$$E = 30,000\text{ ksi} \qquad \nu = 0.3 \qquad \alpha = 6.5\mu/^{o}F$$

$$\Delta T = 40^{o}F$$

**1.36** For a point in plane stress, show

$$\sigma_{xx} = [\varepsilon_{xx} + \nu\varepsilon_{yy}]\frac{E}{(1-\nu^2)} - \frac{E\alpha\,\Delta T}{(1-\nu)}$$

$$\sigma_{yy} = [\varepsilon_{yy} + \nu\varepsilon_{xx}]\frac{E}{(1-\nu^2)} - \frac{E\alpha\,\Delta T}{(1-\nu)} \qquad \textbf{(1.61)}$$

**1.37** For a point in plane stress, show

$$\varepsilon_{zz} = -\left(\frac{\nu}{1-\nu}\right)(\varepsilon_{xx} + \varepsilon_{yy}) + \frac{(1-3\nu)}{(1-\nu)}\alpha\,\Delta T$$

## Section 1.10

**1.38** The stress components at a critical point that is in plane stress due to a force $P$ are

$$\sigma_{xx} = 10P\text{ MPa (T)} \qquad \sigma_{yy} = 20P\text{ MPa (C)} \qquad \tau_{xy} = 5P\text{ MPa}$$

where $P$ is in kilonewtons (kN). The material has a yield stress of 160 MPa as determined in a tension test. If yielding must be avoided, predict the maximum value of the force $P$ by using (a) the maximum shear stress theory and (b) the maximum octahedral shear stress theory.

**1.39** A material has a tensile rupture strength of 18 ksi and a compressive rupture strength of 32 ksi. During use, a component made from this material showed the following stresses on a free surface at a critical point:

$$\sigma_{xx} = 9\text{ ksi (T)} \qquad \sigma_{yy} = 6\text{ ksi (C)} \qquad \tau_{xy} = -4\text{ ksi}$$

Use the modified Mohr's theory to determine the factor of safety.

**1.40** For *plane stress*, show that the von Mises stress representing can be written as

$$\sigma_{\text{von}} = \sqrt{\sigma_{xx}^2 + \sigma_{yy}^2 - \sigma_{xx}\sigma_{yy} + 3\tau_{xy}^2}$$

# 2 One-Dimensional Structural Members

## LEARNING OBJECTIVES

1. To understand how complexities are added to the classical theories of axial members, torsion of circular shafts, and symmetric bending of beams.
2. To understand the concept and use of discontinuity functions in the analysis of one-dimensional structural members subjected to discontinuous loads.

Chair legs, utility poles, drills, bookshelves, diving boards, frame of a building or a car, skeleton of a human being, are among the countless applications of one-dimensional structural members. The theories and formulas for one-dimensional structural members vary in complexity depending upon the loading, material properties, and cross-sectional geometries, many of which we will consider in this chapter.

In Chapter 1, we discussed the relationships of displacements to strains, and strains to stresses. In Section 2.1, we will see the two step process by which stresses are related to external forces and moments. In Section 2.2, we synthesis all these relationships into a logic by which we relate displacements to external forces and moments. In Section 2.3, we use the logic to develop the classical theories of axial members, torsion of circular shafts, and symmetric bending of beams which we saw in the introductory course of mechanics of materials. A synopsis of all three theories is presented in a tabular form to emphasize the commonalities and differences in the derivation of stress and deformation formulas. The limitations and assumptions that are made to get the simplest possible formulas are identified with the intent of dropping them to include complexities in geometry, loading, and materials.

Discontinuous loading on a structural member is fairly common. In the introductory course these discontinuities in loading were accounted for by segmenting the structural member such that in each segment the loading was continuous. Relative deformations of segment ends were found, and the overall deformation was obtained by using the observation that deformation was continuous at the point of loading discontinuity. This approach is algebraically tedious, particularly for beam-bending problems and statically indeterminate structural members. In this chapter the concept of discontinuity functions is introduced and effectively used for analyzing the deformation of statically determinate and indeterminate structural members that are subjected to discontinuous loads.

## 2.1 INTERNAL FORCES AND MOMENTS

**Stress resultant** is another name for internal forces and moments. Stresses being an internal distributed force system can always be replaced by equivalent internal forces and moments for purpose of equilibrium on a free body diagram. The two step process of relating stresses to external forces is emphasized in Figure 2.1.

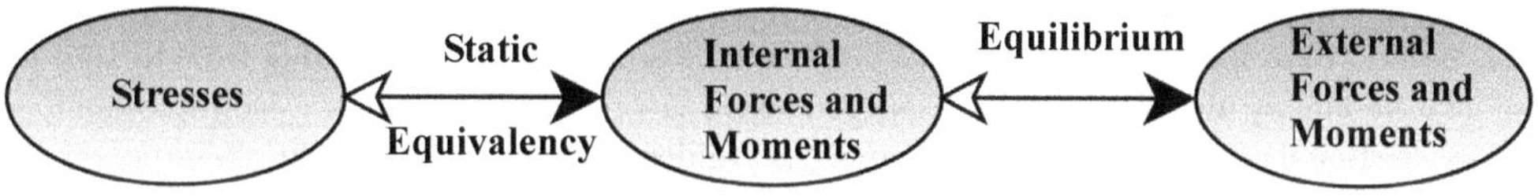


**Figure 2.1** Two step process of relating stresses to external forces and moments

Figure 2.2*a* shows the stresses on a differential element with the outward normal in the *x* direction. Figure 2.2*b* shows the statically equivalent internal forces and moments. The static equivalency of normal stress $\sigma_{xx}$ to the internal forces and moments shown in Figure 2.2 results in the Equation (2.1).The static equivalency of shear stresses to the internal forces and moments shown in Figure 2.2 results in Equation (2.2).

$$N = \int_A \sigma_{xx}\, dA \qquad M_z = -\int_A y\sigma_{xx}\, dA \qquad M_y = -\int_A z\sigma_{xx}\, dA \tag{2.1}$$

$$V_y = \int_A \tau_{xy}\, dA \qquad V_z = \int_A \tau_{xz}\, dA \qquad T = \int_A [y\tau_{xz} - z\tau_{xy}]dA \tag{2.2}$$

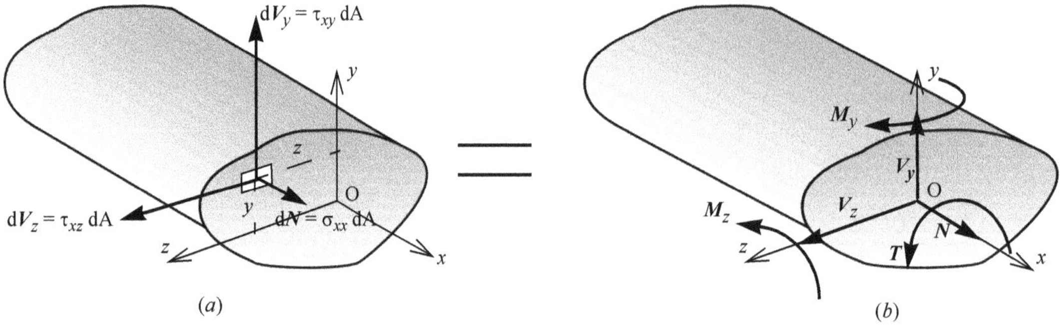


**Figure 2.2** Static equivalency.

For torsion of circular cross sections, the only non-zero stress is shear stress $\tau_{x\theta}$. Figure 2.3 shows that the shear forces can be written as

$$dV_z = (\tau_{x\theta}dA)cos\theta \qquad dV_y = -(\tau_{x\theta}dA)sin\theta \qquad y = \rho cos\theta \qquad z = \rho sin\theta \tag{2.3}$$

Substituting Equation (2.3) into torque expression in Equation (2.2), we obtain the internal torque on circular shafts as:

$$T = \int_A [(\rho cos\theta)(\tau_{x\theta}cos\theta) - (\rho sin\theta)(-\tau_{x\theta}sin\theta)]dA = \int_A \rho\tau_{x\theta}[cos^2\theta + sin^2\theta]dA$$

$$T = \int_A \rho\tau_{x\theta}dA \tag{2.4}$$

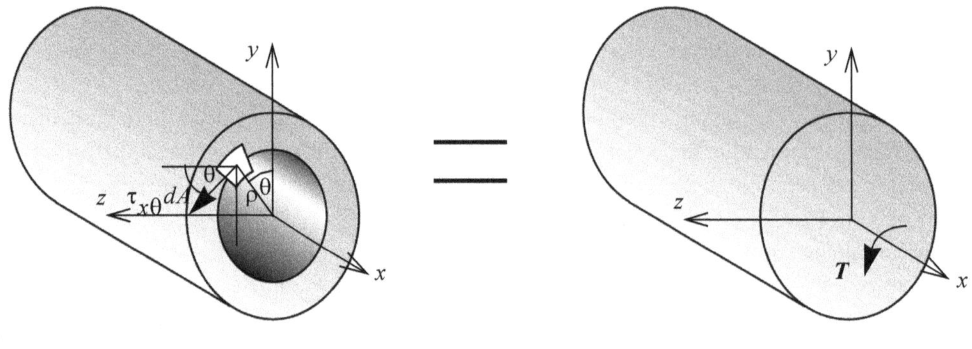


**Figure 2.3** Static equivalency for torsion of circular cross sections.

Stresses are positive or negative according to sign convention discussed in Section 1.2.1, thus the internal forces and moments will be positive or negative according to sign convention discussed in Section 2.3.2.

## 2.1.1 Decoupling axial, bending, and torsion analysis

In statics we saw that any distributed force can be replaced by a force and a moment at any point, or by a single force (and no moment) at a specific point. These specific points help decouple axial from bending, and bending from torsion for purposes of simpler analysis. In the introductory mechanics of materials course we saw that for linear-elastic-isotropic-homogenous material, the *centroid* of the cross section decouples the bending from axial problems. In the same manner, *shear center* decouples the shear forces (hence shear stresses) due to bending from those due to torsion. For more complex material models, the determination of the specific points that decouples axial, bending, and torsion analysis requires more effort.

The **shear center** is a point in space at which the shear stress due to bending can be replaced by statically equivalent internal shear forces and no internal torque.

For linear problems and for non-linear materials we will be able to decouple axial, bending, and torsion analysis. However, for geometric non-linearity, that is, for large deformation it is generally not possible to decouple analysis. Axial forces produce bending displacements and bending forces produce axial displacements as will be seen in Chapter 7.

## EXAMPLE 2.1

A body under applied loads and body forces produces the following displacement field

$$u = 0 \qquad v = Kx(y^2 - z^2) + Kaxz \qquad w = -2Kxyz - Kaxy \tag{E1}$$

where $u$, v, and $w$ are displacements in the $x$, $y$, and $z$ directions, respectively. Figure 2.4 shows the internal forces and moments (stress resultants) on a cross section. Assuming a linear, elastic, isotropic, homogeneous material, determine all the internal forces and moments in terms of modulus of elasticity, Poisson's ratio ν, and the constants $K$ and $a$.

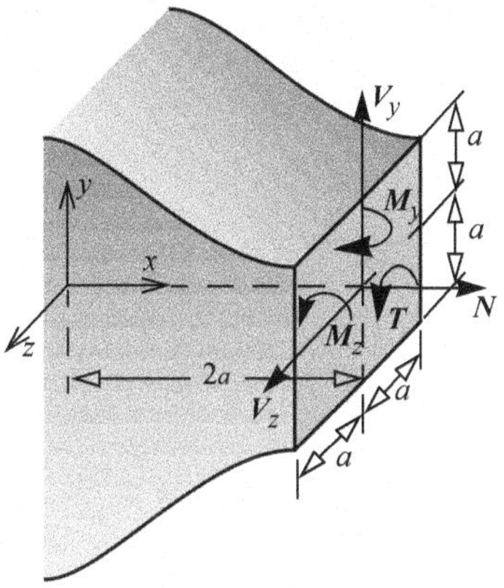


**Figure 2.4** Geometry in Example 2.1.

**PLAN**

The strains can be found from the displacement and evaluated at $x = 2a$. Using the generalized Hooke's law, stresses can be found and then internal forces and moments calculated using Equations (2.1) and (2.2).

**SOLUTION**

1. Strain calculations: From Equation (1.29a) through (1.29d) we obtain the strains as follows and evaluate them at $x = 2a$.

$$\varepsilon_{xx} = \frac{\partial u}{\partial x} = 0 \qquad \varepsilon_{yy} = \left.\frac{\partial v}{\partial y}\right|_{x=2a} = 2Kxy|_{x=2a} = 4Kay \qquad \varepsilon_{zz} = \left.\frac{\partial w}{\partial z}\right|_{x=2a} = -2Kxy|_{x=2a} = -4Kay \tag{E1}$$

$$\gamma_{xy} = \left.\left[\frac{\partial u}{\partial y} + \frac{\partial v}{\partial x}\right]\right|_{x=2a} = K(y^2 - z^2) + Kaz|_{x=2a} = K(y^2 - z^2) + Kaz \tag{E2}$$

$$\gamma_{yz} = \left.\left[\frac{\partial v}{\partial z} + \frac{\partial w}{\partial y}\right]\right|_{x=2a} = [-2Kxz + Kax - 2Kxz - Kax]|_{x=2a} = -8Kaz \tag{E3}$$

$$\gamma_{xz} = \left.\left[\frac{\partial w}{\partial x} + \frac{\partial u}{\partial z}\right]\right|_{x=2a} = [-2Kyz - Kay]|_{x=2a} = -Ky[2z + a] \tag{E4}$$

2. Stress calculations: From Equations (1.42) and (1.46) we obtain the stresses as shown below.

$$\sigma_{xx} = \frac{E[(1-\nu)\varepsilon_{xx} + \nu\varepsilon_{yy} + \nu\varepsilon_{zz}]}{(1-2\nu)(1+\nu)} = \frac{E[\nu(4Kay) + \nu(-4Kay)]}{(1-2\nu)(1+\nu)} = 0 \tag{E5}$$

$$\sigma_{yy} = \frac{E[\nu\varepsilon_{xx} + (1-\nu)\varepsilon_{yy} + \nu\varepsilon_{zz}]}{(1-2\nu)(1+\nu)} = \frac{E[(1-\nu)(4Kay) + \nu(-4Kay)]}{(1-2\nu)(1+\nu)} = \frac{4EKay}{(1+\nu)} \tag{E6}$$

$$\sigma_{zz} = \frac{E[\nu\varepsilon_{xx} + \nu\varepsilon_{yy} + (1-\nu)\varepsilon_{zz}]}{(1-2\nu)(1+\nu)} = \frac{E[(1-\nu)(-4Kay) + \nu(4Kay)]}{(1-2\nu)(1+\nu)} = -\left[\frac{4EKay}{(1+\nu)}\right] \tag{E7}$$

$$\tau_{xy} = G\gamma_{xy} = \frac{EK}{2(1+\nu)}[(y^2 - z^2) + az] \tag{E8}$$

$$\tau_{yz} = G\gamma_{yz} = \frac{E[-8Kaz]}{2(1+\nu)} = -\left[\frac{4KEaz}{(1+\nu)}\right] \tag{E9}$$

$$\tau_{xz} = G\gamma_{yz} = \frac{E[-Ky(2z+a)]}{2(1+\nu)} = -\left[\frac{EKy}{2(1+\nu)}(2z+a)\right] \tag{E10}$$

3. Internal forces and moments calculations: With $\sigma_{xx} = 0$, from Equations (2.1) and (2.2)we can write

$$\textbf{ANS. } N = \int_A \sigma_{xx} dA = 0 \qquad M_y = -\int_A z\sigma_{xx} dA = 0 \qquad M_z = -\int_A y\sigma_{xx} dA = 0 \tag{E11}$$

$$V_y = \int_A \tau_{xy} dA = \frac{EK}{2(1+\nu)}\int_A [(y^2 - z^2) + az] dA = \frac{EK}{2(1+\nu)}\left|\int_A y^2 dA - \int_A z^2 dA + a\int_A z dA\right| = \frac{EK}{2(1+\nu)}[I_{zz} - I_{yy}] \tag{E12}$$

In the above equation we noted that the area moment of inertias are $I_{zz} = \int_A y^2 dA$ and $I_{yy} = \int_A z^2 dA$ and $\int_A z dA = 0$ because $z$ is measured from the centroid of the cross section. Similarly we obtain

$$V_z = \int_A \tau_{xz} dA = -\left[\frac{5KEa}{2(1+\nu)}\right]\int_A y dA = 0 \tag{E13}$$

$$T = \int_A (y\tau_{xz} - z\tau_{xy}) dA = -\left[\frac{EK}{2(1+\nu)}\right]\int_A (2zy^2 + ay^2 + zy^2 - z^3 + az^2) dA = -\left[\frac{EK}{2(1+\nu)}\right]\left[\int_A (3zy^2 - z^3) dA + a(I_{zz} + I_{yy})\right] \tag{E14}$$

The first integral is zero as integrand is an odd function of $z$ and we are integrating in the $z$ direction from $-a$ to $+a$, hence

$$T = -\left[\frac{EKa}{2(1+\nu)}\right](I_{zz} + I_{yy}) \tag{E15}$$

For a square of $2a$ by $2a$ dimension, we know

$$I_{yy} = I_{zz} = (2a)(2a)^3/12 = 4a^4/3 \tag{E16}$$

Substituting the above results in Equations (E12) and (E15) we obtain the results below.

$$\textbf{ANS.} \quad V_z = 0\ ;\ V_z = 0\ ;\ T = -\left[\frac{4EKa^5}{3(1+\nu)}\right]$$

**COMMENTS**

1. The example demonstrates that if the displacement field is known in an elastic body, then all the variables of the mechanics of materials can be found from the displacement function. We will formalize this process into a logic that we will use to develop the classical theories of one dimensional structural members in Section 2.3 and the theory for thin plates in Chapter 5.
2. We could have integrated any of the above integrals directly. For example, we could have written $V_y = \int_A \tau_{xy} dA = \int_{-a}^{a}\int_{-a}^{a} \tau_{xy} dy dz$, substituted $\tau_{xy}$ and performed the integration to get the same results. But using the concept of centroid, area moment of inertia, and symmetric integration of an odd function, we avoided lot of algebraic tedium.

## EXAMPLE 2.2

The normal stress on a cross section is given by $\sigma_{xx} = a + by + cz$, where $a$, $b$, and $c$ are constants. Determine the conditions necessary to obtain the formula below.

$$\sigma_{xx} = \frac{N}{A} - \left(\frac{M_z}{I_{zz}}\right)y - \left(\frac{M_y}{I_{yy}}\right)z \tag{E1}$$

where, A is the cross sectional area, $I_{yy}$ and $I_{zz}$ are the second area moments of inertia about the y and z axis.

**PLAN**

By comparing Equation (E1) to the given stress we can determine the constants $a$, $b$, and $c$. Substituting the given stress into Equation (2.1), we obtain three equations. By looking at the equations we identify the conditions to obtain the constants $a$, $b$, and $c$.

**SOLUTION**

By comparing the given stress to the stress in Equation (E1), we obtain

$$a = \frac{N}{A} \qquad b = -\left(\frac{M_z}{I_{zz}}\right) \qquad c = -\left(\frac{M_y}{I_{yy}}\right) \tag{E2}$$

Substituting the given stress into Equation (2.1) to obtain

$$N = \int_A (a + by + cz) dA = a\int_A dA + b\int_A y dA + c\int_A z dA = aA + by_c + cz_c \tag{E3}$$

$$M_y = -\int_A z(a + by + cz) dA = -a\int_A z dA - b\int_A yz dA - c\int_A z^2 dA = -az_c - bI_{yz} - cI_{yy} \tag{E4}$$

$$M_z = -\int_A y(a + by + cz) dA = -a\int_A y dA - b\int_A y^2 dA - c\int_A yz dA = -ay_c - bI_{zz} - cI_{yz} \tag{E5}$$

where, $y_c$ and $z_c$ are the coordinates of the centroid.

To obtain Equation (E2) from the above equations, the following conditions must be satisfied:

$$\textbf{ANS.}\ y_c = \int_A y dA = 0 \qquad z_c = \int_A z dA \qquad I_{yz} = \int_A yz dA = 0 \tag{E6}$$

**COMMENTS**

1. For $y_c$ and $z_c$ to be zero, the origin must be at the centroid of the cross section. This is the requirement for decoupling the axial from bending problem for a linear-elastic-isotropic-homogenous material cross section.
2. For $I_{yz}$ to be zero, either $y$ or $z$ must be an axis of symmetry for the cross section. This is requirement along with the requirement that loading be in plane of symmetry, results in symmetric bending studied in the introductory mechanics of materials.
3. Equation (E1) represents superposition of stresses due to axial loads and symmetric bending about $y$ and $z$ axis.

4. If the origin is at the centroid but $I_{yz}$ is not zero, then Equations (E3) through (E5) can be solved to obtain

$$a = \frac{N}{A} \qquad b = -\left(\frac{M_z I_{yy} - M_y I_{yz}}{I_{yy}I_{zz} - I_{yz}^2}\right) \qquad c = -\left(\frac{M_y I_{zz} - M_z I_{yz}}{I_{yy}I_{zz} - I_{yz}^2}\right) \tag{E7}$$

Substituting the above constants into the given stress expression we obtain

$$\sigma_{xx} = \frac{N}{A} - \left(\frac{M_z I_{yy} - M_y I_{yz}}{I_{yy}I_{zz} - I_{yz}^2}\right)y - \left(\frac{M_y I_{zz} - M_z I_{yz}}{I_{yy}I_{zz} - I_{yz}^2}\right)z \tag{E8}$$

The above equation represents superposition of stresses due to axial loads and unsymmetrical bending loads.

## EXAMPLE 2.3

The cross-section shown in Figure 2.5 has a uniform thickness $t$. Assume $t \ll a$. The shear stresses in the cross section were found to be as given. (*a*) Replace the shear stresses by equivalent shear forces $V_y$, $V_z$ and torque $T$ acting at the centroid $C$. (*b*) Determine the coordinates of the shear center. Solve the problem in terms variables $K$, $a$, and $t$.

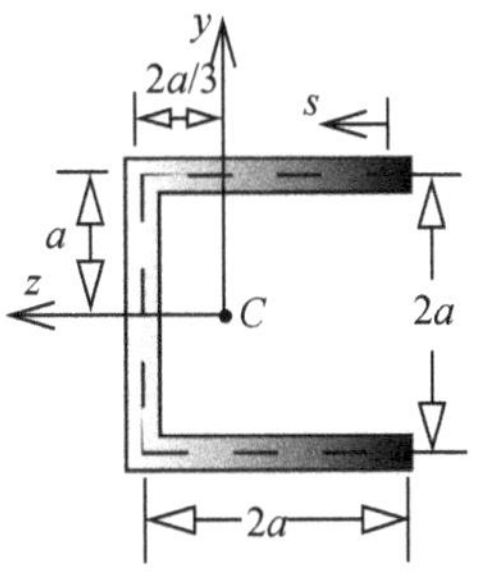


$$\tau_{xy} = 0 \qquad \tau_{xz} = \frac{Ks}{t} \qquad 0 \le s < 2a$$

$$\tau_{xy} = -\frac{K}{2at}(-4a^2 + 6as - s^2) \qquad \tau_{xz} = 0 \qquad 2a < s < 4a$$

$$\tau_{xy} = 0 \qquad \tau_{xz} = \frac{K(s-6a)}{t} \qquad 4a < s \le 6a$$

**Figure 2.5**

**PLAN**

(a) By substituting the stress distribution in Equation (2.2), we obtain the equivalent shear forces $V_y$, $V_z$ and torque $T$. (b) By static equivalency we determine the coordinates of the shear center.

**SOLUTION**

(a) We note the differential area in all segments is $dA = tds$. Substituting the stress distribution in Equation (2.2) we obtain

$$V_y = \int_A \tau_{xy} dA = \int_{2a}^{4a} \frac{-K[(-4a^2 + 6as - s^2)]}{2at} tds = \frac{-K}{2a}\left[\left(-4a^2 s + 3as^2 - \frac{s^3}{3}\right)\Bigg|_{2a}^{4a}\right] = -\left(\frac{14Ka^2}{3}\right) \tag{E1}$$

$$V_z = \int_A \tau_{xz} dA = \int_0^{2a} \frac{Ks}{t} tds_1 + \int_{4a}^{6a} \frac{K(s-6a)}{t} tds_3 = K\left[\left(\frac{s^2}{2}\right)\Bigg|_0^{2a} + \frac{(s-6a)^2}{2}\Bigg|_{4a}^{6a}\right] = K\left[\frac{4a^2}{2} + \frac{0 - 4a^2}{2}\right] = 0 \tag{E2}$$

$$T = \int_A [y\tau_{xz} - z\tau_{xy}]dA = \int_0^{2a} (a)\left[\frac{Ks}{t}\right]tds - \int_{2a}^{4a} \left(\frac{2a}{3}\right)\left[\frac{-K(-4a^2 + 6as - s^2)}{2at}\right]tds + \int_{4a}^{6a} (-a)\left[\frac{K(s-6a)}{t}\right]tds$$

$$T = K\left[a\left(\frac{s^2}{2}\right)\Bigg|_0^{2a} + \left(\frac{2}{3}\right)\left(-4a^2 s + 3as^2 - \frac{s^3}{3}\right)\Bigg|_{2a}^{4a} - a\frac{(s-6a)^2}{2}\Bigg|_{4a}^{6a}\right] = K\left(\frac{64}{9}\right)a^3 \tag{E3}$$

**ANS.** $V_y = -14Ka^2/3$; $V_z = 0$; $T = 64Ka^3/9$ CCW

(b) Figure 2.6 shows two statically equivalent systems. As $V_z = 0$, the shear center is on the $z$ axis as shown. By considering the moments about $C$ for the two systems we obtain

$$T = -e_z(V_y) \qquad \text{or} \qquad e_z = \frac{64Ka^3/9}{14Ka^2/3} = \frac{32}{21}a \tag{E4}$$

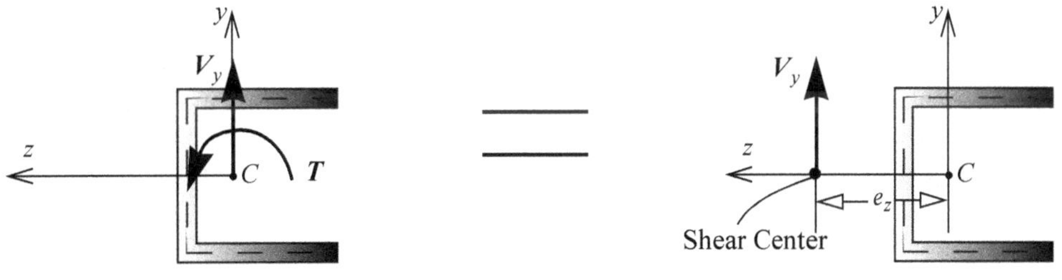


**Figure 2.6** Static equivalency in Example 2.3.

**ANS.** $e_y = 0$; $e_z = 32a/21$

**COMMENTS**

1. Shear center, like the centroid, depends only on the geometry of the cross section. Both points lie on the axis of symmetry.
2. Centroid and shear center are imaginary points in space that are defined to simplify analysis.
3. In this example and the previous example, the stress distributions were given. In actual analysis these stress distributions have to be found from formulas that have been developed by considering axial, bending, and torsion separately. This basic idea that we can develop formulas by considering each type of loading separately was done in the introductory course of mechanics of materials and will be considered again in the next section.

## 2.2 LOGIC IN MECHANICS OF MATERIALS

Displacements, strains, stresses, internal forces and moments, and external forces and moments are the fundamental variables in mechanics of materials. These fundamental variables are logically related as symbolically shown in Figure 2.7. This logic is used for developing both the simple theories discussed in the introductory course on the mechanics of materials and the advanced theories encountered in graduate courses. It is possible to start at any point in the logic and move either clockwise (solid arrows) or counterclockwise (hollow arrows). It is not possible to relate displacement directly to external forces without imposing limitations and making assumptions regarding the geometry of the body, material behavior, and external loading. This is emphasized by the absence of arrows directly linking displacements and external forces in Figure 2.7.

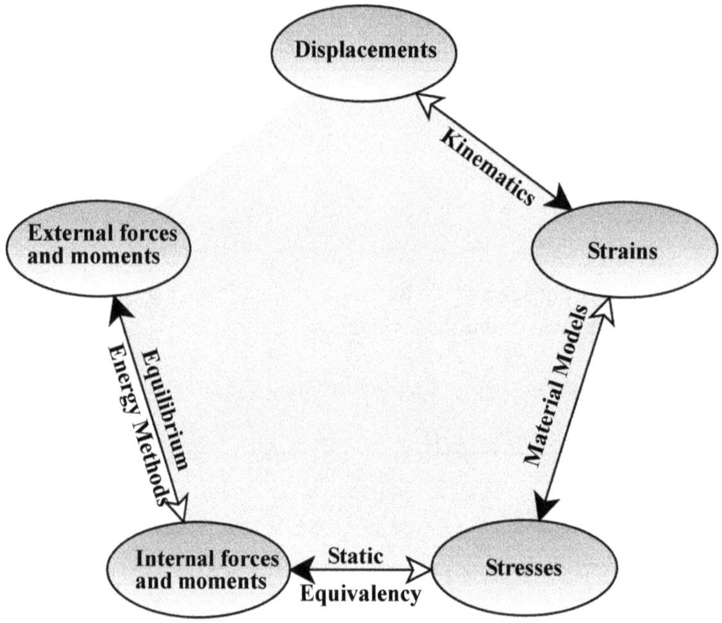


**Figure 2.7** Logic in structural analysis.

The logic shown in Figure 2.7 is intrinsically very modular; that is, *the equations* (kinematics, material models, static equivalency, and equilibrium) *linking any two fundamental variables are mutually independent*. Thus, if complexity is added in one part, then the theory can be re-derived with minimum change. This will be demonstrated many times in this book through examples, in post-text problems, and in Chapter 5 on thin plates. Understanding the modular nature of this logic and learning how complexities can be incorporated to refine basic theories is an important learning objective of this book.

In Chapter 7 we will see that energy methods are an alternative to Newtonian methods of equilibrium equations and can replace the last link in the logic. The modular character of the logic however remains unchanged and complexities can as easily incorporated with energy methods as with equilibrium equations.

## 2.3 CLASSICAL THEORIES OF 1-D STRUCTURAL MEMBERS

This section presents the classical theory of axial members, symmetric bending of beams, and torsion of circular shafts in tabular form to emphasize the *commonalities and differences* in the derivation of various formulas. We will start by reviewing the simplest possible theories, that is, the classical theories discussed in the introductory course on the mechanics of materials. The limitations and assumptions that lead to simplifications will be identified with the intent of dropping them to incorporate complexities as discussed in Section 2.3.3.

Table 2.1 is a synopsis of the derivation the elementary theories. Assumptions and equations with suffixes -A, -B, and -T refer to axial, bending, and torsion, respectively. An assumption without any suffix is applicable to all three theories. This notation is used to facilitate the demonstration that complexities enter the derivation at the same level and are accounted for in nearly a similar manner for all three structural members.

### 2.3.1 Limitations

The following limitations are common to theories about axial members, torsion of circular shafts, and symmetric bending of beams.

1. The length of the member is significantly greater (approximately 10 times) than the greatest dimension in the cross section. Approximation across the cross section is now possible because the region of approximation is small.
2. We are away from regions of stress concentration, where displacements and stresses can be three-dimensional. The results from the simplified theories can be extrapolated into regions of stress concentration using stress concentration factor. In a similar manner, results can be extrapolated into regions near cracks by means of stress intensity factors.
3. The variation of external loads or changes in the cross-sectional area is gradual except in regions of stress concentration. The theory of elasticity shows that this limitation is necessary; otherwise approximations across the cross section would be untenable.
4. The external loads are such that the axial, torsion, and bending problems can be studied individually. This requires not only that the applied loads be in a given direction, but also that the loads pass through specific points associated with the cross section as discussed in Section 2.1.

For torsion and bending, we impose the following additional limitations.

- **Torsion:** We limit ourselves to circular cross sections. This permits us to use arguments of axisymmetry in deducing deformation. This limitation is dropped in Section 6.7 to develop theory of torsion of non-circular shafts.
- **Bending:** We limit ourselves to beam cross sections that have a plane of symmetry to which the loading is restricted. This limitation ensures symmetric bending. It is assumed that the $x$-$y$ plane is the plane of symmetry and that bending is occurring about the $z$ axis (i.e., cross sections rotate about the $z$ axis).

### 2.3.2 Definition of variables and sign conventions

Figure 2.8 shows a slender member with the $x$ axis along the axial direction and $y$ and $z$ in the plane of the cross section. We will use this coordinate system for axial and bending problems. For torsion problems we will use a polar coordinate system ($\rho$, $\theta$, $x$). The figure is symbolic and the theories for axial and bending members are *not* limited to circular cross sections.

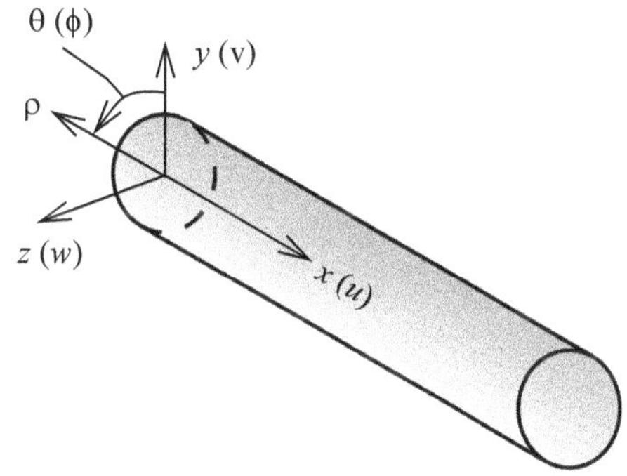


**Figure 2.8** Coordinate systems and deformation variables.

Table 2.1 is a synopsis of the derivation of the elementary classical theories studied in the introductory course on the mechanics of materials. See Vable [2013] for additional details on the development of these theories. Assumptions are made with the motivation to develop the *simplest possible theories.*

The variables used in the table are defined as follows.

1. The displacements $u$, v, and $w$ will be considered to be positive in the positive $x$, $y$, and $z$ direction, respectively.
2. The rotation $\phi$ of the cross section will be considered to be positive counter clockwise with respect to the $x$ axis.
3. The external distributed torque per unit length $t(x)$ is positive counterclockwise with respect to the $x$ axis.
4. The external distributed forces per unit length $p_x(x)$ and $p_y(x)$ are considered to be positive in the positive $x$ and $y$ direction, respectively.
5. $E$ is the modulus of elasticity and $G$ is shear modulus of elasticity.
6. $A$ is the cross sectional area; $I_{zz}$ = second area moment of inertia about $z$; $J$ = polar moment of the area about $x$.
7. $EA$ = axial rigidity; $EI_{zz}$ = bending rigidity; $GJ$ = torsional rigidity.
8. On a free body diagram, the positive internal axial force $N$ is shown as tensile.
9. On a free body diagram, the positive internal shear force $V_y$ is shown in the same direction as positive $\tau_{xy}$.
10. On a free body diagram, the positive internal bending moment $\boldsymbol{M_z}$ is shown to put the positive $y$ face in compression.
11. On a free body diagram, positive internal torque $\boldsymbol{T}$ is shown counterclockwise with respect to the outward normal.

If the sign conventions for internal forces and moments are followed in drawing free body diagrams, the formulas in Table 2.1 will give the right sign for displacements and stress components. If the internal forces and moments are drawn to equilibrate the external forces and moments, the formulas in Table 2.1 should be used only to determine the magnitude of the displacement and stress components; the direction of these components must be done by inspection.

**Table 2.1** Synopsis of Elementary Theories of One-Dimensional Structural Members

| | **Axial** | **Bending** | **Torsion** |
|---|---|---|---|
| | | **Deformations** | |
| **Assumption 1** | Deformations are not functions of time. | | |
| **Assumption 2** | 2-A: Plane sections remain plane and parallel. | 2a-B: Squashing deformation is significantly smaller than deformation due to bending.<br>2b-B: Plane sections before deformation remain plane after deformation.<br>2c-B: A plane perpendicular to the beam axis remains *nearly* perpendicular after deformation. | 2a-T: Plane sections perpendicular to the axis remain plane during deformation.<br>2b-T: All radial lines rotate by equal angle during deformation on a cross section.<br>2c-T: Radial lines remain straight during deformation. |
| | $u = u_0(x)$ **(2.5-A)** | $v = v(x) \quad u = -y\frac{dv}{dx}$ **(2.5a-B)** | $\phi = \phi(x)$ **(2.5-T)** |
| | | **Strains** | |
| **Assumption 3** | The strains are small. | | |
| | $\varepsilon_{xx} = \frac{du_0}{dx}(x)$ **(2.6-A)** | $\varepsilon_{xx} = -y\frac{d^2v}{dx^2}(x)$ **(2.6-B)** | $\gamma_{x\theta} = \rho\frac{d\phi}{dx}(x)$ **(2.6-T)** |
| | | **Stresses** | |
| **Assumption 4** | The material is isotropic. | | |
| **Assumption 5** | There are no thermal or other non-mechanical strains. | | |
| **Assumption 6** | The material is elastic. | | |
| **Assumption 7** | Stress and strain are linearly related. | | |
| Using Hooke's law | $\sigma_{xx} = E\frac{du_0}{dx}(x)$ **(2.7-A)** | $\sigma_{xx} = -Ey\frac{d^2v}{dx^2}(x)$ **(2.7-B)** | $\tau_{x\theta} = G\rho\frac{d\phi}{dx}(x)$ **(2.7-T)** |
| | | **Internal Forces and Moments** | |
| **Static equivalency** | $N = \int_A \sigma_{xx} dA$ **(2.8a-A)** | $N = \int_A \sigma_{xx} dA = 0$ **(2.8a-B)** | $T = \int_A \rho\tau_{x\theta} dA$ **(2.8-T)** |
| | $M_z = -\int_A y\sigma_{xx} dA = 0$ **(2.8b-A)** | $M_z = -\int_A y\sigma_{xx} dA$ **(2.8b-B)** | |
| | $M_y = -\int_A z\sigma_{xx} dA = 0$ **(2.8c-A)** | $V_y = \int_A \tau_{xy} dA$ **(2.8c-B)** | |
| **Sign convention** | 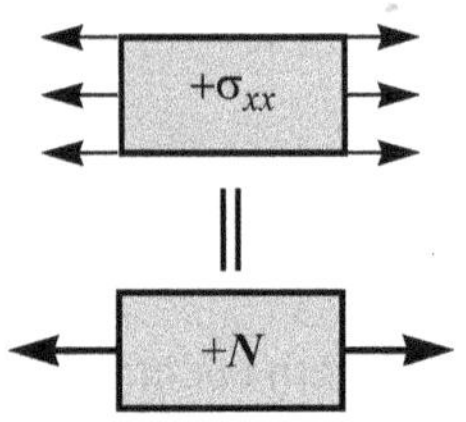 | 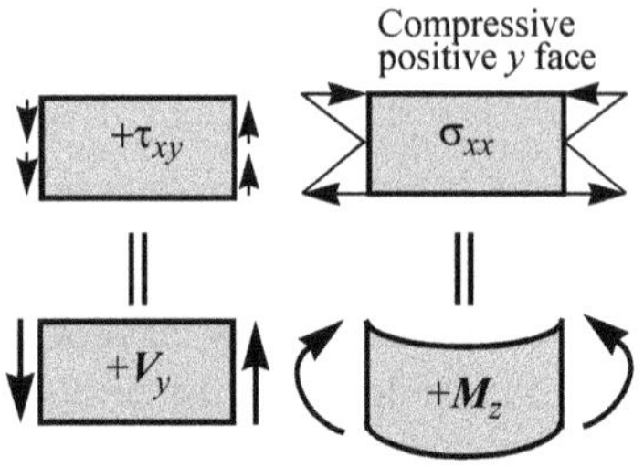 | 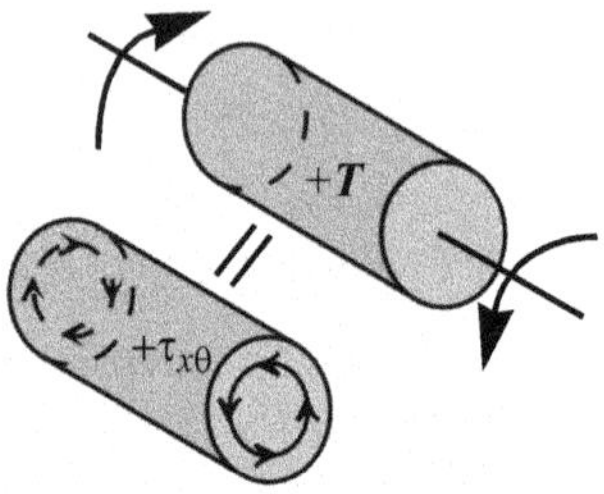 |
| | Substituting stresses into internal forces and moments and noting $du_0/dx$, $d^2v/dx^2$, and $d\phi/dx$ are functions of *x* only, while the integration is with respect to *y* and *z* | | |
| **Origin location** | $\int_A yE\, dA = 0$ **(2.9-A)** | $\int_A yE\, dA = 0$ **(2.9-B)** | |
| | $N = \frac{du_0}{dx}\int_A E\, dA$ **(2.10-A)** | $M_z = \frac{d^2v}{dx^2}\int_A Ey^2\, dA$ **(2.10-B)** | $T = \frac{d\phi}{dx}\int_A G\rho^2 dA$ **(2.10-T)** |
| **Assumption 8** | The material is homogeneous across the cross section. | | |
| **The origin is at the centroid of the cross section.** | $\int_A y\, dA = 0$ **(2.11-A)** | $\int_A y\, dA = 0$ **(2.11-B)** | |
| | $\frac{du_0}{dx} = \frac{N}{EA}$ **(2.12-A)** | $\frac{d^2v}{dx^2} = \frac{M_z}{EI_{zz}}$ **(2.12-B)** | $\frac{d\phi}{dx} = \frac{T}{GJ}$ **(2.12-T)** |

**Table 2.1** Synopsis of Elementary Theories of One-Dimensional Structural Members

| | Axial | Bending | Torsion |
|---|---|---|---|
| | $A = \int_A dA$ | $I_{zz} = \int_A y^2 dA$ | $J = \int_A \rho^2 dA$ |
| | **Stress Formulas** | | |
| | Substituting Equations (2.12-A), (2.12-B), and (2.12-T) into Equations (2.7-A), (2.7-B), and (2.7-T) | | |
| | $\sigma_{xx} = \frac{N}{A}$ **(2.13-A)** | $\sigma_{xx} = -\left(\frac{M_z y}{I_{zz}}\right)$ **(2.13-B)**<br>See Section 3.2.3 for shear stresses in bending. | $\tau_{x\theta} = \frac{T\rho}{J}$ **(2.13-T)** |
| | **Deformation Formulas** | | |
| **Assumption 9**<br>**Assumption 10**<br>**Assumption 11** | The material is homogeneous between $x_1$ and $x_2$.<br>The structural member is not tapered between $x_1$ and $x_2$<br>The external loads do not change with $x$ between $x_1$ and $x_2$. | | |
| | Integrating Equations (2.12-A) and (2.12-T) | | |
| | $u_2 - u_1 = \frac{N(x_2 - x_1)}{EA}$ **(2.14-A)** | See Section 3.4.3 for beam deflection. | $\phi_2 - \phi_1 = \frac{T(x_2 - x_1)}{GJ}$ **(2.14-T)** |
| | **Equilibrium Equations** | | |
| | 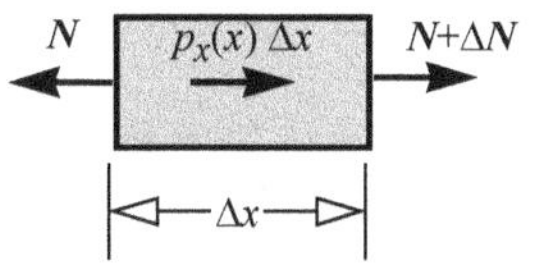  | 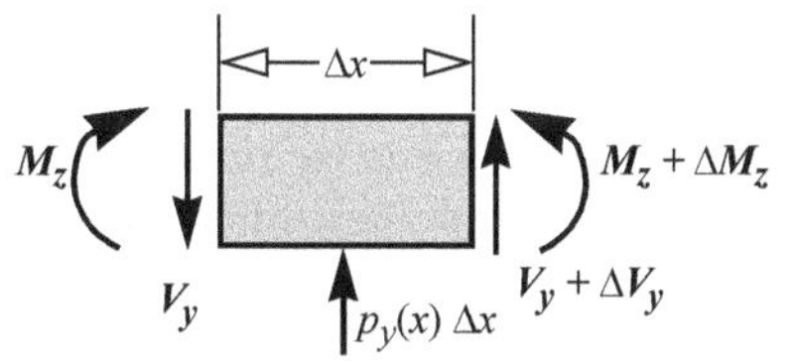  | 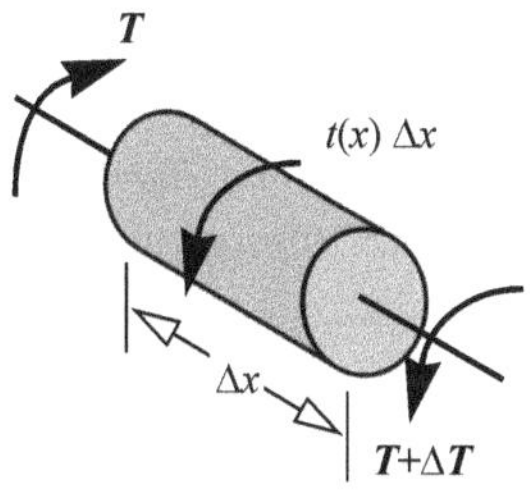  |
| | $\frac{dN}{dx} = -p_x(x)$ **(2.15-A)** | $\frac{dV_y}{dx} = -p_y(x)$ **(2.15a-B)**<br>$\frac{dM_z}{dx} = -V_y$ **(2.15b-B)** | $\frac{dT}{dx} = -t(x)$ **(2.15-T)** |
| | **Differential Equations** | | |
| | Substituting Equations (2.12-A), (2.12-B), and (2.12-T) into Equations (2.15-A), (2.15a-B), (2.15b-B), and (2.15-T) | | |
| | $\frac{d}{dx}\left(EA\frac{du_0}{dx}\right) = -p_x(x)$ **(2.16-A)** | $\frac{d^2}{dx^2}\left(EI_{zz}\frac{d^2 v}{dx^2}\right) = p_y(x)$ **(2.16-B)** | $\frac{d}{dx}\left(GJ\frac{d\phi}{dx}\right) = -t(x)$ **(2.16-T)** |

## 2.3.3 Complexity map

Two observations regarding stress formulas and differential equation on deformations are recorded below before considering the various complexities in examples, problems, and later chapters.

- The formulas for stresses given by Equations (2.13-A), (2.13-B), and (2.13-T) or their equivalent form are obtained by substituting Equations (2.12-A), (2.12-B), and (2.12-T) or their equivalent forms into Equations (2.7-A), (2.7-B), and (2.7-T) or their equivalent forms, irrespective of the complexity included.
- The differential equations given by Equations (2.16-A), (2.16-B), and (2.16-T) or their equivalent form are obtained from by substituting Equations (2.12-A), (2.12-B), and (2.12-T) or their equivalent form into Equations (2.15-A), (2.15a-B), (2.15b-B), and (2.15-T) or their equivalent forms, irrespective of the complexity included.

1. Dropping of Assumption 1 has two impacts. First, time becomes an independent variable like $x$, hence all ordinary derivatives become partial derivatives of $x$. Second, the equilibrium equations in which forces and moments were equated to zero are now equated to the inertial forces. In Example 2.6 we consider vibration of beams. In problem 2.21 we consider the wave equation in axial rods. In problem 2.24 we consider vibration of circular shafts.
2. Dropping of any part of Assumption 2 results in changing the form of deformation equations. Form of deformation equations may also change if some of the limitations are dropped. The flow of development of the new equations remains the same. In Example 2.6, Assumption 2c-B is dropped and shear is incorporated into beam theory. In problem 2.7, the axial and bending deformation are considered together. In Problems 2.8, 2.9, and 2.10 the limitation on beam bending that loading be in plane of symmetry is dropped and equations for unsymmetric bending are developed.
3. Dropping of Assumption 3 incorporates geometric nonlinearity into the theory. Problem 2.26 considers non-linearity in context of axial rods. Example 7.5 considers large strain in symmetric beam bending.
4. Assumption 5 is dropped in problem 2.13 to incorporate effect of temperature on beam bending.

5. Assumption 6 is dropped in problems 2.17 and 2.18 to incorporate elastic-perfectly plastic material model in torsion of circular shafts and symmetric bending of beams.
6. Assumption 7 on material linearity is dropped and formula for stress is developed in Example 2.5 and in problem 2.29. In problem 2.27 stress and deformation formulas are developed for torsion of circular shafts. In Example 7.6 power law is incorporated into symmetric beam bending to develop boundary value problem statement.
7. Assumption 8 on material on material homogeneity across the cross section is dropped in Example 2.4 for beams. Formulas are developed for determining neutral axis, bending normal stress, and the differential equation. In Example 2.7 the formulas for shear stress are developed. In problems 2.11 and 2.12 stress formulas and differentiable equations are developed for laminated axial rods and torsion of laminated circular shafts.
8. Assumption 10 of untapered member is dropped in problem 2.3 for axial rod and in problem 2.4 for torsion of circular shafts.
9. Assumption 11 is dropped in problem 2.33 to incorporate distributed axial force and in problem 2.34 to incorporate distributed torque.

In problem 2.19 the differential equation for a beam on elastic foundation is considered. Solutions for beams on elastic foundations are discussed in details in Chapter 3.

More than one complexity may be added simultaneously as demonstrated in Example 2.6 in which vibration of Timoshenko beams is considered. In problems 2.14 through 2.16 temperature effects in laminated members is considered. In problems 2.31 and 2.32 two non-linear materials make up the structural members. In a similar manner we could create vibrations of laminated axial rods, beams, and shafts. Combinations of complexities is very large, but hopefully the reader by end of this chapter will develop a sense of how these complexities can be added to classical theories of axial rods, bending of symmetric beams, and torsion of circular shafts.

*An important check on these complex theories is that we should be able to reproduce results of classical theories by appropriate substitution of variables.* This check will be performed in the examples presented next.

## EXAMPLE 2.4

Figure 2.9 shows a laminated composite cross section. We assume that all material cross sections are symmetric about the $y$ axis and that the loading in still in the plane of symmetry, thus ensuring symmetric bending about the $z$ axis. All assumptions except Assumption 8 in Table 2.1 are valid. (a) Assuming no axial force, determine the location of the origin, that is location of neutral axis.(b) Obtain a formula relating normal stress $\sigma_{xx}$ and internal bending moment $M_z$. (c) Obtain the fourth order differential equation governing the deflection $v(x)$ of the beam. (d) Show that the results for parts $a$, $b$ and $c$ give the same results as classical beam theory for homogeneous material.

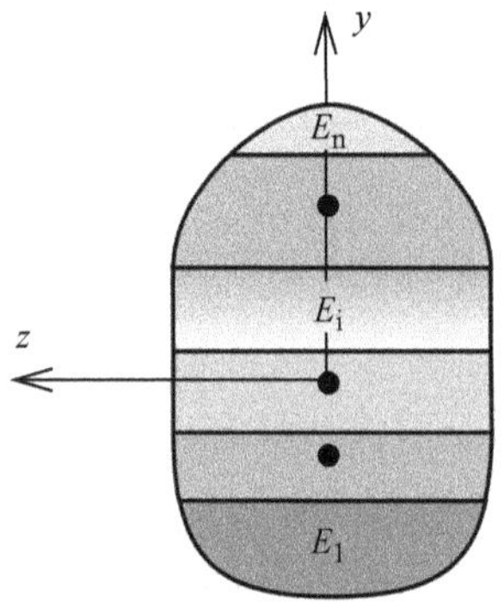


**Figure 2.9** Laminated cross section in Example 2.4.

**PLAN**

All equations before Assumption 8 in Table 2.1 are valid. We write the integrals in Equations (2.9-B) and (2.10-B) as sum of integral over each material to obtain new expressions. We then follow the same steps as in Table 2.1 to obtain the desired results.

**SOLUTION**

(a) In Equation (2.9-B), we substitute $y = \eta - \eta_c$, where $\eta$ is a variable measured from some arbitrary point and $\eta_c$ is the location of the origin $y$ from the arbitrary point. We obtain

$$\int_A yE\,dA = \int_A \eta E\,dA - \int_A \eta_c E\,dA = 0 \quad \text{or} \quad \eta_c = \left[\int_A \eta E dA\right] \Big/ \left[\int_A E dA\right] \tag{E1}$$

The above integrals are written as sum of integral over each material for which $E_i$ is a constant.

$$\textbf{ANS.}\ \eta_c = \frac{\sum_{i=1}^{n}\int_{A_i}\eta E_i dA}{\sum_{i=1}^{n}\int_{A_i}E_i dA} = \frac{\sum_{i=1}^{n}E_i\int_{A_i}\eta dA}{\sum_{i=1}^{n}E_i\int_{A_i}dA} = \frac{\sum_{i=1}^{n}\eta_{ci}E_iA_i}{\sum_{i=1}^{n}E_iA_i} \quad \text{where} \quad \eta_{ci} = \frac{\int_{A_i}\eta dA}{\int_{A_i}dA} = \frac{\int_{A_i}\eta dA}{A_i} \tag{E2}$$

where $\eta_{ci}$ is the location of the centroid of the i$^{th}$ material area.

(b) The integral in Equation (2.10-B) is written as sum of integral over each material for which $E_j$ is a constant.

$$M_z = \frac{d^2v}{dx^2}\int_A Ey^2\,dA = \frac{d^2v}{dx^2}\left[\sum_{j=1}^{n} E_j\int_{A_j} y^2\,dA\right] = \frac{d^2v}{dx^2}\left[\sum_{j=1}^{n} E_j(I_{zz})_j\right] \quad \text{or} \quad \frac{d^2v}{dx^2} = \frac{M_z}{\sum_{j=1}^{n} E_j(I_{zz})_j} \tag{E3}$$

From Equation (2.7-B) we can write the stress for the i$^{th}$ material as

$$(\sigma_{xx})_i = -E_i y(d^2v/dx^2) \tag{E4}$$

Substituting Equation (E3) into the above equation we obtain our result as shown below.

$$\textbf{ANS.}\ (\sigma_{xx})_i = -\left[\frac{E_i y M_z}{\sum_{j=1}^{n} E_j(I_{zz})_j}\right] \tag{E5}$$

(c) The equilibrium equations are unaffected by the non-homogeneity of the cross section. Substituting Equation (E3) into Equations (2.15b-B) and (2.15a-B)and we obtain

$$V_y = -\left\{\frac{dM_z}{dx}\right\} = -\frac{d}{dx}\left\{\frac{d^2v}{dx^2}\left[\sum_{j=1}^{n} E_j(I_{zz})_j\right]\right\} \qquad \frac{dV_y}{dx} = -\left[\frac{d^2}{dx^2}\left\{\frac{d^2v}{dx^2}\left[\sum_{j=1}^{n} E_j(I_{zz})_j\right]\right\}\right] = -p_y(x) \tag{E6}$$

$$\textbf{ANS.}\ \frac{d^2}{dx^2}\left\{\frac{d^2v}{dx^2}\left[\sum_{j=1}^{n} E_j(I_{zz})_j\right]\right\} = p_y \tag{E7}$$

(d) For homogenous material we have

$$E_1 = E_2 = \bullet\ \bullet E_j \bullet = E_n = E \tag{E8}$$

Substituting the above equation into Equations (E2), (E5), and (E7) we obtain the following results

$$\eta_c = \frac{E\sum_{i=1}^{n}\eta_{ci}A_i}{E\sum_{i=1}^{n}A_i} = \frac{\sum_{i=1}^{n}\eta_{ci}A_i}{\sum_{i=1}^{n}A_i} \tag{E9}$$

$$\sigma_{xx} = -\frac{EyM_z}{E\sum_{j=1}^{n}(I_{zz})_j} = -\frac{yM_z}{\sum_{j=1}^{n}(I_{zz})_j} = -\left[\frac{M_z y}{I_{zz}}\right] \tag{E10}$$

$$\frac{d^2}{dx^2}\left\{\frac{d^2v}{dx^2}\left[E\sum_{j=1}^{n}(I_{zz})_j\right]\right\} = \frac{d^2}{dx^2}\left\{EI_{zz}\frac{d^2v}{dx^2}\right\} = p_y \tag{E11}$$

The above equations are the same as that for homogenous material.

**COMMENTS**

1. In this example, the kinematic equations, the Hooke's law, the static equivalency equations, and equilibrium equations did not change. What changed was the evaluation of the integrals generated from static equivalency.
2. The example emphasizes the two observations made in Section 2.3.3 with regard to obtaining stress formulas and differential equations.
3. We will use the results from this example in Example 2.7 where we develop equations for shear stress in laminated beams.

## EXAMPLE 2.5

The stress–strain curve for a material is given by $\sigma = E\varepsilon^n$. Assume the material behavior is same in tension and compression. All assumptions except Assumption 7 in Table 2.1 are valid. (a) For the rectangular cross section shown in Figure 2.10, obtain an equation relating bending normal stress $\sigma_{xx}$ and internal bending moment $M_z$. (b) Obtain the differential equation governing the deflection v($x$) of the beam (c) Show that for $n = 1$ we obtain the result of classical beam theory.

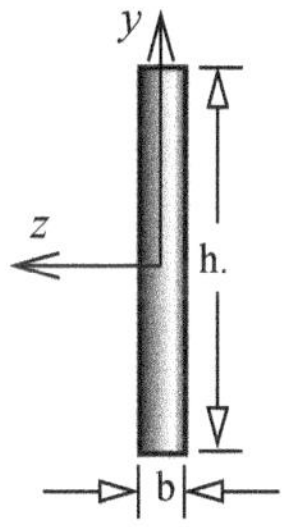


**Figure 2.10** Geometry in Example 2.5.

**PLAN**

All equations before Assumption 7 are valid. We obtain a new stress expression in terms of curvature and substitute it in the equivalent moment equation. We solve for the curvature in terms of moment and substitute it back into the stress equation to obtain the desired result

**SOLUTION**

From Equation (2.6-B) we have

$$\varepsilon_{xx} = -y(d^2v/dx^2) \tag{E1}$$

As the material behavior in tension and compression is the same, we have:

$$\sigma_{xx} = \begin{cases} E\varepsilon_{xx}^n & \varepsilon_{xx} \geq 0 \\ -E(-\varepsilon_{xx})^n & \varepsilon_{xx} \leq 0 \end{cases} \quad \text{or} \quad \sigma_{xx} = \begin{cases} E(-yd^2v/dx^2)^n & y < 0 \\ -E(yd^2v/dx^2)^n & y > 0 \end{cases} \tag{E2}$$

From the symmetry we conclude that the equivalent internal moment in the lower half ($y < 0$ )is same as that in the upper half ($y > 0$ ). The integral in Equation (2.8b-B) can be written as twice the integral in the positive $y$ half as shown below. Noting that curvature is only a function of $x$ we obtain:

$$M_z = -\int_A y\sigma_{xx}dA = -2\left\{\int_0^{h/2} y[-E(yd^2v/dx^2)^n](bdy)\right\} = 2bE(d^2v/dx^2)^n\int_0^{h/2} y^{n+1}dy = 2bE(d^2v/dx^2)^n \left.\frac{y^{n+2}}{n+2}\right|_0^{h/2}$$

$$M_z = 2bE\frac{(h/2)^{n+2}}{n+2}(d^2v/dx^2)^n \quad \text{or} \quad (d^2v/dx^2)^n = \frac{(n+2)M_z}{2b(h/2)^{n+2}E} \tag{E3}$$

Substituting Equation (E3) into Equation (E2), we obtain

$$\textbf{ANS.}\quad \sigma_{xx} = \begin{cases} (n+2)\left(\frac{2^{n+1}}{bh^2}\right)\left(\frac{-y}{h}\right)^n M_z & y < 0 \\ -(n+2)\left(\frac{2^{n+1}}{bh^2}\right)\left(\frac{y}{h}\right)^n M_z & y > 0 \end{cases} \tag{E4}$$

The above equation is desired stress expression.

(b) The equilibrium equations are unaffected by the material non-linearity. Substituting Equation (E3) into Equations (2.15b-B) and (2.15a-B)and we obtain

$$V_y = -\left\{\frac{dM_z}{dx}\right\} = -\frac{d}{dx}\left\{\left[2bE\frac{(h/2)^{n+2}}{n+2}\right]\left(\frac{d^2v}{dx^2}\right)^n\right\} \qquad \frac{dV_y}{dx} = -\left[\frac{d^2}{dx^2}\left\{\left[2bE\frac{(h/2)^{n+2}}{n+2}\right]\left(\frac{d^2v}{dx^2}\right)^n\right\}\right] = -p_y(x)$$

$$\textbf{ANS.}\quad \frac{d^2}{dx^2}\left\{\left[2bE\frac{(h/2)^{n+2}}{n+2}\right]\left(\frac{d^2v}{dx^2}\right)^n\right\} = p_y(x) \tag{E5}$$

(c) Substituting $n = 1$ in Equations (E4) and (E5) we obtain

$$\sigma_{xx} = \begin{cases} (3)\left(\frac{2^2}{bh^2}\right)\left(\frac{-y}{h}\right)M_z & y < 0 \\ -(3)\left(\frac{2^2}{bh^2}\right)\left(\frac{y}{h}\right)^{0.5} M_z & y > 0 \end{cases} \quad \text{or} \quad \sigma_{xx} = -\left(\frac{M_z y}{bh^3/12}\right) \tag{E6}$$

$$\frac{d^2}{dx^2}\left\{\left[2bE\frac{(h/2)^3}{3}\right]\left(\frac{d^2v}{dx^2}\right)\right\} = p_y(x) \quad \text{or} \quad \frac{d^2}{dx^2}\left\{Ebh^3/12\left(\frac{d^2v}{dx^2}\right)\right\} = p_y(x) \tag{E7}$$

Noting that for a rectangular cross section $I_{zz} = bh^3/12$, the above results are the same as that for classical beam theory.

**COMMENTS**

1. In this example the kinematic equations, the static equivalency equations, and equilibrium equations did not change.
2. This example also emphasizes the two observations made in Section 2.3.3 with regard to obtaining stress formulas and differential equations
3. In this example we did not have to find the neutral axis because of geometric and material symmetry. If the geometry is not symmetric about the $z$ axis, then we will have to use Equation (2.8a-B) to determine the location of the neutral axis, that is, the origin of $y$, after which we would proceed as we did in this example.
4. If we were given a beam with loading, then we could make an imaginary cut and relate the internal moment to external loads and then calculate the bending normal stress from the results.
5. For certain loading it may be possible to solve the differential equation by techniques discussed in Sections 2.4 and 2.5. But for most general loadings we will have to use numerical techniques.

## EXAMPLE 2.6

In Timoshenko beams[a] the assumption of planes remaining perpendicular to the axis of the beam is dropped to account for shear by permitting the cross section to rotate by an angle $\psi$ from the vertical symmetric axis $y$. (a) Obtain the differential equations on deflection $v(x,t)$ and angle $\psi(x, t)$ for vibration of symmetric Timoshenko beam by starting with the displacement field given below. Use the following variables: mass density $\rho_m$; shear modulus $G$; modulus of elasticity $E$; distributed load $p_y(x, t)$; cross sectional area A; and second area moment of inertia $I_{zz}$. (b) Show that for $\psi = dv/dx$ and with no dynamic effects we obtain the result of classical beam theory.

$$u = -y\psi(x, t) \qquad v = v(x, t) \tag{2.17a-B}$$

**PLAN**

We follow the logic shown in Figure 2.7. The strain–displacement equations are as before—by differentiating the given displacements, we can find normal strain $\varepsilon_{xx}$ and shear strain $\gamma_{xy}$. Hooke's law is same as before—the normal and shear stress can be found. The static equivalency equations (2.8b-B) and (2.8c-B) are the same as before—by substituting the normal and shear stress, we can obtain new expressions for $M_z$ and $V_y$. In equilibrium equations, the inertial forces must be included.

1. **Strains:** The normal strain and shear strain can be obtained as

$$\varepsilon_{xx} = \frac{\partial u}{\partial x} = -y\frac{\partial \psi}{\partial x} \qquad \gamma_{xy} = \frac{\partial u}{\partial y} + \frac{\partial v}{\partial x} = \frac{\partial v}{\partial x} - \psi \tag{E8}$$

2. **Stresses:** By Hooke's law we obtain the stresses as

$$\sigma_{xx} = E\varepsilon_{xx} = -Ey\frac{\partial \psi}{\partial x} \qquad \tau_{xy} = G\gamma_{xy} = G\left(\frac{\partial v}{\partial x} - \psi\right) \tag{E9}$$

3. **Internal forces and moments:** The stresses can be substituted in Equations (2.8b-B) and (2.8c-B). Noting that $\psi$ and v are functions of $x$ and $t$ only and do not vary across the cross section, and assuming that the material is homogeneous across the cross section ($E$ and $G$ do not vary across the cross section), we obtain

$$M_z = -\int_A y\sigma_{xx} = \int_A Ey^2\frac{\partial \psi}{\partial x}dA = E\frac{\partial \psi}{\partial x}\int_A y^2 dA = EI_{zz}\frac{\partial \psi}{\partial x} \qquad V_y = \int_A \tau_{xy}dA = \int_A G\left(\frac{\partial v}{\partial x} - \psi\right)dA = GA\left(\frac{\partial v}{\partial x} - \psi\right) \tag{E10}$$

4. **Equilibrium:** Figure 2.11 shows a differential element of the beam. The mass of the differential element is $\rho_m A\Delta x$ and the linear acceleration is $\partial^2 v/\partial t^2$. Similarly the mass moment of inertia is $\rho_m I_{zz}\Delta x$ and the angular acceleration is $\partial^2\psi/\partial t^2$. The inertial force and the inertial moment acting on the differential element is shown in Figure 2.11$b$.

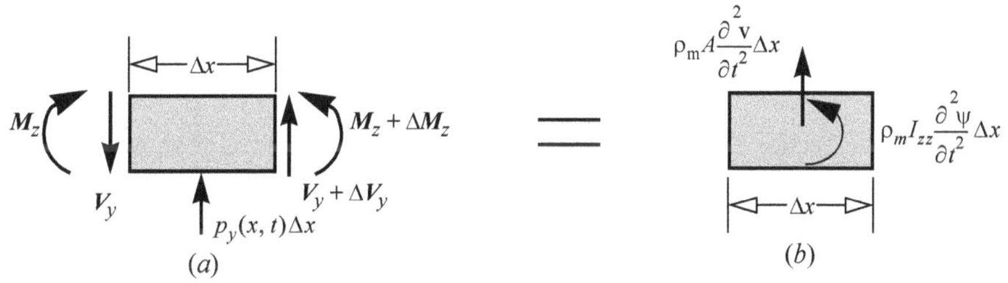


**Figure 2.11** Differential element of a beam with inertial force.

By the force equilibrium and moment equilibrium we obtain.

$$V_y + \Delta V_y - V_y + p_y(x, t)\Delta x = \rho_m A\frac{\partial^2 v}{\partial t^2}\Delta x \tag{E11}$$

$$M_z + \Delta M_z + (V_y + \Delta V_y)\frac{\Delta x}{2} + V_y\frac{\Delta x}{2} - M_z = \rho_m I_{zz}\frac{\partial^2 \psi}{\partial t^2}\Delta x \tag{E12}$$

Dividing by $\Delta x$ and taking the limit $\Delta x \to 0$ we obtain

$$\lim_{\Delta x \to 0}\left(\frac{\Delta V_y}{\Delta x}\right) + p_y(x, t) = \rho_m A\frac{\partial^2 v}{\partial t^2} \quad \text{or} \quad \frac{\partial V_y}{\partial x} + p_y(x, t) = \rho_m A\frac{\partial^2 v}{\partial t^2} \tag{E13}$$

$$\lim_{\Delta x \to 0}\left(\frac{\Delta M_z}{\Delta x} + \Delta V_y\frac{\Delta x}{2}\right) + V_y = \rho_m I_{zz}\frac{\partial^2 \psi}{\partial t^2} \quad \text{or} \quad \frac{\partial M_z}{\partial x} + V_y = \rho_m I_{zz}\frac{\partial^2 \psi}{\partial t^2} \tag{E14}$$

Substituting Equations (E10) and (E10) into Equations (E13) and (E14) we obtain the differential equations below.

$$\text{ANS.}\ \frac{\partial}{\partial x}\left[GA\left(\frac{\partial v}{\partial x} - \psi(x)\right)\right] + p_y(x, t) = \rho_m A\frac{\partial^2 v}{\partial t^2} \qquad \frac{\partial}{\partial x}\left[EI_{zz}\frac{\partial \psi}{\partial x}\right] + GA\left(\frac{\partial v}{\partial x} - \psi(x)\right) = \rho_m I_{zz}\frac{\partial^2 \psi}{\partial t^2} \tag{2.17b-B}$$

(b) To remove the dynamic effects we substitute $\partial^2 v/\partial t^2 = 0$ and $\partial^2\psi/\partial t^2 = 0$ and replace partial derivatives with ordinary derivatives of $x$ to obtain

$$\frac{d}{dx}\left[GA\left(\frac{dv}{dx} - \psi\right)\right] = -p_y \qquad \frac{d}{dx}\left(EI_{zz}\frac{d\psi}{dx}\right) = -GA\left(\frac{dv}{dx} - \psi\right) \tag{2.17c-B}$$

Substituting the second equation into the first equation above we obtain

$$\frac{d^2}{dx^2}\left\{EI_{zz}\frac{d\psi}{dx}\right\} = p_y \tag{E15}$$

Substituting $\psi = dv/dx$ in the above equation we obtain the differential equation of classical beam theory.

**COMMENTS**

1. The change in displacement functions resulted in different strain expressions but the same strain-displacement equations were used. The constitutive equation (Hooke's law) and the equivalency equations did not change, but resulted in different expressions for stresses, and internal shear force and moment. The equilibrium equations changed to account for inertial forces and moments, but the concept of substituting the internal forces and moments in terms of deformation to get differential equation did not change.
2. In classical beam vibration the rotational inertial term is neglected, that is $\rho_m I_{zz}\partial^2\psi/\partial t^2 = 0$. Substituting this into Equation (2.17b-B) and then substituting the result into Equation (2.17b-B) we obtain

$$-\left[\frac{\partial^2}{\partial x^2}\left(EI_{zz}\frac{\partial\psi}{\partial x}\right)\right] + p_y(x,t) = \rho_m A\frac{\partial^2 v}{\partial t^2}$$

Substituting $\psi = \partial v/\partial x$ and assuming $EI_{zz}$ is constant we obtain the classical beam vibration equation below.

$$\frac{\partial^2 v}{\partial t^2} + c^2\frac{\partial^4 v}{\partial x^4} = \frac{p_y(x,t)}{\rho_m A} \qquad \text{where} \qquad c^2 = \frac{EI_{zz}}{\rho_m A} \tag{2.17d-B}$$

3. The three examples show the modular character of the logic shown in Figure 2.7 and how complexities can be added to obtain stress formulas and differential equations.

a. See Fung [1965] or Reddy [1993].

### 2.3.4 Shear stress in thin symmetric beams

Assumption 2c-B implies that the shear strain $\gamma_{xy}$ must be so small that we can neglect it in kinematic equations. By Hooke's law this implies that $\tau_{xy}$ should be small. In beam bending, a check on the validity of the analysis is to compare the maximum shear stress $\tau_{xy}$ with the maximum normal stress $\sigma_{xx}$ for the *entire* beam. If the two stress components are comparable, then the shear strain cannot be neglected in kinematic considerations and we should use Timoshenko beam theory of Example 2.6. The maximum bending normal stress $\sigma_{xx}$ in the beam should be nearly an order of magnitude greater than the maximum bending shear stress $\tau_{xy}$ as shown below.

$$|(\sigma_{xx})_{max}| \gg |(\tau_{xy})_{max}| \tag{2.18-B}$$

Consider a circular cross section that is glued together from six wooden strips as shown in Figure 2.12. The bending load $P$, causes the moment $M_z$ to vary along the length of the beam. Thus from Equation (2.13-B) we know that the magnitude of the normal stress $\sigma_{xx}$ will change along the length of the beam. The shear stress at the glue joints must balance the variation of normal stress $\sigma_{xx}$ along the length of the beam.

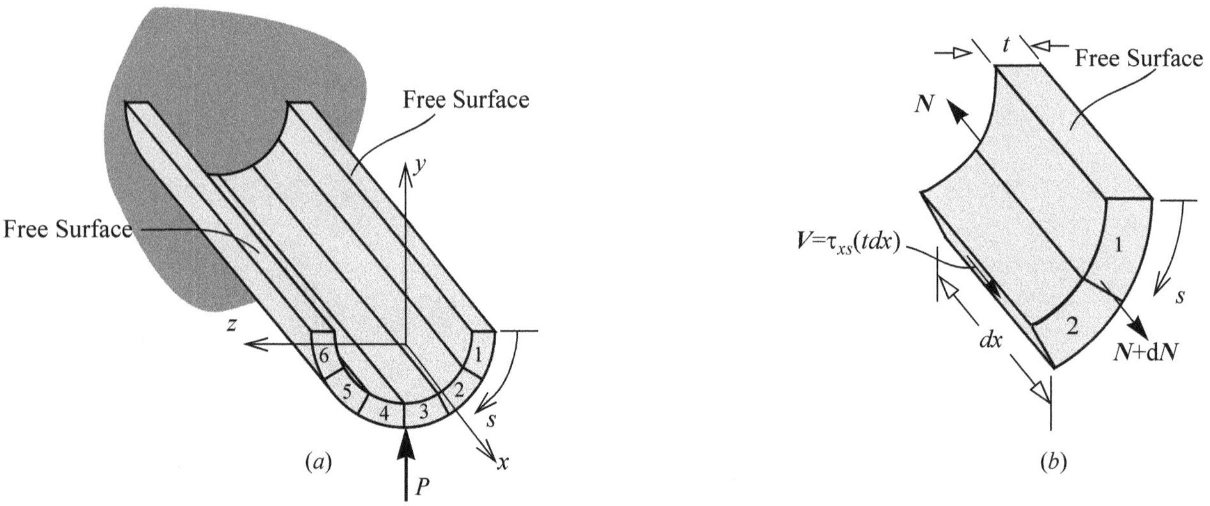


**Figure 2.12** Shear stress in a circular cross section.

The outward normal of the surface will be in a different direction for each glue surface on which we consider the shear stress. We define a tangential coordinate $s$ in the tangent direction to the centerline of the cross section. The outward normal to the glued surface will be in the $s$ direction and the shear stress will be $\tau_{sx}$. Once more, by symmetry of shear stresses, $\tau_{xs} = \tau_{sx}$. At a point, if the $s$ direction and the $y$ direction are the same, then $\tau_{xs}$ will equal plus or minus $\tau_{xy}$. If the $s$ direction and the $z$ direction are the same at a point, then $\tau_{xs}$ will equal plus or minus $\tau_{xz}$.

Figure 2.12*b* shows a differential element of the beam. The shear force that balances the change in axial force $N_s$ is shown on only one surface. The surface on the other end of the free body diagram is assumed to be a free surface; that is, the shear stress is zero on these other surfaces. *The direction of the s coordinate is from the free surface toward the point at which the shear stress is being calculated.*

In a beam cross section, the top, bottom, and side surfaces are always assumed to be surfaces on which the shear stress is zero. We further assume that *the beam is thin perpendicular to the centerline of the cross section* (i.e. $t$ is small). The assumptions imply

- The shear stress normal to the centerline is small because the surfaces on either side of the centerline are free surfaces, and hence can be neglected.
- The shear stress $\tau_{sx}$ is uniform in the direction normal to the centerline.

By equilibrium of forces in the $x$ direction in Figure 2.12*b*, we obtain $(N_s + dN_s) - N_s + \tau_{sx} t\,dx = 0$, or

$$\tau_{sx} t = -\frac{dN_s}{dx} = -\frac{d}{dx}\int_{A_s} \sigma_{xx} dA \qquad \textbf{(2.19-B)}$$

*Area $A_s$ is the area on the cross section that is between the free surface and the point at which the shear stress is being evaluated.*

Substituting Equation (2.13-B) into Equation (2.19-B), and noting that the moment $M_z$ and the area moment of inertia $I_{zz}$ do not vary over the cross section, we obtain

$$\tau_{sx} t = \frac{d}{dx}\left[\frac{M_z}{I_{zz}}\int_{A_s} y\,dA\right] = \frac{d}{dx}\left[\frac{M_z Q_z}{I_{zz}}\right] \qquad \textbf{(2.20-B)}$$

where $Q_z$ is referred to as the first moment of the area $A_s$ and is defined as

$$\boxed{Q_z = \int_{A_s} y\,dA} \qquad \textbf{(2.21-B)}$$

We assume *the beam is not tapered*. This implies that $I_{zz}$ and $Q_z$ are not functions of $x$, and these quantities can be taken outside the derivative sign, we obtain $\tau_{sx} t = (Q_z/I_{zz})(dM_z/dx)$. Substituting Equation (2.15b-B), the relationship between shear force $V_y$ and moment $M_z$, we obtain

$$\tau_{sx} t = -\left(\frac{Q_z V_y}{I_{zz}}\right) \quad \text{or} \quad \boxed{\tau_{sx} = \tau_{xs} = -\left(\frac{V_y Q_z}{I_{zz} t}\right)} \qquad \textbf{(2.22-B)}$$

From Equation (2.21-B), we note that $Q_z$ is the first moment of the area $A_s$ about the $z$ axis. Figure 2.13 shows the area $A_s$ between the top free surface and the point at which shear stress is being found (the *S–S* line). The integral in Equation (2.21-B) is the numerator in the definition of the centroid of the area $A_s$. Analogous to the moment due to a force, the first moment of an area can be found by placing the area $A_s$ at its centroid and finding the moment about the neutral axis. That is, *$Q_z$ is the product of*

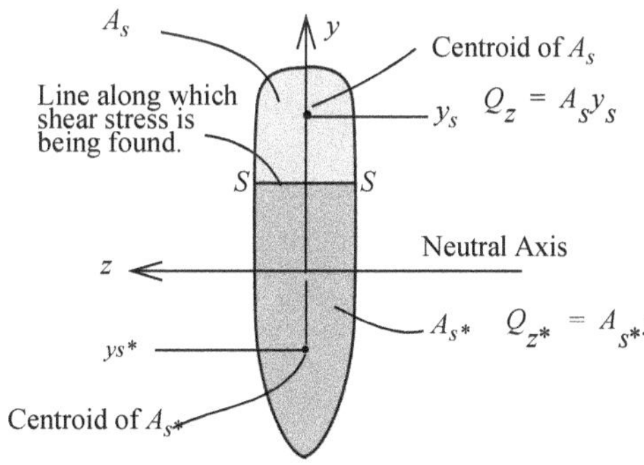


**Figure 2.13** Calculation of $Q_z$.

*area $A_s$ and the distance of the centroid of the area $A_s$ from the neutral axis*, as shown in Figure 2.13. Alternatively, $Q_z$ can be found by using the bottom surface as the free surface, shown as $Q_{z*}$ in Figure 2.13. In the introductory course on the mechanics of materials it was shown that $Q_z$ is *maximum at the neutral axis*; hence bending shear stress is maximum at the neutral axis of a cross section.

We note that since $y$ is measured from the centroid, the first moment of the entire cross-sectional area is zero; hence $Q_z + Q_{z*} = 0$. The equation implies that $Q_z$ and $Q_{z*}$ will have the same magnitude but opposite signs. Thus, we will get the same magnitude of the shear stress whether we use $Q_z$ or $Q_{z*}$ in Equation (2.22-B). But which gives the correct sign (direction)? The answer is that both will give the correct sign, provided the following point is remembered. The $s$ direction in Equation (2.22-B) is measured from the free surface used in the calculation of $Q_z$.

The product of the shear stress and the thickness is called the *shear flow*, designated by the symbol $q$, as shown below.

$$q = \tau_{xs} t \qquad \textbf{(2.23-B)}$$

The units of shear flow $q$ are *force per unit length*. The terminology is from fluid flow in channels, but the term is used extensively for discussing shear stresses in thin cross sections, probably because of the visual image of a flow it conveys in the discussion of shear stress direction.

## EXAMPLE 2.7

(a) For the composite laminated beam in Example 2.4, obtain a formula relating bending shear stress $\tau_{xs}$ and the internal shear force $V_y$. Assume the beam is not tapered. (b) Show that the results for part *a* give the same results as classical beam theory for homogeneous material.

**PLAN**

The equilibrium equation (2.19-B) is independent of the material model. Substituting the normal stress for the laminated composite obtained in Example 2.4 and proceeding as in this section we obtain the required formula.

**SOLUTION**

From the solution in Example 2.4 we have the bending normal stress re-written below in a slightly different form.

$$\sigma_{xx} = -\left[\frac{EyM_z}{\sum_{j=1}^{n} E_j(I_{zz})_j}\right] \tag{E1}$$

where $y$ and $E$ are the coordinates and modulus of elasticity of the material at the point where stress is to be evaluated.
We substitute Equation (E1) into Equation (2.19-B) and note that the moment $M_z$ and the sum of bending rigidities are section properties and hence do not change in the integration over the cross section. We obtain the equations below.

$$\tau_{sx}t = -\frac{d}{dx}\int_{A_s}\sigma_{xx}dA = -\frac{d}{dx}\int_{A_s}\frac{-EyM_z}{\sum_{j=1}^{n} E_j(I_{zz})_j}dA = \frac{d}{dx}\left[\frac{M_z}{\sum_{j=1}^{n} E_j(I_{zz})_j}\int_{A_s} Ey\ dA\right] \tag{E2}$$

where, as before, the area $A_s$ is the area on the cross section that is between the free surface and the point at which the shear stress is being evaluated. We define the integral in the above equation to obtain the equation below.

$$\tau_{sx}t = \frac{d}{dx}\left[\frac{M_z Q_{comp}}{\sum_{j=1}^{n} E_j(I_{zz})_j}\right] \quad \text{where} \quad Q_{comp} = \int_{A_s} Ey\ dA \tag{E3}$$

We can write the integral over the area $A_s$ as the sum of integrals over the materials that are in the area $A_s$ as shown below.

$$Q_{comp} = \sum_{j=1}^{n}\int_{A_j} Ey\ dA = \sum_{j=1}^{n} E_j\int_{A_j} y\ dA = \sum_{j=1}^{n} E_j(Q_z)_j \quad \text{where} \quad (Q_z)_j = \int_{A_j} y\ dA \tag{E4}$$

where, $n_s$ are the number of materials in the area $A_s$.
As the beam is not tapered $Q_{comp}$ and the bending rigidities are not functions of $x$ and can be taken outside the derivative sign. Substituting Equation (2.15b-B) we obtain the desired results shown below.

$$\tau_{sx}t = \frac{Q_{comp}}{\sum_{j=1}^{n} E_j(I_{zz})_j}\frac{dM_z}{dx} = -\frac{Q_{comp}V_y}{\sum_{j=1}^{n} E_j(I_{zz})_j} \tag{E5}$$

**ANS.** $\tau_{sx} = \tau_{xs} = -\dfrac{Q_{comp}V_y}{t\sum_{j=1}^{n} E_j(I_{zz})_j}$

(b) For homogenous material we have

$$E_1 = E_2 = \bullet \bullet E_j \bullet = E_n = E \tag{E6}$$

From Equations (E3) and (E5) we obtain

$$Q_{comp} = E\int_{A_s} y\ dA = EQ_z \tag{E7}$$

$$\tau_{sx} = -\left(\frac{EQ_zV_y}{tE\sum_{j=1}^{n}(I_{zz})_j}\right) = -\left(\frac{V_yQ_z}{I_{zz}t}\right) \tag{E8}$$

The above equation is the same as for homogenous material.

**COMMENTS**

1. Once more the key to solving this example was the recognition that equilibrium equations (2.19-B) and (2.15b-B) do not depend upon the material homogeneity.
2. The calculation of $Q_{comp}$ is similar to that of $Q_z$ described in Figure 2.13, with the difference that now contribution from each material has to be calculated and then summed. See Vable [2013] for additional details.
3. $Q_{comp}$ will be zero if we include the entire cross section. Thus, like $Q_z$, it starts from zero at top increases as the point moves downward, reaching a maximum value at the neutral axis and then it starts decreasing till it reaches a zero value at the bottom. Based on this we conclude that the bending shear stress in each material is maximum at point closest to the neutral axis.

## 2.4 BOUNDARY VALUE PROBLEMS

A **boundary value problem** is a mathematical listing of the differential equations, the boundary conditions, and any other conditions necessary to solve the differential equations.

The differential equations for the axial, bending, and torsion cases are given by Equations (2.16-A), (2.16-B), and (2.16-T), respectively in Table 2.1. The distributed loads $p_x(x)$, $p_y(x)$, and $t(x)$ in the differential equations are known functions. The integration of these distributed loads results in the **particular solution.** The sum of the particular solution and homogeneous solution represents the complete solution of the differential equation. The **homogeneous solution** corresponds to the right hand side of the differential equation set to zero, that is, we set $p_x(x)$, $p_y(x)$, and $t(x)$ to zero in the differential equation to obtain the homogeneous solution by integration. The homogeneous solution contains unknown constants that are determined from the boundary conditions.

The boundary conditions are conditions on the internal forces/moments or displacements/rotations at the two ends (boundary) of the member. To understand the principle of how boundary conditions are determined, we generalize a principle discussed in statics for determining the reaction force and/or moments. Recall that in drawing free body diagrams, we used the following principles for determining reaction forces and moment at the supports.

1. If a point cannot move in a given direction, then a reaction force opposite to the direction acts at the support point.
2. If a line cannot rotate about an axis in a given direction, then a reaction moment opposite to the direction acts at the support.

If we were to make an imaginary cut very close to the support and then draw a free body diagram, we would find that the internal force will equal to the reaction force and the internal moment will equal the reaction moment. Thus, the first observation implies that if a point cannot move (displacement is zero), then the internal force is not known because the reaction force is not known. Similarly, the second observation implies that if a line cannot rotate (rotation is zero) around an axis at the support, then the internal moment is not known because the reaction moment is not known. The reverse is equally true. If structural member's end is free, then the internal force and internal moment will be zero; but the free end can displace and rotate by any amount that is dictated by the loading. Thus, when we specify a value for internal force, then we cannot specify displacement; and when we specify the value for internal moment, then we cannot specify rotation.

### 2.4.1 Axial displacement

The axial displacement $u_0(x)$ can be found from the differential equation (2.16-A) which we rewrite below for convenience.

- Differential equation:

$$\boxed{\frac{d}{dx}\left(EA\frac{du_0}{dx}\right) = -p_x(x)} \tag{2.24-A}$$

Integrating the differential equation twice we obtain:

$$EA\frac{du_0}{dx} = I_1(x) + C_1 \qquad EAu_0 = I_2(x) + C_1x + C_2 \qquad \text{where} \qquad I_1(x) = -\int p_x(x)dx \qquad I_2(x) = \int I_1(x)dx \tag{2.25-A}$$

where, the indefinite integrals can be found as $p_x(x)$ is known; $C_1$ and $C_2$ are constants to be determined from the boundary conditions. The particular solution is $I_2(x)/(EA)$ and the homogeneous solution is $(C_1x + C_2)/(EA)$. In the above integration, we assumed $EA$ is a constant, but if it is not, then it can be easily accounted for by dividing Equation (2.25-A) by $EA$ before integrating.

We note that the internal axial force is: $N = EA(du_0/dx)$. As discussed earlier, it is the nature of boundary conditions that at each end (boundary) of the axial member, either the axial displacement $u_0$ or the internal axial force $N$ must be specified as shown below.

- Boundary conditions: At each end specify

$$\boxed{u_0 \quad \text{or} \quad N} \tag{2.26-A}$$

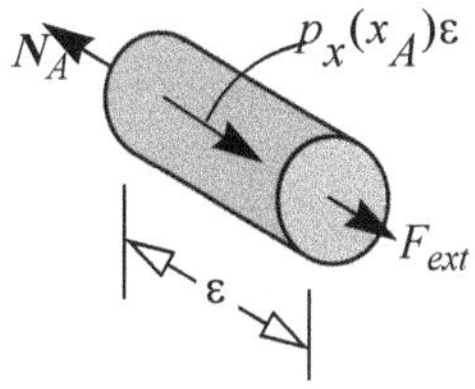


**Figure 2.14** Boundary condition on the internal axial force.

To determine the boundary condition on the internal axial force $N$, we make an imaginary cut an infinitesimal distance ($\varepsilon$) from the end (say point $A$) and draw a free body diagram of the infinitesimal element as in Figure 2.14. The internal axial force is drawn in tension in accordance with our sign convention (see Table 2.1).

By equilibrium of forces and by taking the limit $\varepsilon \to 0$, we obtain the following boundary condition:

$$\lim_{\varepsilon \to 0}(F_{ext} - N_A - p_x(x_A)\varepsilon) = 0 \qquad \text{or} \qquad N_A = F_{ext}$$

The above equation shows that the distributed axial force does not affect the boundary condition on the internal axial force. The magnitude of the internal axial force $N$ at the end of an axial bar is equal to the concentrated external axial force applied at the end, but getting the *right sign* requires drawing the free body diagram with the internal force $N$ drawn in accordance with the sign convention. If $F_{\text{ext}}$ is zero, we can write the boundary condition of zero value of $N$ by inspection.

### 2.4.2 Torsional rotation

The rotation $\phi(x)$ can be found from the differential equation (2.16-T) which we rewrite below for convenience.

- Differential equation:

$$\boxed{\frac{d}{dx}\left(GJ\frac{d\phi}{dx}\right) = -t(x)} \qquad \textbf{(2.27-T)}$$

Integrating the differential equation twice we obtain:

$$GJ\frac{d\phi}{dx} = I_1(x) + C_1 \qquad GJ\phi = I_2(x) + C_1x + C_2 \qquad \text{where} \qquad I_1(x) = -\int t(x)dx \qquad I_2(x) = \int I_1(x)dx \qquad \textbf{(2.28-A)}$$

where, the indefinite integrals can be found as $t(x)$ is known; $C_1$ and $C_2$ are constants to be determined from the boundary conditions. The particular solution is $I_2(x)/(GJ)$ and the homogeneous solution is $(C_1x + C_2)/(GJ)$. In the above integration, we assumed $GJ$ is a constant, but if it is not, then it can be easily accounted for by dividing Equation (2.28-A) by $GJ$ before integrating.

We note that internal torque is: $T = GJ(d\phi/dx)$. As discussed earlier, it is the nature of boundary conditions that at each end (boundary) of the circular shaft, either the rotation $\phi$ or the internal torque $T$ must be specified as shown below.

- Boundary conditions: At each end specify

$$\boxed{\phi \quad \text{or} \quad T} \qquad \textbf{(2.29-T)}$$

To determine the boundary condition on internal torque $T$, we make an imaginary cut an infinitesimal distance ($\varepsilon$) from the end (say point $A$) and draw a free body diagram of the infinitesimal element as shown in Figure 2.15. The internal torque $T$ is drawn in accordance with our sign convention (see Table 2.1)..

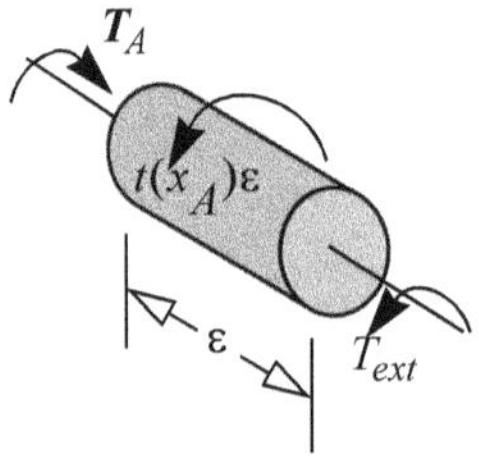


**Figure 2.15** Boundary condition on the internal torque.

By equilibrium of moments and by taking the limit $\varepsilon \to 0$, we obtain the boundary condition

$$\lim_{\varepsilon \to 0}(T_{\text{ext}} - T_A - t(x_A)\varepsilon) = 0 \qquad \text{or} \qquad T_A = T_{\text{ext}}$$

The above equation shows that the distributed torque does not affect the boundary condition on the internal torque. The magnitude of the internal torque $T$ at the end of a shaft is equal to the concentrated external torque applied at the end, but getting the *right sign* requires drawing the free body diagram with the internal torque $T$ drawn in accordance with the sign convention. If $T_{\text{ext}}$ is zero, we can write the boundary condition of zero value of $T$ by inspection.

### 2.4.3 Beam deflection

The beam deflection $v(x)$ can be found from the differential equation (2.16-B) which is written below for convenience.

- Differential equation:

$$\boxed{\frac{d^2}{dx^2}\left(EI_{zz}\frac{d^2v}{dx^2}\right) = p_y(x)} \qquad \textbf{(2.30a-B)}$$

Integrating the differential equation four times we obtain:

$$\frac{d}{dx}\left(EI_{zz}\frac{d^2v}{dx^2}\right) = I_1(x) + C_1 \qquad I_1(x) = \int p_y(x)dx \qquad \textbf{(2.30b-B)}$$

$$EI_{zz}\frac{d^2v}{dx^2} = I_2(x) + C_1x + C_2 \qquad I_2(x) = \int I_1(x)dx \qquad \textbf{(2.30c-B)}$$

$$EI_{zz}\frac{dv}{dx} = I_3(x) + C_1(x^2/2) + C_2x + C_3 \qquad I_3(x) = \int I_2(x)dx \qquad \textbf{(2.30d-B)}$$

$$EI_{zz}v = I_4(x) + C_1(x^3/6) + C_2(x^2/2) + C_3x + C_4 \qquad I_4(x) = \int I_3(x)dx \qquad \textbf{(2.30e-B)}$$

where, the indefinite integrals can be found as $p_y(x)$ is known; $C_1$, $C_2$, $C_3$, $C_4$ are constants to be determined from the boundary conditions. The particular solution is $I_4(x)/(EI_{zz})$ and the homogeneous solution is

$[C_1(x^3/6) + C_2(x^2/2) + C_3x + C_4]/(EI_{zz})$. In the above integration, we assumed $EI_{zz}$ is a constant, but if it is not, then it can be easily accounted for by dividing Equation (2.30c-B) by $EI_{zz}$ before integrating.

We write the following expressions for the internal moment and shear force from Equations (2.12-B) and (2.15b-B)

$$M_z(x) = EI_{zz}\frac{d^2v}{dx^2} \qquad V_y(x) = -\left(\frac{dM_z}{dx}\right) = -\left[\frac{d}{dx}\left(EI_{zz}\frac{d^2v}{dx^2}\right)\right] \tag{2.30f-B}$$

The integration of the differential equation four times resulted in v, $dv/dx$, $M_z$, or $V_y$. The four quantities form two groups for the purpose of determining the boundary conditions.

- *Group 1*: At a boundary point either the deflection v or the internal shear force $V_y$ can be specified, but not both.
- *Group 2*: At a boundary point either the slope $dv/dx$ or the internal bending moment $M_z$ can be specified, but not both.

To generate four boundary conditions, two conditions are specified at each end of the beam. We choose one condition from group 1 and one from group 2; alternatively we can make the following statement.

- Boundary conditions: At each end specify

$$\boxed{(v \quad \text{or} \quad V_y) \quad \text{and} \quad \left(\frac{dv}{dx} \quad \text{or} \quad M_z\right)} \tag{2.30g-B}$$

To determine the boundary condition on the internal moment $M_z$, and the shear force $V_y$, we make an imaginary cut an infinitesimal distance (ε) from the end (say point $A$) and draw a free body diagram of the infinitesimal element as shown in Figure 2.14. The internal shear force and internal bending moment are drawn in accordance with our sign convention (see Table 2.1)

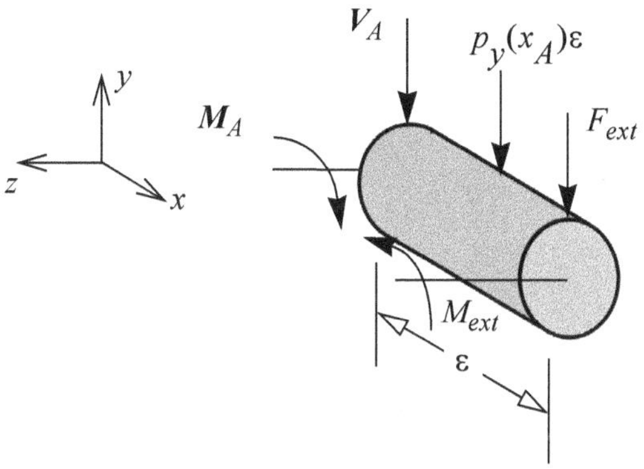


**Figure 2.16** Boundary condition on the internal shear force and bending moment.

By equilibrium of forces and moment and by taking the limit $\varepsilon \to 0$, we obtain the boundary conditions below.

$$\lim_{\varepsilon \to 0}(F_{ext} + V_A - p_y(x_A)\varepsilon) = 0 \quad \text{or} \quad V_A = -F_{ext} \tag{2.31a-B}$$

$$\lim_{\varepsilon \to 0}\left(M_A - M_{ext} + \varepsilon F_{ext} + p_y(x_A)\frac{\varepsilon^2}{2}\right) = 0 \quad \text{or} \quad M_A = M_{ext} \tag{2.31b-B}$$

The above equations show that the distributed force does not affect the boundary condition on either the internal shear force or the bending moment. On a boundary point, the value of the internal shear force $V_y$ and the bending moment $M_z$ is equal in magnitude to the concentrated external force and the concentrated external moment, respectively; but getting the *right sign* requires drawing the free body diagram with the internal shear force $V_y$ and the internal bending moment $M_z$ drawn in accordance with the sign convention. If $F_{ext}$ and $M_{ext}$ are zero, we can write the boundary condition of zero value for shear force and moment by inspection.

A synopsis of the boundary value problems for axial, torsion, and symmetric bending is presented in Table 2.2.

**Table 2.2** Synopsis of Boundary Value Problems

| | Axial | Torsion | Symmetric Bending |
|---|---|---|---|
| Differential equation | $\frac{d}{dx}[EA(du_0/dx)] = -p_x(x)$ | $\frac{d}{dx}[GJ(d\phi/dx)] = -t(x)$ | $\frac{d^2}{dx^2}(EI_{zz}d^2v/dx^2) = p_y(x)$ |
| Boundary conditions at each end | $u_0$ or $N$ | $\phi$ or $T$ | (v or $V_y$) and $\left(\frac{dv}{dx}\text{ or }M_z\right)$ |
| Internal forces/moments | $N = EA(du_0/dx)$ | $T = GJd\phi/dx$ | $M_z = EI_{zz}d^2v/dx^2$ $\quad V_y = -\frac{d}{dx}(EI_{zz}d^2v/dx^2)$ |

## EXAMPLE 2.8

A light pole is subjected to wind pressure that varies quadratically as shown in Figure 2.17. In terms of $E$, $I$, $w$, $L$, and $x$, determine (a) the deflection at the free end and (b) the ground reactions.

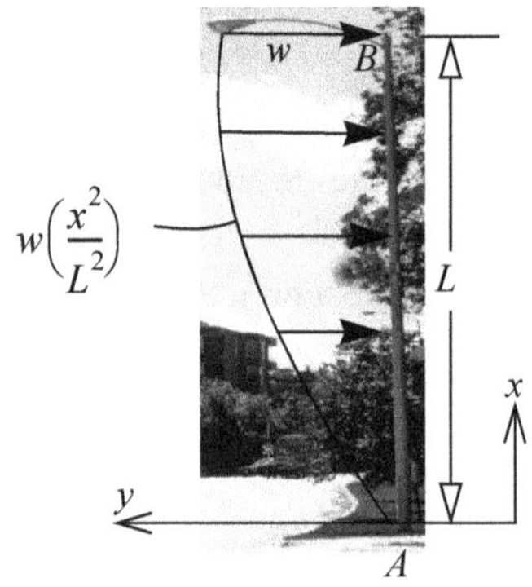


**Figure 2.17** Beam and loading for Example 2.8.

**PLAN**

(a) We substitute $p_y = -w(x^2/L^2)$ into Equation (2.16-B) to obtain the differential equation. The four boundary conditions are: the deflection and the slope at $A$ are zero, and the moment and the shear force at $B$ are zero. We solve the boundary value problem and determine the elastic curve v($x$). By substituting $x = L$ in the elastic curve equation, we obtain the deflection at the free end. (b) By making an imaginary cut just above point $A$ and draw the free body diagram to relate the internal shear force and the internal moment to the reactions at $A$. By substituting $x = 0$ in the expressions for moment and shear force, we obtain the reactions at point $A$.

**SOLUTION**

We write the boundary value problem as discussed in the plan.

- Differential equation

$$\frac{d^2}{dx^2}\left(EI_{zz}\frac{d^2v}{dx^2}\right) = -w\left(\frac{x^2}{L^2}\right) \tag{E1}$$

- Boundary conditions

$$v(0) = 0 \qquad \frac{dv}{dx}(0) = 0 \tag{E2}$$

$$M_z(x = L) = EI_{zz}\frac{d^2v}{dx^2}\bigg|_{x=L} = 0 \qquad V_y(x = L) = -\left[\frac{d}{dx}\left(EI_{zz}\frac{d^2v}{dx^2}\right)\bigg|_{x=L}\right] = 0 \tag{E3}$$

Integrating Equation (E1) four times we obtain

$$\frac{d}{dx}\left(EI_{zz}\frac{d^2v}{dx^2}\right) = -\frac{wx^3}{3L^2} + c_1 \tag{E4}$$

$$EI_{zz}\frac{d^2v}{dx^2} = -\frac{wx^4}{12L^2} + c_1x + c_2 \tag{E5}$$

$$EI_{zz}\frac{dv}{dx} = -\frac{wx^5}{60L^2} + \frac{c_1x^2}{2} + c_2x + c_3 \tag{E6}$$

$$EI_{zz}v = -\frac{wx^6}{360L^2} + \frac{c_1x^3}{6} + \frac{c_2x^2}{2} + c_3x + c_4 \tag{E7}$$

From boundary conditions we obtain

$$\frac{d}{dx}\left(EI_{zz}\frac{d^2v}{dx^2}\right)\bigg|_{x=L} = -\frac{wL^3}{3L^2} + c_1 = 0 \quad \text{or} \quad c_1 = wL/3 \tag{E8}$$

$$EI_{zz}\frac{d^2v}{dx^2}\bigg|_{x=L} = -\frac{wL^4}{12L^2} + c_1x + c_2 = 0 \quad \text{or} \quad c_2 = -(wL^2/4) \tag{E9}$$

$$\frac{dv}{dx}(0) = c_3 = 0 \qquad v(0) = c_4 = 0 \tag{E10}$$

Substituting the constants and simplifying, we obtain

$$v(x) = -\left[\frac{w}{360EI_{zz}L^2}\right](x^6 - 20L^3x^3 + 45L^4x^2) \tag{E11}$$

**(a)** Substituting $x = L$ into Equation (E11), we obtain the deflection at the free end as

$$\textbf{ANS.} \quad v(L) = -\left(\frac{13wL^4}{180EI_{zz}}\right) \tag{E12}$$

(b) The internal moment and shear force can be written as

$$M_z(x) = EI_{zz}\frac{d^2v}{dx^2} = -\frac{wx^4}{12L^2} + \frac{wL}{3}x - \frac{wL^2}{4} \qquad V_y(x) = -\frac{d}{dx}\left(EI_{zz}\frac{d^2v}{dx^2}\right) = \frac{wx^3}{3L^2} - \frac{wL}{3} \tag{E13}$$

We make an imaginary cut just above point $A$ and draw the free body diagram shown in Figure 2.18. By the equilibrium of forces and moments, we obtain

$$M_A = -M_z(0) = -(-wL^2/4) \qquad R_A = -V_y(0) = -(-wL/3) \tag{E14}$$

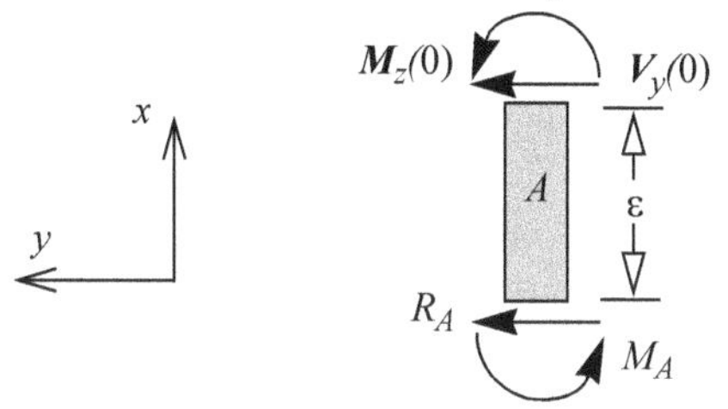


**Figure 2.18** The infinitesimal equilibrium element at $A$ for Example 2.8.

$$\textbf{ANS.} \quad R_A = wL/3 \qquad M_A = wL^2/4$$

## COMMENTS

1. We can check the directions of $R_A$ and $M_A$ by inspection. They are correct because they are the directions necessary for equilibrium of the external distributed force.
2. In drawing the free body diagram in Figure 2.18, the reaction force $R_A$ and the reaction moment $M_A$ can be drawn in any direction, but the internal quantities $V_y$ and $M_z$ must be drawn in accordance with the sign convention. Irrespective of the directions in which $R_A$ and $M_A$ are drawn, the final answer will be as just given. The sign in the equilibrium equations (E14) and (E14) will account for the assumed directions of the reactions.

## EXAMPLE 2.9

In terms of $E$, $I$, $w$, $L$, and $x$, determine (a) the elastic curve and (b) the reaction force at $A$ in Figure 2.19.

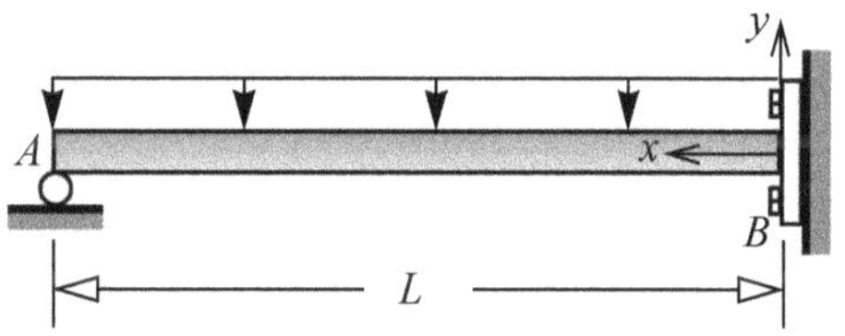


**Figure 2.19** Beam and loading for Example 2.9.

**PLAN**

(a) The two boundary conditions at $A$ are zero deflection and zero moment, and the two boundary conditions at $B$ are zero deflection and zero slope. We can solve the boundary value problem and obtain the elastic curve $v(x)$. (b) After making an imaginary cut just to the right of $A$, we draw a free body diagram and relate the reaction force to the shear force at $x = L$.

**SOLUTION**

We write the boundary value problem as described in the plan.

- Differential equation

$$\frac{d^2}{dx^2}\left(EI_{zz}\frac{d^2v}{dx^2}\right) = -w \tag{E1}$$

- Boundary conditions

$$v(0) = 0 \qquad \frac{dv}{dx}(0) = 0 \qquad v(L) = 0 \qquad EI_{zz}\frac{d^2v}{dx^2}(L) = 0 \tag{E2}$$

Integrating Equation (E1) four times, we obtain

$$\frac{d}{dx}\left(EI_{zz}\frac{d^2v}{dx^2}\right) = -wx + c_1 \tag{E3}$$

$$EI_{zz}\frac{d^2v}{dx^2} = -\left(\frac{wx^2}{2}\right) + c_1x + c_2 \tag{E4}$$

$$EI_{zz}\frac{dv}{dx} = -\left(\frac{wx^3}{6}\right) + c_1\frac{x^2}{2} + c_2x + c_3 \tag{E5}$$

$$EI_{zz}\mathrm{v} = -\left(\frac{wx^4}{24}\right) + c_1\frac{x^3}{6} + c_2\frac{x^2}{2} + c_3x + c_4 \quad \text{(E6)}$$

From boundary conditions we obtain the constants as shown below.

$$\frac{d\mathrm{v}}{dx}(0) = c_3 = 0 \qquad \mathrm{v}(0) = c_4 = 0 \quad \text{(E7)}$$

$$EI_{zz}\frac{d^2\mathrm{v}}{dx^2}(L) = -\left(\frac{wL^2}{2}\right) + c_1x + c_2 \quad \text{or} \quad c_1L + c_2 = wL^2/2 \quad \text{(E8)}$$

$$EI_{zz}\mathrm{v}(L) = -\left(\frac{wL^4}{24}\right) + c_1\frac{x^3}{6} + c_2\frac{x^2}{2} = 0 \quad \text{or} \quad c_1L^2/2 + c_2L = wL^3/6 \quad \text{(E9)}$$

Solving for the constants we obtain

$$c_1 = 5wL/8 \qquad c_2 = -wL^2/8 \quad \text{(E10)}$$

Substituting the constants and simplifying, we obtain the elastic curve below.

$$\textbf{ANS.}\quad \mathrm{v}(x) = -\frac{w}{48EI_{zz}}(2x^4 - 5Lx^3 + 3L^2x^2) \quad \text{(E11)}$$

We make an imaginary cut at an infinitesimal distance right of point $A$ and draw the free body diagram shown in Figure 2.20. By force equilibrium in the $y$ direction, we obtain

$$R_A = V_A = V_y(L) \quad \text{(E12)}$$

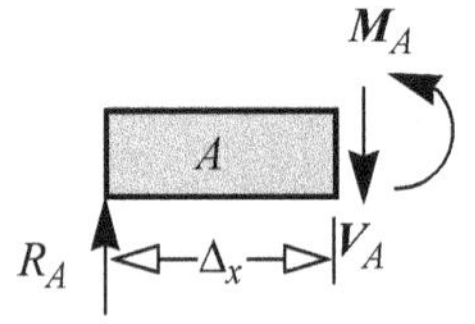


**Figure 2.20** The infinitesimal equilibrium element at $A$ in Example 2.9.

Substituting Equations (E3) and (E10) into Equation (2.30f-B), we obtain

$$V_y(x) = -\frac{d}{dx}\left(EI_{zz}\frac{d^2\mathrm{v}}{dx^2}\right) = wx - \frac{5wL}{8} \quad \text{(E13)}$$

Substituting Equation (E13) into Equation (E12), we obtain the reaction at $A$ as

$$\textbf{ANS.}\quad R_A = 3wL/8$$

**COMMENTS**

1. Using $R_A$ as reaction at support $A$, we could have determined the internal moment and written the boundary value problem using the second-order differential equation (2.12-B) as shown below.

$$\text{Differential equation: } \boldsymbol{M}_z = EI_{zz}\frac{d^2\mathrm{v}}{dx^2} = R_A(L-x) - w\frac{(L-x)^2}{2} \quad \text{(E14)}$$

$$\text{Boundary conditions: } \mathrm{v}(0) = 0 \qquad \frac{d\mathrm{v}}{dx}(0) = 0 \qquad \mathrm{v}(L) = 0 \quad \text{(E15)}$$

The above boundary value problem could be solved to obtain the same results as we did with a fourth order system. Though the algebra is slightly less, this second-order differential equation approach is very problem dependent and does not have the same methodical approach to all problems that the fourth-order differential equation provides. The second-order differential equation approach would be difficult to implement for complex loading such as in Example 2.8.

## 2.5 DISCONTINUITY FUNCTIONS

A structural member usually has point loads applied to it. Internal forces and moments jump by the values of external point forces and moments as one crosses the point loads from one side to the other. In the introductory course on mechanics of materials, the displacements and the rotation calculations had to be conducted for each segment separately, and the continuity conditions on displacement and rotations had to be imposed *explicitly* to solve the problem.

The discontinuity functions permit us to model the discontinuities in loads such that a single expression for distributed load can be used over the entire length of the structural member. The single expression for distributed load can be substituted into the differential equations and the boundary value problem solved as we did in Section 2.4. The continuity conditions are *implicitly* satisfied in the solution of boundary value problem. The solution is independent of whether the problem is statically determinate or indeterminate. Once the displacement function is obtained, then strains, stresses, internal forces and moments can be found.

### 2.5.1 Definitions

Consider a distributed load $p$ and an equivalent load $P = p\varepsilon$ as shown in Figure 2.21. Suppose we now let the intensity of the distributed load continuously increase to infinity, while we decrease the length over which the distributed force is applied to zero in such a manner that the area $p\varepsilon$ remains a finite quantity; then we obtain a concentrated force $P$ applied at $x = a$.

Mathematically, this is stated as follows:

$$P = \lim_{p \to \infty} \lim_{\varepsilon \to 0} (p\varepsilon) = P\langle x - a\rangle^{-1} \tag{2.32}$$

The function $\langle x - a\rangle^{-1}$ is called the Dirac delta function, or just the **delta function**. The delta function is zero except in an infinitesimal region near $a$. As $x$ tends toward $a$, the delta function tends to infinity, but the area under the function is equal to one. The delta function is defined as follows:

$$\langle x - a\rangle^{-1} = \begin{cases} 0 & x \neq a \\ \infty & x \to a \end{cases} \qquad \int_{(a-\varepsilon)}^{(a+\varepsilon)} \langle x - a\rangle^{-1} dx = 1 \tag{2.33a}$$

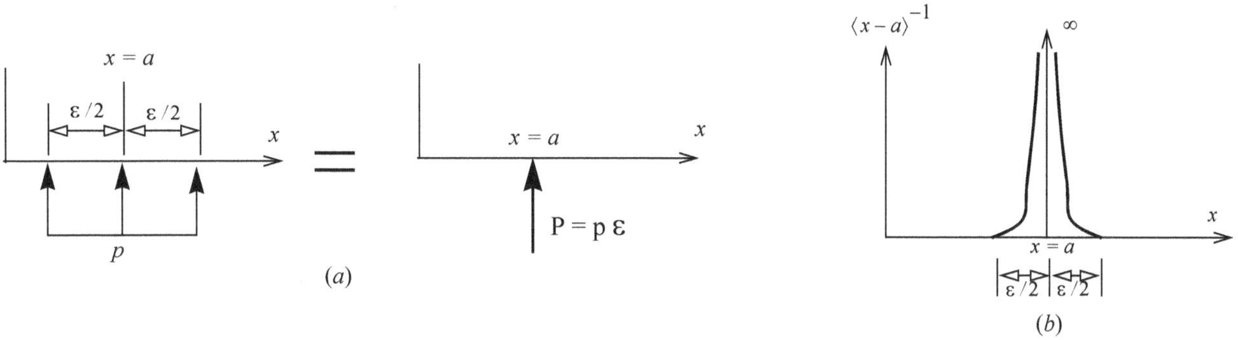


**Figure 2.21** The delta function.

Consider the integral of the delta function. The lower limit of minus infinity emphasizes that the point is before $a$. We write the integral as the sum of three integrals:

$$\int_{-\infty}^{x} \langle x - a\rangle^{-1} dx = \int_{-\infty}^{(a-\varepsilon)} \langle x - a\rangle^{-1} dx + \int_{(a-\varepsilon)}^{(a+\varepsilon)} \langle x - a\rangle^{-1} dx + \int_{(a+\varepsilon)}^{x} \langle x - a\rangle^{-1} dx$$

The first and the third integrals are zero because the delta function is zero at all points in the interval of integration, while the second integral is equal to 1, per Equation (2.33a). Thus, the integral is zero before $a$ and equal to one after $a$. The integral is called the **step function** as defined in Equation (2.33b) and shown in Figure 2.22*a*.

$$\langle x - a\rangle^{0} = \int_{-\infty}^{x} \langle x - a\rangle^{-1} dx = \begin{Bmatrix} 0 & x < a \\ 1 & x > a \end{Bmatrix} \tag{2.33b}$$

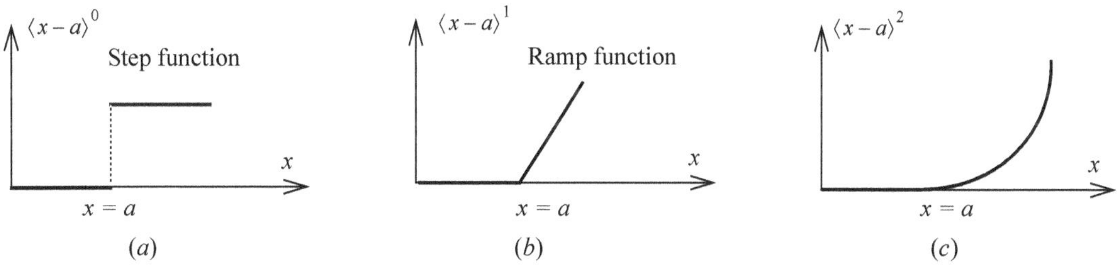


**Figure 2.22** Discontinuity functions.

Now consider the integral of the step function. We can write the integral as sum of two integrals.

$$\int_{-\infty}^{x} \langle x - a\rangle^{0} dx = \int_{-\infty}^{a} \langle x - a\rangle^{0} dx + \int_{a}^{x} \langle x - a\rangle^{0} dx$$

The first integral is zero because the step function is zero at all points in the interval of integration, while the second integral value is $(x - a)$. The second integral is called the *ramp function* shown in Figure 2.22*b*. Proceeding in this manner; we can mathematically define an entire class of functions shown in Equation (2.34).

$$\boxed{\langle x - a\rangle^{n} = \begin{Bmatrix} 0 & x \leq a \\ (x - a)^{n} & x > a \end{Bmatrix} \qquad n \geq 0} \tag{2.34}$$

We can also generate the integral formula of Equation (2.35) from Equation (2.34):

$$\int_{-\infty}^{x} \langle x - a\rangle^{n} dx = \frac{\langle x - a\rangle^{n+1}}{n+1} \qquad n \geq 0 \tag{2.35}$$

One more function, called the *doublet function* is defined in Equation (2.36).

$$\langle x - a\rangle^{-2} = \begin{cases} 0 & x \neq a \\ \infty & x \to a \end{cases} \quad \text{and} \quad \int_{-\infty}^{x} \langle x - a\rangle^{-2} dx = \langle x - a\rangle^{-1} \tag{2.36}$$

The delta function and the doublet function become infinite at $x = a$. Alternatively stated, these functions are singular at $x = a$ and are referred to as **singularity functions**. The entire class of functions $\langle x - a\rangle^n$ for positive and negative $n$ are called the **discontinuity functions.**

*Note that the discontinuity functions are zero if the argument is negative.* By differentiating Equations (2.33b), (2.35), and (2.36) we obtain the formulas for differentiation. The integral formulas are inverse of the differentiation formulas.

$$\frac{d\langle x-a\rangle^{-1}}{dx} = \langle x-a\rangle^{-2} \qquad \frac{d\langle x-a\rangle^{0}}{dx} = \langle x-a\rangle^{-1} \qquad \frac{d\langle x-a\rangle^{n}}{dx} = n\langle x-a\rangle^{n-1} \text{for } n \ge 1$$
$$\int\langle x-a\rangle^{-2}dx = \langle x-a\rangle^{-1} \qquad \int\langle x-a\rangle^{-1}dx = \langle x-a\rangle^{0} \qquad \int\langle x-a\rangle^{n-1}dx = \frac{\langle x-a\rangle^{n}}{n} \text{ for } n \ge 1 \tag{2.37}$$

## 2.5.2 Axial displacement

Figure 2.23*a* shows a template of an axial bar in which a concentrated force is applied at $x = a$. If we make an imaginary cut in the region $x < a$, the internal axial force $N = 0$. If we make an imaginary cut in the region $x > a$, the internal axial force $N = -F$. We can use the discontinuity function to write the internal force and substitute it into Equation (2.15-A) to obtain the two equations shown as template equations in Figure 2.23*a*. If the point force on the actual member is in the direction shown in Figure 2.23, the internal force and distributed force are as given. If the point force is opposite to that shown in Figure 2.23*a*, the signs are reversed in the template equations.

For statically determinate problems and for problems for which $p_x$ is a simple function, it may be possible to determine the internal axial force $N$ and integrate Equation (2.15-A) to get obtain the axial displacements. But starting with the differential equations (2.16-A) and solving the boundary value problem as we did in Section 2.4, is a general approach that covers both the statically determinate and indeterminate problems.

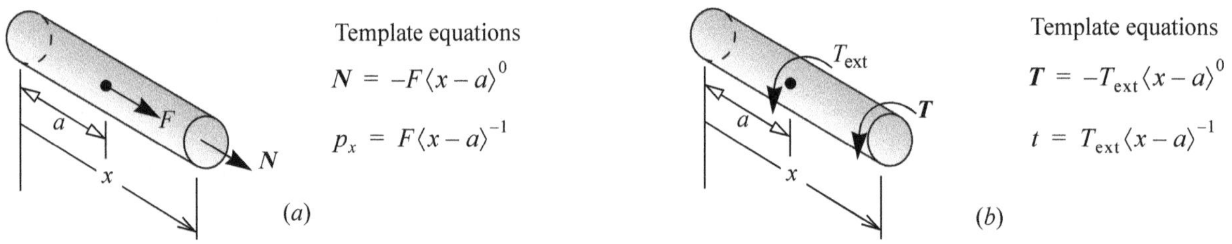

a
x
F
N
Template equations
$N = -F\langle x-a\rangle^0$
$p_x = F\langle x-a\rangle^{-1}$
(a)
$T_{ext}$
a
x
T
Template equations
$T = -T_{ext}\langle x-a\rangle^0$
$t = T_{ext}\langle x-a\rangle^{-1}$
(b)


**Figure 2.23** Template and template equations for use of discontinuity functions in (*a*) Axial members. (*b*) Torsion of shafts.

## 2.5.3 Torsional rotation

Figure 2.23*b* shows a template of a shaft on which a concentrated torque is applied at $x = a$. If we make an imaginary cut in the region $x < a$, the internal torque $T = 0$. If we make an imaginary cut in the region $x > a$, the internal torque is $T = -T_{ext}$. We can use the discontinuity function to write the internal torque and substitute it into Equation (2.15-T) to obtain the two equations shown as template equations in Figure 2.23*b*. If the concentrated torque $T_{ext}$ on the actual shaft is in the direction shown in Figure 2.23*b*, the internal torque and the distributed torque are as given. If the concentrated torque is opposite to that shown in Figure 2.23*b*, the signs are reversed in the template equations

The use of discontinuity functions in torsion is similar to that for axial displacement and will not be discussed further.

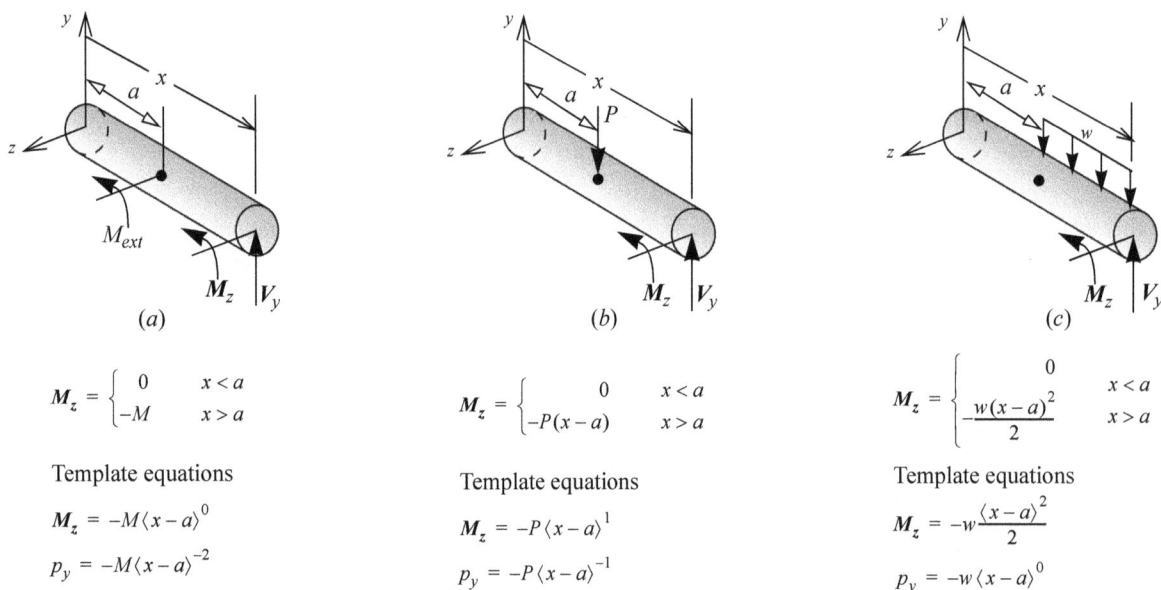

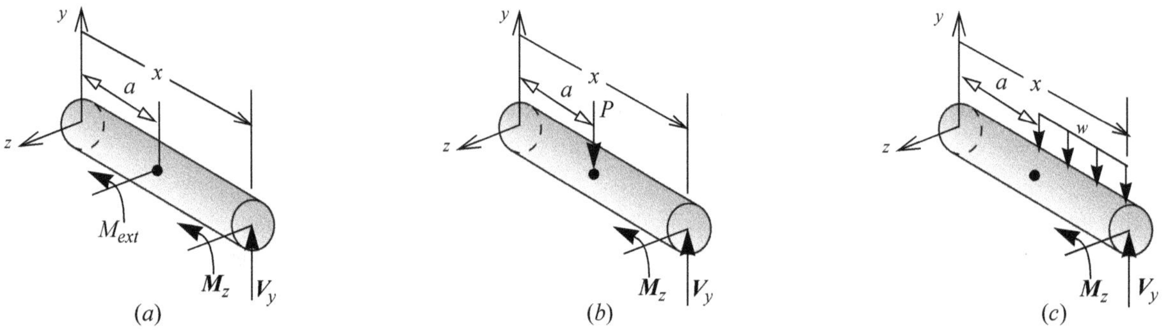


(a)
$$M_z = \begin{cases} 0 & x < a \\ -M & x > a \end{cases}$$
Template equations
$$M_z = -M\langle x-a\rangle^0$$
$$p_y = -M\langle x-a\rangle^{-2}$$

(b)
$$M_z = \begin{cases} 0 & x < a \\ -P(x-a) & x > a \end{cases}$$
Template equations
$$M_z = -P\langle x-a\rangle^1$$
$$p_y = -P\langle x-a\rangle^{-1}$$

(c)
$$M_z = \begin{cases} 0 & x < a \\ -\dfrac{w(x-a)^2}{2} & x > a \end{cases}$$
Template equations
$$M_z = -w\frac{\langle x-a\rangle^2}{2}$$
$$p_y = -w\langle x-a\rangle^0$$

**Figure 2.24** Templates for the use of the discontinuity functions in beams.

### 2.5.4 Beam deflection

Figure 2.24 shows a free body diagram of a beam in which a concentrated force, moment, or the start of distributed force occurs at $x = a$. If we make an imaginary cut in the region $x < a$, then the internal bending moment $M_z = 0$. If we make an imaginary cut in region $x > a$, then the internal bending moment $M_z$ can be found by means of the moment equilibrium, as shown in Figure 2.24. The moment can then be written by using the definition of discontinuity functions. By substituting Equation (2.15a-B) into Equation (2.15b-B), we obtain Equation (2.38-B).

$$\frac{d^2 M_z}{dx^2} = p_y(x) \tag{2.38-B}$$

By substituting the moment expressions for each case obtained in Figure 2.24 into Equation (2.38-B), we obtain an expression for the distributed force for each case as shown in Figure 2.24.

The template equations are associated with the coordinate system shown in Figure 2.24. If the coordinate system is changed, corresponding changes must be made in the template equations. If the external forces and moments are in the directions shown on the template free body diagram, then the expressions for $M_z$ and $p_y$ will be in accordance with the template equations. If the external forces and moments are opposite, then the signs will have to be reversed in the template equations.

**EXAMPLE 2.10**

The column shown in Figure 2.25 has a specific weight of $\gamma = 0.1$ lb/in$^3$, modulus of elasticity of $E = 4000$ ksi, and cross-sectional area $A = 100$ in$^2$. Determine (a) the movement of the rigid plate at $C$ and (b) the reaction force at $A$.

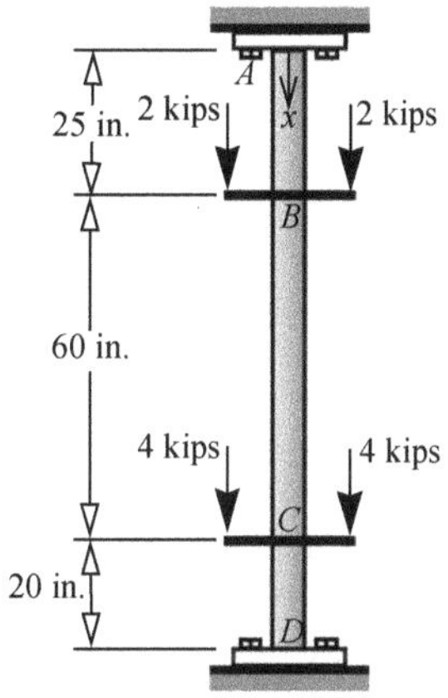


**Figure 2.25** Column for Example 2.10.

**PLAN**

The specific weight results in a distributed force per unit length of $\gamma A$. To this we add the loads at $B$ and $C$ using discontinuity function to get $p_x(x)$. We write the differential equation (2.16-A) and zero displacements at $A$ and $D$ as the boundary conditions to complete the boundary value problem statement. Solving the boundary value problem we obtain $u(x)$. We find $u(x= 85)$ to get the movement of the plate at $C$ and $N(x = 0)$, the internal axial force, which is equal to the reaction force $R_A$.

**SOLUTION**

The weight per unit length is $p = \gamma A = 10$ lb/inch in the positive direction of $x$. The concentrated forces at $B$ and $C$ are in the increasing direction of $x$ as in Figure 2.23. The total distributed force is $p_x = [10 + 4000\langle x-25\rangle^{-1} + 8000\langle x-85\rangle^{-1}]$ lb/in. The boundary value problem can be written as

- Differential equation

$$\frac{d}{dx}[EA(du/dx)] = -[10 + 4000\langle x-25\rangle^{-1} + 8000\langle x-85\rangle^{-1}] \tag{E1}$$

- Boundary conditions

$$u(0) = 0 \qquad u(105) = 0 \tag{E2}$$

Integrating Equation (E1) twice we obtain

$$EA(du/dx) = -[10x + 4000\langle x-25\rangle^{0} + 8000\langle x-85\rangle^{0}] + c_1 \tag{E3}$$

$$EAu = -[5x^2 + 4000\langle x-25\rangle^{1} + 8000\langle x-85\rangle^{1}] + c_1 x + c_2 \tag{E4}$$

From boundary conditions and the fact that discontinuity functions with negative arguments are zero, we obtain

$$EAu(0) = -[0 + 4000\langle -25\rangle^{1} + 8000\langle -85\rangle^{1}] + c_1(0) + c_2 \quad \text{or} \quad c_2 = 0 \tag{E5}$$

$$EAu(105) = -[5(105)^2 + 4000\langle 105-25\rangle^{1} + 8000\langle 105-85\rangle^{1}] + c_1(105) = 0$$

$$-[5(105)^2 + 4000(80) + 8000(20)] + c_1(105) = 0 \qquad \text{or} \qquad c_1 = 5096.4 \tag{E6}$$

Substituting Equations (E5) and (E6) and the value of $EA = 4000(10^3)(100) = 400(10^6)$ into Equation (E4), we obtain

$$400(10^6)u = -[5x^2 + 4000\langle x-25\rangle^1 + 8000\langle x-85\rangle^1] + (5096.4)x \quad \text{(E7)}$$

(a) Substituting $x = 85$ in Equation (E7), we obtain:

$$400(10^6)u_c = -[5(85)^2 + 4000\langle 60\rangle^1 + 8000\langle 0\rangle^1] + (5096.4)(85)$$

**ANS.** $u_c = 0.393(10^{-3})$ in

(b) Substituting $x = 0$ in Equation (E4) and using Equation (E6), we obtain the internal force at $A$ as follows:

$$N_A = EA(du/dx)|_{x=0} = -[10(0) + 4000\langle -25\rangle^0 + 8000\langle -85\rangle^0] + c_1 = c_1 = 5096.4 \text{ lb} \quad \text{(E8)}$$

By inspection, the reaction force will be upward, and its magnitude will be equal to $N_A$. Thus we obtain the following answer.

**ANS.** $R_A = 5096$ lb upward

## COMMENTS

1. We did not draw any free body diagram to find the internal axial force because Equation (2.16-A) is an equilibrium equation.
2. We determined the direction of the reaction force at $A$ by inspection. For more complex loading we must draw a free body diagram after making an infinitesimal cut away from $A$ and write the equilibrium equation.
3. Without the use of the discontinuity functions the solution would be tedious. We would have to solve a statically indeterminate problem that required us to find the relative displacement for each segment by integration.

## EXAMPLE 2.11

Use the discontinuity functions for the three templates shown in Figure 2.26 to write the moment and distributed force expressions.

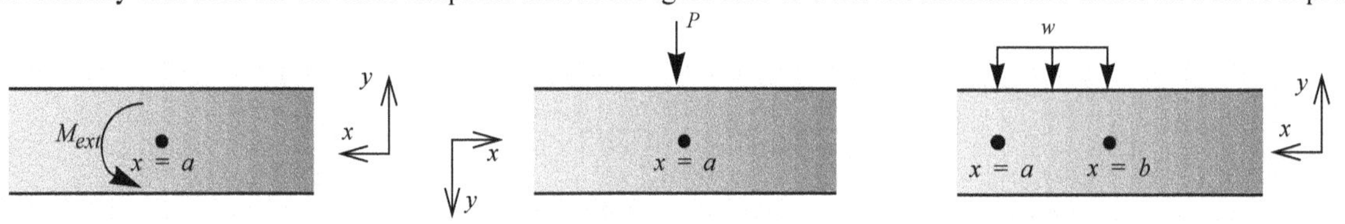


**Figure 2.26** The three cases of Example 2.11.

**PLAN**

For cases I and II, we can make an imaginary cut after $x = a$ and draw the shear force and bending moment in accordance with the sign convention. By equilibrium we can obtain the moment expression and rewrite it by using the discontinuity functions. By differentiating twice, we can obtain the distributed force expression. For case III, we can write the expression for the distributed force by using the discontinuity functions and integrate twice to obtain the moment expression.

**SOLUTION**

**Case I:** We make an imaginary cut at $x > a$, and draw the free body diagram as shown in Figure 2.27*a*. The moment and shear force are drawn in accordance with our sign convention.

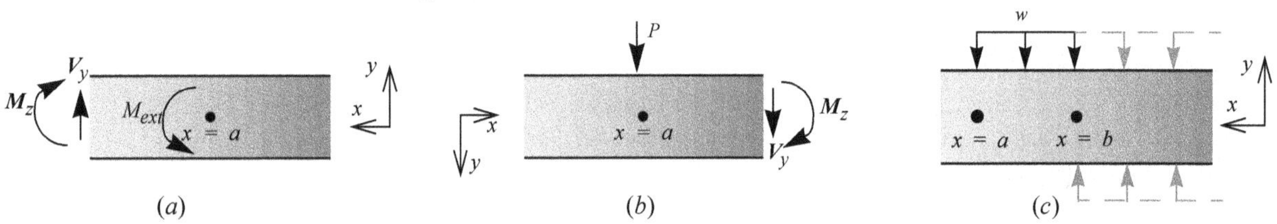


**Figure 2.27** Free body diagrams for Example 2.11: (a) case I and (b) case II.(c) case III

By moment equilibrium in Figure 2.27(a) we obtain Equation (E1).

$$M_z = M_{ext} \quad \text{(E1)}$$

Equation (E1) is valid only after $x > a$. We use the step function to write the moment expression. By differentiating twice as indicated in Equation (2.38-B), we obtain the distributed force expression.

**ANS.** $M_z = M_{ext}\langle x-a\rangle^0 \quad p_y = M_{ext}\langle x-a\rangle^{-2}$

**Case II:** We make an imaginary cut at $x > a$, and draw the free body diagram as shown in Figure 2.27*b*. The moment and shear force are drawn in accordance with our sign convention. By moment equilibrium in Figure 2.27*b* we obtain

$$M_z = P(x-a) \quad \text{(E2)}$$

Equation (E2) is valid only after $x > a$. We can use the ramp function to write the moment expression. By differentiating twice as indicated in Equation (2.38-B), we obtain the distributed force expression.

**ANS.** $M_z = P\langle x-a\rangle^1 \quad p_y = P\langle x-a\rangle^{-1}$

**Case III:** The distributed force is in the negative $y$ direction; its start can be represented by the step function at $x = a$. The end of the distributed force can also be represented by a step function, with an opposite sign used at the start to obtain

$$\textbf{ANS. } p_y = -w\langle x-a\rangle^0 + w\langle x-b\rangle^0 \quad \text{(E3)}$$

Substituting Equation (E3) into Equation (2.38-B) and integrating twice, we obtain the moment expression as shown below.

$$\textbf{ANS. } M_z = -w\langle x-a\rangle^2/2 + w\langle x-b\rangle^2/2 \quad \text{(E4)}$$

## COMMENTS

1. The three cases presented could be part of a beam with more complex loading. But the contribution for each load would be calculated as shown in this example.
2. In obtaining Equation (E4) we did not write integration constants, since at this stage we were seeking only expressions for use in the calculation of displacements. However, when we integrate for displacements, we will write integration constants that will be determined from the boundary conditions.
3. In case III we did not have to draw the free body diagram. This is an advantage in problems where the distributed load changes character over the length of the beam.

## EXAMPLE 2.12

Determine the equation of the elastic curve in Figure 2.28 in terms of *E*, *I*, *L*, *P*, and *x*.

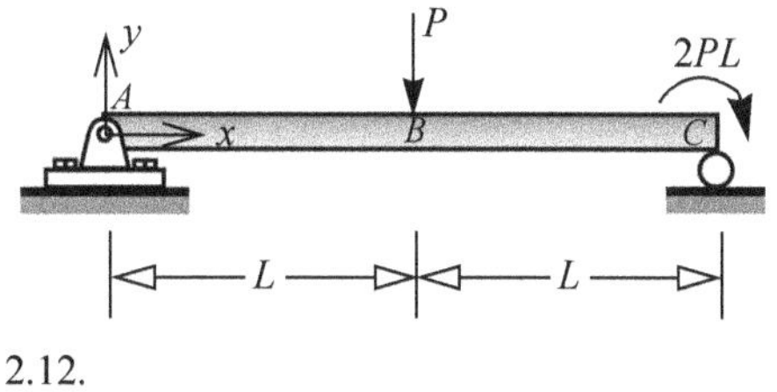


**Figure 2.28** Beam and loading for Example 2.12.

**PLAN**

The distributed load as per the template in Figure 2.24 is $p_y = -P\langle x-a\rangle^{-1}$. We write the differential equation (2.16-B) using the discontinuity functions. The boundary conditions at support *A* are deflection and moment are zero and at support *B* the deflection is zero and the moment will equal to plus or minus $2PL$ with sign determined from a free body diagram. We write and solve the boundary value problem to obtain the elastic curve v(*x*).

**SOLUTION**

We draw a free body diagram after making an imaginary cut just left of *C* as shown in Figure 2.29. By moment equilibrium with ε going to zero we obtain our moment boundary condition as

$$\lim_{\varepsilon\to 0}[M_z(2L) - \varepsilon R_C + 2PL] = 0 \qquad \text{or} \qquad M_z(2L) = -2PL \quad \text{(E1)}$$

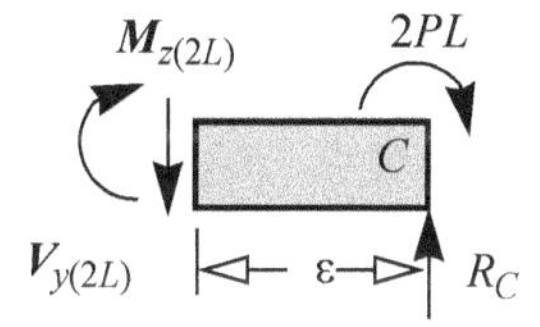


**Figure 2.29** Free body diagram in Example 2.12.

We write the boundary value problem as shown below.

- Differential equation

$$\frac{d^2}{dx^2}\left(EI_{zz}\frac{d^2v}{dx^2}\right) = -P\langle x-L\rangle^{-1} \quad \text{(E2)}$$

- Boundary conditions

$$v(0) = 0 \qquad M_z(0) = EI_{zz}\frac{d^2v}{dx^2}(0) = 0 \quad \text{(E3)}$$

$$v(2L) = 0 \qquad M_z(2L) = EI_{zz}\frac{d^2v}{dx^2}(2L) = -2PL \quad \text{(E4)}$$

Integrating Equation (E2) we obtain

$$\frac{d}{dx}\left(EI_{zz}\frac{d^2v}{dx^2}\right) = -P\langle x-L\rangle^0 + c_1 \quad \text{(E5)}$$

$$EI_{zz}\frac{d^2v}{dx^2} = -P\langle x-L\rangle^1 + c_1x + c_2 \quad \text{(E6)}$$

$$EI_{zz}\frac{dv}{dx} = -\frac{P}{2}\langle x-L\rangle^2 + \frac{c_1x^2}{2} + c_2x + c_3 \quad \text{(E7)}$$

$$EI_{zz}v = -\frac{P}{6}\langle x-L\rangle^3 + \frac{c_1x^3}{6} + \frac{c_2x^2}{2} + c_3x + c_4 \quad \text{(E8)}$$

From boundary conditions we obtain

$$v(0) = c_2 = 0 \qquad EI_{zz}\frac{d^2v}{dx^2}(0) = c_4 = 0 \tag{E9}$$

$$EI_{zz}\frac{d^2v}{dx^2}(2L) = -P\langle L\rangle^1 + c_1(2L) + c_2 = -2PL \qquad \text{or} \qquad c_1 = -P/2 \tag{E10}$$

$$v(2L) = -\frac{P}{6}\langle L\rangle^3 + \frac{c_1(2L)^3}{6} + c_3(2L) = 0 \qquad \text{or} \qquad c_3 = \left(\frac{5PL^2}{12}\right) \tag{E11}$$

Substituting Equations (E9), (E10), and (E11) into Equation (E8) and simplifying we obtain the elastic curve as

$$\textbf{ANS.} \quad v(x) = -\left(\frac{P}{12EI_{zz}}\right)(2\langle x-L\rangle^3 + x^3 - 5L^2x) \tag{E12}$$

## COMMENTS

1. Note the external moment at $C$ enters the analysis through the boundary condition in Equation (E4) and not through the distributed load. If we had tried to incorporate it in distributed load as $-2PL\langle x-2L\rangle^{-2}$, then this term would always be zero as $x$ is always less than $2L$ and we would have essentially neglected the moment.
2. Without the discontinuity functions, we would have had to write two differential equations, one applicable in $AB$ and another in $BC$. We would then have had to use the continuity of the displacements and slope at $B$ to evaluate the two extra constants. The use of the discontinuity functions implicitly accounts for the continuity conditions by using a single moment expression for the entire beam.
3. We can easily obtain the deflection expressions for regions $AB$ and $BC$ from Equations (E12) and definition of discontinuity function as shown below.

$$v(x) = \frac{P}{12EI_{zz}}[x^3 - 5L^2x] \quad 0 \le x \le L \qquad \text{and} \qquad v(x) = \frac{P}{12EI_{zz}}[(2(x-L)^3 + x^3 - 5L^2x)] \quad L \le x \le 2L$$

4. *Dimension Check:* We note that all terms on the right-hand side of Equations (E12) have the dimension of length to the cubic power. Thus the expression in Equation (E11) is dimensionally homogeneous.

## EXAMPLE 2.13

A beam with a bending rigidity of $EI = 42{,}000$ kN · m$^2$ is shown in Figure 2.30. Determine (a) the deflection at point $B$ and (b) the maximum internal bending moment.

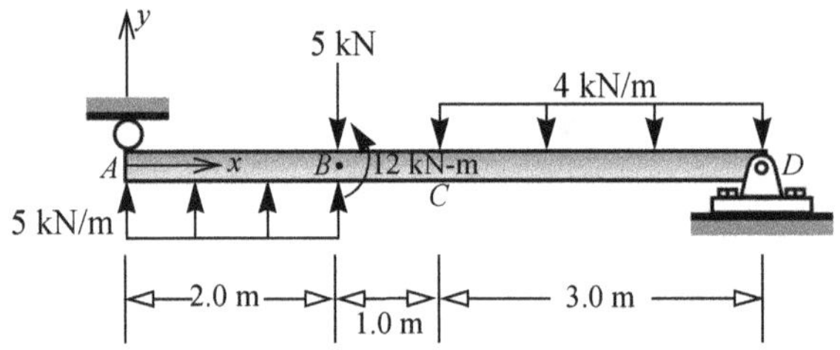


**Figure 2.30** Beam and loading for Example 2.13

**PLAN**

We write the distributed force expression in terms of the discontinuity functions using templates in Figure 2.24 We use Equation (2.16-B) and the boundary conditions at $A$ and $D$ of zero deflection and moment to write the boundary value problem, which we solve to obtain the elastic curve v($x$). (a) The deflection at $B$ is v($x$ = 2). (b) The equations of internal bending moment and shear force will be obtained in the determination of v($x$). We find the location $x_m$ where shear force is zero. Substituting $x_m$ in the moment equation will give us the maximum moment.

**SOLUTION**

We note that the distributed force in section $AB$ is positive, starts at zero, and ends at $x = 2$. The distributed force in section $CD$ is negative, starts at $x = 3$, and is over the rest of the beam. We use the template equations in Figure 2.24 to write the distributed force expression as shown in Equation (E1).

$$p_y = 5 - 5\langle x-2\rangle^0 - 4\langle x-3\rangle^0 - 5\langle x-2\rangle^{-1} - 12\langle x-2\rangle^{-2} \tag{E1}$$

We write the boundary value problem as shown below.

- Differential equation

$$\frac{d^2}{dx^2}\left(EI_{zz}\frac{d^2v}{dx^2}\right) = 5 - 5\langle x-2\rangle^0 - 4\langle x-3\rangle^0 - 5\langle x-2\rangle^{-1} - 12\langle x-2\rangle^{-2} \tag{E2}$$

- Boundary conditions

$$v(0) = 0 \qquad EI_{zz}\frac{d^2v}{dx^2}(0) = 0 \tag{E3}$$

$$v(6) = 0 \qquad EI_{zz}\frac{d^2v}{dx^2}(6) = 0 \tag{E4}$$

Integrating Equation (E2) four time, we obtain

$$\frac{d}{dx}\left(EI_{zz}\frac{d^2v}{dx^2}\right) = 5x - 5\langle x-2\rangle^1 - 4\langle x-3\rangle^1 - 5\langle x-2\rangle^0 - 12\langle x-2\rangle^{-1} + c_1 \tag{E5}$$

$$EI_{zz}\frac{d^2v}{dx^2} = \frac{5}{2}x^2 - \frac{5}{2}\langle x-2\rangle^2 - 2\langle x-3\rangle^2 - 5\langle x-2\rangle^1 - 12\langle x-2\rangle^0 + c_1x + c_2 \tag{E6}$$

$$EI_{zz}\frac{dv}{dx} = \frac{5}{6}x^3 - \frac{5}{6}\langle x-2\rangle^3 - \frac{2}{3}\langle x-3\rangle^3 - \frac{5}{2}\langle x-2\rangle^2 - 12\langle x-2\rangle^1 + c_1\frac{x^2}{2} + c_2x + c_3 \tag{E7}$$

$$EI_{zz}v = \frac{5}{24}x^4 - \frac{5}{24}\langle x-2\rangle^4 - \frac{2}{12}\langle x-3\rangle^4 - \frac{5}{6}\langle x-2\rangle^3 - 6\langle x-2\rangle^2 + c_1\frac{x^3}{6} + c_2\frac{x^2}{2} + c_3x + c_4 \tag{E8}$$

From boundary conditions we obtain

$$EI_{zz}\frac{d^2v}{dx^2}(0) = c_2 = 0 \qquad v(0) = c_4 = 0 \tag{E9}$$

$$EI_{zz}\frac{d^2v}{dx^2}(6) = \frac{5}{2}(6)^2 - \frac{5}{2}\langle 4\rangle^2 - 2\langle 3\rangle^2 - 5\langle 4\rangle^1 - 12\langle 4\rangle^0 + c_1(6) = 0 \qquad \text{or} \qquad c_1 = 0 \tag{E10}$$

$$\frac{5}{24}(6)^4 - \frac{5}{24}\langle 4\rangle^4 - \frac{2}{12}\langle 3\rangle^4 - \frac{5}{6}\langle 4\rangle^3 - 6\langle 4\rangle^2 + c_3(6) = 0 \qquad \text{or} \qquad c_3 = -\frac{323}{36} = -8.97 \tag{E11}$$

Substituting the constants and simplifying, we obtain the elastic curve below.

$$v(x) = \frac{1}{72EI_{zz}}[15x^4 - 15\langle x-2\rangle^4 - 12\langle x-3\rangle^4 - 60\langle x-2\rangle^3 - 432\langle x-2\rangle^2 - 646x] \tag{E12}$$

(a) Substituting $x = 2$ into Equation (E12), we obtain the deflection at point $B$ as follows:

$$v(2) = \frac{1}{72(42)(10^3)}[15(2)^4 - 15\langle 0\rangle^4 - 12\langle -1\rangle^4 - 60\langle 0\rangle^3 - 432\langle 0\rangle^2 - 646(2)] \tag{E13}$$

**ANS.** $v(2) = -0.35$ mm

Substituting the zero values of $c_1$ and $c_2$ into Equation (E6), we obtain the internal moment and shear force as

$$M_z(x) = EI_{zz}\frac{d^2v}{dx^2}(x) = \frac{5}{2}x^2 - \frac{5}{2}\langle x-2\rangle^2 - 2\langle x-3\rangle^2 - 5\langle x-2\rangle^1 - 12\langle x-2\rangle^0 \tag{E14}$$

$$V_y = -\frac{d}{dx}\left(EI_{zz}\frac{d^2v}{dx^2}\right) = -5x + 5\langle x-2\rangle^1 + 4\langle x-3\rangle^1 + 5\langle x-2\rangle^0 + 12\langle x-2\rangle^{-1} \tag{E15}$$

The shear force in each segment can be written using the above equation and definition of discontinuity functions as

$$V_{AB} = -5x \qquad V_{BC} = -5x + 5(x-2) + 5 = -5 \qquad V_{CD} = -5 + 4(x-3) \tag{E16}$$

The shear force can only be zero in interval $CD$. Setting $V_{CD}$ to zero we can determine the location $x_m$ where shear force will be zero and moment will be maximum as shown below.

$$-5 + 4(x_m - 3) = 0 \qquad \text{or} \qquad x_m = 5/4 + 3 = 4.25 \text{ m} \tag{E17}$$

Substituting the value of $x_m$ into Equation (E14) we obtain the maximum moment as

$$M_{max} = M_z(4.25) = \frac{5}{2}(4.25)^2 - \frac{5}{2}(2.25)^2 - 2(1.25)^2 - 5(2.25) - 12 = 6.125 \text{ kN-m} \tag{E18}$$

**ANS.** $M_{max} = 6.125$ kN-m

**COMMENTS**

1. This problem would be algebraically very tedious to solve without the discontinuity functions.
2. As shear force varies linearly, its maximum value will exist at end of intervals shown in Equation (E16).
3. Maximum bending normal stress can be found from the maximum bending moment if we know the cross section geometry. Similarly, we can find the maximum bending shear stress from the maximum shear force.
4. The above comments emphasize that once the displacement function is known, all mechanics variables can be calculated.

## 2.6 CLOSURE

The logic shown in Figure 2.7 emphasizes that the *kinematic equations* between displacements and strains, the *constitutive equations* between strains and stresses, the *static equivalency* equations between stresses and internal forces/moments, and the *equi-*

*librium equations* between internal forces/moments and external forces/moments can be treat as independent modules, permitting one to add complexities to one module without changing the fundamental equations in the other modules. Thus, dynamics effects and elastic foundation effects change the equilibrium equations without changing the kinematics equations, constitutive equations, and equivalency equations. In the Timoshenko beam, we changed the kinematic equations but not the equations in other modules. In composite laminated structural members, we drop the assumption of material homogeneity across the cross section that we make in the module of static equivalency, but there are no other changes. To incorporate material non-linearity or inelastic strains, we drop the assumptions related to Hooke's law, thus changing the constitutive equations but nothing else. In unsymmetric bending, we drop the restrictions on the geometry of the cross sections of symmetric beams, thus changing the kinematic equations without changing the fundamental equations in other modules.

In Chapter 5 on Thin Plates, we will follow a similar path of developing first the simplest possible theory using logic shown in Figure 2.7 and then add complexities one at a time. In Chapter 7 on Energy Methods, we replace the entire module of equilibrium equations with energy method equations but do not change the equations of the other modules. An understanding of the modular character of the logic, and how complexities are incorporated not only will help you with the remaining chapters in this book but will also give you the capability of adding complexities to structures not covered here.

In this chapter we also saw the discontinuity functions and their applications in solving boundary value problems. In Chapter 3 on *Influence Functions* we will use these discontinuity functions for solving the boundary value problems with more complex loadings and beams on elastic foundations.

## 2.7 SYNOPSIS OF EQUATIONS

| | Axial | Symmetric Bending | Torsion |
|---|---|---|---|
| Displacements | $u = u_o(x)$ | $v = v(x) \qquad u = -y\dfrac{dv}{dx}$ | $\phi = \phi(x)$ |
| Strain | $\varepsilon_{xx} = \dfrac{du_o}{dx}(x)$ | $\varepsilon_{xx} = -y\dfrac{d^2v}{dx^2}(x)$ | $\gamma_{x\theta} = \rho\dfrac{d\phi}{dx}(x)$ |
| Stress | $\sigma_{xx} = E\dfrac{du_o}{dx}(x)$ | $\sigma_{xx} = -Ey\dfrac{d^2v}{dx^2}(x)$ | $\tau_{x\theta} = G\rho\dfrac{d\phi}{dx}(x)$ |
| Internal forces and moments | $N = \int_A \sigma_{xx}dA$<br>$M_y = 0 \qquad M_z = 0$ | $N = 0$<br>$M_z = -\int_A y\sigma_{xx}dA \qquad V_y = \int_A \tau_{xy}dA$ | $T = \int_A \rho\tau_{x\theta}dA$ |
| Origin | $\int_A ydA = 0$ | $\int_A ydA = 0$ | |
| Deformation formulas | $\dfrac{du_o}{dx} = \dfrac{N}{EA}$<br>$u_2 - u_1 = \dfrac{N(x_2 - x_1)}{EA}$ | $\dfrac{d^2v}{dx^2} = \dfrac{M_z}{EI_{zz}}$ | $\dfrac{d\phi}{dx} = \dfrac{T}{GJ}$<br>$\phi_2 - \phi_1 = \dfrac{T(x_2 - x_1)}{GJ}$ |
| Stress formulas | $\sigma_{xx} = \dfrac{N}{A}$ | $\sigma_{xx} = -\left(\dfrac{M_z y}{I_{zz}}\right) \qquad \tau_{xs} = -\left(\dfrac{V_y Q_z}{I_{zz}t}\right)$ | $\tau_{x\theta} = \dfrac{T\rho}{J}$ |
| Equilibrium | $\dfrac{dN}{dx} = -p_x(x)$ | $\dfrac{dV_y}{dx} = -p_y(x) \qquad \dfrac{dM_z}{dx} = -V_y$ | $\dfrac{dT}{dx} = -t(x)$ |
| Differential equations | $\dfrac{d}{dx}\left(EA\dfrac{du_o}{dx}\right) = -p_x(x)$ | $\dfrac{d^2}{dx^2}\left(EI_{zz}\dfrac{d^2v}{dx^2}\right) = p_y(x)$ | $\dfrac{d}{dx}\left(GJ\dfrac{d\phi}{dx}\right) = -t(x)$ |
| Boundary conditions at each end | $u_0$ or $N$ | $\phi$ or $T$ | ($v$ or $V_y$) and $\left(\dfrac{dv}{dx}\text{ or }M_z\right)$ |
| Torsion non-circular bars | $T = \int_A [y\tau_{xz} - z\tau_{xy}]dA$ | | |
| Discontinuity functions | $\dfrac{d\langle x-a\rangle^{-1}}{dx} = \langle x-a\rangle^{-2}$<br>$\int\langle x-a\rangle^{-2}dx = \langle x-a\rangle^{-1}$ | $\dfrac{d\langle x-a\rangle^{0}}{dx} = \langle x-a\rangle^{-1}$<br>$\int\langle x-a\rangle^{-1}dx = \langle x-a\rangle^{0}$ | $\dfrac{d\langle x-a\rangle^{n}}{dx} = n\langle x-a\rangle^{n-1}$ for $n \ge 1$<br>$\int\langle x-a\rangle^{n-1}dx = \dfrac{\langle x-a\rangle^{n}}{n}$ for $n \ge 1$ |

# PROBLEMS

In problems below: $A$ is the cross-sectional area; $I_{yy}$, $I_{zz}$, and $I_{yz}$ are the second area moment of inertias of the cross section; $J$ is the polar moment of the cross section; $E$ is modulus of elasticity; $G$ is the shear modulus of elasticity; ν is the Poisson's ration; $\rho_m$ is the mass density of the material; α is the coefficient of thermal expansion; displacements, internal forces and moments, and external distributed forces are as described in Section 2.3.2. All assumptions in Table 2.1 are valid except one or two that are part of the problem.

## Section 2.1

**2.1** The circular cross-section with center at $O$ shown in Fig. P2.1 has a uniform thickness $t$. Assume $t \ll a$. The shear stress in the cross section was found to be $\tau_{x\theta} = K(\cos\theta - \cos\alpha)$. (a) Replace the shear stress by equivalent shear forces $V_y$, $V_z$ and torque $\boldsymbol{T}$ acting at point O. (b) Determine the coordinates of the shear center. Solve the problem in terms variables $K$, $a$, $t$, and $\alpha$.

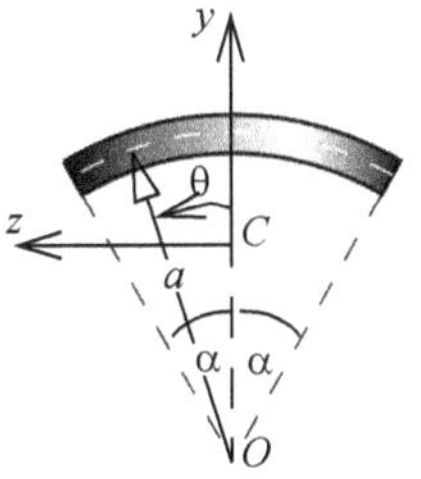


**Fig. P2.1**

**2.2** The cross-section shown in Fig. P2.2 has a uniform thickness t. Assume $t \ll a$. The shear stresses in the cross section were found to be

$$\tau_{xy} = 0 \qquad \tau_{xz} = K[as_1 - s_1^2]/2 \qquad 0 \le s_1 < a$$
$$\tau_{xy} = 0 \qquad \tau_{xz} = 0 \qquad 0 < s_2 < 2a$$
$$\tau_{xy} = 0 \qquad \tau_{xz} = K[2as_3 - s_3^2]/2 \qquad 0 < s_3 \le 2a$$

(a) Replace the shear stresses by equivalent shear forces $V_y$, $V_z$ and torque $\boldsymbol{T}$ acting at the centroid C. (b) Determine the coordinates of the shear center. Solve the problem in terms variables $K$, $a$, and $t$.

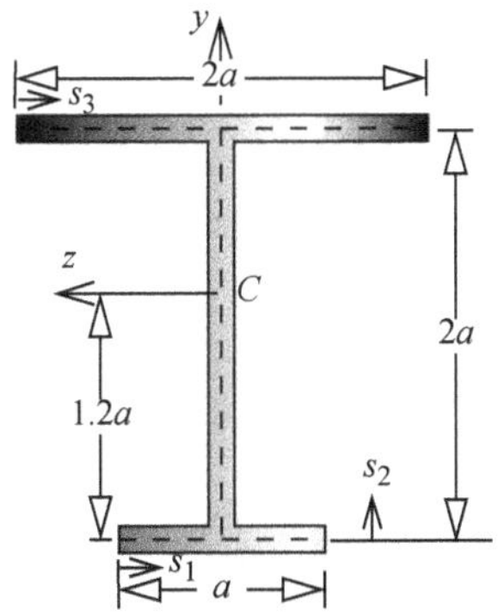


**Fig. P2.2**

## Sections 2.2-2.3

**2.3** The tapered bar shown in Fig. P2.3 has a cross-sectional area that varies with $x$ as $A = K(4L - 3x)$. Determine the elongation of the bar in terms of $P$, $L$, $E$, and $K$.

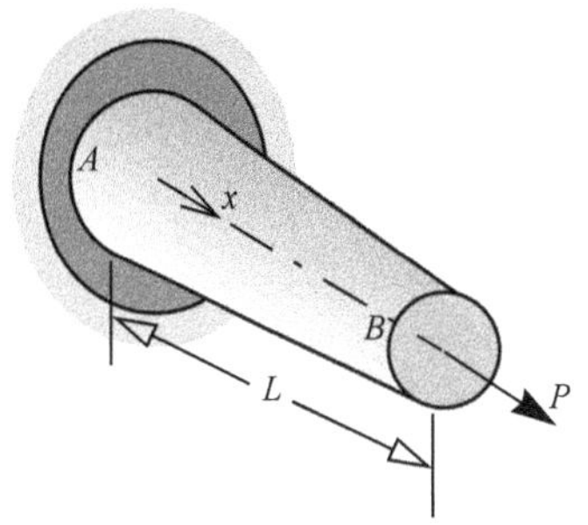


**Fig. P2.3**

**2.4** The radius of the tapered shaft shown in Fig. P2.4 varies as $R = (r/L)(2L - 0.25x)$. In terms of $T_{ext}$, $L$, $G$, and $r$, determine (a) the rotation of section at $B$ and (b) the maximum shear stress in the shaft.

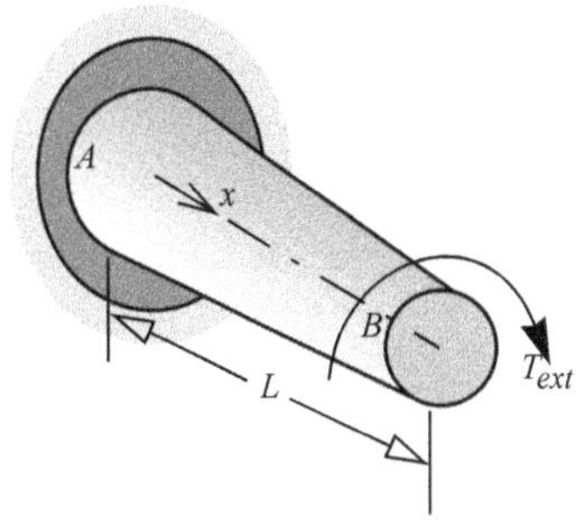


**Fig. P2.4**

**2.5** The displacements $u$, v, and $w$ of a point on a cross section of a noncircular shaft are

$$u = \psi(y,z)\frac{d\phi}{dx} \qquad v = -xz\frac{d\phi}{dx} \qquad w = xy\frac{d\phi}{dx}$$

where $d\phi/dx$ is the rate of twist and is considered to be *constant*, and $\psi(x, y)$, called the warping function, describes the movement of points out of the plane of the cross section.

Show that the shear stresses for the noncircular shaft are given by

$$\tau_{xy} = G\left(\frac{\partial\psi}{\partial y} - z\right)\frac{d\phi}{dx} \qquad \tau_{xz} = G\left(\frac{\partial\psi}{\partial z} + y\right)\frac{d\phi}{dx}$$

**2.6** The displacements for a shaft with an elliptical cross section under torsion are

$$u = \phi Kyz \qquad v = -\phi xz \qquad w = \phi xy$$

where the constant $K$ is related to the semi-axis of the ellipse $a$-$b$ as $K = (b^2 - a^2)/(b^2 + a^2)$. The angle of twist per unit length $\phi$ may be treated as constant for this problem. The axis of the shaft is in the $x$ direction. Determine the relationship between the internal torque $\boldsymbol{T}$ and $\phi$ on a *cross section* in terms of the shear modulus $G$, $a$, and $b$. Also find the stress components in terms of the internal torque $\boldsymbol{T}$.

**2.7** The displacement in the $x$ direction of a beam cross section is given by $u(x) = u_0 - y(dv/(dx))$. Assume that the Assumptions 3 to 8 in Table 2.1 are valid show that

$$N = EA\frac{du_0}{dx} - EAy_c\frac{d^2v}{dx^2} \quad \text{and} \quad M_z = -EAy_c\frac{du_0}{dx} + EI_{zz}\frac{d^2v}{dx^2}$$

where $y_c$ is the coordinate of the centroid of the cross section measured from some arbitrary origin. If $y$ is measured from the centroid of the cross section, show

$$\sigma_{xx} = \frac{N}{A} - \frac{M_z y}{I_{zz}}$$

**2.8** In unsymmetric bending the limitation that cross section has a plane of symmetry and the loading is in the plane of symmetry is dropped.

The displacement in the $x$ direction of a beam cross section is given by $u(x) = -y(dv/dx) - z(dw/dx)$. Assume there is no axial force, $y$ and $z$ are measured from the centroid of the cross section. Show that

$$\sigma_{xx} = -\left(\frac{I_{yy}M_z - I_{yz}M_y}{I_{yy}I_{zz} - I_{yz}^2}\right)y - \left(\frac{I_{zz}M_y - I_{yz}M_z}{I_{yy}I_{zz} - I_{yz}^2}\right)z$$

**2.9** For unsymmetric bending of beams described in problem 2.8 show that the differential equation for untapered beams is as given below.

$$\frac{d^4v}{dx^4} = \frac{1}{E}\left(\frac{I_{yy}p_y - I_{yz}p_z}{I_{yy}I_{zz} - I_{yz}^2}\right) \qquad \frac{d^4w}{dx^4} = \frac{1}{E}\left(\frac{I_{zz}p_z - I_{yz}p_y}{I_{yy}I_{zz} - I_{yz}^2}\right)$$

**2.10** For unsymmetric bending of beams, using the results in problem 2.8, show that for thin untapered beam the shear stress is as given below.

$$\tau_{sx}t = -\left(\frac{I_{yy}Q_z - I_{yz}Q_y}{I_{yy}I_{zz} - I_{yz}^2}\right)V_y - \left(\frac{I_{zz}Q_y - I_{yz}Q_z}{I_{yy}I_{zz} - I_{yz}^2}\right)V_z$$

**2.11** Figure 2.9 on page 44 shows a laminated composite cross section of an axial rod. Show the axial stress in the i$^{th}$ material $(\sigma_{xx})_i$ and the differential equation governing the displacement $u_o(x)$ of the rod are as shown below.

$$(\sigma_{xx})_i = \frac{NE_i}{\sum_{j=1}^{n} E_jA_j} \qquad \frac{d}{dx}\left[\left(\sum_{j=1}^{n} E_jA_j\right)\frac{du_o}{dx}\right] = -p_x(x)$$

**2.12** Fig. P2.12 shows a laminated shaft in which all materials are securely bonded together. Show the torsional shear stress in the i$^{th}$ material $(\tau_{x\theta})_i$ and the differential equation governing the shaft rotation $\phi(x)$ are as shown below.

$$(\tau_{x\theta})_i = \frac{G_i\rho T}{\sum_{j=1}^{n} G_jJ_j} \qquad \frac{d}{dx}\left[\left\{\sum_{j=1}^{n} G_jJ_j\right\}\frac{d\phi}{dx}\right] = -t(x)$$

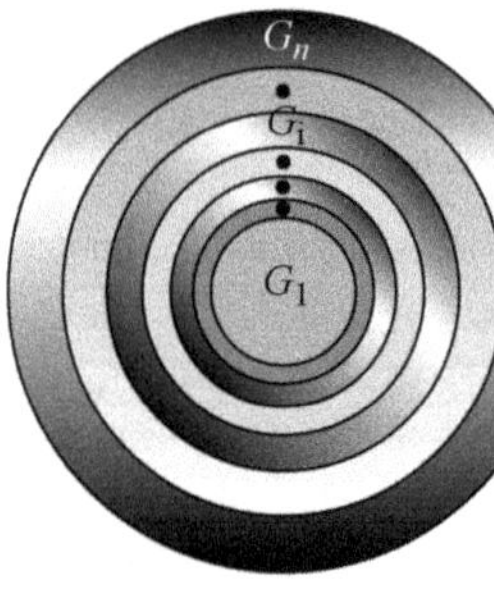


**Fig. P2.12**

**2.13** A symmetric beam is subjected to transverse load $p_y$ and its temperature is increased by $\Delta T(x, y)$. The temperature distribution is such that it produces no axial force at a cross section. Starting with the displacement in Equation (2.5a-B), show the bending normal stress $\sigma_{xx}$ and the differential equation governing the deflection of the beam are

$$\sigma_{xx} = -\frac{(M_z - m_T)y}{I_{zz}} - E\alpha\Delta T \qquad \frac{d^2}{dx^2}\left[EI_{zz}\frac{d^2v}{dx^2}\right] = p_y(x) - \frac{d^2m_T}{dx^2}$$

where, $m_T = E\alpha\int_A y\Delta T(x, y)dA$ is the thermal moment.

**2.14** Fig. P2.14 shows a laminated cross section of an axial member that is subjected to a uniform temperature change across the cross section $\Delta T = T_0$. Assume $E_1 = 2E$, $E_2 = E$, $\alpha_1 = \alpha$, and $\alpha_2 = 2\alpha$. Determine the maximum axial stress on the cross section in both materials in terms $N$, $E$, $\alpha$, $h$, and $T_0$.

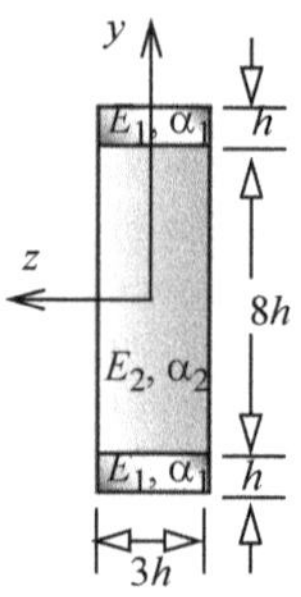


**Fig. P2.14**

**2.15** The axial rod shown in Fig. P2.15 has the cross section and material properties given in problem 2.14. Assume the change in temperature $T_0$ is uniform across the length and the cross section. Determine the maximum axial stress in both materials in terms $E$, $\alpha$, $h$, and $T_0$

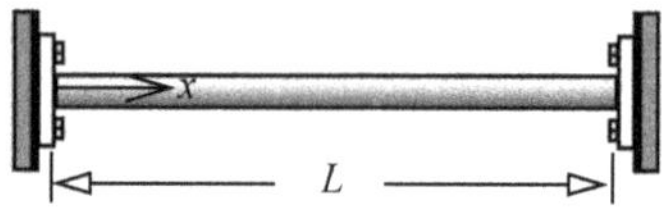


**Fig. P2.15**

**2.16** Fig. P2.14 shows a laminated cross section of a symmetric beam that is subjected to a linearly varying temperature change across the cross section $\Delta T = T_0(y/h)$. Assume $E_1 = 2E$, $E_2 = E$, $\alpha_1 = \alpha$, and $\alpha_2 = 2\alpha$. Determine the magnitude of the maximum bending normal stress in both materials in terms $M_z$, $E$, $\alpha$, $h$, and $T_0$.

**2.17** The solid, elastic–perfectly plastic shaft shown in Fig. P2.17 has a plastic zone 0.5 inch deep in section $AB$. The shear yield stress of the shaft is 30,000 psi, and a shear modulus of 12,000 ksi. Determine (a) the magnitude of the applied torque $T_{ext}$, (b) the rotation of section $C$..

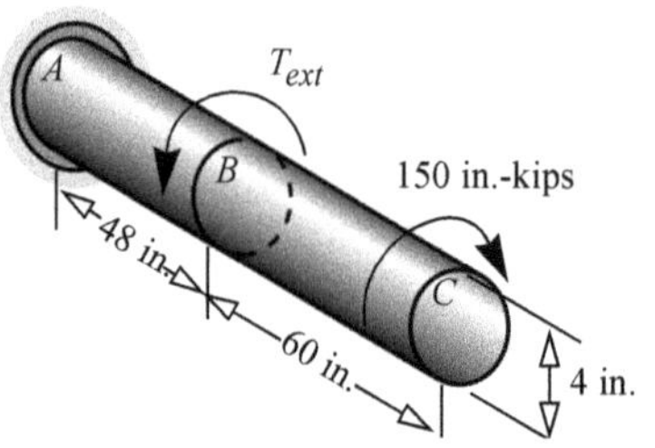


**Fig. P2.17**

**2.18** A beam made from an elastic–perfectly plastic material has a yield stress of 30 ksi. Determine the internal bending moment if the elastic–plastic boundary is 0.5 inch from the top and bottom on the rectangular cross section shown in Fig. P2.18

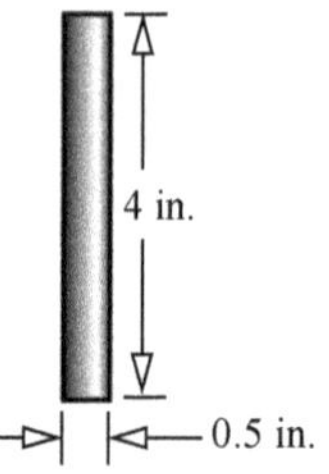


**Fig. P2.18**

**2.19** A beam resting on an elastic foundation has a distributed spring force that depends upon the deflections at a point. Show the differential equation governing the deflection of the beam is:

$$\frac{d^2}{dx^2}\left(EI_{zz}\frac{d^2v}{dx^2}\right) + kv = p_y$$

where $k$ is the spring constant per unit length called the foundation modulus.

**2.20** Determine the elongation of the rotating bar shown in Fig. P2.20 in terms of the rotating speed $\omega$, density $\rho_m$, length $L$, and cross-sectional area $A$.

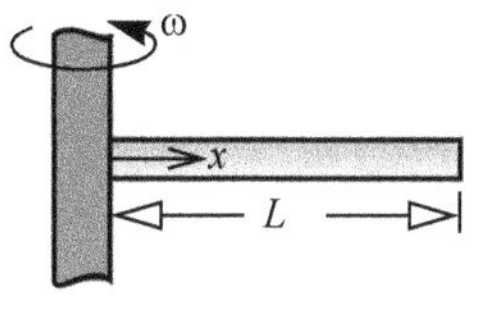


**Fig. P2.20**

**2.21** Derive the *wave equation* below by considering the dynamic equilibrium of an axial rod's differential element

$$\frac{\partial^2 u}{\partial t^2} = c^2\left(\frac{\partial^2 u}{\partial x^2}\right) \qquad \text{where} \qquad c = \sqrt{E/\rho_m} \qquad \textbf{(2.39-A)}$$

$c$ is the velocity of sound propagation in the material.

**2.22** Show that the functions $f(x - ct)$ and $g(x + ct)$ satisfy Equation (2.39-A).

**2.23** Show the following solution satisfies Equation (2.17d-B).

$$v(x, t) = G(x)H(t)$$
$$G(x) = A\cos(\omega x) + B\sin(\omega x) + C\cosh(\omega x) + D\sinh(\omega x)$$
$$H(t) = E\cos(c\omega^2)t + D\sin(c\omega^2)t$$

**2.24** Derive the differential equation for torsional vibration of shafts below by considering the dynamic equilibrium of the differential element of a circular shaft.

$$\frac{\partial^2 \phi}{\partial t^2} = c^2\frac{\partial^2 \phi}{\partial x^2} \qquad \text{where} \qquad c = \sqrt{G/\rho_m} \qquad \textbf{(2.39-T)}$$

**2.25** Show the following solution satisfies Equation (2.39-T)

$$\phi = [A\cos(\omega x/c) + B\sin(\omega x/c)][C\cos(\omega t) + D\sin(\omega t)]$$

where $A$, $B$, $C$, and $D$ are constants to be determined from the boundary conditions and the initial conditions, and $\omega$ is the frequency of vibration.

**2.26** The strain displacement relationship for a large axial strain is given by $\varepsilon_{xx} = du_0/dx + (1/2)(du_0/dx)^2$. All assumptions except Assumption 3 in Table 2.1 are valid. Show that

$$\frac{du_0}{dx} = \sqrt{1 + \frac{2N}{EA}} - 1 \qquad \sigma_{xx} = \frac{N}{A}$$

**2.27** A circular solid shaft of radius $R$ is made from a nonlinear material that has a shear stress–shear strain relationship given by $\tau = G\gamma^{0.4}$. All assumptions except Assumption 7 in Table 2.1 are valid. Show that the maximum shear stress and relative rotation are as given below.

$$\tau_{max} = 0.5411\frac{T}{R^3} \qquad \phi_2 - \phi_1 = 0.2154(x_2 - x_1)\left(\frac{T}{GR^{3.4}}\right)^{2.5}$$

**2.28** Determine the magnitude of the maximum torsional shear stress and rotation of section at $C$ for the shaft and loading shown in Fig. P2.28. The shaft cross section and material are the same as in problem 2.27.Use $G = 100$ MPa.

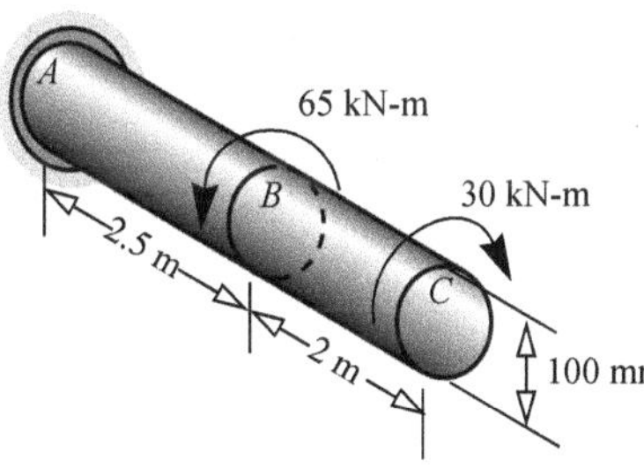


**Fig. P2.28**

**2.29** The beam cross section shown in Figure 2.33 is made from a material that has a stress–strain curve given by $\sigma = 400\varepsilon^{0.4}$ ksi. Determine (a) the location of the neutral axis and (b) the bending normal stress in terms of $y$ and the internal moment $M_z$.

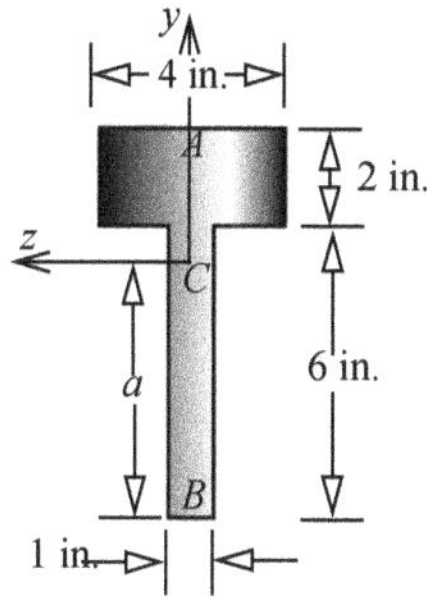


**Fig. P2.29**

**2.30** Determine the maximum bending normal stress for the two beams and loading shown in Figure 2.30. The beam cross section and material are the same as in problem 2.29.

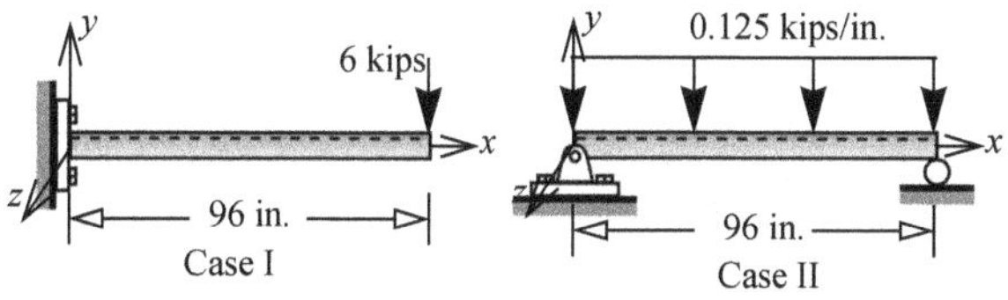


**Fig. P2.30**

**2.31** A composite circular shaft cross section is made from two nonlinear materials as shown in Fig. P2.31 The stress–strain relationships for materials 1 and 2 are given by $\tau_1 = G\gamma^{0.5}$ and $\tau_2 = 2G\gamma^{0.5}$. Determine the maximum torsional shear stress in both materials and rotation of the section at $B$ in terms of $T_{ext}$, $L$, $G$, and $R$.

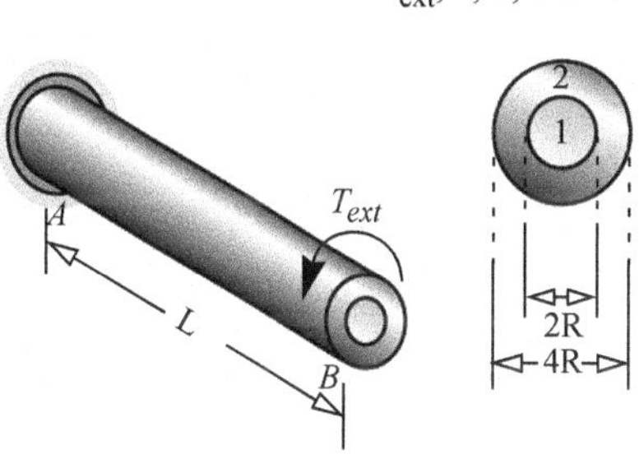


**Fig. P2.31**

**2.32** A composite beam cross section is made from two non-linear materials as shown in Fig. P2.32. The stress–strain relationships for materials 1 and 2 are given by $\sigma_1 = E\varepsilon^{0.5}$ and $\sigma_2 = 2E\varepsilon^{0.5}$.Determine the magnitude of the maximum bending normal stress in both materials in terms of $M_z$, $E$, and $h$.

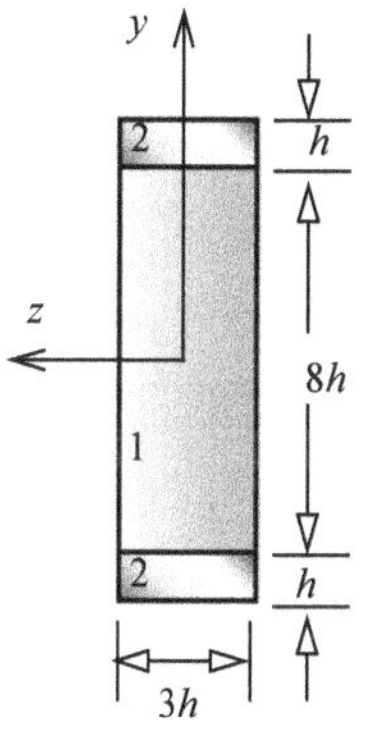


**Fig. P2.32**

## Section 2.4

**2.33** The frictional force per unit length on a cast iron pipe being pulled from the ground varies quadratically as shown in Fig. P2.33. Determine the force $F$ needed to pull the pipe out of ground and the elongation of the pipe before the pipe slips in terms of the modulus of elasticity $E$, cross-sectional area $A$, length $L$, and maximum value of frictional force $f_{max}$.

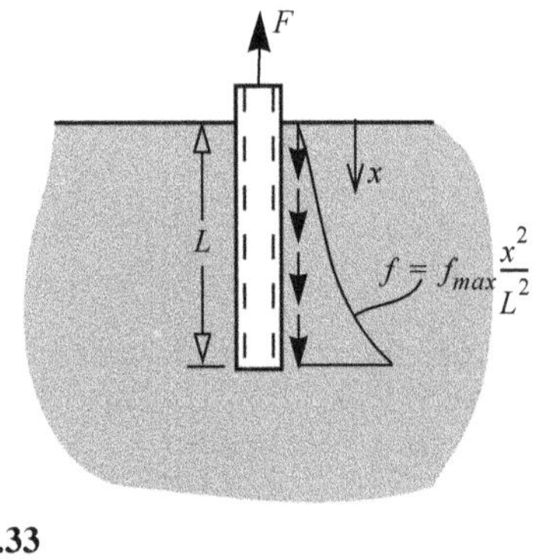


**Fig. P2.33**

**2.34** The external torque on the drill bit shown in Fig. P2.34 varies quadratically to a maximum intensity of $q(x^2/L^2)$ in.-lb/in, as shown in Fig. P2.34. If the drill bit diameter is $d$, its length $L$, and modulus of rigidity $G$, determine (a) the maximum shear stress on the drill bit and (b) the relative rotation of the end of the drill bit with respect to the chuck.

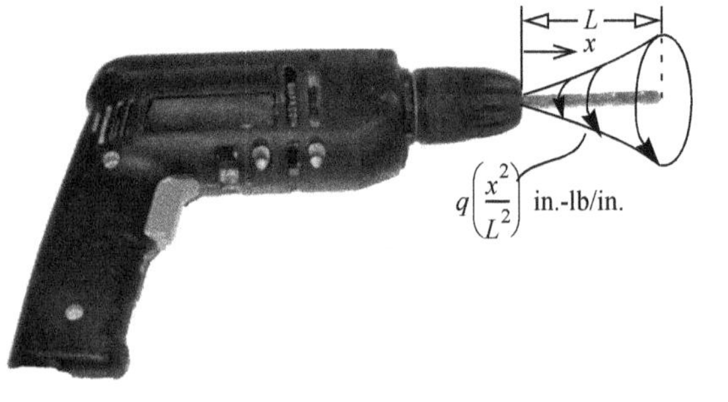


**Fig. P2.34**

**2.35** In terms of $w$, $L$, $E$, and $I$, determine the deflection and slope at $x = L$ of the beam shown in Fig. P2.35.

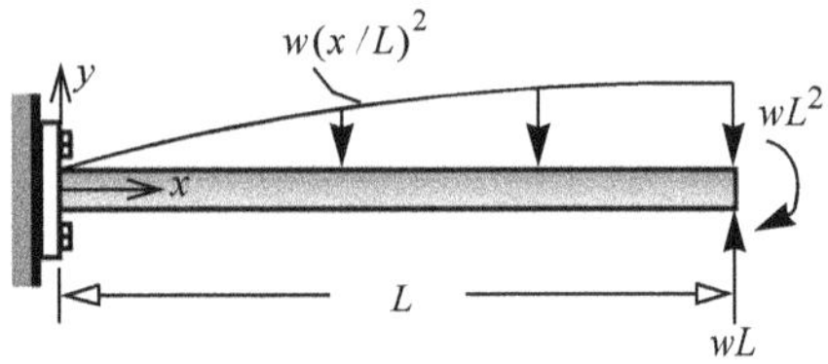


**Fig. P2.35**

**2.36** Determine the reaction and moment at the left wall on the beam shown in Fig. P2.36, and also the slope at $x = L$.

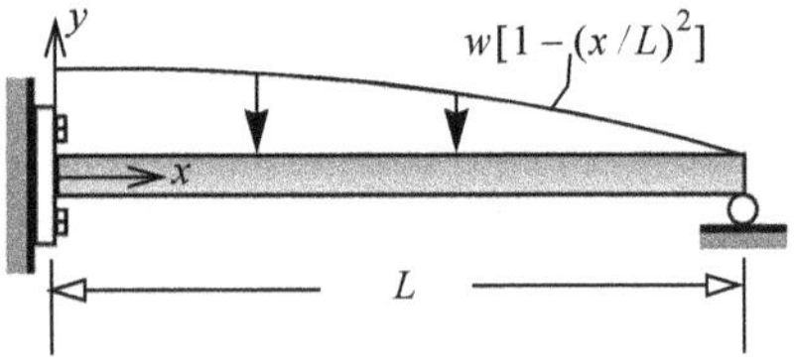


**Fig. P2.36**

**2.37** Determine the deflection and moment reaction at $x = L$ in terms of $E$, $I$, $w$, and $L$ for the beam shown in Fig. P2.37.

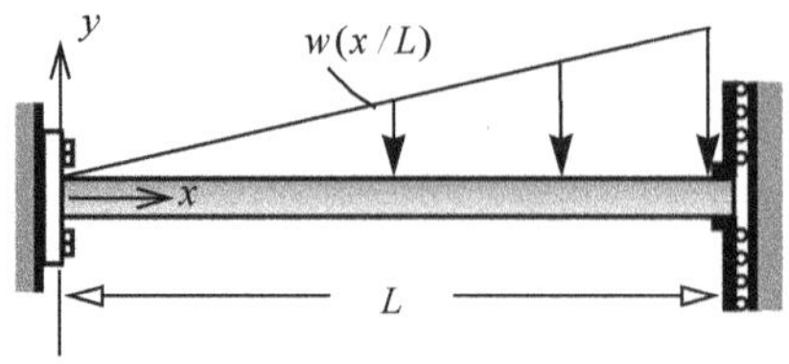


**Fig. P2.37**

**2.38** Determine the elastic curve and the reactions at $A$ for the beam shown in Fig. P2.38.

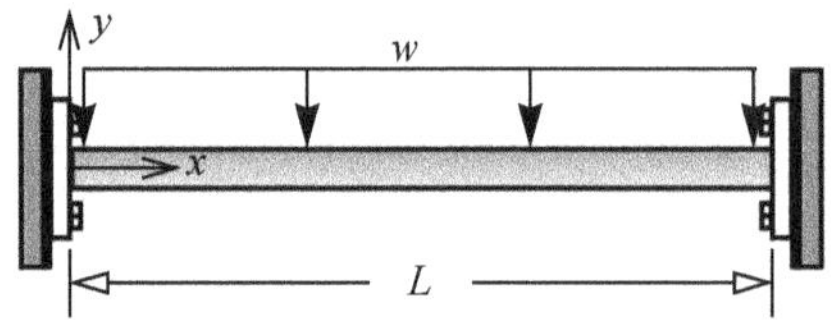


**Fig. P2.38**

**2.39** A linear spring that has a spring constant $K$ is attached to a beam at one end as shown in Fig. P2.39. In terms of $w$, $E$, $I$, $L$, and $K$. Use the discontinuity functions to write the boundary value problem, but do not integrate or solve.

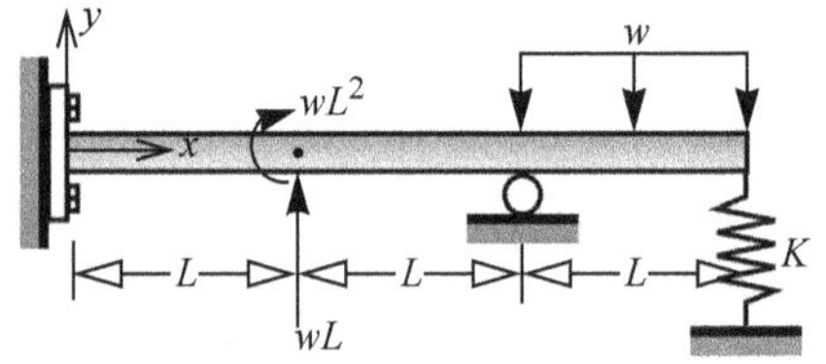


**Fig. P2.39**

## Section 2.5

**2.40** Fig. P2.40 shows a rectangular steel ($E$ = 30,000 ksi, ν = 0.25) bar of 0.5 in thickness, there is a gap of 0.01 in between the section at $D$ and a rigid wall before the forces are applied. Assuming that the applied forces are sufficient to close the gap, determine (a) the movement of section $C$ with respect to the left wall and (b) the change in the depth $d$ of segment $CD$. Use the discontinuity function to solve the problem.

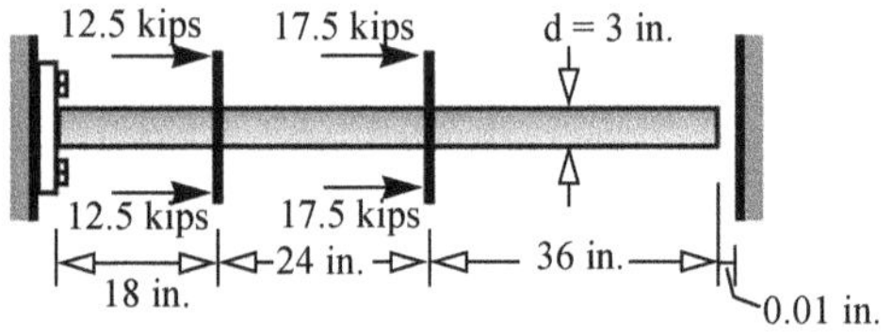


**Fig. P2.40**

**2.41** A solid circular steel ($G$ = 12,000 ksi, $E$ = 30,000 ksi) shaft of 4-inch diameter is loaded as shown in Fig. P2.41. Determine the maximum shear stress in the shaft. Use the discontinuity functions to solve the problem.

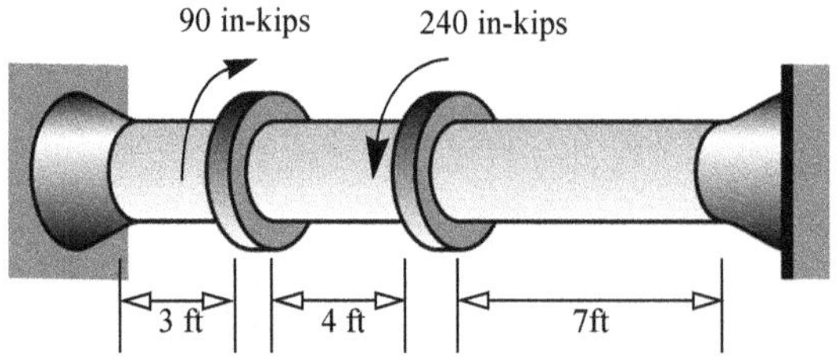


**Fig. P2.41**

**2.42** A uniform distributed torque of $q$ in . lb/in is applied to the entire shaft as shown in Fig. P2.42. In addition to the distributed torque, a concentrated torque of $T = 3\,qL$ in . lb is applied at section $B$. Let the shear modulus be $G$ and radius of the shaft be $r$. Determine, in terms of $q$, $L$, $G$, and $r$, (a) the rotation of section at $B$ and (b) the magnitude of maximum torsional shear stress in the shaft.

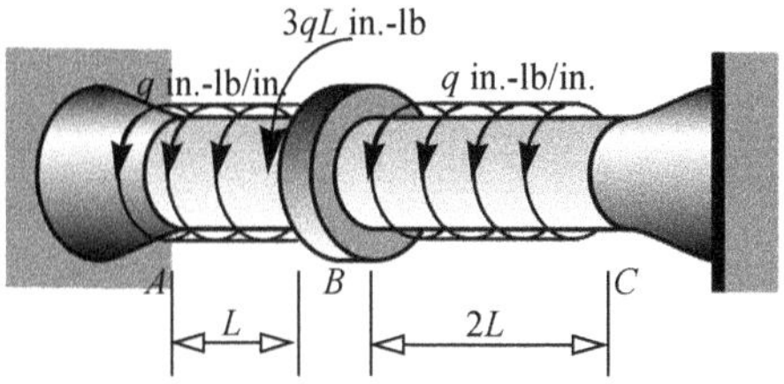


**Fig. P2.42**

**2.43** Determine the elastic curve and the reactions at $A$ for the beam shown in Fig. P2.43.

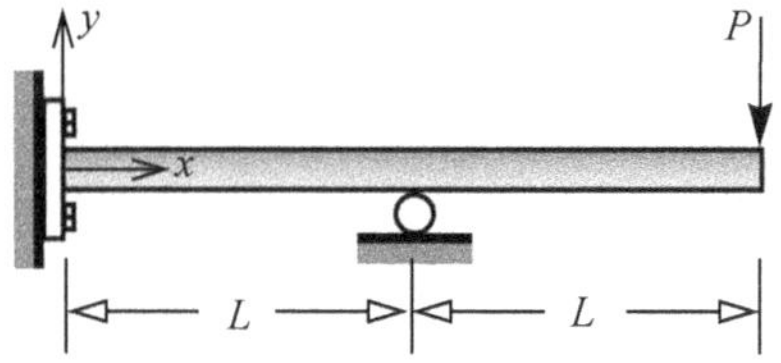


**Fig. P2.43**

**2.44** For the beam and loading shown in Fig. P2.44, in terms of $w$, $L$, $E$, and $I$, determine (a) the equation of the elastic curve and (b) the deflection at $x = L$.

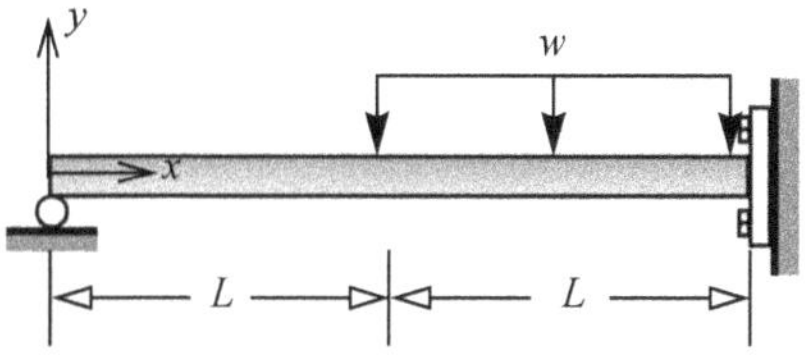


**Fig. P2.44**

**2.45** For the beam and loading shown in Fig. P2.45, in terms of $w$, $L$, $E$, and $I$, determine (a) the equation of the elastic curve and (b) the deflection at $x = L$.

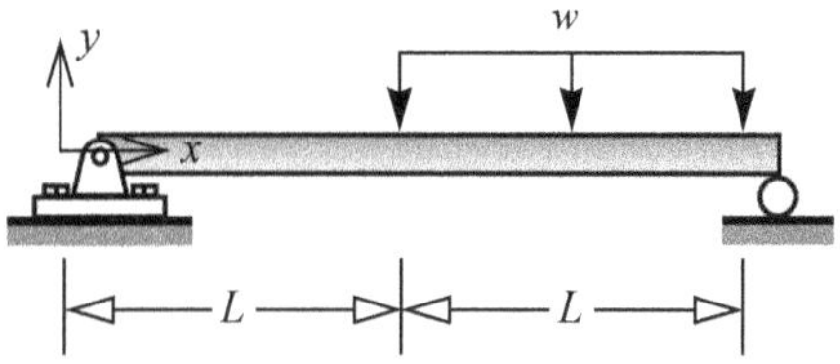


**Fig. P2.45**

**2.46** For the beam and loading shown in Fig. P2.46, in terms of $w$, $L$, $E$, and $I$, determine (a) the equation of the elastic curve and (b) the deflection at $x = L$.

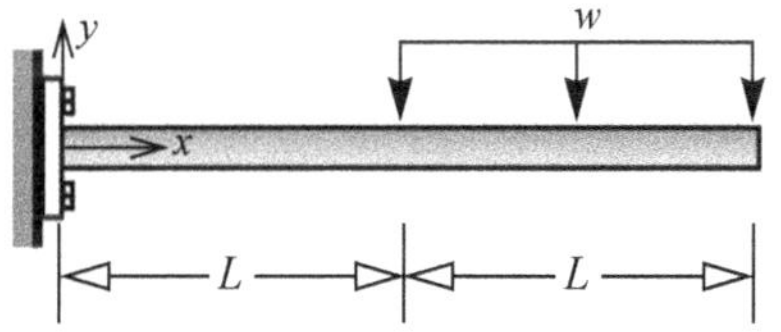


**Fig. P2.46**

**2.47** Determine the deflection of the beam at point $C$ in terms of $E$, $I$, $w$, and $L$ for the beam shown in Fig. P2.47.

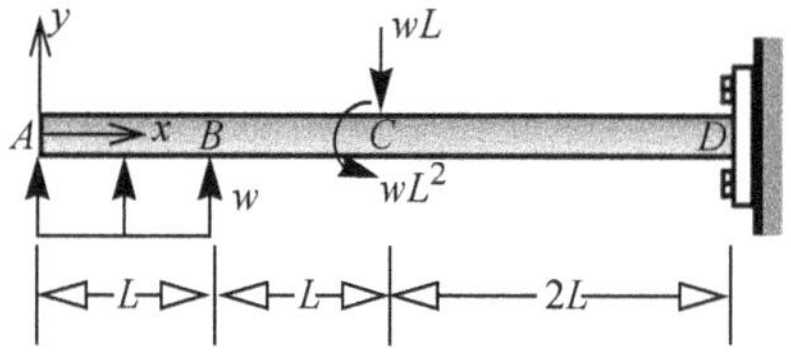


**Fig. P2.47**

**2.48** Determine the deflection of the beam at point $C$ in terms of $E$, $I$, w, and $L$ for the beam shown in Fig. P2.48.

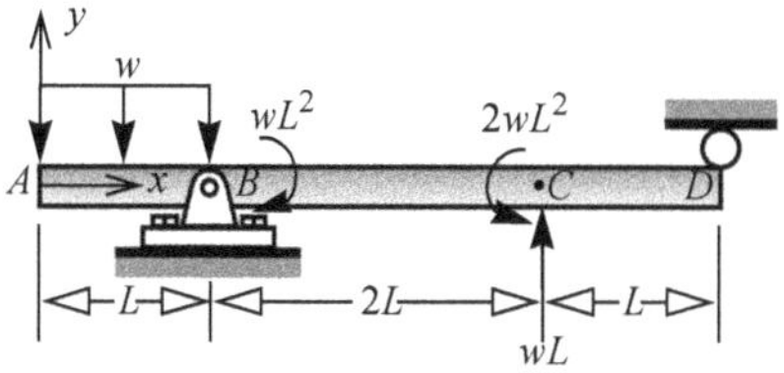


**Fig. P2.48**

# 3 Influence Functions

## LEARNING OBJECTIVE

1. Understand the concept of influence functions and its applications to classical beams and beams on elastic foundations.

*Green's function* was the name given to the function George Green introduced in 1828 in the solution of Poisson's equation governing the potential of an electric field. As the applications of Green's concept grew, the name *Green's function* was replaced by *influence function*. *Fundamental solution* is the name used for influence function when the region of interest is infinite. Fundamental solutions are used in the formulation of *boundary integral equations* as a solution to boundary value problems. The numerical solution of boundary integral equations is called the *Boundary Element Method*, or *BEM*.

In this chapter we develop the concept of influence functions and fundamental solutions and apply it to classical beams and beams on elastic foundations with arbitrary loading.

## 3.1 BASIC CONCEPTS

Definitions and the mathematical equations for solving problems of classical beams and beams on elastic foundations are introduced in this section.

### 3.1.1 Definitions

A **source point** ($\xi$) is a point in the material at which a disturbance is placed. A **field point** ($x$) is a point in the material where the impact of disturbance is evaluated. **Influence function** $\mathcal{G}(x, \xi)$ relates a value of a variable at the field point ($x$) to the unit value of the disturbance placed at the source point ($\xi$). In beam bending applications, the points at which we place concentrated force or moment are the source points, and points at which we evaluate beam deflection, slope, shear force, and bending moment are the field points.

Influence functions associated with infinite bodies are called **fundamental solutions**. The influence function is said to be **singular** if it becomes infinite at the source point. The disturbance associated with a singular influence function is called a **singularity**. The most common representation of singularity is the delta function. Figure 2.21 shows singularity in the delta function. When the singularity is represented by the delta function it is called the **source singularity**. A **doublet singularity** is created by placing two source singularities of equal and opposite magnitude at a distance that shrinks to zero while the source strength tends to infinity such that the product of source strength and distance results in a finite value. Point force is a source singularity and point moment is a doublet singularity.

In the next section we develop equations and notations that are applicable to classical beams, infinite and finite beams on elastic foundations which we will study later.

### 3.1.2 Force (source) singularity influence functions in beams

The forcing function on the right hand side of the differential equation is replaced by a delta function and the boundary value problem is solved to obtain the influence function in an application. The differential equation governing symmetric beam deflec-

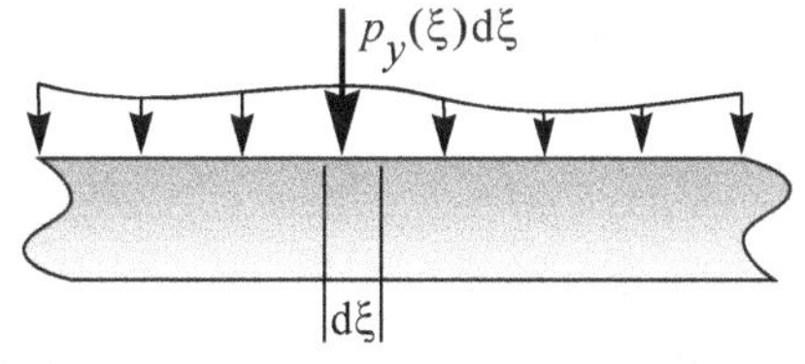


**Figure 3.1** Force acting on a differential length of a beam.

tion is a fourth order ordinary differential equation. The forcing function is the distributed load $p_y$ in the $y$-direction. We replace $p_y$ by the delta function and solve the boundary value problem to obtain the influence function $\mathcal{G}(x, \xi)$ that represents the deflection at field point $x$ due to a unit value of concentrated force placed at source point $\xi$. Multiplying the influence function by the magnitude of force $P$ gives us the displacement v(x) due to force $P$ as

$$\boxed{\mathrm{v}(x) = \mathcal{G}(x, \xi)P} \tag{3.1}$$

In beam bending, $p_y(\xi)d\xi$ is the force acting on the differential length at source point $\xi$ as shown in Figure 3.1. By distributing the force over the length of the beam we obtain the deflection at any point $x$ as

$$\boxed{\mathrm{v}(x) = \int_{Length} \mathcal{G}(x, \xi)p_y(\xi)d\xi} \tag{3.2a}$$

We obtain the slope, moment, and shear force by successive differentiation of Equation (3.2a) as shown below.

$$\boxed{\psi(x) = \frac{d\mathrm{v}}{dx} = \int_{Length} \mathcal{G}_1(x, \xi)p_y(\xi)d\xi \qquad \text{where} \qquad \mathcal{G}_1(x, \xi) = \frac{\partial \mathcal{G}}{\partial x}} \tag{3.2b}$$

$$\boxed{M_z(x) = EI\frac{d^2\mathrm{v}}{dx^2} = \int_{Length} \mathcal{G}_2(x, \xi)p_y(\xi)d\xi \qquad \text{where} \qquad \mathcal{G}_2(x, \xi) = EI\frac{\partial \mathcal{G}_1}{\partial x}} \tag{3.2c}$$

$$\boxed{V_y(x) = -\left(\frac{dM_z}{dx}\right) = \int_{Length} \mathcal{G}_3(x, \xi)p_y(\xi)d\xi \qquad \text{where} \qquad \mathcal{G}_3(x, \xi) = -\left(\frac{\partial \mathcal{G}_2}{\partial x}\right)} \tag{3.2d}$$

The influence functions introduced above are the: the slope $\mathcal{G}_1(x, \xi)$, the internal bending moment $\mathcal{G}_2(x, \xi)$, and the internal shear force $\mathcal{G}_3(x, \xi)$ at field point $x$ due to a unit value of force placed at source point $\xi$.

In solving the fourth order differential equation to obtain $\mathcal{G}(x, \xi)$, we have to use the boundary conditions to evaluate the integration constants. The *influence function is valid for a given set of boundary conditions*. In Section 3.2.1 we will obtain $\mathcal{G}(x, \xi)$ for a simply supported beam. We can then use Equation (3.2a) for a simply supported beam with any loading $p_y$. We evaluate the above integrals analytically when $p_y$ is a simple function. When $p_y$ is obtained from experimental data or represented by a complicated function then we will evaluate the integral numerically.

### 3.1.3 Moment (doublet) singularity influence functions in beams

A point moment is a couple of two equal and opposite forces placed at $\xi$ and $\xi$+$\varepsilon$ as shown in Figure 3.2. By superposition we can write Equation (3.1) as:

$$\mathrm{v}(x) = \mathcal{G}(x, \xi)P_1 + \mathcal{G}(x, \xi+\varepsilon)P_2 = -\mathcal{G}(x, \xi)P + \mathcal{G}(x, \xi+\varepsilon)P$$

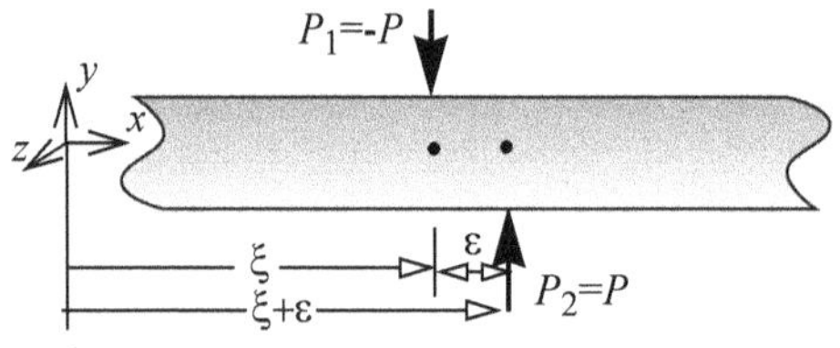


**Figure 3.2** Moment singularity

Expanding the second term by Taylor series we obtain

$$\mathrm{v}(x) = -\mathcal{G}(x, \xi)P + \left[\mathcal{G}(x, \xi) + \frac{\partial \mathcal{G}}{\partial \xi}\varepsilon + \frac{1}{2!}\frac{\partial^2 \mathcal{G}}{\partial \xi^2}\varepsilon^2 + \bullet\ \bullet\right]P = P\varepsilon\left[\frac{\partial \mathcal{G}}{\partial \xi} + \frac{1}{2!}\frac{\partial^2 \mathcal{G}}{\partial \xi^2}\varepsilon + \bullet\ \bullet\right]$$

We now let $\varepsilon \to 0$ and $P \to \infty$ such that $\lim_{\varepsilon \to 0} \lim_{P \to \infty} (P\varepsilon) = M$ to obtain

$$\boxed{\mathrm{v}(x) = M\left[\frac{\partial \mathcal{G}}{\partial \xi}\right] = \mathcal{H}(x, \xi)M \qquad \mathcal{H}(x, \xi) = \frac{\partial \mathcal{G}}{\partial \xi}} \tag{3.3a}$$

Equation (3.3a) is independent of the support conditions and gives the influence function associated with a point moment. Note that $M$ is *positive counterclockwise* with respect to the $z$-axis.

Once more, by successive differentiation of Equation (3.3a), we obtain the slope, the internal bending moment, and the internal shear force as given below

$$\boxed{\psi(x) = \frac{d\mathrm{v}}{dx} = \mathcal{H}_1(x, \xi)M \qquad \text{where} \qquad \mathcal{H}_1(x, \xi) = \frac{\partial \mathcal{H}}{\partial x}} \tag{3.3b}$$

$$\boxed{M_z(x) = EI\frac{d^2\mathrm{v}}{dx^2} = \mathcal{H}_2(x, \xi)M \qquad \text{where} \qquad \mathcal{H}_2(x, \xi) = EI\frac{\partial \mathcal{H}_1}{\partial x}} \tag{3.3c}$$

$$\boxed{V_y(x) = -\left(\frac{\mathrm{d}M_z}{\mathrm{d}x}\right) = \mathcal{H}_3(x,\xi)M \qquad \text{where} \qquad \mathcal{H}_3(x,\xi) = -\left(\frac{\partial \mathcal{H}_2}{\partial x}\right)} \tag{3.3d}$$

The influence functions introduced above are: the displacement $\mathcal{H}(x, \xi)$, the slope $\mathcal{H}_1(x, \xi)$, the internal bending moment $\mathcal{H}_2(x, \xi)$, and the internal shear force $\mathcal{H}_3(x, \xi)$ at field point $x$ due to a unit value of moment placed at source point $\xi$.

### 3.1.4 Numerical integration

The integration in Equation (3.2a) through (3.3d) can be done analytically when the loading $p_y$ is simple, or numerically if the loading is defined by experimental data or represented by a complicated function.
Consider the integral

$$I = \int_a^b f(\xi)d\xi \tag{3.4a}$$

In numerical integration, the above integral is approximated as

$$I \cong \sum_{i=1}^{N} w_i f(\xi_i) \tag{3.4b}$$

where, $\xi_i$ are called the **base points**; $w_i$ are called the **weights**, and $N$ is the number of segments into which the integral interval ($b$-$a$) is divided into. Alternatively stated, the interval ($b$-$a$) has $N$+1 base points and the function $f(\xi)$ is evaluated at these base points. There are many numerical integration schemes that been have developed to reduce the error of approximation. These integration schemes differ from one another in the location of the base points $\xi_i$ and the value of the weights $w_i$. **Gauss-quadrature** is the most popular integration scheme but we will use the simplest numerical integration scheme called the **Trapezoidal rule** to demonstrate the concepts.

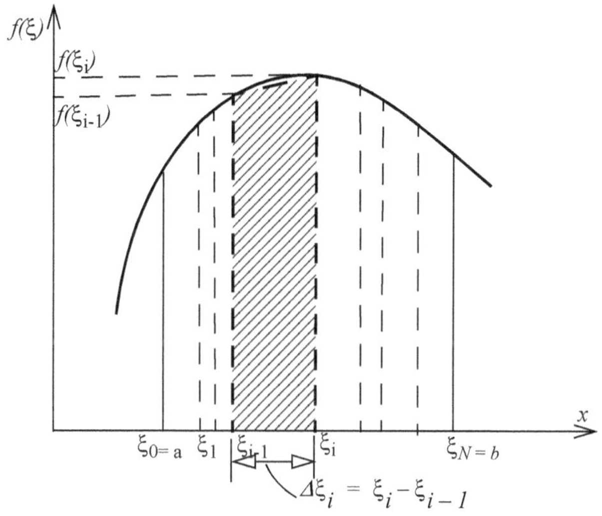


**Figure 3.3** Integration by trapezoidal rule.

The integral in Equation (3.4a) represents the area under the curve defined by the function $f(\xi)$. We can approximate the function by a straight line between two successive base points as shown in Figure 3.3. The straight line approximation of the function creates an area of a trapezoid, hence the name of the rule. The sum of the areas of the trapezoids is the value of the integral as shown below.

$$I \cong \sum_{i=1}^{N} \frac{\Delta\xi_i}{2}[f(\xi_i) + f(\xi_{i-1})] = \frac{\Delta\xi_1}{2} f(\xi_0) + \sum_{i=1}^{N-1} \frac{(\Delta\xi_i + \Delta\xi_{i+1})}{2}[f(\xi_i)] + \frac{\Delta\xi_N}{2} f(\xi_N) \tag{3.4c}$$

Comparing Equations (3.4b) and (3.4c) we see that $w_0 = \Delta\xi_1/2$, $w_N = \Delta\xi_N/2$, and $w_i = (\Delta\xi_i + \Delta\xi_{i+1})/2$ for $i = 1$ to $N$-1. For further simplification, we assume the base points are equally spaced—$\Delta\xi_i = \Delta\xi$ for all $i$'s. Equation (3.4c) simplifies as

$$\boxed{I = \left[\frac{f(\xi_0) + f(\xi_N)}{2} + \sum_{i=1}^{N-1} f(\xi_i)\right](\Delta\xi)} \tag{3.4d}$$

### 3.1.5 Non-dimensional variables

Higher accuracies can be obtained in numerical calculations by use of non-dimensional variables. Non-dimensionalizing also simplifies equations for programming and makes all variables unit less.

Variables with curved bars will refer to the non-dimensional variable. We start with non-dimensionalizing coordinates with respect to some characteristic length $L_o$ and distributed load $p_o$ as shown in Equation (3.5a), (3.5b), and (3.5c). We note: $p_o L_o$ has the dimension of force; $p_o L_o^2$ has the dimension of moment; $p_o L_o^3/EI$ has the dimension of slope; $p_o L_o^4/EI$ has the dimension of deflection, where $EI$ is the bending rigidity of the beam.

Equations (3.5e) through (3.5i) are obtained using the above observations. Equation (3.1) can be written as

$$[\hat{v}(\hat{x})](p_oL_o^4/EI) = \mathcal{G}(x,\xi)[\hat{P}p_oL_o] \quad \text{or} \quad \hat{v}(\hat{x}) = [\mathcal{G}(x,\xi)(EI/L_o^3)]\hat{P} \quad \text{or}$$

$$\hat{v}(\hat{x}) = \hat{\mathcal{G}}(\hat{x},\hat{\xi})\hat{P}$$

where, $\hat{\mathcal{G}}(\hat{x},\hat{\xi})$ in the above equation is given by Equation (3.6a). In a similar manner the remaining equations in Table 3.1 can be derived (see problems 3.1-3.2).

**Table 3.1** List of non-dimensional variables

| Original variable | Non-dimensionalized variable | | Original variable | Non-dimensionalized variable | |
|---|---|---|---|---|---|
| $x$ | $\hat{x} = x/L_o$ | **(3.5a)** | $\mathcal{G}$ | $\hat{\mathcal{G}} = \mathcal{G}[EI/L_o^3]$ | **(3.6a)** |
| $\xi$ | $\hat{\xi} = \xi/L_o$ | **(3.5b)** | $\mathcal{G}_1 = \partial\mathcal{G}/\partial x$ | $\hat{\mathcal{G}}_1 = \mathcal{G}_1[EI/L_o^2] = \partial\hat{\mathcal{G}}/\partial\hat{x}$ | **(3.6b)** |
| $p_y$ | $\hat{p}_y = p_y/p_o$ | **(3.5c)** | $\mathcal{G}_2$ | $\hat{\mathcal{G}}_2 = \mathcal{G}_2/L_o = \partial\hat{\mathcal{G}}_1/\partial\hat{x}$ | **(3.6c)** |
| $P$ | $\hat{P} = P/(p_oL_o)$ | **(3.5d)** | $\mathcal{G}_3$ | $\hat{\mathcal{G}}_3 = \mathcal{G}_3 = \partial\hat{\mathcal{G}}_2/\partial\hat{x}$ | **(3.6d)** |
| v | $\hat{v} = v[EI/(p_oL_o^4)]$ | **(3.5e)** | $\mathcal{H}$ | $\hat{\mathcal{H}} = \mathcal{H}[EI/L_o^2]$ | **(3.6e)** |
| $\psi = dv/dx$ | $\hat{\psi} = \psi[EI/(p_oL_o^3)]$ | **(3.5f)** | $\mathcal{H}_1 = \partial\mathcal{H}/\partial x$ | $\hat{\mathcal{H}}_1 = \mathcal{H}_1[EI/L_o] = \partial\hat{\mathcal{H}}/\partial\hat{x}$ | **(3.6f)** |
| $M$ | $\hat{M} = M/(p_oL_o^2)$ | **(3.5g)** | $\mathcal{H}_2$ | $\hat{\mathcal{H}}_2 = \mathcal{H}_2 = \partial\hat{\mathcal{H}}_1/\partial\hat{x}$ | **(3.6g)** |
| $M_z = EI(d^2v/dx^2)$ | $\hat{M}_z = M_z/(p_oL_o^2)$ | **(3.5h)** | $\mathcal{H}_3$ | $\hat{\mathcal{H}}_3 = \mathcal{H}_3L_o = d\hat{\mathcal{H}}_2/d\hat{x}$ | **(3.6h)** |
| $V_y = -\frac{d}{dx}(EId^2v/dx^2)$ | $\hat{V}_y = V_y/(p_oL_o)$ | **(3.5i)** | | | |

## 3.2 INFLUENCE FUNCTIONS FOR CLASSICAL BEAMS

The differential equation governing deflection of a classical beam with constant bending rigidity is given by

$$EI\frac{d^4v}{dx^4} = p_y \tag{3.7}$$

To determine the influence function we replace $p_y(x)$ by the delta function $\langle x-\xi\rangle^{-1}$ and v($x$) by $G(x,\xi)$ to obtain

$$EI\frac{d^4\mathcal{G}}{dx^4} = \langle x-\xi\rangle^{-1} \tag{3.8}$$

The homogeneous $\mathcal{G}_h(x,\xi)$ and the particular solution $\mathcal{G}_p(x,\xi)$ to the above equation are

$$\mathcal{G}_h(x,\xi) = c_1\frac{x^3}{6} + c_2\frac{x^2}{2} + c_3x + c_4 \qquad \mathcal{G}_p(x,\xi) = \frac{1}{6EI}\langle x-\xi\rangle^3 \tag{3.9}$$

The solution to Equation (3.8) is

$$\mathcal{G}(x,\xi) = \mathcal{G}_h(x,\xi) + \mathcal{G}_p(x,\xi) = \frac{1}{6EI}\langle x-\xi\rangle^3 + c_1\frac{x^3}{6} + c_2\frac{x^2}{2} + c_3x + c_4 \tag{3.9a}$$

The constants c's will be determined from the boundary conditions for a specific beam. Once the four constants are determined, the influence function $\mathcal{G}(x,\xi)$ will be defined.

### 3.2.1 Influence functions for simply supported beam

We will develop the solution for a simply supported beam shown in Figure 3.4*a* to elaborate the determination of constants c's in Equation (3.9a) from the boundary conditions.

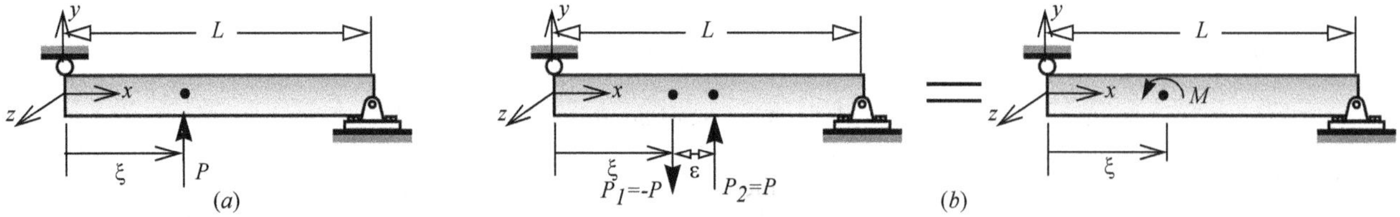


**Figure 3.4** Simply Supported Beam. (*a*) Source singularity or point force. (*b*) Doublet singularity or point moment.

Noting that deflection and moment are zero at each end of the beam shown in Figure 3.4*a*, the boundary conditions on $\mathcal{G}(x, \xi)$ are:

$$\mathcal{G}(0, \xi) = 0 \qquad EI\frac{d^2\mathcal{G}}{dx^2}(0, \xi) = 0 \qquad \mathcal{G}(L, \xi) = 0 \qquad EI\frac{d^2\mathcal{G}}{dx^2}(L, \xi) = 0 \tag{3.10}$$

Using the four boundary conditions in Equation (3.10) we obtain the integration constants as

$$c_1 = -\frac{(L-\xi)}{EIL} \qquad c_2 = 0 \qquad c_3 = -\frac{(L-\xi)^3}{6EIL} + \frac{(L-\xi)L}{6EI} \qquad c_4 = 0 \tag{3.11}$$

Substituting Equation (3.11) into Equation (3.9a) we obtain:

$$\boxed{\mathcal{G}(x, \xi) = \frac{1}{6EI}\left[\langle x-\xi\rangle^3 - \frac{(L-\xi)}{L}x^3 - \frac{(L-\xi)^3}{L}x + (L-\xi)Lx\right]} \tag{3.12a}$$

Figure 3.4*b* shows a doublet singularity obtained by placing two equal and opposite forces and replacing the couple by the moment. The influence function associated with the point moment can be found from Equation (3.3a) as shown below:

$$\boxed{\mathcal{H}(x, \xi) = \frac{\partial \mathcal{G}}{\partial \xi} = \frac{1}{6EI}\left[-3\langle x-\xi\rangle^2 + \frac{x^3}{L} + \frac{3(L-\xi)^2}{L}x - Lx\right]} \tag{3.12b}$$

The above influence functions $\mathcal{G}(x, \xi)$ and $\mathcal{H}(x, \xi)$ incorporates the boundary conditions of simple support. These functions can now be used for finding solutions of simply supported beams with any loading. Table 3.2 shows all the influence functions that are applicable to simply supported beam and are obtained by appropriate differentiation of Equations (3.12a) and (3.12b) and by appropriate non-dimensioanlizing the influence functions.

**Table 3.2** Influence functions for a simply supported beam

| Influence functions | | Non-dimensionalized form of influence functions | |
|---|---|---|---|
| $\mathcal{G} = \frac{1}{6EI}\left[\langle x-\xi\rangle^3 - \frac{(L-\xi)}{L}x^3 - \frac{(L-\xi)^3}{L}x + (L-\xi)Lx\right]$ | **(3.13a)** | $\widehat{\mathcal{G}} = \frac{1}{6}[\langle \widehat{x} - \widehat{\xi}\rangle^3 - (1-\widehat{\xi})\widehat{x}^3 - (1-\widehat{\xi})^3\widehat{x} + (1-\widehat{\xi})\widehat{x}]$ | **(3.14a)** |
| $\mathcal{G}_1 = \frac{1}{6EI}\left[3\langle x-\xi\rangle^2 - 3\frac{(L-\xi)}{L}x^2 - \frac{(L-\xi)^3}{L} + (L-\xi)L\right]$ | **(3.13b)** | $\widehat{\mathcal{G}}_1 = \frac{1}{6}[3\langle \widehat{x} - \widehat{\xi}\rangle^2 - 3(1-\widehat{\xi})\widehat{x}^2 - (1-\widehat{\xi})^3 + (1-\widehat{\xi})]$ | **(3.14b)** |
| $\mathcal{G}_2 = \left[\langle x-\xi\rangle^1 - \frac{(L-\xi)}{L}x\right]$ | **(3.13c)** | $\widehat{\mathcal{G}}_2 = [\langle \widehat{x} - \widehat{\xi}\rangle^1 - (1-\widehat{\xi})\widehat{x}]$ | **(3.14c)** |
| $\mathcal{G}_3 = -\left[\langle x-\xi\rangle^0 - \frac{(L-\xi)}{L}\right]$ | **(3.13d)** | $\widehat{\mathcal{G}}_3 = -[\langle \widehat{x} - \widehat{\xi}\rangle^0 - (1-\widehat{\xi})]$ | **(3.14d)** |
| $\mathcal{H} = \frac{1}{6EI}\left[-3\langle x-\xi\rangle^2 + \frac{x^3}{L} + \frac{3(L-\xi)^2}{L}x - Lx\right]$ | **(3.13e)** | $\widehat{\mathcal{H}} = \frac{1}{6}[-3\langle \widehat{x} - \widehat{\xi}\rangle^2 + \widehat{x}^3 + 3(1-\widehat{\xi})^2\widehat{x} - \widehat{x}]$ | **(3.14e)** |
| $\mathcal{H}_1 = \frac{1}{6EI}\left[-6\langle x-\xi\rangle^1 + 3\frac{x^2}{L} + \frac{3(L-\xi)^2}{L} - L\right]$ | **(3.13f)** | $\widehat{\mathcal{H}}_1 = \frac{1}{6}[-3\langle \widehat{x} - \widehat{\xi}\rangle^1 + 3\widehat{x}^2 + 3(1-\widehat{\xi})^2 - 1]$ | **(3.14f)** |
| $\mathcal{H}_2 = \left[-\langle x-\xi\rangle^0 + \frac{x}{L}\right]$ | **(3.13g)** | $\widehat{\mathcal{H}}_2 = [-\langle \widehat{x} - \widehat{\xi}\rangle^0 + \widehat{x}]$ | **(3.14g)** |
| $\mathcal{H}_3 = \left[\langle x-\xi\rangle^{-1} - \frac{1}{L}\right]$ | **(3.13h)** | $\widehat{\mathcal{H}}_3 = [\langle \widehat{x} - \widehat{\xi}\rangle^{-1} - 1]$ | **(3.14h)** |

## EXAMPLE 3.1

Obtain the elastic curve v(x) for the beam and loading shown in Figure 3.5 using the influence functions in Table 3.2.

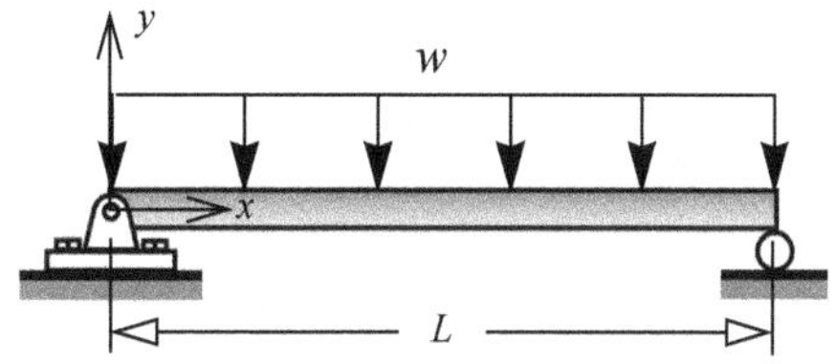


**Figure 3.5** Beam in Example 3.1.

**PLAN**

We can substitute $p_y = -w$ and Equation (3.13a) into Equation (3.2a) and integrate to obtain v(x).

**SOLUTION**

Substituting $p_y = -w$ and Equation (3.13a) into Equation (3.2a) we obtain:

$$v(x) = \int_0^L \mathcal{G}(x, \xi)p_y(\xi)d\xi = \int_0^L \left(\frac{-w}{6EI}\right)\left[\langle x-\xi\rangle^3 - \frac{(L-\xi)}{L}x^3 - \frac{(L-\xi)^3}{L}x + (L-\xi)Lx\right]d\xi$$

$$v(x) = \left(\frac{-w}{6EI}\right)\left[-\frac{\langle x-\xi\rangle^4}{4} + \frac{(L-\xi)^2}{2L}x^3 + \frac{(L-\xi)^4}{4L}x - \frac{(L-\xi)^2}{2}Lx\right]\Bigg|_0^L = \left(\frac{-w}{24EI}\right)[-\langle x-L\rangle^4 + \langle x\rangle^4 - 2Lx^3 - L^3x + L^3x] \quad \text{(E1)}$$

Noting that for $x < L$ the term $\langle x-L\rangle^4$ is zero, while for $x > 0$ the term $\langle x\rangle^4 = x^4$, we obtain our result.

**ANS.** $v(x) = \left(\frac{-w}{24EI}\right)[x^4 - 2Lx^3 + L^3x]$

## COMMENTS

1. The internal bending moment can be found from the solution as $M_z = -w[x^2 - xL]/2$. We note that v($x$) and $M_z$ are zero at $x = 0$ and $x = L$, thus satisfying the boundary conditions as expected. We also note that $EI(\partial^4 v/\partial x^4) = -w$, thus the solution satisfies the differential equation and hence the boundary value problem. This validates the methodology of using influence function for finding deflections of beams.
2. We could have obtained the solution from second or fourth order boundary value problem for the simple loading in this example. Examples 3.3 and 3.4 demonstrate the use of influence function for more complex loadings.

## EXAMPLE 3.2

Obtain the elastic curve v($x$) for the beam shown in Figure 3.6$a$ using the influence functions in Table 3.2.

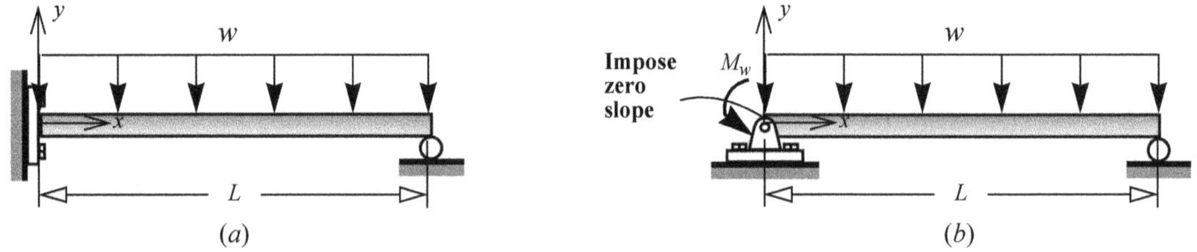


**Figure 3.6** ($a$) Original beam. ($b$) Equivalent beam

**PLAN**

The beam in Figure 3.6$a$ differs from the simply supported beam in two ways: it has a reaction moment and the slope at the end is zero as shown in Figure 3.6$b$. We write the solution using superposition of a point moment $M_w$ and a distributed load. By imposing the zero slope condition we obtain the value of $M_w$ and the corresponding solution.

**SOLUTION**

By superposition of Equation (3.2a) and Equation (3.3a) we obtain

$$v(x) = \mathcal{H}(x, 0)M_w + \int_0^L \mathcal{G}(x, \xi)p_y(\xi)d\xi \quad \text{(E1)}$$

Using the solution of Example 3.1 for the integral and Equation (3.13e) with $\xi = 0$ we obtain

$$v(x) = \frac{M_w}{6EI}\left[-3\langle x-0\rangle^2 + \frac{x^3}{L} + \frac{3(L-0)^2}{L}x - Lx\right] + \left(-\frac{w}{24EI}\right)[x^4 - 2Lx^3 + L^3x]$$

$$v(x) = \frac{M_w}{6EI}\left[-3x^2 + \frac{x^3}{L} + 2Lx\right] - \frac{w}{24EI}[x^4 - 2Lx^3 + L^3x] \quad \text{(E2)}$$

We take the derivative of Equations (E2) and set it equal to zero at $x = 0$ to obtain $M_w$ as shown below.

$$\frac{dv}{dx}(0) = \left\{\frac{M_w}{6EI}\left[-6x + \frac{x^2}{L} + 2L\right] - \frac{w}{24EI}[4x^3 - 6Lx^2 + L^3]\right\}\Bigg|_{x=0} = \frac{M_w}{6EI}[2L] - \frac{w}{24EI}[L^3] = 0$$

$$M_w = \frac{wL^2}{8} \quad \text{(E3)}$$

Substituting Equation (E3) into Equation (E2) we obtain the solution

$$v(x) = \frac{wL^2}{48EI}\left[-3x^2 + \frac{x^3}{L} + 2Lx\right] - \frac{w}{24EI}[x^4 - 2Lx^3 + L^3x] = \frac{w}{48EI}[-3L^2x^2 + x^3L + 2L^3x] - \frac{w}{48EI}[2x^4 - 4Lx^3 + 2L^3x] \quad \text{(E4)}$$

**ANS.** $v(x) = -\frac{wx^2}{48EI}[2x^2 - 5xL + 3L^2]$

## COMMENTS

1. This example shows the adaptation of influence function associated with simply supported beams for solutions of indeterminate and beams with other boundary condition than those associated with simply supported beams.
2. We could have found the influence function from the boundary conditions and used it to find the solution. See problem 3.4.

## EXAMPLE 3.3

A 10 ft. simply supported beam shown in Figure 3.7 is loaded with a distributed force whose values vary as shown in Table 3.3. The beam has a rectangular cross section with depth of 8 in. in the $y$ direction and width of 2 in. in the $z$ direction. The modulus of elasticity for of the beam material is 8,100 ksi. Determine: (a) The deflection and bending moment at the mid section of the beam. (b) The maximum deflection and bending normal stress in the beam.

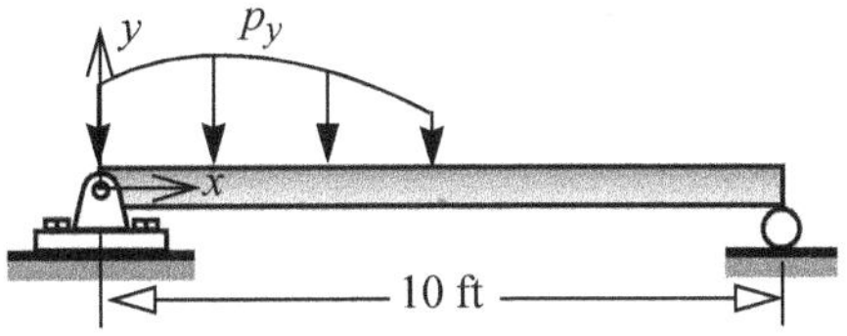


**Figure 3.7** Beam and loading in Example 3.3.

**Table 3.3** Distributed load values in Example 3.3.

| $x$ (ft.) | 0 | 1 | 2 | 3 | 4 | 5 | 6 | 7 | 8 | 9 | 10 |
|---|---|---|---|---|---|---|---|---|---|---|---|
| $p_y(x)$ (lb/ft) | -300 | -310 | -306 | -288 | -256 | -210 | 0 | 0 | 0 | 0 | 0 |

**PLAN**

Using $L_0 = 10$ ft and the maximum value of the distributed load of $p_0 = 310$ lb/ft., we write the non-dimensionalized form of Equation (3.2a) for deflection and internal moment. (*a*) We use numerical integration to obtain displacement and bending moment at mid point. (*b*) Using the procedure of part *a* we determine the deflection and moment at each base point and plot it to determine the maximum deflection and moment in the beam. Using flexure formula we find the maximum bending normal stress.

**SOLUTION**

The second area moment of inertia and bending rigidity can be found as:

$$I = \frac{1}{12}(2)(8^3) = 85.33 \text{ in.}^4 \qquad EI = (8100)(10^3)(85.33) = 691.2(10^6) \text{ lbs-in}^2 \tag{E1}$$

The non-dimensionalized form of Equation (3.2a) for displacement and internal bending moment can be written. The integrals are evaluated by trapezoidal rule and the formula for equally spaced base points written using Equation (3.4d) as shown below.

$$\hat{v}(\hat{x}) = \int_0^1 \hat{G}(\hat{x}, \hat{\xi})\hat{p}_y(\hat{\xi})d\hat{\xi} = \left[\sum_{i=1}^{N-1}\hat{G}_i\hat{p}_i + 0.5\{\hat{G}_0\hat{p}_0 + \hat{G}_N\hat{p}_N\}\right](\Delta\hat{\xi}) \tag{E2}$$

$$\hat{M}_z(\hat{x}) = \int_0^1 \hat{G}_2(\hat{x}, \hat{\xi})\hat{p}_y(\hat{\xi})d\hat{\xi} = \left[\sum_{i=1}^{N-1}(\hat{G}_2)_i\hat{p}_i + 0.5\{(\hat{G}_2)_0\hat{p}_0 + (\hat{G}_2)_N\hat{p}_N\}\right](\Delta\hat{\xi}) \tag{E3}$$

where, $\hat{G}(\hat{x}, \hat{\xi})$ and $\hat{G}_2(\hat{x}, \hat{\xi})$ are given by Equations (3.14a) and (3.14c). Equations (E2) and (E3) are evaluated using spreadsheet at the mid point of $\hat{x} = 0.5$ as shown in Table 3.4 and explained below.

**Table 3.4** Solution at midpoint using spreadsheet in Example 3.3

| | *A* | *B* | *C* | *D* | *E* | *F* | *G* | *H* | *I* |
|---|---|---|---|---|---|---|---|---|---|
| 1 | | x = | 0.5 | | | | | | |
| 2 | $\xi$ ft. | Py lb/ft | $\hat{\xi}$ | $\hat{p}_y$ | $\langle\hat{x}-\hat{\xi}\rangle$ | $\hat{G}$ | $\hat{G}_i\hat{p}_i$ | $\hat{G}_2$ | $\hat{G}_{2i}\hat{p}_i$ |
| 3 | 0 | -300 | 0 | -0.9677 | 0.5 | 0.0000E+00 | 0.0000E+00 | 0.0000E+00 | 0.0000E+00 |
| 4 | 1 | -310 | 0.1 | -1.0000 | 0.4 | 6.1667E-03 | -6.1667E-03 | -5.0000E-02 | 5.0000E-02 |
| 5 | 2 | -306 | 0.2 | -0.9871 | 0.3 | 1.1833E-02 | -1.1681E-02 | -1.0000E-01 | 9.8710E-02 |
| 6 | 3 | -288 | 0.3 | -0.9290 | 0.2 | 1.6500E-02 | -1.5329E-02 | -1.5000E-01 | 1.3935E-01 |
| 7 | 4 | -256 | 0.4 | -0.8258 | 0.1 | 1.9667E-02 | -1.6241E-02 | -2.0000E-01 | 1.6516E-01 |
| 8 | 5 | -210 | 0.5 | -0.6774 | 0 | 2.0833E-02 | -1.4113E-02 | -2.5000E-01 | 1.6935E-01 |
| 9 | 6 | 0 | 0.6 | 0.0000 | 0 | 1.9667E-02 | 0.0000E+00 | -2.0000E-01 | 0.0000E+00 |
| 10 | 7 | 0 | 0.7 | 0.0000 | 0 | 1.6500E-02 | 0.0000E+00 | -1.5000E-01 | 0.0000E+00 |
| 11 | 8 | 0 | 0.8 | 0.0000 | 0 | 1.1833E-02 | 0.0000E+00 | -1.0000E-01 | 0.0000E+00 |
| 12 | 9 | 0 | 0.9 | 0.0000 | 0 | 6.1667E-03 | 0.0000E+00 | -5.0000E-02 | 0.0000E+00 |
| 13 | 10 | 0 | 1 | 0.0000 | 0 | 6.9389E-18 | 0.0000E+00 | -5.5511E-17 | 0.0000E+00 |
| 14 | | | | | | | -6.3530E-03 | | 6.2258E-02 |

(i) The values of base points and distributed load given in Table 3.3 are copied into columns *A* and *B* starting from row 3.
(ii) The base points are divided by the length of the beam of 10 ft. and the distributed load values are divided by the maximum value of 310 lb/ft to obtain the entries in columns *C* and *D* from row 3 onwards.
(iii) In cell *C1* we enter the non-dimensionalized value of the point at which we wish to find the displacement and bending moment. Subsequently, we will need to just change value in *C1* to obtain displacement and internal moment at any point.
(iv) The discontinuity function $\langle\hat{x} - \hat{\xi}\rangle$ is evaluated in column *E* by comparing entries of row 3 onwards to row 1 value in column *C* and calculating the difference if it results in a positive value.

(v) Equations (3.14a) and (3.14c) are entered into cells *F3* and *H3* and copied into the rest of the column, giving us the values of the influence functions $\widehat{G}$ and $\widehat{G}_2$ at each base point.

(vi) The product of column *D* and *F* is written in column *H* and the product *D* and *H* is written in column *I*. In cells *G14* and *I14* the sum of the entries in *G* and *I* column are written as per Equations (E2) and (E3).

From values in cell *H14* and *I14* we obtain the non dimensionalized deflection and bending moment as:

$$\widehat{v} = -6.3530(10^{-3}) \qquad \widehat{M}_z = 62.258(10^{-3}) \tag{E4}$$

We note that the length is $L = 120$ in , the distributed load is $p_o = 310/12 = 25.833$ lb/in .Substituting these values and Equations (E1) and (E4) into Equation (3.5e) and Equation (3.5h) we obtain the deflection and bending moment at mid-point as:

$$v(x=5) = \frac{p_o L^4}{EI}\widehat{v} = \frac{(25.833)(120^4)}{691.2(10^6)}[-6.3530(10^{-3})] = -49.235(10^{-3}) \text{ in} \tag{E5}$$

$$M_z(x=5) = p_o L^2 \widehat{M}_z = (25.833)(120^2)(62.258)(10^{-3}) = 23160 \text{ in-lb} \tag{E6}$$

**ANS.** $v(x=5) = -0.0492$ in. ; $M_z(x=5) = 25272$ in.-lb

(b) If we change the value in cell *C1* sequentially from 0 to 1 in steps of 0.1 then we will obtain the values of non-dimensional deflection and bending moment at each base point. These values can be plotted as shown in Figure 3.8.

From spreadsheet and Figure 3.8 we obtain the maximum values of non-dimensionalized displacement and moment as

$$\widehat{v}_{max} = \widehat{v}(x=0.5) = -6.3530(10^{-3}) \qquad \widehat{M}_{max} = \widehat{M}_z(x=0.4) = 67.935(10^{-3}) \tag{E7}$$

Substituting $L = 120$ in. , $p_o = 25.833$ lb/in. and Equations (E1) and (E7) into Equation (3.5e) and Equation (3.5h) we obtain the maximum déflection and bending moment as

$$v_{max} = \frac{p_o L^4}{EI}\widehat{v}_{max} = \frac{(25.833)(120^4)}{691.2(10^6)}[-6.3530(10^{-3})] = -49.235(10^{-3}) \text{ in} \tag{E8}$$

$$M_{max} = p_o L^2 \widehat{M}_{max} = (25.833)(120^2)(67.935)(10^{-3}) = 25272 \text{ in-lb} \tag{E9}$$

Noting that $y_{max} = \pm 4$ in we obtain the maximum bending normal stress as

$$(\sigma_{xx})_{max} = -\left[\frac{M_{max} y_{max}}{I}\right] = -\left[\frac{(25272)(\pm 4)}{85.33}\right] = \mp 1184.7 \text{ psi} \tag{E10}$$

**ANS.** $v_{max} = -0.0492$ in. ;$(\sigma_{xx})_{max} = 1185$ psi (C) or (T)

## COMMENTS

1. This example demonstrates the basic concept of using influence functions with numerical integration to calculate values of displacements, moments and stresses from experimentally measured values of distributed load.

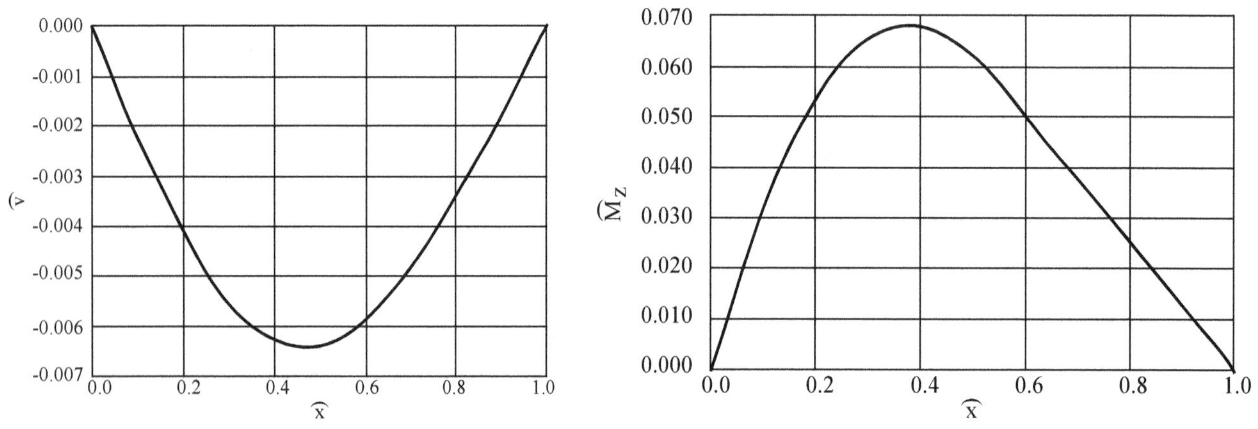


**Figure 3.8** Plot of non-dimensional deflection and bending moment in Example 3.3.

2. The maximum values of deflection and moment were at base points $x = 6$ ft and $x = 4$ ft. Base point location for maximum values is an artifact of our methodology.
3. We could evaluate deflection and moment at any point and not just at base points. Thus, we could evaluate these quantities at additional closely spaced points in the vicinity of the base point where we obtained the maximum. The improvement in accuracy will be marginal as our trapezoidal integration scheme is too simple. If we want higher accuracy it would be better to use higher order integration scheme such as Gauss quadrature.
4. It is a good practice to test the spread sheet (or a computer program) with a known solution such as in Example 3.1 with an assumed value for *w*.

## EXAMPLE 3.4

A 10 ft. simply supported beam shown in Figure 3.9 has a distributed force that varies as shown. The beam has a rectangular cross section with depth of 8 inch in the $y$ direction and width of 2 inch in the $z$-direction. The modulus of elasticity for of the beam material is 8,100 ksi. Determine: (a) The deflection and bending moment at the mid section of the beam. (b) The maximum deflection and bending normal stress in the beam.

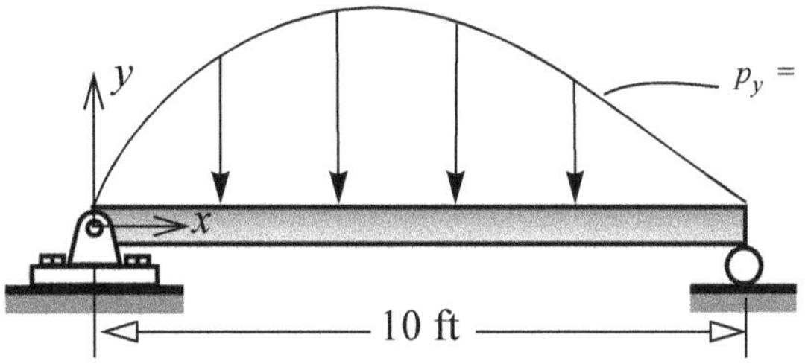


**Figure 3.9** Beam and loading in Example 3.4.

**PLAN**

We generate a table for distributed load and then solve the problem the same way as in Example 3.3.

**SOLUTION**

We generate a table using an increment of 1 ft between base points as shown in Table 3.5.

**Table 3.5** Value of distributed load in Example 3.4

| $x$(ft) | 0 | 1 | 2 | 3 | 4 | 5 | 6 | 7 | 8 | 9 | 10 |
|---|---|---|---|---|---|---|---|---|---|---|---|
| $p_y(x)$ (lb/ft) | 0.000 | -83.883 | -144.371 | -179.800 | -191.253 | -181.959 | -156.585 | -120.523 | -79.233 | -37.691 | 0.000 |

We follow the steps outlined in Example 3.3 using the same spreadsheet with column for distributed load replaced by those in Table 3.5. We use the maximum value of $p_y$ as $p_o$, i.e. $p_o = 191.26$ lb/ft $= 15.938$ lb/in . We obtain the non dimensionalized deflection and bending moment at mid point as:

$$\widehat{v} = -9.9254(10^{-3}) \qquad \widehat{M_z} = 98.584(10^{-3}) \tag{E1}$$

Substituting $L = 120$ in , $p_o = 15.938$ lb/in , $EI = 691.2(10^6)$ lbs-in$^2$ and (E1) into Equation (3.5e) and Equation (3.5h) we obtain the deflection and bending moment as

$$v(x=5) = \frac{p_o L^4}{EI}\widehat{v} = \frac{(15.938)(120^4)}{691.2(10^6)}[-9.9254(10^{-3})] = -47.457(10^{-3}) \text{ in.} \tag{E2}$$

$$M_z(x=5) = p_o L^2 \widehat{M_z} = (25.833)(120^2)(-98.584)(10^{-3}) = 22625 \text{ in.-lb} \tag{E3}$$

**ANS.** $v(x=5) = -0.0475$ in. ; $M_z(x=5) = 22,625$ in.-lb

Spreadsheet shows the maximum values of non-dimensionalized displacement and moment are at mid point and are given by Equation (E1); hence maximum displacement and moment are given by Equations (E2) and (E3).

Noting that $y_{max} = \pm 4$ in we obtain the maximum bending normal stress as

$$(\sigma_{xx})_{max} = -\left[\frac{M_{max}y_{max}}{I}\right] = -\left[\frac{(22625)(\pm 4)}{85.33}\right] = \mp 1060.6 \text{ psi} \tag{E4}$$

**ANS.** $v_{max} = -0.0475$ in ; $(\sigma_{xx})_{max} = 1060.6$ psi (C) or (T)

### COMMENTS

1. The example demonstrates the use of numerical integration with influence function for beams with complex distributed loads.
2. We did not make effective use of the fact that distributed load was a known function. We could have evaluated the distributed load at many more base points to improve accuracy. If we use increments of 0.25 ft between base points and solve for non-dimensionalized displacements and stress then we will obtain

$$\widehat{v} = -9.9254(10^{-3}) \quad \text{and} \quad \widehat{M_z} = 97.838(10^{-3}) \tag{E5}$$

3. Comparing Equations (E1) and (E5) we see no change in displacement up to 5 significant figures and a decrease in moment of 0.75%. Other points may show greater or smaller changes.
4. The number of base points increased from 11 to 41 when we changed the increment from 1 ft to 0.25 ft. If we try to find displacement and moments at each base point to find maximum values then we realize that the process is becoming tedious and it would be advantageous to write a computer program to do the job.
5. An alternative to the scheme used in Example 3.3 is as follows. Plot the distributed load from the values given in Table 3.3 and then fit a curve through the data points in the least square sense. Now we can create a more extensive table with many more closely spaced base points for the distributed load. This will overcome some of the shortcoming of using trapezoidal rule for integration.

## 3.3 BEAMS ON ELASTIC FOUNDATIONS

Railroad tracks rest on elastic support made of cross ties and ballast of crushed rocks, long steel pipes rest on earth or a series of periodic supports, load bearing walls in buildings have beams supported periodically by studs that rest on foundation beams, are some examples of beams on elastic foundations. The simplest model for incorporating the elastic foundation effect is the *Winkler* model in which the foundation resistance is assumed proportional to the beam deflection. This linear model works well for small deflection—the basic assumption in our theory.

We represent the **modulus of foundation** by $k$, which is *spring constant per unit length* and has the dimension of *force per length squared.* The foundation exerts a force per unit length of $kv$ downwards to resist the upward positive deflection. We take a small differential element of a beam resting on the elastic foundation as shown in Figure 3.10*a*.

In the infinitesimal length we can assume $p_y$ is a constant and replace it by an equivalent force as shown in Figure 3.10*b*.

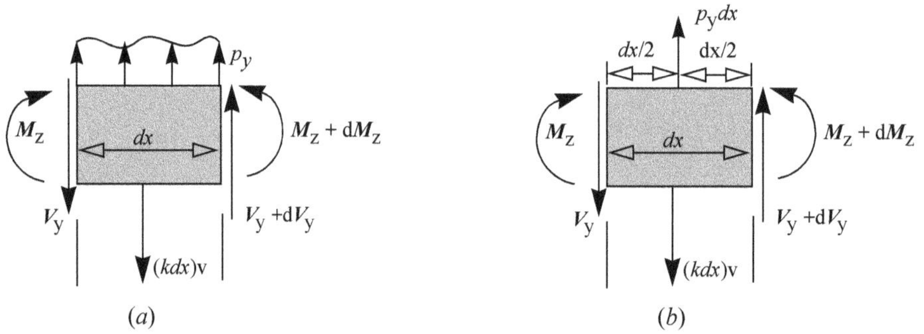


**Figure 3.10** Differential element of a beam on elastic foundation.

By force equilibrium in the $y$-direction and moment equilibrium about the midpoint in Figure 3.10*b* we obtain

$$dV_y + p_y dx - (kdx)v = 0 \qquad \text{or} \qquad \frac{dV_y}{dx} - kv = -p_y \tag{3.15a}$$

$$M_z + dM_z + (V_y + dV_y)\frac{dx}{2} + V_y\frac{dx}{2} - M_z = 0 \qquad \text{or} \qquad V_y = -\frac{dM_z}{dx} \tag{3.15b}$$

In the above equation we neglected the second order term ($dV_y dx$). Substituting $M_z(x) = EI(d^2v/dx^2)$ and Equation (3.15b) into Equation (3.15a) we obtain the differential equation governing deflection of the beam resting on elastic foundation.

$$\boxed{\frac{d^2}{dx^2}\left(EI_{zz}\frac{d^2v}{dx^2}\right) + kv = p_y} \tag{3.16}$$

## 3.4 FUNDAMENTAL SOLUTIONS

The fundamental solution is the influence function associated with an infinite beam on elastic foundation. To simplify calculations, we will place the concentrated force at the origin as shown in Figure 3.11 and later generalize the solution by translating the origin to an arbitrary location $\xi$. To keep focus on the concepts in this section, the mathematical details of the derivation of the fundamental solutions are separated and given in Section 3.8. We substitute $p_y = P\langle x\rangle^{-1}$ into Equations (3.16) to obtain

$$\frac{d^2}{dx^2}\left(EI\frac{d^2v}{dx^2}\right) + kv = P\langle x\rangle^{-1} \tag{3.17a}$$

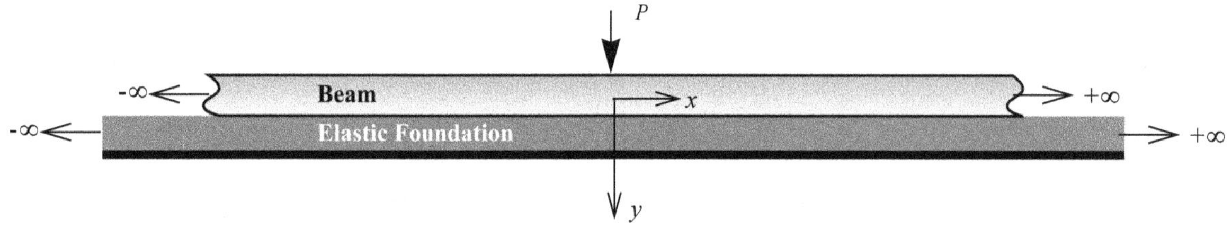


**Figure 3.11** Infinite beam on elastic foundation with a concentrated force.

The disturbance introduced by the concentrated force must die out at infinity. Thus, shear force and bending moment must go to zero at infinity as shown below

$$\lim_{|x| \to \infty}\left[\frac{d}{dx}\left(EI\frac{d^2v}{dx^2}\right)\right] \to 0 \qquad \lim_{|x| \to \infty}\left[EI_{zz}\frac{d^2v}{dx^2}\right] \to 0 \tag{3.17b}$$

In addition, displacements and slope must remain bounded at infinity as shown below.

$$\lim_{|x| \to \infty} [\mathrm{v}] < \infty \qquad \lim_{|x| \to \infty} \left[\frac{d\mathrm{v}}{dx}\right] < \infty \tag{3.17c}$$

Equations (3.17a) through (3.17c) represent the boundary value problem we need to solve. Integrating Equation (3.17a) from minus infinity to plus infinity, noting the integral of delta function is one and using Equation (3.17b) we obtain

$$\left.\frac{d}{dx}\left(EI_{zz}\frac{d^2\mathrm{v}}{dx^2}\right)\right|_{-\infty}^{\infty} + \int_{-\infty}^{\infty} k\mathrm{v}dx = P\int_{-\infty}^{\infty} \langle x\rangle^{-1} dx \qquad \text{or} \qquad \int_{-\infty}^{\infty} k\mathrm{v}dx = P \tag{3.17d}$$

Equation (3.17d) represents static equilibrium as the left hand side is the total foundation resistance which should equal to the applied force $P$. We use the symmetry in the problem to reduce the algebra by considering only $x > 0$ and later generalizing the solution to include $x < 0$. By symmetry we obtain that the slope at the origin is zero, which replaces the condition on moment at infinity. The displacement will be symmetric and Equation (3.17d) can be written as:

$$2\int_0^{\infty} k\mathrm{v}dx = P \tag{3.17e}$$

$$\frac{\partial \mathrm{v}}{\partial x}(0) = 0 \tag{3.18a}$$

In the development that follows we shall assume that the bending rigidity $EI_{zz} = EI$ is a constant. For $x > 0$, the solution to the above described boundary value problem as shown in Equation (3.41) is

$$\mathrm{v}(x) = \left(\frac{P\beta}{2k}\right)e^{-\beta x}[\cos\beta x + \sin\beta x] \qquad x > 0 \tag{3.18b}$$

where the parameter β is

$$\boxed{\beta = \left(\frac{k}{4EI}\right)^{1/4}} \tag{3.18c}$$

We generalize the solution by taking the absolute value of $x$ and extend our solution to negative values of $x$. To further generalize our solution, we consider the concentrated force at $x = \xi$ rather than the origin. This is a simple translation of the origin in an infinite beam and the impact is that we write $x - \xi$ in place of $x$. Noting the definition of fundamental solution we rewrite Equation (3.18b) as

$$\boxed{\mathrm{v}(x) = \mathcal{G}(x, \xi)P \qquad \mathcal{G}(x, \xi) = \left(\frac{\beta}{2k}\right)e^{-\beta|x-\xi|}[\cos\beta|x-\xi| + \sin\beta|x-\xi|]} \tag{3.19}$$

Consider the following derivative of an absolute quantity

$$|x-\xi| = \begin{cases} (x-\xi) & x > \xi \\ -(x-\xi) & x < \xi \end{cases} \qquad \therefore \qquad \frac{\partial}{\partial x}|x-\xi| = \begin{cases} 1 & x > \xi \\ -1 & x < \xi \end{cases} \tag{3.20}$$

We introduce the definition of *sgn function* below which makes the derivative of absolute functions compact.

$$\boxed{sgn(x-\xi) = \begin{cases} 1 & x > \xi \\ -1 & x < \xi \end{cases}} \tag{3.21}$$

We note the following identities using Equation (3.21)

$$\frac{\partial}{\partial x}|x-\xi| = -\left[\frac{\partial}{\partial \xi}|x-\xi|\right] = sgn(x-\xi) \tag{3.22}$$

$$\boxed{\frac{\partial \mathcal{G}}{\partial x} = -\left(\frac{\partial \mathcal{G}}{\partial \xi}\right)} \tag{3.23}$$

Equation (3.23) permit us to change the derivative from $x$ to $\xi$. *The above identity holds only for fundamental solutions.* The above identity is not valid for influence function in Equation (3.12a), which is associated with a beam of finite length.

By taking the appropriate derivatives of $\mathcal{G}(x, \xi)$ with respect to $x$ we obtain expressions for fundamental solutions associated with the slope, bending moment, and shear force. These expressions are given by Equations (3.25b), (3.25c), and (3.25d) in Table 3.6. The details of the derivation of these equations are given in Section 3.8.

Now the fundamental solution associated with concentrated moment is given by Equation (3.3a). Using the identity of Equation (3.23), we conclude that

$$\mathcal{H}(x, \xi) = -G_1(x, \xi) \tag{3.24}$$

The expression for $\mathcal{H}(x, \xi)$ is shown in Equation (3.25e). We may then take the derivatives of $\mathcal{H}(x, \xi)$ with respect to $x$ to obtain expressions for expressions for fundamental solutions associated with the slope, bending moment, and shear force. These expressions are given by Equations (3.25f), (3.25g), and (3.25h). The details of the derivation of these equations are given in Section 3.8.

**Table 3.6** Fundamental solutions for an infinite beam on elastic foundation.I

| Fundamental Solutions | | Non-dimensionalized form of Fundamental Solutions | |
|---|---|---|---|
| $\mathcal{G} = \frac{\beta}{2k} e^{-\beta\lvert x-\xi\rvert}[\cos\beta\lvert x-\xi\rvert + \sin\beta\lvert x-\xi\rvert]$ | **(3.25a)** | $\widehat{\mathcal{G}} = \left(\frac{1}{8\widehat{\beta}^3}\right) e^{-\widehat{\beta}\lvert\widehat{x}-\widehat{\xi}\rvert}[\cos\widehat{\beta}\lvert\widehat{x}-\widehat{\xi}\rvert + \sin\widehat{\beta}\lvert\widehat{x}-\widehat{\xi}\rvert]$ | **(3.26a)** |
| $\mathcal{G}_1 = -\left(\frac{\beta^2}{k}\right) sgn(x-\xi) e^{-\beta\lvert x-\xi\rvert} \sin\beta\lvert x-\xi\rvert$ | **(3.25b)** | $\widehat{\mathcal{G}}_1 = -\left(\frac{1}{4\widehat{\beta}^2}\right) sgn(\widehat{x}-\widehat{\xi}) e^{-\widehat{\beta}\lvert\widehat{x}-\widehat{\xi}\rvert} \sin\widehat{\beta}\lvert\widehat{x}-\widehat{\xi}\rvert$ | **(3.26b)** |
| $\mathcal{G}_2 = \frac{1}{4\beta} e^{-\beta\lvert x-\xi\rvert}(\sin\beta\lvert x-\xi\rvert - \cos\beta\lvert x-\xi\rvert)$ | **(3.25c)** | $\widehat{\mathcal{G}}_2 = \left(\frac{1}{4\widehat{\beta}}\right) e^{-\widehat{\beta}\lvert\widehat{x}-\widehat{\xi}\rvert}(\sin\widehat{\beta}\lvert\widehat{x}-\widehat{\xi}\rvert - \cos\widehat{\beta}\lvert\widehat{x}-\widehat{\xi}\rvert)$ | **(3.26c)** |
| $\mathcal{G}_3 = -\left(\frac{1}{2}\right) sgn(x-\xi) e^{-\beta\lvert x-\xi\rvert} \cos\beta\lvert x-\xi\rvert$ | **(3.25d)** | $\widehat{\mathcal{G}}_3 = -\left(\frac{1}{2}\right) sgn(\widehat{x}-\widehat{\xi}) e^{-\widehat{\beta}\lvert\widehat{x}-\widehat{\xi}\rvert} \cos\widehat{\beta}\lvert\widehat{x}-\widehat{\xi}\rvert$ | **(3.26d)** |
| $\mathcal{H} = sgn(x-\xi)\left(\frac{\beta^2}{k}\right) e^{-\beta\lvert x-\xi\rvert} \sin\beta\lvert x-\xi\rvert$ | **(3.25e)** | $\widehat{\mathcal{H}} = \left(\frac{1}{4\widehat{\beta}^2}\right) sgn(\widehat{x}-\widehat{\xi}) e^{-\widehat{\beta}\lvert\widehat{x}-\widehat{\xi}\rvert} \sin\widehat{\beta}\lvert\widehat{x}-\widehat{\xi}\rvert$ | **(3.26e)** |
| $\mathcal{H}_1 = -\left(\frac{\beta^3}{k}\right) e^{-\beta\lvert x-\xi\rvert}(\sin\beta\lvert x-\xi\rvert - \cos\beta\lvert x-\xi\rvert)$ | **(3.25f)** | $\widehat{\mathcal{H}}_1 = -\left(\frac{1}{4\widehat{\beta}}\right) e^{-\widehat{\beta}\lvert\widehat{x}-\widehat{\xi}\rvert}(\sin\widehat{\beta}\lvert\widehat{x}-\widehat{\xi}\rvert - \cos\widehat{\beta}\lvert\widehat{x}-\widehat{\xi}\rvert)$ | **(3.26f)** |
| $\mathcal{H}_2 = -\left(\frac{1}{2}\right) sgn(x-\xi) e^{-\beta\lvert x-\xi\rvert} \cos\beta\lvert x-\xi\rvert$ | **(3.25g)** | $\widehat{\mathcal{H}}_2 = -\left(\frac{1}{2}\right) sgn(\widehat{x}-\widehat{\xi}) e^{-\widehat{\beta}\lvert\widehat{x}-\widehat{\xi}\rvert} \cos\widehat{\beta}\lvert\widehat{x}-\widehat{\xi}\rvert$ | **(3.26g)** |
| $\mathcal{H}_3 = -\left(\frac{\beta}{2}\right) e^{-\beta\lvert x-\xi\rvert}[\cos\beta\lvert x-\xi\rvert + \sin\beta\lvert x-\xi\rvert]$ | **(3.25h)** | $\widehat{\mathcal{H}}_3 = -\left(\frac{\widehat{\beta}}{2}\right) e^{-\widehat{\beta}\lvert\widehat{x}-\widehat{\xi}\rvert}[\cos\widehat{\beta}\lvert\widehat{x}-\widehat{\xi}\rvert + \sin\widehat{\beta}\lvert\widehat{x}-\widehat{\xi}\rvert]$ | **(3.26h)** |

The non-dimensionalized forms of fundamental solutions in Table 3.6 are obtained using variables defined in Table 3.1 and the following.

$$\widehat{\beta} = \beta L_o \qquad \widehat{k} = kL_o/p_o \tag{3.27}$$

It should be noted that the *sgn function* is a constant, hence its derivative is zero for $x > \xi$ and $x < \xi$ but neither the function nor it's derivative exists at $x = \xi$. The equations in Table 3.6 are valid for $x > \xi$ and $x < \xi$ but not at $x = \xi$, unless the application of the formula show that the variable is continuous at $x = \xi$. It also follows that *during integration it will be necessary to consider the regions $x > \xi$ and $x < \xi$ separately.*

### 3.4.1 Some properties of fundamental solutions associated with force singularity

From (3.25a) through (3.25d) we note

- $\mathcal{G}(x, \xi)$ and $\mathcal{G}_2(x, \xi)$ are even functions about the source point $\xi$.
- $\mathcal{G}_1(x, \xi)$ and $\mathcal{G}_3(x, \xi)$ are an odd function about the source point $\xi$.

Thus, the displacement is an even function of $(x - \xi)$. The odd derivatives of displacement with respect to $x$ are odd functions and even derivatives are even functions, i.e.,

$$\mathrm{v}(-x+\xi) = \mathrm{v}(x-\xi) \qquad \frac{d\mathrm{v}}{dx}(-x+\xi) = -\frac{d\mathrm{v}}{dx}(x-\xi) \qquad M_z(-x+\xi) = M_z(x-\xi) \qquad V_y(-x+\xi) = -V_y(x-\xi) \tag{3.28}$$

We further note that amplitude of all variables decreases exponentially as we move away from the point of application of the force. Thus, the magnitude of the maximum values of the variables is just before or just after the point of application of the force. Substituting $x = \xi + \varepsilon$ and letting $\varepsilon$ go to zero we obtain the magnitude of the maximum values as shown below:

$$|\mathrm{v}_{max}| = P\beta/(2k) \qquad |(d\mathrm{v}/dx)_{max}| = P\beta^2/k \qquad |(M_z)_{max}| = P/4\beta \qquad |(V_y)_{max}| = P/2 \tag{3.29}$$

Figure 3.12 shows the variation of the four variables non-dimensionalized with respect to their maximum values for the range of $-2\pi < \beta x < 2\pi$.

Figure 3.12 shows the following:

1. Displacement and moment are even functions of $x$ and slope and shear force are odd functions about the source, consistent with Equation (3.28).
2. The disturbance introduced by the concentrated force at the origin dies out rapidly as one move away from the source. This is emphasized by all four variables.

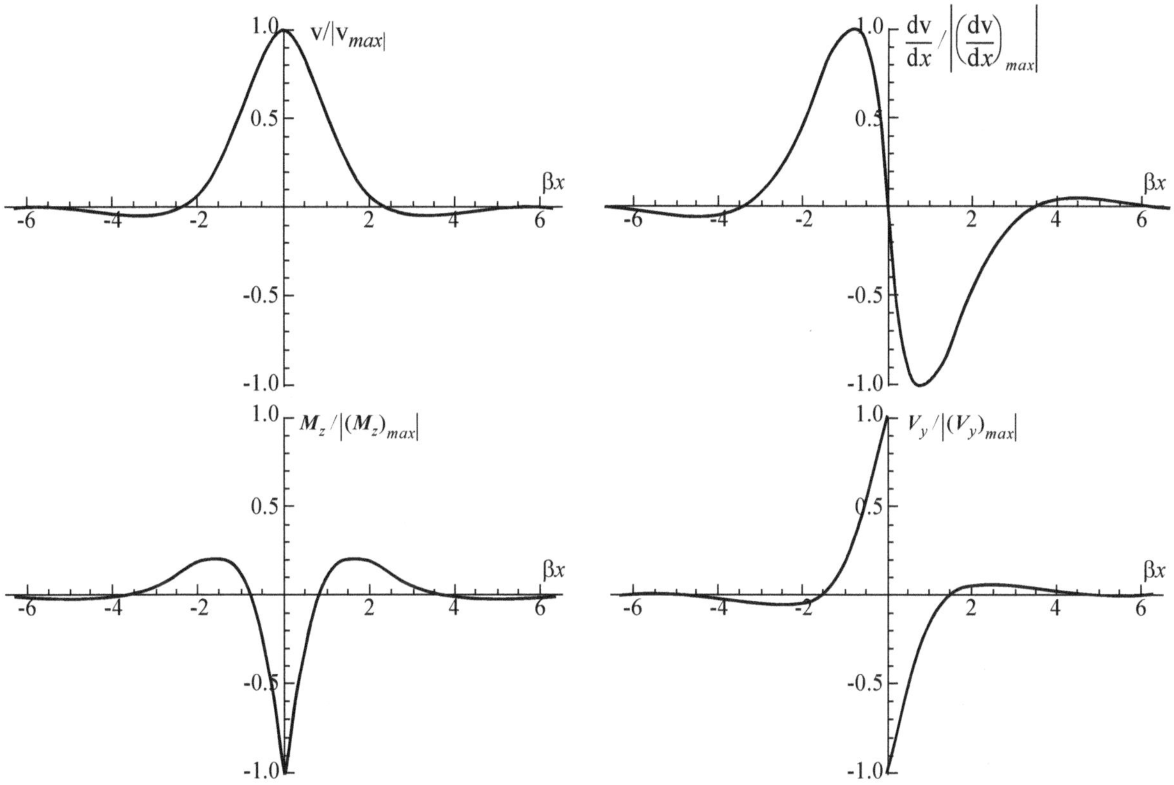


**Figure 3.12** Plot of non-dimensionalized variables in an infinite beam with concentrated force.

3. Displacement, slope and moment are continuous at the origin but the shear force jumps by the value of applied load *P*. This is as expected as the beam does not break or form kinks, hence the continuity of these variables. There is no external moment at the origin, hence no jump in the internal moment but there is a concentrated force *P*, hence the jump in internal shear force.

In a similar manner we can obtain properties associated with moment singularity (See problem 3.17).

## EXAMPLE 3.5

Obtain the displacement and internal bending moment function for a point on an infinite beam on elastic foundation under a uniform load shown in Figure 3.13.

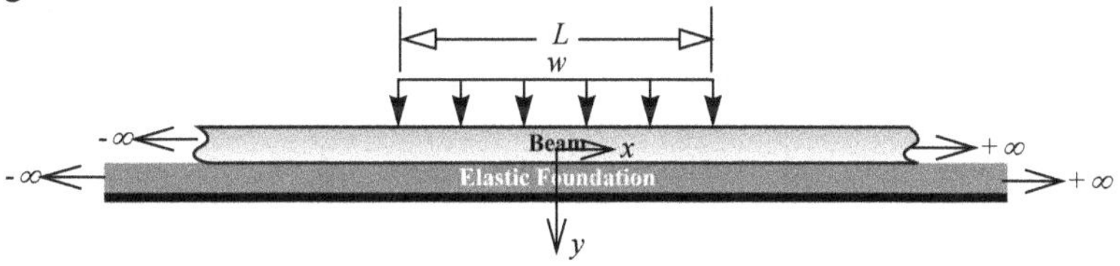


**Figure 3.13** Beam on elastic foundation in Example 3.5.

**PLAN**

We can substitute $p_y = w$ and Equation (3.25a) into Equation (3.2a) and its equivalent form for moment. We can write the integral as a sum of two integral between $-L/2 \le \xi < x - \varepsilon$ and $x + \varepsilon < \xi \le L/2$ to define the sign function. Using formulas of Equations (3.49) and Equations (3.50) we can perform the integration and let $\varepsilon \to 0$ to obtain v($x$) and $M_z$. The moment can also be found directly from v($x$).

**SOLUTION**

We can substitute $p_y = w$ and Equation (3.19) into Equation (3.2a) to obtain the following

$$\mathrm{v}(x) = \int_{-L/2}^{L/2} G(x,\xi)p_y(\xi)d\xi = \lim_{\varepsilon \to 0}\left[\int_{-L/2}^{x-\varepsilon} G(x,\xi)w\,d\xi + \int_{x+\varepsilon}^{L/2} G(x,\xi)w\,d\xi\right]$$

$$\mathrm{v}(x) = \left(\frac{w\beta}{2k}\right)\lim_{\varepsilon \to 0}\left[\int_{-L/2}^{x-\varepsilon} e^{-\beta|x-\xi|}[\cos\beta|x-\xi| + \sin\beta|x-\xi|]d\xi + \int_{x+\varepsilon}^{L/2} e^{-\beta|x-\xi|}[\cos\beta|x-\xi| + \sin\beta|x-\xi|]\ d\xi\right] \quad \text{(E1)}$$

Substituting Equations (3.49) and Equations (3.50) into the above equation we obtain:

$$\mathrm{v}(x) = \left(\frac{w\beta}{2k}\right)\lim_{\varepsilon \to 0}\left\{\left[\frac{1}{\beta}sgn(x-\xi)e^{-\beta|x-\xi|}\cos\beta|x-\xi|\right]\Bigg|_{\xi=-L/2}^{\xi=x-\varepsilon} + \left[\frac{1}{\beta}sgn(x-\xi)e^{-\beta|x-\xi|}\cos\beta|x-\xi|\right]\Bigg|_{\xi=x+\varepsilon}^{\xi=L/2}\right\}$$

$$\mathrm{v}(x) = \left(\frac{w}{2k}\right)\lim_{\varepsilon\to 0}\left\{\begin{matrix} sgn(\varepsilon)e^{-\beta|\varepsilon|}cos\,\beta|\varepsilon| - sgn(x+L/2)e^{-\beta|x+L/2|}cos\,\beta|x+L/2| \\ + sgn(x-L/2)e^{-\beta|x-L/2|}cos\,\beta|x-L/2| - sgn(-\varepsilon)e^{-\beta|\varepsilon|}cos\,\beta|\varepsilon| \end{matrix}\right\} \tag{E2}$$

**ANS.** $\mathrm{v}(x) = [w/(2k)]\{2 - sgn(x+L/2)e^{-\beta|x+L/2|}cos(\beta|x+L/2| + sgn(x-L/2)e^{-\beta|x-L/2|}cos\,\beta|x-L/2|)\}$

**Method I for finding $M_z$**

We can write the moment expression as:

$$M_z(x) = \int_{-L/2}^{L/2} G_2(x,\xi)p_y(\xi)d\xi = \lim_{\varepsilon\to 0}\left[\int_{-L/2}^{x-\varepsilon} G_2(x,\xi)w\,d\xi + \int_{x+\varepsilon}^{L/2} G_2(x,\xi)w\,d\xi\right] \tag{E3}$$

Substituting Equation (3.25c) we obtain

$$M_z(x) = \left(\frac{w}{4\beta}\right)\lim_{\varepsilon\to 0}\left[\int_{-L/2}^{x-\varepsilon} e^{-\beta|x-\xi|}(sin\,\beta|x-\xi| - cos\,\beta|x-\xi|)d\xi + \int_{x+\varepsilon}^{L/2} e^{-\beta|x-\xi|}(sin\,\beta|x-\xi| - cos\,\beta|x-\xi|)d\xi\right] \tag{E4}$$

Substituting Equations (3.49) and Equations (3.50) into the above equation we obtain:

$$M_z(x) = \left(\frac{w}{4\beta}\right)\lim_{\varepsilon\to 0}\left\{\left[\frac{1}{\beta}sgn(x-\xi)e^{-\beta|x-\xi|}sin\,\beta|x-\xi|\right]\Bigg|_{\xi=-L/2}^{\xi=x-\varepsilon} + \left[\frac{1}{\beta}sgn(x-\xi)e^{-\beta|x-\xi|}sin\,\beta|x-\xi|\right]\Bigg|_{\xi=x+\varepsilon}^{\xi=L/2}\right\}$$

$$M_z(x) = \left(\frac{w}{4\beta^2}\right)\lim_{\varepsilon\to 0}\left\{\begin{matrix} sgn(\varepsilon)e^{-\beta|\varepsilon|}sin\,\beta|\varepsilon| - sgn(x+L/2)e^{-\beta|x+L/2|}sin\,\beta|x+L/2| \\ + sgn(x-L/2)e^{-\beta|x-L/2|}sin\,\beta|x-L/2| - sgn(-\varepsilon)e^{-\beta|\varepsilon|}sin\,\beta|\varepsilon| \end{matrix}\right\} \tag{E5}$$

**ANS.** $M_z(x) = [w/(4\beta^2)]\{-sgn(x+L/2)e^{-\beta|x+L/2|}sin\,\beta|x+L/2| + sgn(x-L/2)e^{-\beta|x-L/2|}sin\,\beta|x-L/2|\}$

**Method II for finding $M_z$**

To find $M_z$ from the derivatives of v(x) we rewrite Equations (3.47) and Equations (3.48) using the identity in Equation (3.23) as:

$$\frac{\partial}{\partial x}[e^{-\beta|x-\xi|}cos\,\beta|x-\xi|] = -\beta\,sgn(x-\xi)e^{-\beta|x-\xi|}[cos\,\beta|x-\xi| + sin\,\beta|x-\xi|] \tag{E6}$$

$$\frac{\partial}{\partial x}[e^{-\beta|x-\xi|}sin\,\beta|x-\xi|] = -\beta\,sgn(x-\xi)e^{-\beta|x-\xi|}[-cos\,\beta|x-\xi| + sin\,\beta|x-\xi|] \tag{E7}$$

The functions in Equations (E6) and (E7) can be evaluated at $\xi = -L/2$ and $\xi = L/2$ to yield the formulas we need. Noting that the sign function is a constant for the purpose of integration, we can take the derivatives of v($x$), use Equations (E6) and (E7),to obtain

$$\frac{d\mathrm{v}}{dx} = \left(\frac{w}{2k}\right)\left\{\begin{matrix} \beta[sgn(x+L/2)]^2e^{-\beta|x+L/2|}[cos\,\beta|x+L/2| + sin\,\beta|x+L/2|] \\ -\beta[sgn(x-L/2)]^2e^{-\beta|x-L/2|}[cos\,\beta|x-L/2| + sin\,\beta|x-L/2|] \end{matrix}\right\}$$

$$\frac{d\mathrm{v}}{dx} = \left(\frac{w\beta}{2k}\right)\{e^{-\beta|x+L/2|}[cos\,\beta|x+L/2| + sin\,\beta|x+L/2|] - e^{-\beta|x-L/2|}[cos\,\beta|x-L/2| + sin\,\beta|x-L/2|]\} \tag{E8}$$

$$\frac{d^2\mathrm{v}}{dx^2} = \left(\frac{w\beta}{2k}\right)\{-2\beta\,sgn(x+L/2)e^{-\beta|x+L/2|}sin\,\beta|x+L/2| + 2\beta\,sgn(x-L/2)e^{-\beta|x-L/2|}sin\,\beta|x-L/2|\} \tag{E9}$$

Substituting $k = 4\beta^4EI$ from Equation (3.18c) and Equation (E9) into moment curvature relationship $M_z(x) = EI(d^2\mathrm{v}/dx^2)$ we obtain the answer for moment function obtained in Method I.

**COMMENTS**

1. The discontinuity in the sign function dictated that we write the integral in two parts and approach the discontinuity in the limit.
2. In the next example the solution from this example will be used to find the maximum deflection and bending normal stress.

## EXAMPLE 3.6

A very long rectangular beam with a modulus of elasticity of 30,000 ksi rest on an elastic foundation of modulus of 2 ksi. The beam cross section and loading are as shown in Figure 3.14. Determine the maximum deflection, maximum bending normal stress, and the maximum foundation force per unit length acting on the beam.

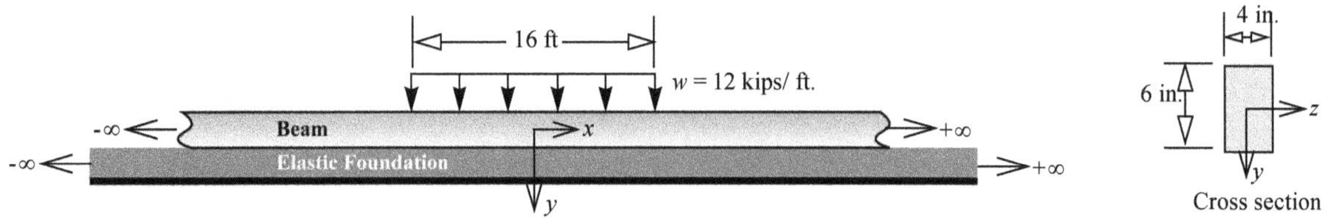


**Figure 3.14** Beam on elastic foundation in Example 3.6.

**PLAN**

The maximum deflection and moment will be at the origin and its value can be found from the solution of deflection and moment function in Example 3.5. The maximum moment can be found from the flexure formula. The maximum force per unit length exerted on the beam will be the maximum deflection multiplied by the foundation modulus.

**SOLUTION**

The maximum deflection and moment will be under the load, i.e., for $-L/2 \leq x \leq L/2$. In this region we note that $x + L/2$ is always positive and $x - L/2 = -(L/2 - x)$ is always negative. We can substitute for the sign function and remove the absolute values from the solution obtained in Example 3.5 to obtain

$$v(x) = \left(\frac{w}{2k}\right)[2 - e^{-\beta(x+L/2)}cos\beta(x+L/2) - e^{-\beta(L/2-x)}cos\beta(L/2-x)] \qquad -L/2 \leq x \leq L/2 \qquad (E1)$$

$$M_z(x) = -\left(\frac{w}{4\beta^2}\right)[e^{-\beta(x+L/2)}sin\beta(x+L/2) + e^{-\beta(L/2-x)}sin\beta(L/2-x)] \qquad -L/2 \leq x \leq L/2 \qquad (E2)$$

We substitute $x = 0$ in the above equations to obtain the maximum values for displacement and moment as:

$$v_{max} = v(0) = \left(\frac{w}{k}\right)[1 - e^{-(\beta L/2)}cos(\beta L/2)] \qquad M_{max} = M_z(0) = -\left(\frac{w}{2\beta^2}\right)[e^{-(\beta L/2)}sin(\beta L/2)] \qquad (E3)$$

The area moment of inertia is: $I = (4)(6^3)/12 = 72 \text{ in.}^4$.The parameters $\beta$ and $\beta L/2$

$$\beta = \left(\frac{k}{4EI}\right)^{1/4} = \left[\frac{2}{4(30000)(72)}\right]^{1/4} = 0.021935 \text{ in.}^{-1} \qquad \frac{\beta L}{2} = \frac{0.021935(16)(12)}{2} = 2.1057 \qquad (E4)$$

Substituting Equation (E4), $w = 1$ kips/in. and $k = 2$ ksi into Equation (E3) we obtain:

$$v_{max} = [1 - e^{-2.1057}cos(2.1057)]/2 = 0.531 \text{ in.} \qquad (E5)$$

$$M_{max} = -\frac{1}{2[0.021935]^2}e^{-2.1057}sin(2.1057) = -108.86 \text{ in.-kips} \qquad (E6)$$

The maximum force per unit length exerted is:

$$F_{max} = kv_{max} = (2)(0.531) = 1.062 \text{ kips/in} = 12.74 \text{ kips/ft} \qquad (E7)$$

The maximum bending normal stress will be at ($y_{max} = \pm 3$ in.) and can be found as shown below

$$(\sigma_{xx})_{max} = -\left[\frac{M_{max}y_{max}}{I}\right] = -\left[\frac{(-108.86)(\pm 3)}{72}\right] = \pm 4.536 \text{ ksi} \qquad (E8)$$

**ANS.** $v_{max} = 0.531$ in. $\quad F_{max} = 12.74$ kips/ft $\quad (\sigma_{xx})_{max} = 4.536$ ksi (T) or (C)

## EXAMPLE 3.7

Figure 3.15 shows a very long rectangular beam with a modulus of elasticity of 30,000 ksi that rests on an elastic foundation of modulus of 2 ksi. The beam is subjected to a transverse load of $p_y = 300e^{-(x/10)}sin(\pi x/10)$ lb/ft and has the cross section shown. Determine the deflection and bending moment at the mid section of the beam.

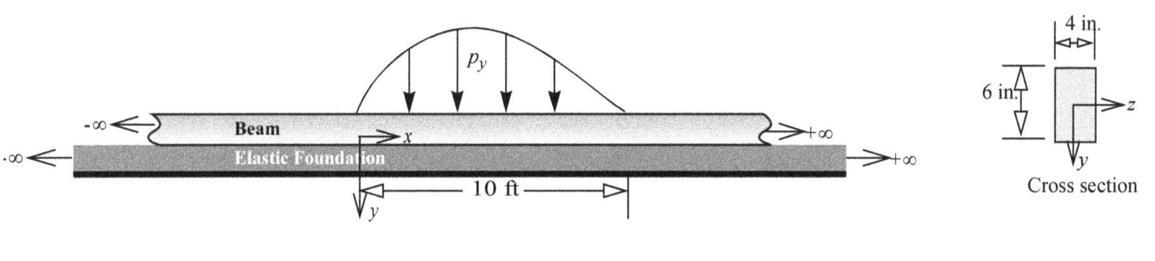


**Figure 3.15**

**PLAN**

The loading on the beam is same as in Example 3.4. We replace the influence functions associated with simply supported beams by the fundamental solutions associated with beams on elastic foundation and determine the deflection and moment in the same way as in Example 3.4.

**SOLUTION**

We non-dimensionalize the variables using $L_o = 10$ ft and the maximum value of distributed load $p_o = 191.25$ lb/ft. The area moment of inertia, bending rigidity, the parameters $\beta$ and its non-dimensional form can be found as

$$I = (4)(6^3)/12 = 72 \text{ in.}^4 \qquad EI = 2.160(10^6) \text{ kips-in.}^2 = 2.160(10^9) \text{ lbs-in.}^2$$

$$\beta = (k/4EI)^{\frac{1}{4}} = [2/\{4(2.160)(10^6)\}]^{\frac{1}{4}} = 0.021935 \text{ in.}^{-1} \qquad \hat{\beta} = 0.021935(10)(12) = 2.6321 \qquad (E1)$$

The integrals for displacements and moment can be evaluated using trapezoidal rule with equally spaced base points given by Equations (E2) and (E3) in Example 3.3. But now $\widehat{G}(\widehat{x}, \widehat{\xi})$ and $\widehat{G}_2(\widehat{x}, \widehat{\xi})$ are given by Equations (3.26a) and (3.26c). Equations (E2) and (E3) in Example 3.3 are evaluated using spreadsheet at the midpoint of $\widehat{x} = 0.5$ as shown in Table 3.7 and explained below.

(i) The given distributed load is calculated at base points as shown in columns *A* and *B* starting from row 3.

(ii) The base points are divided by the length of the beam of 10 ft. and the distributed load values are divided by the maximum value of 191.25 lb/ft to obtain the entries in columns *C* and *D* from row 3 onwards.

(iii) In cell *C1* we enter the non-dimensionalized value of the point at which we wish to find the displacement and bending moment. Subsequently, we will need to just change value in *C1* to obtain displacement and internal moment at any point.

(iv) The difference $\widehat{x} - \widehat{\xi}$ is evaluated in column *E* and the $sgn(\widehat{x} - \widehat{\xi})$ function evaluated in column *F*. It should be noted that the sign function does not appear in fundamental solutions $\widehat{G}$ or $\widehat{G}_2$. If shear force is to be evaluated then it will be needed in the formula for $\widehat{G}_3$.

(v) Equations (3.26a) and (3.26c) are entered into cells *F3* and *H3* and copied into the rest of the column, giving us the values of the fundamental solutions $\widehat{G}$ and $\widehat{G}_2$ at each base point.

The product of column *D* and *F* is written in column *H* and the product *D* and *H* is written in column *I*. In cell *G14* and *I14* the sum of the entries in *G* and *I* column are written as per Equations (E2) and (E3).

From values in cell *I14* and *K14* we obtain the non dimensionalized deflection and bending moment as:

$$\widehat{v} = 3.4177(10^{-3}) \qquad \widehat{M}_z = 20.774(10^{-3}) \tag{E2}$$

We obtain the deflection and bending moment at mid-point as from the non-dimensional values as:

$$v(x=5) = \frac{p_o L^4}{EI}\widehat{v} = \frac{(191.25/12)(120^4)}{2.160(10^9)}[3.4177(10^{-3})] = 5.229(10^{-3}) \text{ in.} \tag{E3}$$

$$M_z(x=5) = p_o L^2 \widehat{M}_z = (191.25/12)(120^2)(20.774)(10^{-3}) = 47677 \text{ in.-lb} \tag{E4}$$

**ANS.** $v(x=5) = 0.052$ in. ; $M_z(x=5) = 47677$ in.-lb

**Table 3.7** Solution at midpoint using spreadsheet in Example 3.7

| | A | B | C | D | E | F | G | H | I | J | K |
|---|---|---|---|---|---|---|---|---|---|---|---|
| 1 | | | 0.5 | | | | | | | | |
| 2 | $\xi$ ft. | $p_y$ lb/ft | $\widehat{\xi}$ | $\widehat{p}_y$ | $\widehat{x} - \widehat{\xi}$ | $sgn(\widehat{x} - \widehat{\xi})$ | $\widehat{\beta}\lvert\widehat{x} - \widehat{\xi}\rvert$ | $\widehat{G}$ | $\widehat{G}_i\widehat{p}_i$ | $\widehat{G}_2$ | $\widehat{G}_{2i}\widehat{p}_i$ |
| 3 | 0 | 0.00 | 0 | 0.0000 | 0.5 | 1 | 1.3161E+00 | 2.2422E-03 | 0.0000E+00 | 1.8232E-02 | 0.0000E+00 |
| 4 | 1 | 83.88 | 0.1 | 0.4386 | 0.4 | 1 | 1.0529E+00 | 3.2623E-03 | 1.4308E-03 | 1.2387E-02 | 5.4329E-03 |
| 5 | 2 | 144.37 | 0.2 | 0.7549 | 0.3 | 1 | 7.8964E-01 | 4.4010E-03 | 3.3222E-03 | 2.5895E-04 | 1.9547E-04 |
| 6 | 3 | 179.80 | 0.3 | 0.9401 | 0.2 | 1 | 5.2643E-01 | 5.5353E-03 | 5.2038E-03 | -2.0319E-02 | -1.9102E-02 |
| 7 | 4 | 191.25 | 0.4 | 1.0000 | 0.1 | 1 | 2.6321E-01 | 6.4575E-03 | 6.4575E-03 | -5.1491E-02 | -5.1491E-02 |
| 8 | 5 | 181.96 | 0.5 | 0.9514 | 0.0 | -1 | 0.0000E+00 | 6.8546E-03 | 6.5214E-03 | -9.4979E-02 | -9.0364E-02 |
| 9 | 6 | 156.59 | 0.6 | 0.8187 | -0.1 | -1 | 2.6321E-01 | 6.4575E-03 | 5.2870E-03 | -5.1491E-02 | -4.2158E-02 |
| 10 | 7 | 120.52 | 0.7 | 0.6302 | -0.2 | -1 | 5.2643E-01 | 5.5353E-03 | 3.4882E-03 | -2.0319E-02 | -1.2805E-02 |
| 11 | 8 | 79.23 | 0.8 | 0.4143 | -0.3 | -1 | 7.8964E-01 | 4.4010E-03 | 1.8233E-03 | 2.5895E-04 | 1.0728E-04 |
| 12 | 9 | 37.69 | 0.9 | 0.1971 | -0.4 | -1 | 1.0529E+00 | 3.2623E-03 | 6.4291E-04 | 1.2387E-02 | 2.4411E-03 |
| 13 | 10 | 0.00 | 1 | 0.0000 | -0.5 | -1 | 1.3161E+00 | 2.2422E-03 | 1.5852E-19 | 1.8232E-02 | 1.2890E-18 |
| 14 | | | | | | | | | 3.4177E-03 | | -2.0774E-02 |

**COMMENT**

1. This example demonstrates that the methodology of solving a beam bending problem is the same for classical beam and beam on elastic foundation. The change between the applications is captured by the influence function, which can be defined in a sub-program.

## 3.5 FINITE BEAMS

We consider a finite beam *AB* resting on elastic foundation in Figure 3.16*a*. There is no loss of generality to assume the origin at the center of the beam. In Figure 3.16*b* we consider the finite beam *AB* as part of an infinite beam with a force and a moment applied at each end. These forces are in the positive *y*-direction and the moment is positive in the counter-clockwise direction with respect to the *z* axis. These forces and moments are unknown quantities to be determined as described later. These forces and moments represent jumps in the shear force and bending moments as we move from the inside of the finite beam to the outside into the infinite beam.

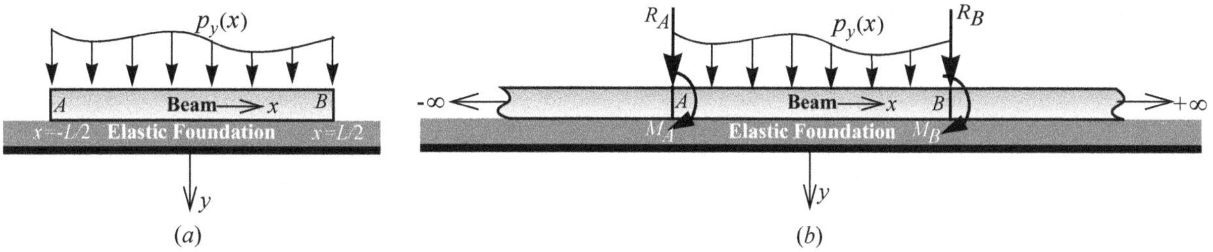


**Figure 3.16** Beams on Elastic Foundations. (a) Finite beam. (b) Part of an infinite beam.

Using the fundamental solutions in Table 3.6 we can write the following by superposition.

$$\mathrm{v}(x) = R_A\mathcal{G}(x,-L/2) + M_A\mathcal{H}(x,-L/2) + R_B\mathcal{G}(x,L/2) + M_B\mathcal{H}(x,L/2) + \int_{-L/2}^{L/2}\mathcal{G}(x,\xi)p_y(\xi)d\xi \tag{3.30a}$$

$$\frac{\partial \mathrm{v}}{\partial x}(x) = R_A\mathcal{G}_1(x,-L/2) + M_A\mathcal{H}_1(x,-L/2) + R_B\mathcal{G}_1(x,L/2) + M_B\mathcal{H}_1(x,L/2) + \int_{-L/2}^{L/2}\mathcal{G}_1(x,\xi)p_y(\xi)d\xi \tag{3.30b}$$

$$M_z(x) = R_A\mathcal{G}_2(x,-L/2) + M_A\mathcal{H}_2(x,-L/2) + R_B\mathcal{G}_2(x,L/2) + M_B\mathcal{H}_2(x,L/2) + \int_{-L/2}^{L/2}\mathcal{G}_2(x,\xi)p_y(\xi)d\xi \tag{3.30c}$$

$$V_y(x) = R_A\mathcal{G}_3(x,-L/2) + M_A\mathcal{H}_3(x,-L/2) + R_B\mathcal{G}_3(x,L/2) + M_B\mathcal{H}_3(x,L/2) + \int_{-L/2}^{L/2}\mathcal{G}_3(x,\xi)p_y(\xi)d\xi \tag{3.30d}$$

In the above equations, all quantities are known except forces and moments at $A$ and $B$ in Figure 3.16*b*. Applying the four boundary conditions that are applicable to the finite beam we will generate 4 equations in 4 unknowns that can be solved. The equations above then gives us the complete solution at any point $x$.

In order to be able to evaluate the *sgn* functions, if the left boundary at $A$ is considered then $x = -L/2+\varepsilon$ and if the right boundary at $B$ is considered then $x = L/2-\varepsilon$. We then let $\varepsilon \to 0$. Table 3.8 shows the values of the fundamental solutions at the two boundaries for convenience.

**Table 3.8** Fundamental solutions values on the boundary.

| Left side $x = -L/2+\varepsilon$ | | Right Side $x = L/2-\varepsilon$ | |
|---|---|---|---|
| $\xi = -L/2$ | $\xi = L/2$ | $\xi = -L/2$ | $\xi = L/2$ |
| $\mathcal{G} = \frac{\beta}{2k}$ | $\mathcal{G} = \frac{\beta}{2k}e^{-\beta L}(\cos\beta L + \sin\beta L)$ | $\mathcal{G} = \frac{\beta}{2k}e^{-\beta L}[\cos\beta L + \sin\beta L]$ | $\mathcal{G} = \frac{\beta}{2k}$ |
| $\mathcal{G}_1 = 0$ | $\mathcal{G}_1 = \frac{\beta^2}{k}e^{-\beta L}\sin\beta L$ | $\mathcal{G}_1 = -\frac{\beta^2}{k}e^{-\beta L}\sin\beta L$ | $\mathcal{G}_1 = 0$ |
| $\mathcal{G}_2 = -\left(\frac{1}{4\beta}\right)$ | $\mathcal{G}_2 = \frac{1}{4\beta}e^{-\beta L}(\sin\beta L - \cos\beta L)$ | $\mathcal{G}_2 = \frac{1}{4\beta}e^{-\beta L}(\sin\beta L - \cos\beta L)$ | $\mathcal{G}_2 = -\left(\frac{1}{4\beta}\right)$ |
| $\mathcal{G}_3 = -\left(\frac{1}{2}\right)$ | $\mathcal{G}_3 = \frac{1}{2}e^{-\beta L}\cos\beta L$ | $\mathcal{G}_3 = -\frac{1}{2}e^{-\beta L}\cos\beta L$ | $\mathcal{G}_3 = \frac{1}{2}$ |
| $\mathcal{H} = 0$ | $\mathcal{H} = -\frac{\beta^2}{k}e^{-\beta L}\sin\beta L$ | $\mathcal{H} = \frac{\beta^2}{k}e^{-\beta L}\sin\beta L$ | $\mathcal{H} = 0$ |
| $\mathcal{H}_1 = \frac{\beta^3}{k}$ | $\mathcal{H}_1 = -\frac{\beta^3}{k}e^{-\beta L}(\sin\beta L - \cos\beta L)$ | $\mathcal{H}_1 = -\frac{\beta^3}{k}e^{-\beta L}(\sin\beta L - \cos\beta L)$ | $\mathcal{H}_1 = \left(\frac{\beta^3}{k}\right)$ |
| $\mathcal{H}_2 = -\left(\frac{1}{2}\right)$ | $\mathcal{H}_2 = \frac{1}{2}e^{-\beta L}\cos\beta L$ | $\mathcal{H}_2 = -\frac{1}{2}e^{-\beta L}\cos\beta L$ | $\mathcal{H}_2 = \frac{1}{2}$ |
| $\mathcal{H}_3 = -\left(\frac{\beta}{2}\right)$ | $\mathcal{H}_3 = -\left(\frac{\beta}{2}\right)e^{-\beta L}[\cos\beta L + \sin\beta L]$ | $\mathcal{H}_3 = -\left(\frac{\beta}{2}\right)e^{-\beta L}[\cos\beta L + \sin\beta L]$ | $\mathcal{H}_3 = -\left(\frac{\beta}{2}\right)$ |

In practice, the algebra needed to solve for the four unknowns can be tedious. So we will consider simplifications that will let us obtain analytical solutions, and in case of complicated distribution of $p_y(x)$ we could get numerical solutions using spreadsheets as we have done in several examples. In developing these solutions we will assume the following:

*Beam is continuously attached to the elastic foundation irrespective of the loads and boundary conditions.*

The above assumption is necessary to develop analytical solutions. If beam is not attached continuously then the negative values of deflection v at a point indicates the beam and the foundation separates at that point. It is the responsibility of the designer to ensure the mathematical model actually simulates the real life condition and thus should make a decision if negative values of deflection v are acceptable or not.

### 3.5.1 Symmetric loading

We assume that the beam has symmetric boundary conditions and loading, that is $p_y(x) = p_y(-x) = p_s(x)$. With these simplifications we conclude that

$$R_B = R_A = R_s \quad \text{and} \quad M_B = -M_A = -M_s \tag{3.31a}$$

Substituting Equation (3.31a) into Equations (3.30a) through (3.30d) we obtain

$$\mathrm{v}(x) = R_s[\mathcal{G}(x,-L/2)+\mathcal{G}(x,L/2)] + M_s[\mathcal{H}(x,-L/2)-\mathcal{H}(x,L/2)] + \int_{-L/2}^{L/2}\mathcal{G}(x,\xi)p_s(\xi)d\xi \tag{3.32a}$$

$$\frac{\partial \mathrm{v}}{\partial x}(x) = R_s[\mathcal{G}_1(x,-L/2)+\mathcal{G}_1(x,L/2)] + M_s[\mathcal{H}_1(x,-L/2)-\mathcal{H}_1(x,L/2)] + \int_{-L/2}^{L/2}\mathcal{G}_1(x,\xi)p_s(\xi)d\xi \tag{3.32b}$$

$$M_z(x) = R_s[\mathcal{G}_2(x,-L/2)+\mathcal{G}_2(x,L/2)] + M_s[\mathcal{H}_2(x,-L/2)-\mathcal{H}_2(x,L/2)] + \int_{-L/2}^{L/2}\mathcal{G}_2(x,\xi)p_s(\xi)d\xi \tag{3.32c}$$

$$V_y(x) = R_s[\mathcal{G}_3(x,-L/2)+\mathcal{G}_3(x,L/2)] + M_s[\mathcal{H}_3(x,-L/2)-\mathcal{H}_3(x,L/2)] + \int_{-L/2}^{L/2}\mathcal{G}_3(x,\xi)p_s(\xi)d\xi \tag{3.32d}$$

We now need only two boundary conditions at one of the ends to obtain $R_s$ and $M_s$ as demonstrated in Example 3.8.

### 3.5.2 Asymmetric loading and boundary conditions

We assume that the beam has asymmetric boundary conditions and the loading, that is, $p_y(x) = -p_y(-x) = p_a(x)$. With these simplifications we conclude that

$$R_B = -R_A = -R_a \quad \text{and} \quad M_B = M_A = M_a \tag{3.32e}$$

Substituting Equation (3.32e) into Equations (3.30a) through (3.30d) we obtain

$$\mathrm{v}(x) = R_a[\mathcal{G}(x,-L/2)-\mathcal{G}(x,L/2)] + M_a[\mathcal{H}(x,-L/2)+\mathcal{H}(x,L/2)] + \int_{-L/2}^{L/2}\mathcal{G}(x,\xi)p_a(x)d\xi \tag{3.33a}$$

$$\frac{\partial \mathrm{v}}{\partial x}(x) = R_a[\mathcal{G}_1(x,-L/2)-\mathcal{G}_1(x,L/2)] + M_a[\mathcal{H}_1(x,-L/2)+\mathcal{H}_1(x,L/2)] + \int_{-L/2}^{L/2}\mathcal{G}_1(x,\xi)p_a(x)d\xi \tag{3.33b}$$

$$M_z(x) = R_a[\mathcal{G}_2(x,-L/2)-\mathcal{G}_2(x,L/2)] + M_a[\mathcal{H}_2(x,-L/2)+\mathcal{H}_2(x,L/2)] + \int_{-L/2}^{L/2}\mathcal{G}_2(x,\xi)p_a(x)d\xi \tag{3.33c}$$

$$V_y(x) = R_a[\mathcal{G}_3(x,-L/2)-\mathcal{G}_3(x,L/2)] + M_a[\mathcal{H}_3(x,-L/2)+\mathcal{H}_3(x,L/2)] + \int_{-L/2}^{L/2}\mathcal{G}_3(x,\xi)p_a(x)d\xi \tag{3.33d}$$

### 3.5.3 General loading and boundary conditions

Figure 3.17 shows how a beam under arbitrary distributed loads and boundary conditions can be analyzed as a sum of symmetric and an asymmetric problem. The variables for the symmetric and asymmetric can be related to the arbitrary distributed loading as shown below.

$$\begin{aligned} p_s(x) &= [p_y(x)+p_y(-x)]/2 \qquad P_s = [P_1+P_2]/2 \qquad M_s = [M_1+M_2]/2 \\ p_a(x) &= [p_y(x)-p_y(-x)]/2 \qquad P_a = [P_1-P_2]/2 \qquad M_a = [M_1-M_2]/2 \end{aligned} \tag{3.34a}$$

In writing the above equation we made use of the fact that any function can be written as a sum of a symmetric function and asymmetric function as shown for the forces $P_1$ and $P_2$ below. In a similar manner other variables can be written.

$$P_1 = (P_1+P_2)/2+(P_1-P_2)/2 = P_s+P_a \quad \text{and} \quad P_2 = (P_1+P_2)/2-(P_1-P_2)/2 = P_s-P_a \tag{3.34b}$$

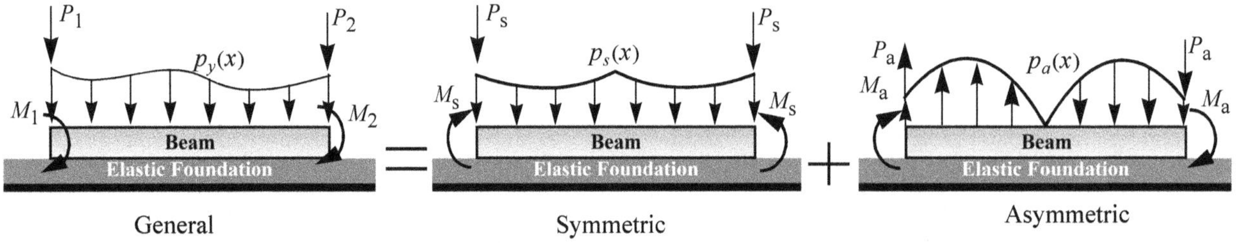


**Figure 3.17** General boundary loads.

The solutions of symmetric and asymmetric problem are found as described in Sections 3.5.1and 3.5.2. By superposition we obtain the solutions to the general problem. In other words

$$R_A = R_s+R_a \qquad R_B = R_s-R_a \qquad M_A = M_s+M_a \qquad M_B = -M_s+M_a \tag{3.35}$$

Substituting the above into Equations (3.30a) through (3.30d) we obtain the complete solution of the original problem, with $p_y$ being the original distributed load in the equations. We demonstrate this approach in the next two examples. This approach will be of greater use if we create a table of solutions for symmetric and asymmetric problems.

## EXAMPLE 3.8

A finite beam on elastic foundation is loaded as shown in Figure 3.18. (*a*) In terms of $E$, $I$, $\beta$, $k$, and $L$ determine the unknown force $R_s$ and moment $M_s$ in Equations (3.32a) through (3.32d). (*b*) Make a plot of the deflection, slope, internal bending moment, and internal shear force across the beam assuming $k = 2$ ksi, $P = 1.5$ kips and $L = 10$ ft.

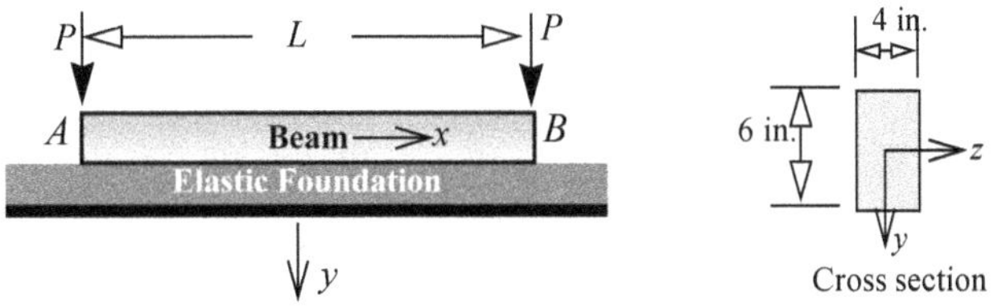


**Figure 3.18** Finite beam under end forces.

**PLAN**

(*a*) We make an imaginary cut at $x = L/2 - \varepsilon$ and draw the free body diagram to get the boundary condition on moment and shear force. Noting that $p_y(x) = 0$, we generate two equations using Equations (3.32c) and (3.32d) and solve for $R_s$ and $M_s$. (*b*) From Equations (3.30a) through (3.30d), we obtain deflection, slope, bending moment, and shear force across the beam as a function of $x$, which we then plot.

**SOLUTION**

(*a*) Figure 3.19 shows a free body diagram after making an imaginary cut at an infinitesimal distance from the end of the beam. The internal bending moment and shear force are drawn as per our sign convention.

By equilibrium of force and moment we obtain

$$M_z(L/2 - \varepsilon) = 0 \qquad V_y(L/2 - \varepsilon) = P \tag{E1}$$

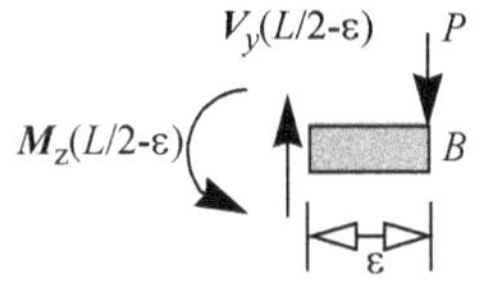


**Figure 3.19** Free body diagram at beam end.

From Table 3.8 we have

$$\mathcal{G}_2(L/2 - \varepsilon, -L/2) = e^{-\beta L}(\sin\beta L - \cos\beta L)/4\beta \qquad \mathcal{G}_2(L/2 - \varepsilon, L/2) = -(1/4\beta) \tag{E2}$$

$$\mathcal{H}_2(L/2 - \varepsilon, -L/2) = -e^{-\beta L}\cos\beta L/2 \qquad \mathcal{H}_2(L/2 - \varepsilon, L/2) = 1/2 \tag{E3}$$

$$\mathcal{G}_3(L/2 - \varepsilon, -L/2) = -e^{-\beta L}\cos\beta L/2 \qquad \mathcal{G}_3(L/2 - \varepsilon, L/2) = 1/2 \tag{E4}$$

$$\mathcal{H}_3(L/2 - \varepsilon, -L/2) = -\beta e^{-\beta L}(\sin\beta L + \cos\beta L)/2 \qquad \mathcal{H}_3(L/2 - \varepsilon, L/2) = -\beta/2 \tag{E5}$$

Substituting $p_y(\xi) = 0$ and Equations (3.32c) and (3.32d) into Equation (E1) we obtain

$$R_s[e^{-\beta L}(\sin\beta L - \cos\beta L) - 1]/4\beta - M_s[e^{-\beta L}\cos\beta L + 1]/2 = 0 \tag{E6}$$

$$R_s[1 - e^{-\beta L}\cos\beta L]/2 + \beta M_s[1 - e^{-\beta L}(\sin\beta L + \cos\beta L)]/2 = P \tag{E7}$$

Solving Equations (E6) and (E7) we obtain

$$\textbf{ANS. } R_s = P\frac{4(1 + e^{-\beta L}\cos\beta L)}{1 - e^{-2\beta L} + 2e^{-\beta L}\sin\beta L} \qquad M_s = -\left(\frac{2P}{\beta}\right)\frac{1 - e^{-\beta L}(\sin\beta L - \cos\beta L)}{(1 - e^{-2\beta L} + 2e^{-\beta L}\sin\beta L)} \tag{E8}$$

(*b*) From the given information we obtain the following values for the various parameters

$$I = (4)(6^3)/12 = 72 \text{ in.}^4 \qquad EI = 2.160(10^6) \text{ kips-in.}^2 = 2.160(10^9) \text{ lbs-in.}^2 \tag{E9}$$

$$\beta = (k/4EI)^{1/4} = [2/\{4(2.160)(10^6)\}]^{1/4} = 0.021935 \text{ in.}^{-1} = 0.26321 \text{ ft.}^{-1} \qquad \beta L = 2.6321 \tag{E10}$$

Substituting $P = 1.5$ kips and the above parameters into Equation (E8) we obtain

$$R_s = 5.2801 \text{ kips} \qquad M_s = -9.6547 \text{ ft-kips} \tag{E11}$$

The values of deflection, slope, shear force, and moment were calculated on a spreadsheet after substituting Equation (E11) into Equations (3.32a) through (3.32d). The location and the maximum values of these quantities are shown in Table 3.9.

**Table 3.9** Maximum values and the location in Example 3.8.

| | $v_{max}$ | $(dv/dx)_{max}$ | $(M_z)_{max}$ | $(V_y)_{max}$ |
|---|---|---|---|---|
| Location $x$ | ±5 ft. | ±5 ft. | 0 | ±5 ft. |
| Value | $2.264(10^{-3})$ ft. | $0.7160(10^{-3})$ rads. | 2.5778 ft.-kips | ∓1.5 kips |

The values deflection, slope, shear force, and moment were non-dimensionalized with the maximum values shown in Table 3.9 and plotted as a function of $x/L$ and are shown in Figure 3.20.

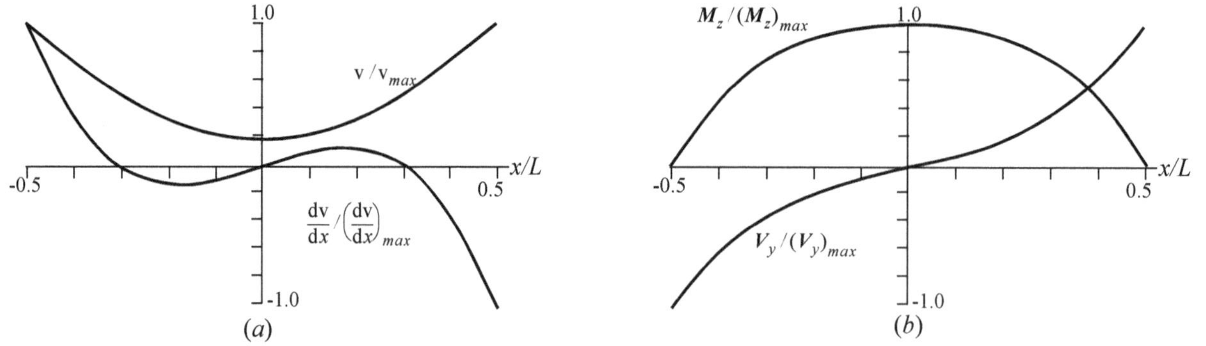


**Figure 3.20** Plots of (*a*) displacement and slope (*b*) Internal moment and shear force.

## COMMENTS

1. Figure 3.20 shows that deflection and moment are symmetric about the origin. Slope and shear force are anti-symmetric hence zero at the origin
2. Figure 3.20 shows that the moment at each end is zero, shear force at left end is the -*P* and +P at right end, meeting the boundary conditions.

## EXAMPLE 3.9

A finite beam on elastic foundation is loaded as shown in Figure 3.18. (a) In terms of $E$, $I$, $\beta$, $k$, and $L$ determine the unknown forces $R_A$, $R_B$ and moments $M_A$, $M_B$ in Equations (3.32a) through (3.32d). (b) Make a plot of the deflection, slope, internal bending moment, and internal shear force across the beam assuming the beam cross-section and material properties of the beam and foundation are the same as in Example 3.8.

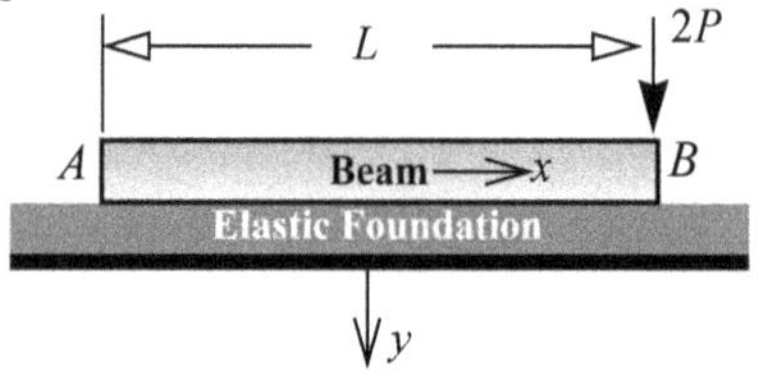


**Figure 3.21** Finite beam under end forces in Example 3.9.

**PLAN**

The problem can be written as a symmetric problem with load $P$ at each end and an asymmetric problem with loads $P$ in opposite directions at the end. The symmetric problem was solved in Example 3.8. The asymmetric problem can be solved and superposed to get the complete solution.

**SOLUTION**

Figure 3.22 shows the superposition of symmetric and asymmetric problem.

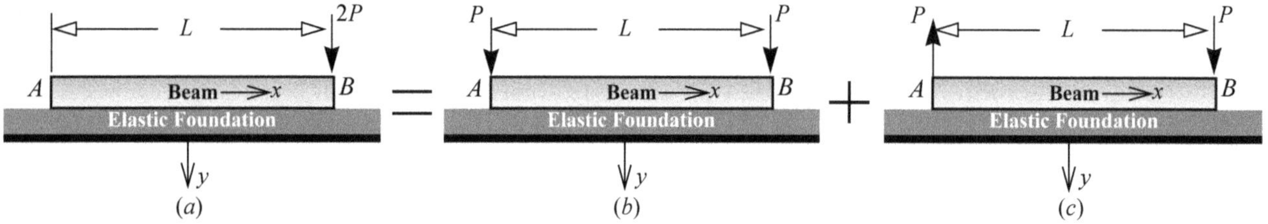


**Figure 3.22** Superposition in Example 3.9. (*a*) End load (*b*) Symmetric loads (*c*) Asymmetric loads.

From Example 3.8 we have the following for the symmetric problem

$$\mathcal{G}_2(L/2-\varepsilon, -L/2) = e^{-\beta L}(\sin\beta L - \cos\beta L)/4\beta \qquad \mathcal{G}_2(L/2-\varepsilon, L/2) = -(1/4\beta) \tag{E1}$$

$$\mathcal{H}_2(L/2-\varepsilon, -L/2) = -e^{-\beta L}\cos\beta L/2 \qquad \mathcal{H}_2(L/2-\varepsilon, L/2) = 1/2 \tag{E2}$$

$$\mathcal{G}_3(L/2-\varepsilon, -L/2) = -e^{-\beta L}\cos\beta L/2 \qquad \mathcal{G}_3(L/2-\varepsilon, L/2) = 1/2 \tag{E3}$$

$$\mathcal{H}_3(L/2-\varepsilon, -L/2) = -\beta e^{-\beta L}(\sin\beta L + \cos\beta L)/2 \qquad \mathcal{H}_3(L/2-\varepsilon, L/2) = -\beta/2 \tag{E4}$$

$$R_s = P\frac{4(1+e^{-\beta L}\cos\beta L)}{1-e^{-2\beta L}+2e^{-\beta L}\sin\beta L} \qquad M_s = -\left(\frac{2P}{\beta}\right)\frac{1-e^{-\beta L}(\sin\beta L - \cos\beta L)}{(1-e^{-2\beta L}+2e^{-\beta L}\sin\beta L)} \tag{E5}$$

$$R_s = 5.2801 \text{ kips} \qquad M_s = -9.6547 \text{ ft-kips} \tag{E6}$$

Substituting $x = L/2-\varepsilon$, $p_y(\xi) = 0$, Equations (E2) through (E5), and Equations (3.33c) and (3.33d) into Equation (E1) we obtain the following as $\varepsilon \to 0$.

$$R_a[e^{-\beta L}(\sin\beta L - \cos\beta L) + 1]/4\beta + M_a[1 - e^{-\beta L}\cos\beta L]/2 = 0 \tag{E7}$$

$$-R_a[1+e^{-\beta L}\cos\beta L]/2 - \beta M_a[1+e^{-\beta L}(\sin\beta L+\cos\beta L)]/2 = P \quad \text{(E8)}$$

Solving Equations (E7) and (E8) we obtain

$$R_a = -4P\frac{[1-e^{-\beta L}\cos\beta L]}{1-2e^{-\beta L}\sin\beta L - e^{-2\beta L}} \qquad M_a = \frac{2P}{\beta}\frac{[e^{-\beta L}(\sin\beta L-\cos\beta L)+1]}{1-2e^{-\beta L}\sin\beta L - e^{-2\beta L}} \quad \text{(E9)}$$

Substituting $\beta = 0.26321$ ft.$^{-1}$, $\beta L = 2.6321$ and $P = 1.5$ kips into the above equation we obtain

$$R_a = -6.8962 \text{ kips} \qquad M_a = 13.5323 \text{ ft-kips} \quad \text{(E10)}$$

By superposition we obtain

$$R_A = R_s + R_a = 5.2801 + (-6.8962) = -1.6106 \text{ kips} \qquad R_B = R_s - R_a = 5.2801 - (-6.8962) = 12.1763 \text{ kips}$$
$$M_A = M_s + M_a = -9.6547 + 13.5323 = 3.8726 \text{ ft-kips} \qquad M_B = -M_s + M_a = 9.6547 + 13.5323 = 23.187 \text{ ft-kips} \quad \text{(E11)}$$

**ANS.** $R_A = -1.6106$ kips $\quad R_B = 12.1763$ kips $\quad M_A = 3.8726$ ft-kips $\quad M_B = 23.187$ ft-kips

The values of deflection, slope, shear force, and moment were calculated on a spreadsheet after substituting Equation (E11) into Equations (3.32a) through (3.32d). The location and the maximum values of these quantities are shown in Table 3.10.

**Table 3.10** Maximum values and the location in Example 3.8.

| | $v_{max}$ | $(dv/dx)_{max}$ | $(M_z)_{max}$ | $(V_y)_{max}$ |
|---|---|---|---|---|
| Location $x$ | 5 ft. | 5 ft. | 2 2 ft | 5 ft. |
| Value | $5.617(10^{-3})$ ft. | $1,431(10^{-3})$ rads. | 3.52 ft.-kips | 2 kips |

The values deflection, slope, shear force, and moment were non-dimensionalized with the maximum values shown in Table 3.10 and plotted as a function of $x/L$ and are shown in Figure 3.23.

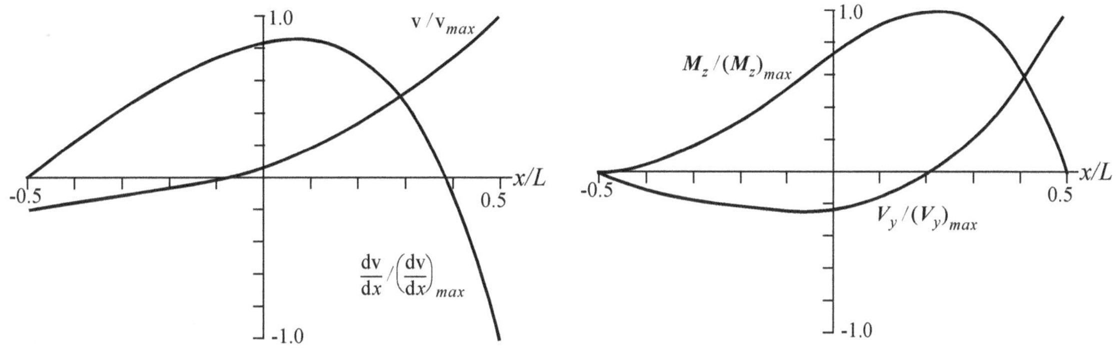


**Figure 3.23**

**COMMENTS**

1. The example demonstrates the solution to any loading can be constructed as a superposition of solution to symmetric and asymmetric loading.
2. The graph shows that deflection is negative at $x/L < -0.1$. If the assumption that the beam is not continuously attached to elastic foundation then the beam would separate from the elastic foundation and the above analysis will not be valid. A possibility would be to redefine the length of beam as $0.6L$ and redo do the analysis iteratively till only positive values of deflection are obtained.

The above comment emphasizes that the theoretical prediction should be checked to see if it is even possible and a further check should be to conduct an experiment to validate the theoretical model.

## 3.6 CLOSURE

The concept of influence function makes it possible to develop techniques by which we can analyze beam deformation and stresses for complex loading or loading obtained experimentally. Influence functions intrinsically satisfy the boundary conditions. However, by replacing supports with appropriate reaction forces and / or moments we can also analyze beams with different boundary conditions.

We developed the concept of fundamental solutions by extending the concept of influence functions to infinite beams on elastic foundations. The boundary conditions now required that any disturbance due to force or moment must die out at infinity. We then interpreted the finite beam as part of the infinite beam with unknown fictitious force and moment applied at the boundary of the finite beam. We considered any loading and boundary conditions as a sum of symmetric and asymmetric problem, which helped reduce the algebraic tedium to the solution of two unknowns. The unknowns were determined by satisfying the boundary conditions at one end of the finite beam.

We were able to find analytical solutions when loading was simple for both the classical beams and beams on elastic foundation. We used numerical integration when loading was complex or in data form such as those that will be obtained experimentally. We saw that for numerical integration we should non dimensionalize the variables. We used trapezoidal rule to demonstrate the concept. If higher accuracies are needed then the trapezoidal rule should be replaced by Gauss Quadrature.

This is a stand alone chapter and does not impact subsequent chapters. However, the concepts of influence functions and fundamental solutions are used extensively in science and engineering literature.

## 3.7 SYNOPSIS OF EQUATIONS

### Influence Functions

**Point force**

$$v(x) = \int_L \mathcal{G}(x,\xi)p_y(\xi)d\xi \quad \psi(x) = \frac{dv}{dx} = \int_L \mathcal{G}_1(x,\xi)p_y(\xi)d\xi\ ;\ M_z(x) = EI\frac{d^2v}{dx^2} = \int_L \mathcal{G}_2(x,\xi)p_y(\xi)d\xi\ ;$$

$$V_y(x) = -\left(\frac{dM_z}{dx}\right) = \int_L \mathcal{G}(x,\xi)p_y(\xi)d\xi\ ;\ \mathcal{G}_1(x,\xi) = \frac{\partial \mathcal{G}}{\partial x} \quad \mathcal{G}_2(x,\xi) = EI\frac{\partial \mathcal{G}_1}{\partial x} \quad \mathcal{G}_3(x,\xi) = -\left(\frac{\partial \mathcal{G}_2}{\partial x}\right)$$

**Point moment**

$$v(x) = \mathcal{H}(x,\xi)M \quad \psi(x) = \frac{dv}{dx} = \mathcal{H}_1(x,\xi)M \quad M_z(x) = EI\frac{d^2v}{dx^2} = \mathcal{H}_2(x,\xi)M \quad V_y(x) = -\left(\frac{dM_z}{dx}\right) = \mathcal{H}_3(x,\xi)M$$

$$\mathcal{H}_1(x,\xi) = \frac{\partial \mathcal{H}}{\partial x} \quad \mathcal{H}_3(x,\xi) = -\left(\frac{\partial \mathcal{H}_2}{\partial x}\right) \quad \mathcal{H}_3(x,\xi) = -\left(\frac{\partial \mathcal{H}_2}{\partial x}\right) \quad \mathcal{H}_3(x,\xi) = -\left(\frac{\partial \mathcal{H}_2}{\partial x}\right)$$

### Numerical Integration

**Equally spaced base points**

$$I = \left[\frac{f(\xi_0)+f(\xi_N)}{2} + \sum_{i=1}^{N-1} f(\xi_i)\right](\Delta\xi)$$

**Non-dimensional variables**

$$\hat{x} = x/L_o \quad \hat{\xi} = \xi/L_o \quad \hat{p}_y = p_y/p_o \quad \hat{P} = P/(p_oL_o)$$

$$\hat{v} = v[EI/(p_oL_o^4)] \quad \hat{\psi} = \psi[EI/(p_oL_o^3)] \quad \hat{M} = M/(p_oL_o^2) \quad \hat{V}_y = V_y/(p_oL_o)$$

$$\hat{\mathcal{G}} = \mathcal{G}[EI/L_o^3] \quad \hat{\mathcal{G}}_1 = \mathcal{G}_1[EI/L_o^2] = \partial\hat{\mathcal{G}}/\partial\hat{x} \quad \hat{\mathcal{G}}_2 = \mathcal{G}_2/L_o = \partial\hat{\mathcal{G}}_1/\partial\hat{x} \quad \hat{\mathcal{G}}_3 = \mathcal{G}_3 = \partial\hat{\mathcal{G}}_2/\partial\hat{x}$$

$$\hat{\mathcal{H}} = \mathcal{H}[EI/L_o^2] \quad \hat{\mathcal{H}}_1 = \mathcal{H}_1[EI/L_o] = \partial\hat{\mathcal{H}}/\partial\hat{x} \quad \hat{\mathcal{H}}_2 = \mathcal{H}_2 = \partial\hat{\mathcal{H}}_1/\partial\hat{x} \quad \hat{\mathcal{H}}_3 = \mathcal{H}_3L_o = d\hat{\mathcal{H}}_2/d\hat{x}$$

### Influence functions for classical beams

**General solution**

$$\mathcal{G}(x,\xi) = \frac{1}{6EI}\langle x-\xi\rangle^3 + c_1\frac{x^3}{6} + c_2\frac{x^2}{2} + c_3x + c_4$$

**Simply supported beam**

$$\mathcal{G}(x,\xi) = \frac{1}{6EI}\left[\langle x-\xi\rangle^3 - \frac{(L-\xi)}{L}x^3 - \frac{(L-\xi)^3}{L}x + (L-\xi)Lx\right];\ \mathcal{H}(x,\xi) = \frac{\partial\mathcal{G}}{\partial\xi} = \frac{1}{6EI}\left[-3\langle x-\xi\rangle^2 + \frac{x^3}{L} + \frac{3(L-\xi)^2}{L}x - Lx\right]$$

### Beam on elastic foundations

$$;\frac{\partial\mathcal{G}}{\partial x} = -\left(\frac{\partial\mathcal{G}}{\partial\xi}\right);\ sgn(x-\xi) = \begin{cases} 1 & x>\xi \\ -1 & x<\xi \end{cases}$$

**Infinite beams**

$$\mathcal{D}_1(x,\xi) = \frac{\partial}{\partial\xi}[e^{-\beta|x-\xi|}\cos\beta|x-\xi|] = \beta\, sgn(x-\xi)e^{-\beta|x-\xi|}[\cos\beta|x-\xi| + \sin\beta|x-\xi|]$$

$$\mathcal{D}_2(x,\xi) = \frac{\partial}{\partial\xi}[e^{-\beta|x-\xi|}\sin\beta|x-\xi|] = \beta\, sgn(x-\xi)e^{-\beta|x-\xi|}[-\cos\beta|x-\xi| + \sin\beta|x-\xi|]$$

$$\mathcal{I}_1(x,\xi) = \int e^{-\beta|x-\xi|}\sin\beta|x-\xi|d\xi = sgn(x-\xi)e^{-\beta|x-\xi|}[\cos\beta|x-\xi| + \sin\beta|x-\xi|]/2\beta$$

$$\mathcal{I}_2(x,\xi) = \int e^{-\beta|x-\xi|}\cos\beta|x-\xi|d\xi = sgn(x-\xi)e^{-\beta|x-\xi|}[\cos\beta|x-\xi| - \sin\beta|x-\xi|]/2\beta$$

**Point force**

$$\mathcal{G}(x,\xi) = \beta e^{-\beta|x-\xi|}[\cos\beta|x-\xi| + \sin\beta|x-\xi|]/(2k) \quad \mathcal{G}_1 = -[\beta^2 sgn(x-\xi)e^{-\beta|x-\xi|}\sin\beta|x-\xi|]/k$$

$$\mathcal{G}_2 = e^{-\beta|x-\xi|}(\sin\beta|x-\xi| - \cos\beta|x-\xi|)/4\beta \quad \mathcal{G}_3 = -sgn(x-\xi)e^{-\beta|x-\xi|}\cos\beta|x-\xi|/2$$

**Point moment**

$$\mathcal{H} = sgn(x-\xi)\beta^2e^{-\beta|x-\xi|}\sin\beta|x-\xi|/k \quad \mathcal{H}_1 = -\beta^3e^{-\beta|x-\xi|}(\sin\beta|x-\xi| - \cos\beta|x-\xi|)/k$$

$$\mathcal{H}_2 = -sgn(x-\xi)e^{-\beta|x-\xi|}\cos\beta|x-\xi|/2 \quad \mathcal{H}_3 = -\beta e^{-\beta|x-\xi|}[\cos\beta|x-\xi| + \sin\beta|x-\xi|]/2$$

**Non-dimensionalized form**

$$\hat{\mathcal{G}} = \left[e^{-\hat\beta|\hat x-\hat\xi|}(\cos\hat\beta|\hat x-\hat\xi| + \sin\hat\beta|\hat x-\hat\xi|)\right]/8\hat\beta^3 \quad \hat{\mathcal{G}}_1 = -\left[sgn(\hat x-\hat\xi)e^{-\hat\beta|\hat x-\hat\xi|}\sin\hat\beta|\hat x-\hat\xi|\right]/4\hat\beta^2$$

$$\hat{\mathcal{G}}_2 = \left[e^{-\hat\beta|\hat x-\hat\xi|}(\sin\hat\beta|\hat x-\hat\xi| - \cos\hat\beta|\hat x-\hat\xi|)\right]/4\hat\beta \quad \hat{\mathcal{G}}_3 = -\left[sgn(\hat x-\hat\xi)e^{-\hat\beta|\hat x-\hat\xi|}\cos\hat\beta|\hat x-\hat\xi|\right]/2$$

$$\hat{\mathcal{H}} = \left[sgn(\hat x-\hat\xi)e^{-\hat\beta|\hat x-\hat\xi|}\sin\hat\beta|\hat x-\hat\xi|\right]/4\hat\beta^2 \quad \hat{\mathcal{H}}_1 = -\left[e^{-\hat\beta|\hat x-\hat\xi|}(\sin\hat\beta|\hat x-\hat\xi| - \cos\hat\beta|\hat x-\hat\xi|)\right]/4\hat\beta$$

$$\hat{\mathcal{H}}_2 = -\left[sgn(\hat x-\hat\xi)e^{-\hat\beta|\hat x-\hat\xi|}\cos\hat\beta|\hat x-\hat\xi|\right]/2 \quad \hat{\mathcal{H}}_3 = -\left[\hat\beta e^{-\hat\beta|\hat x-\hat\xi|}(\cos\hat\beta|\hat x-\hat\xi| + \sin\hat\beta|\hat x-\hat\xi|)\right]/2$$

**Finite beams**

$$v(x) = R_A\mathcal{G}(x,-L/2) + M_A\mathcal{H}(x,-L/2) + R_B\mathcal{G}(x,L/2) + M_B\mathcal{H}(x,L/2) + \int_{-L/2}^{L/2}\mathcal{G}(x,\xi)p_y(\xi)d\xi$$

At $x = -L/2+\varepsilon$, $\xi = -L/2$

$$\mathcal{G} = \beta/2k \quad \mathcal{G}_1 = 0 \quad \mathcal{G}_2 = -1/4\beta \quad \mathcal{G}_3 = -1/2;\ \mathcal{H} = 0 \quad \mathcal{H}_1 = \beta^3/k \quad \mathcal{H}_2 = -1/2 \quad \mathcal{H}_3 = -\beta/2$$

At $x = -L/2+\varepsilon$, $\xi = L/2$

$$\mathcal{G} = \beta e^{-\beta L}(\cos\beta L + \sin\beta L)/2k \quad \mathcal{G}_1 = \beta^2e^{-\beta L}\sin\beta L/k \quad \mathcal{G}_2 = e^{-\beta L}(\sin\beta L - \cos\beta L)/4\beta \quad \mathcal{G}_3 = e^{-\beta L}\cos\beta L/2$$

$$\mathcal{H} = -\beta^2e^{-\beta L}\sin\beta L/k \quad \mathcal{H}_1 = -\beta^3e^{-\beta L}(\sin\beta L - \cos\beta L)/k \quad \mathcal{H}_2 = e^{-\beta L}\cos\beta L/2 \quad \mathcal{H}_3 = -\beta e^{-\beta L}[\cos\beta L + \sin\beta L]/2$$

At $x = L/2-\varepsilon$, $\xi = -L/2$

$$\mathcal{G} = \beta e^{-\beta L}[\cos\beta L + \sin\beta L]/2k \quad \mathcal{G}_1 = -\beta^2e^{-\beta L}\sin\beta L/k \quad \mathcal{G}_2 = e^{-\beta L}(\sin\beta L - \cos\beta L)/4\beta \quad \mathcal{G}_3 = -e^{-\beta L}\cos\beta L/2$$

$$\mathcal{H} = \beta^2e^{-\beta L}\sin\beta L/k \quad \mathcal{H}_1 = -\beta^3e^{-\beta L}(\sin\beta L - \cos\beta L)/k \quad \mathcal{H}_2 = -e^{-\beta L}\cos\beta L/2 \quad \mathcal{H}_3 = (-\beta e^{-\beta L}[\cos\beta L + \sin\beta L])/2$$

At $x = L/2-\varepsilon$, $\xi = L/2$

$$\mathcal{G} = \beta/2k \quad \mathcal{G}_1 = 0 \quad \mathcal{G}_2 = -1/4\beta \quad \mathcal{G}_3 = 1/2;\ \mathcal{H} = 0 \quad \mathcal{H}_1 = \beta^3/k \quad \mathcal{H}_2 = 1/2 \quad \mathcal{H}_3 = -\beta/2$$

## 3.8 MATHEMATICAL DETAILS OF THE DERIVATION OF FUNDAMENTAL SOLUTIONS

In this section the details of the derivation of equations introduced in Section 3.4 are presented.

### 3.8.1 Fundamental solution for displacement

For $x > 0$, and assuming $EI$ is a constant, Equation (3.18a) can be write as

$$\frac{d^4 v}{dx^4} + \frac{k}{EI} v = 0 \tag{3.36}$$

We substitute $v = Ae^{\lambda x}$ to obtain

$$A\left(\lambda^4 + \frac{k}{EI}\right)e^{\lambda x} = 0 \quad \text{or} \quad \lambda^4 + \frac{k}{EI} = 0 \tag{3.37a}$$

The four roots of Equation (3.37a) are

$$\lambda_1 = \beta(1+i) \qquad \lambda_2 = -\beta(1+i) \qquad \lambda_3 = \beta(1-i) \qquad \lambda_4 = -\beta(1-i) \tag{3.37b}$$

where, $i = \sqrt{-1}$ and $\beta$ is given by Equation (3.18c). The 3rd and 4th roots are complex conjugate roots 1 and 2. Thus, the solution of Equation (3.36) is

$$v = \boldsymbol{Re}\{A_1 e^{\beta(1+i)x} + A_2 e^{-\beta(1+i)x}\} \tag{3.38a}$$

where, $A_1$ and $A_2$ are complex constants and $\boldsymbol{Re}\{\bullet\}$ implies the real part of the term in the brackets. Later on we will also need the imaginary part of the term in bracket which is symbolized by $\boldsymbol{Im}\{\bullet\}$. We note that the derivative with respect to $x$ is a real quantity and can be moved inside the bracket with respect to the operators $\boldsymbol{Re}\{\bullet\}$ and $\boldsymbol{Im}\{\bullet\}$. We further note that any real variable, such as $\beta$ and $sgn(x-\xi)$, can be taken outside the operators $\boldsymbol{Re}\{\bullet\}$ and $\boldsymbol{Im}\{\bullet\}$ in subsequent derivations. We thus obtain

$$\frac{\partial v}{\partial x} = \boldsymbol{Re}\left\{\frac{\partial}{\partial x}(A_1 e^{\beta(1+i)x} + A_2 e^{-\beta(1+i)x})\right\} = \boldsymbol{Re}\{\beta(1+i)(A_1 e^{\beta(1+i)x} - A_2 e^{-\beta(1+i)x})\} \tag{3.38b}$$

From the requirement that displacements and slopes must be bounded as $x \to \infty$ [Equation (3.17c)] we conclude

$$A_1 = 0 \tag{3.38c}$$

The condition that slope must be zero at the origin [Equation (3.18a)] yields the following

$$\boldsymbol{Re}\{\beta(1+i)(A_2)\} = 0 \tag{3.38d}$$

The equilibrium condition [Equation (3.17e)] yields

$$2\int_0^\infty k\boldsymbol{Re}\{A_2 e^{-\beta(1+i)x}\}dx = 2k\boldsymbol{Re}\left\{\frac{-A_2}{\beta(1+i)}e^{-\beta(1+i)x}\Big|_0^\infty\right\} = \frac{2k}{\beta}\boldsymbol{Re}\left\{\frac{A_2(1-i)}{(1-i^2)}\right\} = \frac{k}{\beta}\boldsymbol{Re}\{A_2(1-i)\} = P \tag{3.38e}$$

We represent $A_2 = B_1 + iB_2$, where, $B_1$ and $B_2$ are real constants. Substituting $A_2$ into Equations (3.38d) and (3.38e) we obtain the following

$$\beta\boldsymbol{Re}\{(1+i)(B_1+iB_2)\} = \beta\boldsymbol{Re}\{(B_1-B_2)+i(B_1+B_2)\} = 0 \quad \text{or} \quad B_1 - B_2 = 0 \tag{3.38f}$$

$$(k/\beta)\boldsymbol{Re}\{(B_1+iB_2)(1-i)\} = (k/\beta)\boldsymbol{Re}\{(B_1+B_2)-i(B_1-B_2)\} = P \quad \text{or} \quad B_1 + B_2 = P\beta/k \tag{3.38g}$$

Solving the above two equations we obtain

$$B_1 = B_2 = P\beta/(2k) \quad \text{and} \quad A_2 = P\beta(1+i)/(2k) \tag{3.38h}$$

Substituting Equations (3.38c) and (3.38h) into Equation (3.38a) we obtain the displacement as

$$v = \left(\frac{P\beta}{2k}\right)\boldsymbol{Re}\{(1+i)e^{-\beta(1+i)x}\} \qquad x > 0 \tag{3.39}$$

We now make use of the following identities governing two complex functions $f$ and $g$

$$\boldsymbol{Re}\{fg\} = \boldsymbol{Re}\{f\}\boldsymbol{Re}\{g\} - \boldsymbol{Im}\{f\}\boldsymbol{Im}\{g\} \qquad \boldsymbol{Im}\{fg\} = \boldsymbol{Re}\{f\}\boldsymbol{Im}\{g\} + \boldsymbol{Im}\{f\}\boldsymbol{Re}\{g\} \tag{3.40a}$$

We also note that

$$e^{-\beta(1+i)x} = e^{-\beta x}(\cos\beta x - i\sin\beta x) \qquad Re\{e^{-\beta(1+i)x}\} = e^{-\beta x}\cos\beta x \qquad Im\{e^{-\beta(1+i)x}\} = -e^{-\beta x}\sin\beta x \tag{3.40b}$$

Using the above three equations we obtain from Equation (3.39) the following

$$v = \left(\frac{P\beta}{2k}\right)[\boldsymbol{Re}\{(1+i)\}\boldsymbol{Re}\{e^{-\beta(1+i)x}\} - \boldsymbol{Im}\{(1+i)\}\boldsymbol{Im}\{e^{-\beta(1+i)x}\}] = \left(\frac{P\beta}{2k}\right)e^{-\beta x}[\cos\beta x + \sin\beta x] \qquad x > 0 \tag{3.41}$$

We generalize the solution in Equation (3.39) by taking the absolute value of $x$ and extend our solution to negative values of $x$. We translate the origin to arbitrary $\xi$ to obtain

$$\boxed{\mathrm{v} = \frac{P\beta}{2k}\boldsymbol{Re}\{(1+i)e^{-\beta(1+i)|x-\xi|}\}} \tag{3.42a}$$

### 3.8.2 Fundamental solutions for derivatives of displacement

The derivatives yield the following equations.

$$\frac{\partial \mathrm{v}}{\partial x} = \left(\frac{P\beta}{2k}\right)\boldsymbol{Re}\{-\beta(1+i)^2 sgn(x-\xi)e^{-\beta(1+i)|x-\xi|}\} = -\left(\frac{P\beta^2}{2k}\right)sgn(x-\xi)\boldsymbol{Re}\{(1+i)^2 e^{-\beta(1+i)|x-\xi|}\} \text{ or}$$

$$\frac{\partial \mathrm{v}}{\partial x} = -\left(\frac{P\beta^2}{2k}\right)sgn(x-\xi)\boldsymbol{Re}\{(2i)e^{-\beta(1+i)|x-\xi|}\} = -\left(\frac{P\beta^2}{2k}\right)sgn(x-\xi)[\boldsymbol{Re}\{2i\}\boldsymbol{Re}\{e^{-\beta(1+i)|x-\xi|}\} - \boldsymbol{Im}\{2i\}\boldsymbol{Im}\{e^{-\beta(1+i)|x-\xi|}\}]$$

$$\boxed{\frac{\partial \mathrm{v}}{\partial x} = -(P\beta^2/k)sgn(x-\xi)e^{-\beta|x-\xi|}sin\beta|x-\xi| = \boldsymbol{G}_1(x,\xi)P} \tag{3.43}$$

$$\frac{\partial^2 \mathrm{v}}{\partial x^2} = \left(\frac{P\beta^3}{2k}\right)[sgn(x-\xi)]^2\boldsymbol{Re}\{(1+i)(2i)e^{-\beta(1+i)|x-\xi|}\} = -\left(\frac{P\beta^3}{k}\right)\boldsymbol{Re}\{(1-i)e^{-\beta(1+i)|x-\xi|}\} \tag{3.43a}$$

$$\frac{\partial^3 \mathrm{v}}{\partial x^3} = \left(\frac{P\beta^4}{k}\right)sgn(x-\xi)\boldsymbol{Re}\{(1-i^2)e^{-\beta(1+i)|x-\xi|}\} = \left(\frac{2P\beta^4}{k}\right)sgn(x-\xi)\boldsymbol{Re}\{e^{-\beta(1+i)|x-\xi|}\} \tag{3.43b}$$

Noting that $EI = k/(4\beta^4)$ and Equations (3.43a) and (3.43b), we can write the expressions for moment and shear force as

$$\boldsymbol{M}_z = EI\frac{\partial^2 \mathrm{v}}{\partial x^2} = -\left(\frac{P}{4\beta}\right)\boldsymbol{Re}\{(1-i)e^{-\beta(1+i)|x-\xi|}\} \qquad \boldsymbol{V}_z = -EI\frac{\partial^3 \mathrm{v}}{\partial x^3} = -\left(\frac{P}{2}\right)sgn(x-\xi)\boldsymbol{Re}\{e^{-\beta(1+i)|x-\xi|}\} \tag{3.43c}$$

Using Equations (3.40a), (3.40a) and (3.40b) we obtain expressions for moment and shear force as

$$\boxed{\boldsymbol{M}_z = -(P/4\beta)e^{-\beta|x-\xi|}[cos\beta|x-\xi| - sin\beta|x-\xi|] = \boldsymbol{G}_2(x,\xi)P} \tag{3.44}$$

$$\boxed{\boldsymbol{V}_z = -(P/2)sgn(x-\xi)e^{-\beta|x-\xi|}cos\beta|x-\xi| = \boldsymbol{G}_3(x,\xi)P} \tag{3.45}$$

The expressions for $\boldsymbol{G}_1(x,\xi)$, $\boldsymbol{G}_2(x,\xi)$, and $\boldsymbol{G}_3(x,\xi)$ above reported in Table 3.6.

### 3.8.3 Derivatives and integrals of fundamental solutions with respect to source point

Using Equations (3.23) and Equations we obtain:

$$\boldsymbol{\mathcal{D}}(x,\xi) = \frac{\partial e^{-\beta(1+i)|x-\xi|}}{\partial \xi} = -\left\{\frac{\partial e^{-\beta(1+i)|x-\xi|}}{\partial x}\right\} = \beta(1+i)sgn(x-\xi)e^{-\beta(1+i)|x-\xi|} \tag{3.46}$$

Equating the real and imaginary parts we obtain the following two formulas

$$\boxed{\boldsymbol{\mathcal{D}}_1(x,\xi) = \frac{\partial}{\partial \xi}[e^{-\beta|x-\xi|}cos\beta|x-\xi|] = \beta sgn(x-\xi)e^{-\beta|x-\xi|}[cos\beta|x-\xi| + sin\beta|x-\xi|]} \tag{3.47}$$

$$\boxed{\boldsymbol{\mathcal{D}}_2(x,\xi) = \frac{\partial}{\partial \xi}[e^{-\beta|x-\xi|}sin\beta|x-\xi|] = \beta sgn(x-\xi)e^{-\beta|x-\xi|}[-cos\beta|x-\xi| + sin\beta|x-\xi|]} \tag{3.48}$$

Consider the integral

$$\boldsymbol{\mathcal{I}}(x,\xi) = \int e^{-\beta(1+i)|x-\xi|}d\xi = -\frac{sgn(x-\xi)}{\beta(1+i)}e^{-\beta(1+i)|x-\xi|} = -\frac{sgn(x-\xi)(1-i)}{\beta(1-i^2)}e^{-\beta(1+i)|x-\xi|} = -\frac{(1-i)sgn(x-\xi)e^{-\beta(1+i)|x-\xi|}}{2\beta}$$

Equating the real and imaginary parts we obtain the following two formulas

$$\boxed{\boldsymbol{\mathcal{I}}_1(x,\xi) = \int e^{-\beta|x-\xi|}sin\beta|x-\xi|d\xi = \frac{1}{2\beta}sgn(x-\xi)e^{-\beta|x-\xi|}[cos\beta|x-\xi| + sin\beta|x-\xi|]} \tag{3.49}$$

$$\boxed{\boldsymbol{\mathcal{I}}_2(x,\xi) = \int e^{-\beta|x-\xi|}cos\beta|x-\xi|d\xi = \frac{1}{2\beta}sgn(x-\xi)e^{-\beta|x-\xi|}[cos\beta|x-\xi| - sin\beta|x-\xi|]} \tag{3.50}$$

# Problems

## Section 3.1

**3.1** Starting with $\mathcal{G}_1 = \partial\mathcal{G}/\partial x$ show that $\widehat{\mathcal{G}}_1 = \partial\widehat{\mathcal{G}}/\partial\widehat{x}$, where $\mathcal{G}$, $\widehat{\mathcal{G}}$, $\mathcal{G}_1$ and $\widehat{\mathcal{G}}_1$ are defined in Table 3.1.

**3.2** Starting with $v(x) = \mathcal{H}(x, \xi)M$ show that $\widehat{\mathcal{H}} = \mathcal{H}[EI/L^2]$. and $\widehat{\mathcal{H}}_1 = \partial\widehat{\mathcal{H}}/\partial\widehat{x}$, where $\mathcal{H}$, $\widehat{\mathcal{H}}$, $\mathcal{H}_1$ and $\widehat{\mathcal{H}}_1$ are defined in Table 3.1.

## Section 3.2

**3.3** Obtain the elastic curve v(*x*) for the beam and loading shown in Fig. P3.3 using Equation (3.13a).

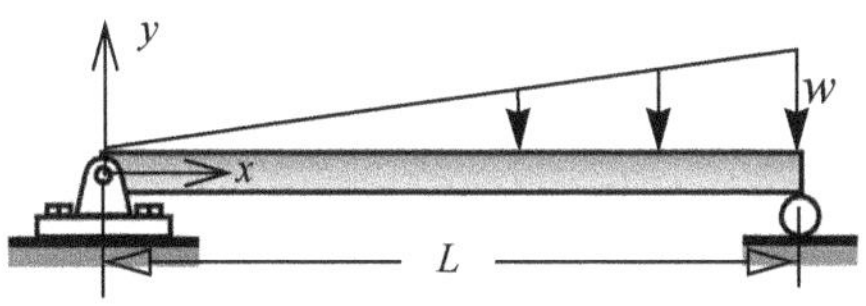


**Fig. P3.3**

**3.4** Obtain the influence function for point load for beam shown in Figure 3.6*a* on page 76 and use it to obtain the elastic curve v(*x*) for the beam and the given load.

**3.5** Obtain the influence functions for point load for the beam shown in Fig. P3.5. and from it obtain the influence function for point moment.

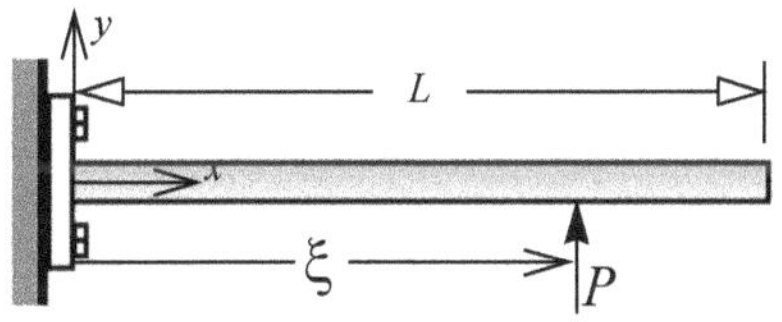


**Fig. P3.5**

**3.6** Obtain the elastic curve v(*x*) for the beam and loading shown in Fig. P3.6 using the influence functions obtained in problem 3.5.

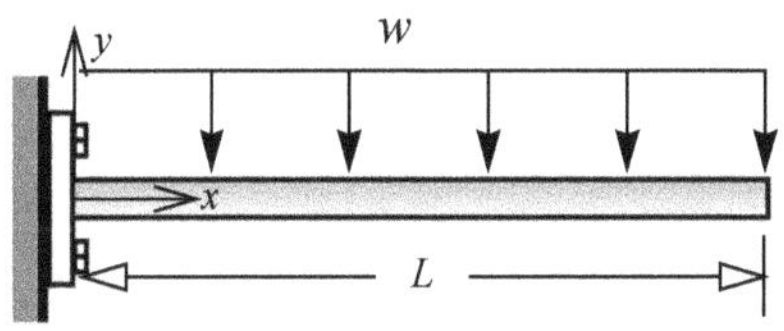


**Fig. P3.6**

**3.7** Obtain the elastic curve v(*x*) for the beam and loading shown in Figure P3.7 using the influence function of problem 3.5 and solution of problem 3.6.

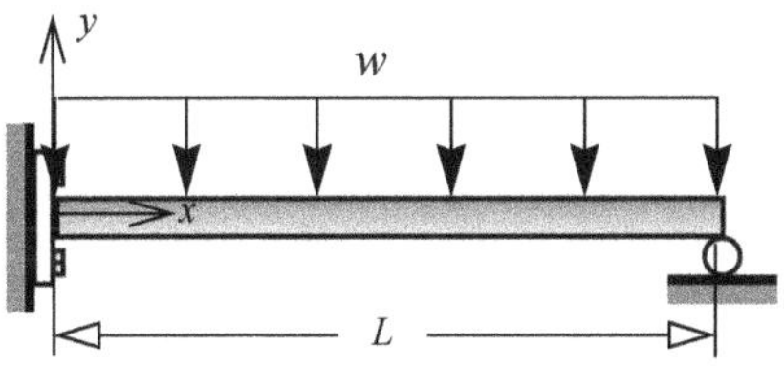


**Figure P3.7**

**3.8** Obtain the elastic curve v(*x*) for the beam and loading shown in Figure P3.8 using the influence function of problem 3.5.

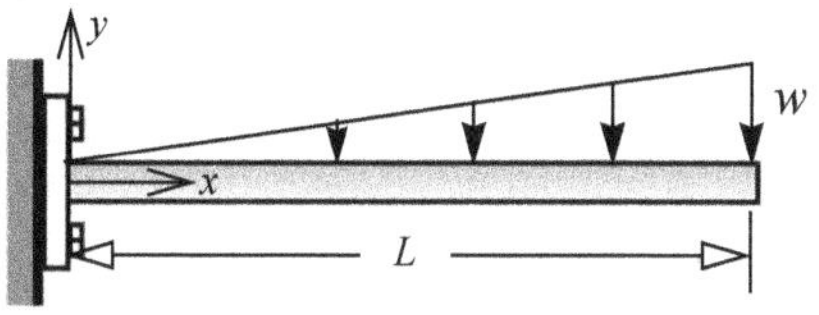


**Figure P3.8**

**3.9** Starting with the solution of $\mathcal{G}$ in problem 3.5, obtain $\mathcal{G}_1$, $\mathcal{G}_2$, and $\mathcal{G}_3$.

**3.10** Starting with the solution of $\mathcal{H}$ in problem 3.5, obtain $\mathcal{H}_1$, $\mathcal{H}_2$, and $\mathcal{H}_3$.

**3.11** Starting with the solution of $\mathcal{G}$ in problem 3.5 and solution of $\mathcal{G}_1$, $\mathcal{G}_2$, and $\mathcal{G}_3$. in problem 3.9, obtain $\widehat{\mathcal{G}}$, $\widehat{\mathcal{G}}_1$, $\widehat{\mathcal{G}}_2$, and $\widehat{\mathcal{G}}_3$.

**3.12** Starting with the solution of $\mathcal{H}$ in problem 3.5 and solution of $\mathcal{H}_1$, $\mathcal{H}_2$, and $\mathcal{H}_3$. in problem 3.10, obtain $\widehat{\mathcal{H}}$, $\widehat{\mathcal{H}}_1$, $\widehat{\mathcal{H}}_2$, and $\widehat{\mathcal{H}}_3$.

**3.13** A 10 ft. simply supported beam shown in Fig. P3.13 has a distributed force that varies as

$$p_y = -500e^{-(x/10)}[1 - cos(2\pi x/10)] \text{ lb/ft.}.$$

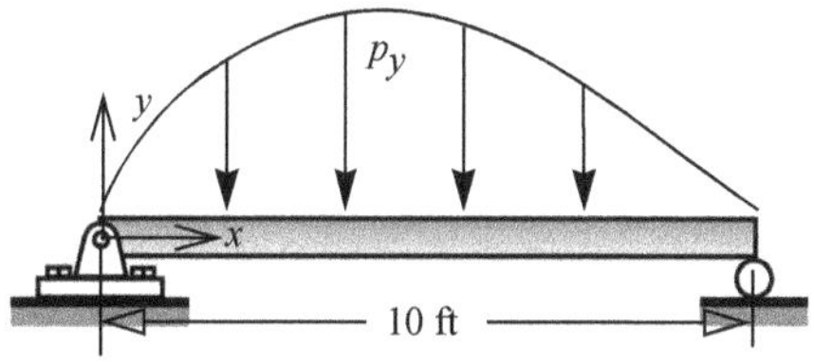


**Fig. P3.13**

The beam has a rectangular cross section with depth of 8 in. in the *y* direction and width of 2 in. in the *z*-direction. The modulus of elasticity for of the beam material is 8,100 ksi. Determine (a) the deflection and bending moment at the mid section of the beam. (b) the maximum deflection and bending normal stress in the beam. Use 10 equal intervals for trapezoidal rule and the maximum value of $p_y$ as $p_o$ and *L*=10 ft. for non-dimensionalizing. Test your spread sheet or program with solution of Example 3.1.

**3.14** A 10 ft. cantilever beam shown in Fig. P3.14 has a distributed force that varies as $p_y = -500e^{-(x/10)}[1 - cos(2\pi x/10)]$ lb/ft. The beam has a rectangular cross section with depth of 10 in. in the *y* direction and width of 2 in. in the *z*-direction. The modulus of elasticity for of the beam material is 8,100 ksi. Determine (a) the deflection and bending moment at the mid section of the beam. (b) the maximum deflection and bending normal stress in the beam. Use 10 equal intervals for trapezoidal rule and the maximum value of $p_y$ as $p_o$ and *L*=10 ft. for non-dimensionalizing. Test your spread sheet or program with solution of problem 3.6.

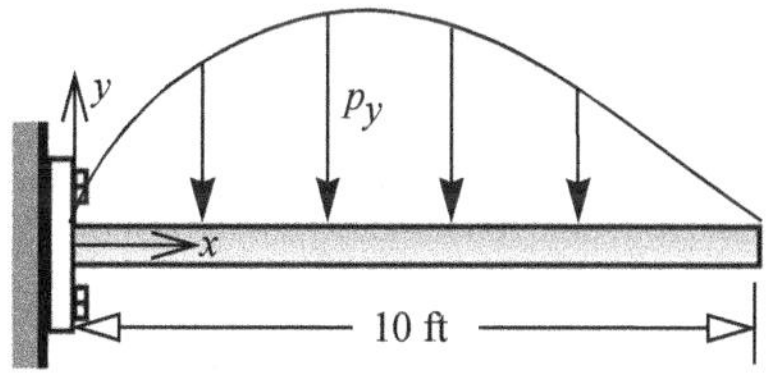


**Fig. P3.14**

## Sections 3.3-3.4

**3.15** By substitution into the differential equation given by Equation (3.17a) show that

$v = e^{\beta x}[A\cos\beta x + B\sin\beta x] + e^{-\beta x}[C\cos\beta x + D\sin\beta x]$ is the solution for all $x > 0$.

**3.16** Starting with Equation (3.19), show that Equation (3.23) is valid.

**3.17** Determine the maximum values for deflection, slope, bending moment, and shear force associated with moment singularity. Make plots similar to Figure 3.8 for moment singularity and comment on the odd and even properties of deflection, slope, bending moment, and shear force about the singular point.

**3.18** A very long rectangular beam with a modulus of elasticity of 30,000 ksi rest on an elastic foundation of modulus of 2 ksi as shown in Fig. P3.18.The beam has a depth of 6 in. in the y-direction and a width of 2 in. in the z-direction and $P$= 64 kips.

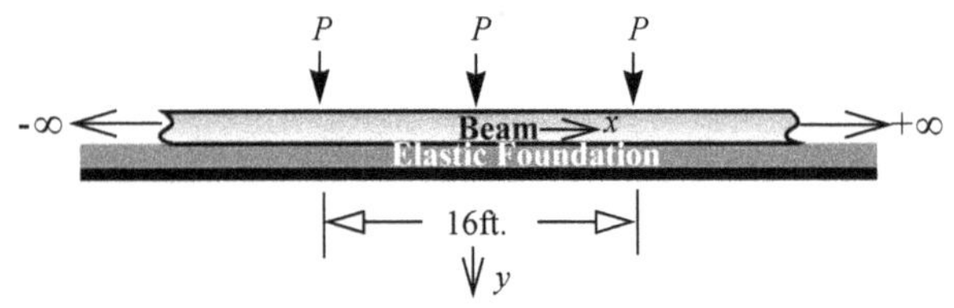


**Fig. P3.18**

For a load of P= 64 kips, determine (a) the deflection at $x = 0$. (b) the maximum bending normal stress. (c) the shear force just to the right of the origin. (c) the shear force at $x = 4$ ft

## Section 3.5

**3.19** A finite beam on elastic foundation is loaded as shown in Fig. P3.19. (a) In terms of $E$, $I$, $\beta$, $k$, $L$, and $M$ determine the unknown force $R_S$ and moment $M_S$ in Equations (3.32a) through (3.32d). (b) the deflection slope, bending moment, and shear force at $x = 0$ assuming $k = 2$ ksi, $\beta = 0.021935$ in.$^{-1}$, $M = 2$ ft-kips and $L = 10$ ft.

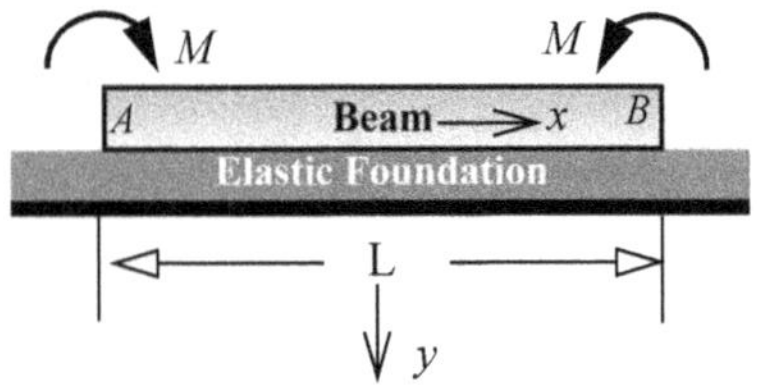


**Fig. P3.19**

**3.20** A finite beam on elastic foundation is loaded as shown in Fig. P3.20. Determine the deflection slope, bending moment, and shear force at $x = 0$ assuming $k = 2$ ksi, $\beta = 0.021935$ in.$^{-1}$, $M = 2$ ft-kips and $L = 10$ ft.

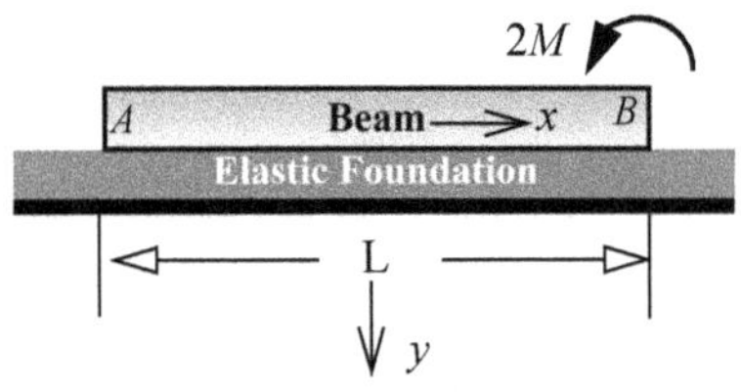


**Fig. P3.20**

# 4 Stability

## LEARNING OBJECTIVE

1. Understand the concept of buckling of columns.

**Stability,** also called **buckling**, is an instability of equilibrium in structures that occurs from compressive loads or stresses. A structure or its components may fail due to buckling at loads that are far smaller than those that produce material strength failure. Very often buckling is a catastrophic failure.

Instability due to compressive inplane stresses of thin plates and shells is called **local buckling**. Instability of a structural member due to inplane loads is called **structural buckling**. One dimensional structural member that can bend due to axial loads is called a **column**. In this chapter we restrict ourselves to study of buckling of columns.

Leonard Euler '''(1707–1782) was the first to study buckling of pin connected columns and the formula he developed is named after him. His work on column buckling was unappreciated during his time by practicing engineers as it did not accurately predict failure in the structural members then in use. End effects, imperfections, stresses exceeding yield stress, shorter column lengths, are some of the complications that have been added to Euler's work to improve accuracy of predictions and we shall see these in this chapter. For all the refinements, the Euler buckling formula is still used three centuries later for column design and is still valid for long columns with pin-supported ends. Such is the power of logical thinking.

## 4.1 EULER BUCKLING

Figure 4.1*a* shows a simply supported column that is axially loaded with a force *P*. We shall initially assume that bending is about the *z* axis as our equations on beam deflection were developed with just this assumption. We shall relax this assumption at the end to generate the formula for a critical buckling load.

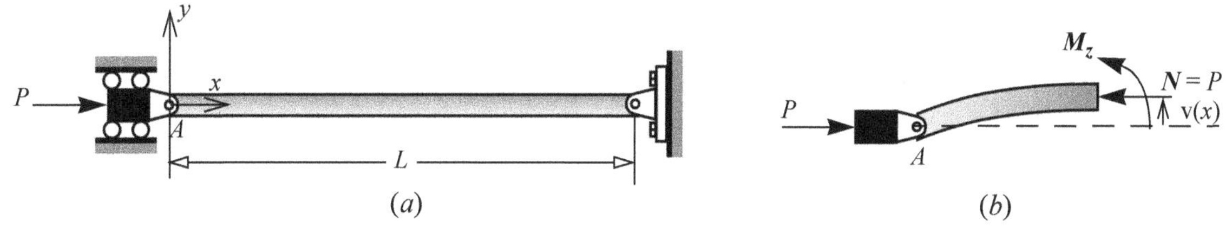


**Figure 4.1** Simply supported column.

Let the bending deflection at any location *x* be given by v(*x*), as shown in Figure 4.1*b*. An imaginary cut is made at some location *x*, and the internal bending moment is drawn according to our sign convention. The internal axial force *N* will be equal to *P*. By balancing the moment at point *A* we obtain $M_z + P\mathrm{v} = 0$. Substituting the moment–curvature relationship $M_z = EI_{zz}(d^2\mathrm{v}/dx^2)$, we obtain the differential equation:

$$EI_{zz}\frac{d^2\mathrm{v}}{dx^2} + P\mathrm{v} = 0 \qquad \textbf{(4.1a)}$$

If buckling can occur about any axis and not just the *z* axis, as we initially assumed, then the subscripts *zz* in the area moment of inertia should be dropped. The boundary value problem can be written as shown by Equations (4.1b) and (4.1c).

Differential Equation $\frac{d^2v}{dx^2} + \lambda^2 v = 0$ where $\lambda = \sqrt{\frac{P}{EI}}$ **(4.1b)**

Boundary Conditions $v(0) = 0 \qquad v(L) = 0$ **(4.1c)**

The solution to the differential equation, is

$$v(x) = A\cos\lambda x + B\sin\lambda x \tag{4.2}$$

From the boundary conditions we obtain two equations which in matrix form can be written as

$$\begin{bmatrix} 1 & 0 \\ \cos\lambda L & \sin\lambda L \end{bmatrix} \begin{Bmatrix} A \\ B \end{Bmatrix} = \begin{Bmatrix} 0 \\ 0 \end{Bmatrix} \tag{4.3}$$

For a nontrivial solution—that is, when $A$ and $B$ are not both zero—the condition is that the determinant of the matrix must be zero. This yield

$$\sin\lambda L = 0 \tag{4.4}$$

Above equation is the *characteristic equation,* or the *buckling equation.* The roots of the characteristic equation are $\lambda L = n\pi$. Substituting for $\lambda$ and solving for $P$, we obtain

$$P_n = \frac{n^2\pi^2 EI}{L^2}, \qquad n = 1, 2, 3, \ldots \tag{4.5}$$

Equation (4.5) represents the values of load $P$ (the eigenvalues) at which buckling would occur. What is the lowest value of $P$ at which buckling will occur? Clearly, for the lowest value of $P$, $n$ should equal 1 in Equation (4.5). Furthermore minimum value of $I$ should be used. The critical buckling load is

$$\boxed{P_{cr} = \frac{\pi^2 EI}{L^2}} \tag{4.6}$$

$P_{cr}$, the **critical buckling load,** is also called **Euler load.**

- *Buckling will occur about the axis that has minimum area moment of inertia.*

### 4.1.1 Buckling modes

From Equation (4.3) we see that $A = 0$. The solution for v can be written as

$$v = B\sin\left(n\pi\frac{x}{L}\right) \tag{4.7}$$

Equation (4.7) represents the buckled mode (eigenvectors). The importance of each buckled mode shape can be appreciated by examining Figure 4.2. If buckled mode 1 is prevented from occurring by installing a restraint (or support), then the column would buckle at the next higher mode at critical load values that are higher than those for the lower modes. Point $I$ on the deflection curves describing the mode shapes has two attributes: it is an inflection point and the magnitude of deflection at this point is zero. Recall that the curvature $d^2v/dx^2$ at an inflection point is zero. Hence the internal moment $M_z$ at this point is zero. If roller supports are put at any other points than the inflection points $I$, as predicted by Equation (4.7), then the boundary-value problem (see Problem 4.9) will have different eigenvalues (critical loads) and eigenvectors (mode shapes).

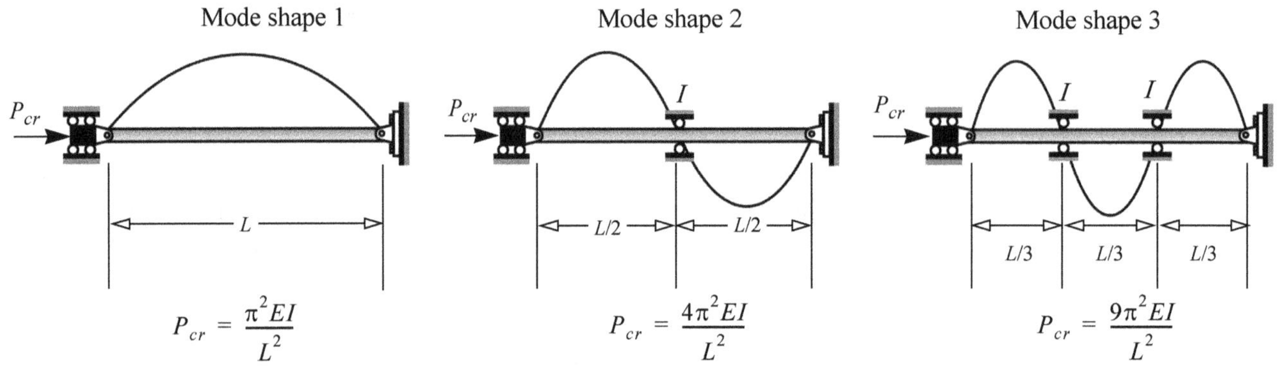


**Figure 4.2** Importance of buckled modes.

In many situations it may not be possible to put roller supports in order to change a mode to a higher critical buckling load. But buckling modes and buckling loads can also be changed by using elastic supports as demonstrate in Example 4.1. Water tanks that supply water to a community has columns with elastic ring supports to increase the critical buckling load.

### 4.1.2 Effects of end conditions

Equation (4.6) is applicable only to simply supported columns. However, the process used to obtain the formula can be used for other types of supports. Table 4.1 shows the critical elements in the derivation process and the results for three other supports. The formula for critical loads for all cases shown in Table 4.1 can be written as

$$P_{cr} = \frac{\pi^2 EI}{L_{eff}^2} \tag{4.8}$$

where $L_{eff}$ is the effective length of the column. The effective length for each case is given in the last row of Table 4.1. This definition of effective length will permit us to extend results that will be derived in Section 4.2 for simply supported imperfect columns to imperfect columns with the supports shown in cases 2 through 4 in Table 4.1.

**Table 4.1 Buckling of columns with different supports**

| | Case 1 | Case 2 | Case 3 | Case 4[a] |
|---|---|---|---|---|
| | 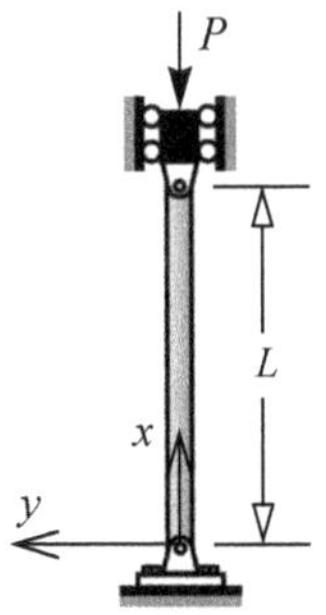 | 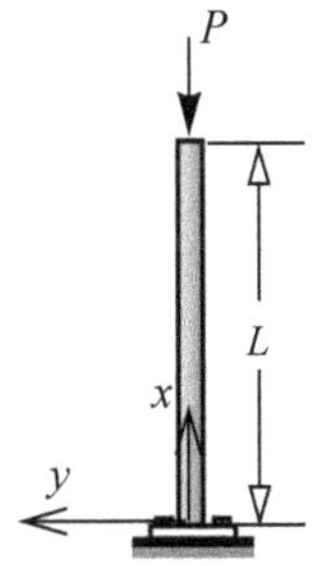 | 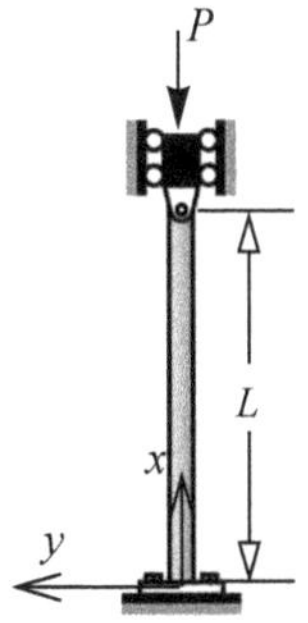 | 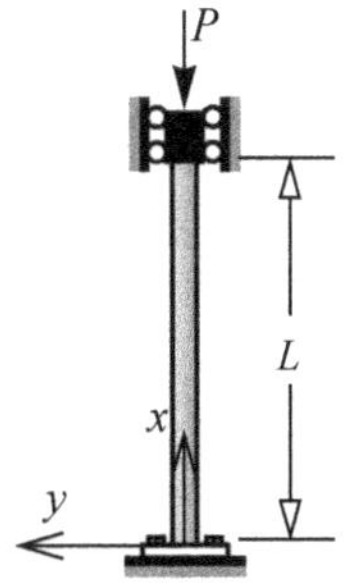 |
| | **Pinned at both ends** | **One end fixed, other end free** | **One end fixed, other end pinned** | **Fixed at both ends** |
| Differential equation | $EI\frac{d^2 v}{dx^2} + Pv = 0$ | $EI\frac{d^2 v}{dx^2} + Pv = Pv(L)$ | $EI\frac{d^2 v}{dx^2} + Pv = R_B(L-x)$ | $EI\frac{d^2 v}{dx^2} + Pv = R_B(L-x) + M_B$ |
| Boundary conditions | $v(0) = 0$<br>$v(L) = 0$ | $v(0) = 0$ $\quad \frac{dv}{dx}(0) = 0$ | $v(0) = 0$ $\quad \frac{dv}{dx}(0) = 0$<br>$v(L) = 0$ | $v(0) = 0$ $\quad \frac{dv}{dx}(0) = 0$<br>$v(L) = 0$ $\quad \frac{dv}{dx}(L) = 0$ |
| Characteristic equation $\lambda = \sqrt{P/(EI)}$ | $\sin \lambda L = 0$ | $\cos \lambda L = 0$ | $\tan \lambda L = \lambda L$ | $2(1 - \cos \lambda L) - \lambda L \sin \lambda L = 0$ |
| Critical load $P_{cr}$ | $\frac{\pi^2 EI}{L^2}$ | $\frac{\pi^2 EI}{4L^2} = \frac{\pi^2 EI}{(2L)^2}$ | $\frac{20.13EI}{L^2} = \frac{\pi^2 EI}{(0.7L)^2}$ | $\frac{4\pi^2 EI}{L^2} = \frac{\pi^2 EI}{(0.5L)^2}$ |
| Effective length $L_{eff}$ | L | $2L$ | $0.7L$ | $0.5L$ |

a. $R_B$ and $M_B$ are the force and moment reactions.

### 4.1.3 Classification of columns

In Equation (4.6), $I$ can be replaced by $Ar^2$, where $A$ is the cross-sectional area and $r$ is the minimum radius of gyration. We obtain

$$\sigma_{cr} = \frac{P_{cr}}{A} = \frac{\pi^2 E}{(L_{eff}/r)^2} \tag{4.9}$$

where $L_{eff}/r$ is the **slenderness ratio** and $\sigma_{cr}$ is the compressive axial stress just before the column would buckle.

Equation (4.9) is valid only in the elastic region—that is, if $\sigma_{cr} < \sigma_{yield}$. If $\sigma_{cr} > \sigma_{yield}$, then elastic failure will be due to stress exceeding the material strength. Thus $\sigma_{cr} = \sigma_{yield}$ defines the failure envelope for a column. Figure 4.3 shows the failure envelopes for steel, aluminum, and wood. As nondimensional variables are used in the plots in Figure 4.3, these plots can also be used for metric units. Note that the slenderness ratio is defined using effective lengths; hence these plots are applicable to columns with different supports.

The failure envelopes in Figure 4.3 show that as the slenderness ratio increases, *the failure due to buckling will occur at stress values significantly lower than the yield stress.* This underscores the importance of buckling in the design of members under compression.

The failure envelopes, as shown in Figure 4.3, depend only on the material property and are applicable to columns of different lengths, shapes, and types of support. These failure envelopes are used for classifying columns as short or long. Short column design is based on using yield stress as the failure stress. Long column design is based on using critical buckling stress as the failure stress. The slenderness ratio at point $A$ for each material is used for separating short columns from long columns for that material. Point $A$ is the intersection point of the straight line representing elastic material failure and the hyperbola curve representing buckling failure.

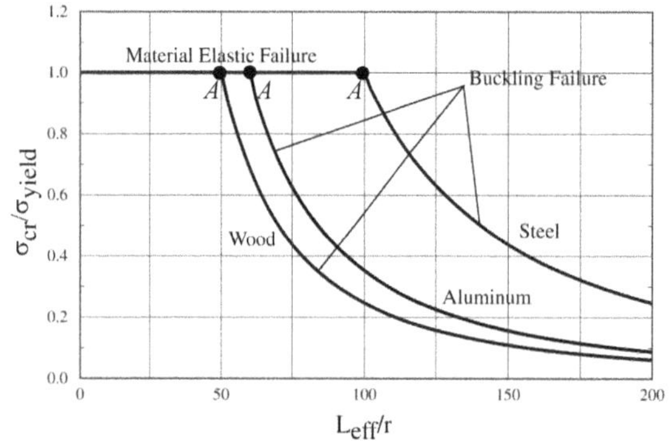


**Figure 4.3** Failure envelopes for Euler columns.

Intermediate column is a third classification used if the critical stress is between yield stress and ultimate stress. The tangent modulus theory of buckling accounts for it by replacing the modulus of elasticity by the tangent modulus of elasticity, that is,

$$P_{cr} = \frac{\pi^2 E_t I}{L_{eff}^2} \tag{4.10}$$

where $E_t$ is the tangent modulus, which depends on the stress level $P_{cr}/A$. Using an iterative procedure and Equation (4.10), the critical buckling load can be determined.

## EXAMPLE 4.1

A linear spring is attached at the free end of a column, as shown in Figure 4.4. Assume that bending about the $y$ axis is prevented. (a) Determine the characteristic equation for this buckling problem. Show that the critical load $P_{cr}$ for (b) $k = 0$ and (c) $k = \infty$ is as given in Table 4.1 for cases 2 and 3, respectively.

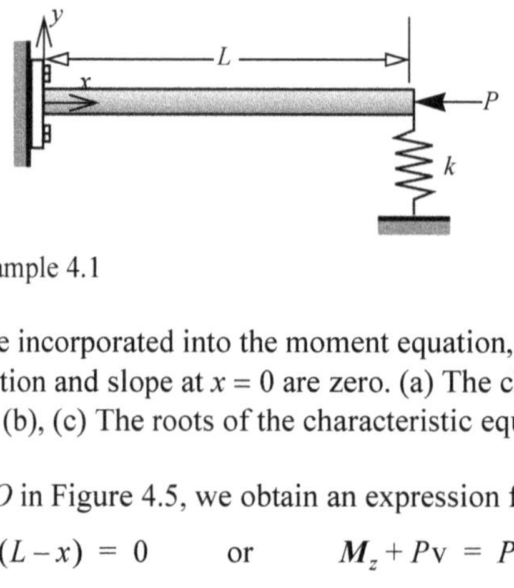


**Figure 4.4** Column with elastic support in Example 4.1

**PLAN**

The spring exerts a spring force $kv_L$ must be incorporated into the moment equation, and hence into the differential equation. The boundary conditions are that the deflection and slope at $x = 0$ are zero. (a) The characteristic equation will be generated while solving the boundary-value problem. (b), (c) The roots of the characteristic equation for the two cases will give $P_{cr}$.

**SOLUTION**

(a) By equilibrium of moment about point $O$ in Figure 4.5, we obtain an expression for moment $M_z$,

$$M_z - P(v_L - v) + kv_L(L - x) = 0 \quad \text{or} \quad M_z + Pv = Pv_L - kv_L(L - x) \tag{E1}$$

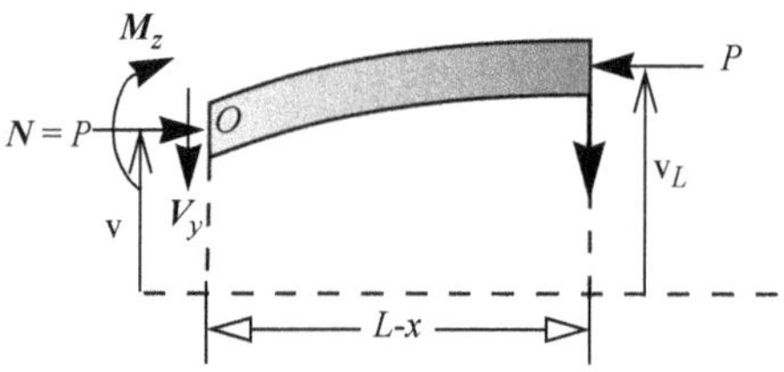


**Figure 4.5** Free-body diagram in Example 4.1.

Substituting the moment–curvature relationship, we obtain

$$EI\frac{d^2v}{dx^2} + Pv = Pv_L - kv_L(L - x) \tag{E2}$$

We can write the boundary value problem as shown below.

Differential equation: $$\frac{d^2v}{dx^2} + \lambda^2 v = \lambda^2 v_L - \frac{kv_L}{EI}(L - x) \qquad \text{(E3)}$$

Boundary Conditions: $$v(0) = 0 \qquad \frac{dv}{dx}(0) = 0 \qquad \text{(E4)}$$

The homogeneous solution $v_H$ to Equation (E3) is given by Equation (4.2). The particular solution is

$$v_P = v_L - \frac{kv_L}{\lambda^2 EI}(L - x) \qquad \text{(E5)}$$

Thus the total solution can be written as

$$v(x) = v_H + v_P = A\cos\lambda x + B\sin\lambda x + v_L - \frac{kv_L}{\lambda^2 EI}(L - x) \qquad \text{(E6)}$$

From boundary conditions we obtain

$$v(0) = A\cos(0) + B\sin(0) + v_L - \frac{kv_L}{\lambda^2 EI}(L - 0) = 0 \qquad \text{or} \qquad A = \left(\frac{kL}{\lambda^2 EI} - 1\right)v_L \qquad \text{(E7)}$$

$$\frac{dv}{dx}(0) = -\lambda A\sin(0) + B\lambda\cos(0) + \frac{kv_L}{\lambda^2 EI} = 0 \qquad \text{or} \qquad B = -\frac{k}{\lambda^3 EI}v_L \qquad \text{(E8)}$$

Substituting the values of $A$ and $B$ into Equation (E6), we obtain

$$v(x) = \left[\left(\frac{kL}{\lambda^2 EI} - 1\right)\cos\lambda x - \frac{k}{\lambda^3 EI}\sin\lambda x + 1 - \frac{k}{\lambda^2 EI}(L - x)\right]v_L \qquad \text{(E9)}$$

Substituting $x = L$ into Equation (E6), we obtain

$$v(L) = \left[\left(\frac{kL}{\lambda^2 EI} - 1\right)\cos\lambda L - \frac{k}{\lambda^3 EI}\sin\lambda L + 1 - 0\right]v_L = v_L \qquad \text{(E10)}$$

Since $v_L$ is a common factor, Equation (E10) can be simplified to the following *characteristic equation:*

$$\textbf{ANS.}\quad \left(\frac{kL}{\lambda^2 EI} - 1\right)\cos\lambda L - \frac{k}{\lambda^3 EI}\sin\lambda L = 0 \qquad \text{(E11)}$$

(b) Substituting $k = 0$ into Equation (E10), we obtain $\cos\lambda L = 0$, which is the characteristic equation for case 2 in Table 4.1. Thus the $P_{cr}$ value corresponding to the smallest root will be as given in Table 4.1 for case 2.

(c) We rewrite Equation (E10) as

$$\tan\lambda L = \lambda L - \frac{\lambda^3 EI}{k} \qquad \text{(E12)}$$

As $k$ tends to infinity, the second term tends to zero and we obtain $\tan\lambda L = \lambda L$, which is the characteristic equation for case 3 in Table 4.1. Thus the $P_{cr}$ value corresponding to the smallest root will be as given in Table 4.1 for case 3.

**COMMENTS**

1. This example shows that a spring could simulate an imperfect support that provides some restraint to deflection. The restraining effect is more than zero (free end) but not as much as a roller support.
2. The spring could also represent other beams that are pin connected at the top end. These pin-connected beams provide elastic restraint to deflection but no restraint to the slope. If the beams were welded rather than pin connected, then we would have to include a torsional spring also at the end.
3. The example demonstrates that the critical buckling loads can be changed by installing elastic restraints.

## EXAMPLE 4.2

Determine the maximum deflection of the column shown in Figure 4.6 in terms of the modulus of elasticity $E$, the length of the column $L$, the area moment of inertia $I$, the axial force $P$, and the intensity of the distributed force $w$.

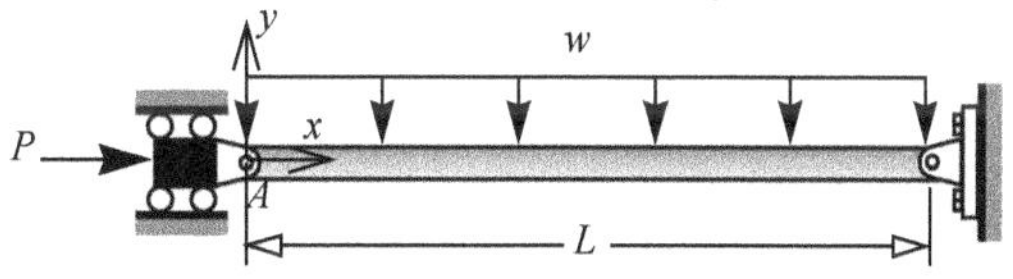


**Figure 4.6** Buckling of beam with distributed load in Example 4.2.

**PLAN**

The moment from the distributed load can be added to the moment for case 1 in Table 4.1 and the differential equation written. The boundary conditions are that the deflection at $x = 0$ and $x = L$ is zero. The boundary-value problem can be solved, and the deflection at $x = L/2$ evaluated to obtain the maximum deflection.

**SOLUTION**

The reaction force in the $y$ direction is half the total load $wL$ acting on the beam. An imaginary cut at some location $x$ can be made and the free-body diagram of the left part drawn as shown in Figure 4.7. By balancing the moment at point $O$, we obtain an expression for the moment $M_z$,

$$M_z + P\text{v}(x) - \frac{wL}{2}x + \frac{wx^2}{2} = 0 \qquad \text{or} \qquad EI_{zz}\frac{d^2\text{v}}{dx^2} + P\text{v} = \frac{wL}{2}x - \frac{wx^2}{2} \tag{E1}$$

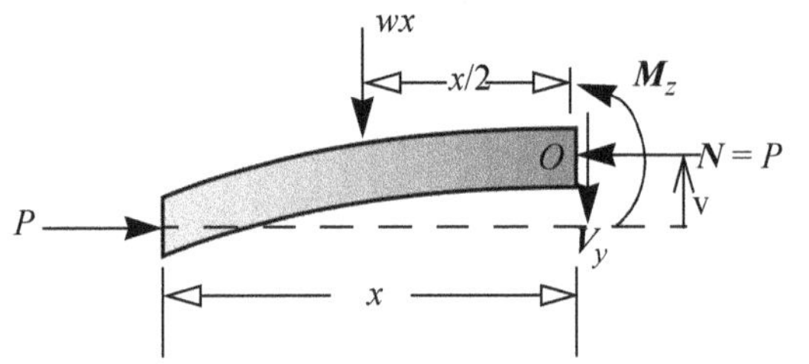


**Figure 4.7** Free-body diagram in Example 4.2.

The zero-deflection boundary conditions at either end are written to complete the statement of the boundary-value problem.

Differential equation:
$$\frac{d^2\text{v}}{dx^2} + \lambda^2\text{v} = \frac{wLx}{2EI} - \frac{wx^2}{2EI} \tag{E2}$$

Boundary Conditions:
$$\text{v}(0) = 0 \qquad \text{v}(L) = 0 \tag{E3}$$

To find the particular solution, we substitute $\text{v}_P = a + bx + cx^2$ into Equation (E2) and simplify,

$$2c + \lambda^2(a + bx + cx^2) = \frac{wLx}{2EI} - \frac{wx^2}{2EI} \qquad \text{or} \qquad (2c + \lambda^2 a) + \left(\lambda^2 b - \frac{wL}{2EI}\right)x + \left(\lambda^2 c + \frac{w}{2EI}\right)x^2 = 0 \tag{E4}$$

If Equation (E4) is to be valid for any value of $x$, then each of the terms in parentheses must be zero and we obtain the values of constants $a$, $b$, and $c$,

$$c = -\frac{w}{2\lambda^2 EI} \qquad b = \frac{wL}{2\lambda^2 EI} \qquad a = -\frac{2c}{\lambda^2} = \frac{w}{\lambda^4 EI} \tag{E5}$$

Hence the particular solution is

$$\text{v}_P = \frac{w}{\lambda^4 EI} + \frac{wL}{2\lambda^2 EI}x - \frac{w}{2\lambda^2 EI}x^2 \tag{E6}$$

The homogeneous solution $\text{v}_H$ to Equation (E2) is given by Equation (4.2). Thus the total solution can be written as

$$\text{v}(x) = \text{v}_H + \text{v}_P = A\cos\lambda x + B\sin\lambda x + \frac{w}{\lambda^4 EI} + \frac{wL}{2\lambda^2 EI}x - \frac{w}{2\lambda^2 EI}x^2 \tag{E7}$$

Substituting $x = 0$ into Equation (E7) and using Equation (E3), we obtain

$$\text{v}(0) = A\cos(0) + B\sin(0) + \frac{w}{\lambda^4 EI} + 0 - 0 = 0 \qquad \text{or} \qquad A = -\frac{w}{\lambda^4 EI} \tag{E8}$$

Substituting $x = L$ into Equation (E7) and using Equation (E3), we obtain

$$\text{v}(L) = A\cos\lambda L + B\sin\lambda L + \frac{w}{\lambda^4 EI} + \frac{wL^2}{2\lambda^2 EI} - \frac{wL^2}{2\lambda^2 EI} = 0 \qquad \text{or} \qquad -\frac{w}{\lambda^4 EI}\cos\lambda L + B\sin\lambda L + \frac{w}{\lambda^4 EI} = 0 \tag{E9}$$

Since $\sin\lambda L = 2\sin(\lambda L/2)\cos(\lambda L/2)$ and $1 - \cos\lambda L = 2\sin^2(\lambda L/2)$ the above equation can be simplified as

$$B = -\frac{w}{\lambda^4 EI}\frac{1 - \cos\lambda L}{\sin\lambda L} = -\frac{w}{\lambda^4 EI}\tan\left(\frac{\lambda L}{2}\right) \tag{E10}$$

By symmetry the maximum deflection will occur at midpoint. Substituting $x = L/2$, $A$ and $B$ into Equation (E7), we obtain

$$\text{v}_{\max} = \text{v}\left(\frac{L}{2}\right) = A\cos\left(\frac{\lambda L}{2}\right) + B\sin\left(\frac{\lambda L}{2}\right) + \frac{w}{\lambda^4 EI} + \frac{wL^2}{4\lambda^2 EI} - \frac{wL^2}{8\lambda^2 EI}$$

$$\text{v}_{\max} = \frac{w}{\lambda^4 EI}\left[-\cos\left(\frac{\lambda L}{2}\right) - \tan\left(\frac{\lambda L}{2}\right)\sin\left(\frac{\lambda L}{2}\right)\right] + \frac{w}{\lambda^4 EI} + \frac{wL^2}{8\lambda^2 EI} \tag{E11}$$

Equation (E11) can be simplified by substituting the tangent function in terms of the sine and cosine functions to obtain

$$\text{v}_{\max} = -\frac{w}{\lambda^4 EI}\left[\sec\left(\frac{\lambda L}{2}\right) - 1\right] + \frac{wL^2}{8\lambda^2 EI} \tag{E12}$$

Substituting for $\lambda$, the maximum deflection can be written as

**ANS.** $$v_{max} = -\frac{wEI}{P^2}\left[sec\left(\frac{L}{2}\sqrt{\frac{P}{EI}}\right) - 1\right] + \frac{wL^2}{8P}$$

**COMMENTS**

1. In Equation (E11), as $\lambda L \to \pi$, the secant function tends to infinity and the maximum displacement becomes unbounded, which means the column becomes unstable. $\lambda L = \pi$ corresponds to the Euler buckling load of Equation (4.6). Thus the *transverse distributed load does not change the critical buckling load* of a column.
2. However the failure mode can be significantly affected by the transverse distributed load. The maximum normal stress will be the sum of axial stress and maximum bending normal stress, $\sigma_{max} = P/A + M_{max}y_{max}/I$. The maximum bending moment will be at $x = L/2$ and can be found from Equation (E1) as $M_{max} = wL^2/8 - Pv_{max}$. Substituting and simplifying gives the maximum normal stress:

$$\sigma_{max} = \frac{P}{A} + \frac{wEy_{max}}{P}\left[sec\left(\frac{L}{2}\sqrt{\frac{P}{EI}}\right) - 1\right] \tag{E13}$$

By equating the maximum normal stress to the yield stress, we obtain a failure envelope, which clearly depends on the value of $w$.

## EXAMPLE 4.3

In terms of $\alpha$ the thermal coefficient of expansion, $r$ is the radius of gyration, and length L, determine the critical change of temperature at which the beam shown in Figure 4.8 will buckle. Assume a uniform increase of temperature.

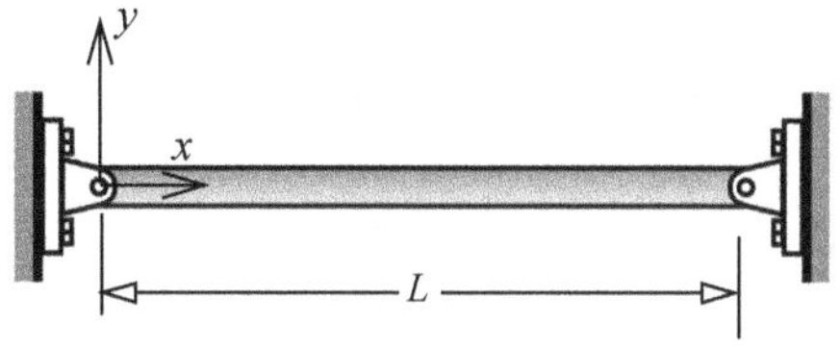


**Figure 4.8** Column buckling from temperature change in Example 4.3.

**PLAN**

A reaction axial force will develop as temperature is increased, which can be equated to the critical buckling load to obtain the critical buckling temperature.

**SOLUTION**

From Hooke's law we have

$$\varepsilon_{xx} = \frac{\sigma_{xx}}{E} + \alpha\Delta T = 0 \quad \text{or} \quad \sigma_{xx} = -E\alpha\Delta T \quad \text{or} \quad (\sigma_{xx})_{cr} = E\alpha\Delta T_{crit}\ (C) \tag{E1}$$

The stress at critical buckling load to obtain

$$(\sigma_{xx})_{cr} = \frac{P_{cr}}{A} = \frac{\pi^2 EI}{AL^2} = \frac{\pi^2 EAr^2}{AL^2} = \frac{\pi^2 E}{(L/r)^2} \tag{E2}$$

Equating the critical stress in the two equations we obtain the critical temperature at buckling.

$$E\alpha\Delta T_{crit} = \frac{\pi^2 E}{(L/r)^2} \tag{E3}$$

**ANS.** $$\Delta T_{crit} = \frac{\pi^2}{\alpha(L/r)^2}$$

**COMMENTS**

1. The derivation above presupposes that the critical buckling temperature is small enough that the mechanical properties do not change and the thermal strain is still a linear function of temperature change.
2. We could use effective length of $0.5L$ in the critical temperature formula for a column with built in both ends (case 4 in Table 4.1). But we cannot extend it to case 2 and 3, because the displacement at one end is not constraint, and temperature increase will simply cause expansion without introducing thermal stress.

## 4.2 IMPERFECT COLUMNS

In the development of the theory for axial members and the symmetric bending of beams, we obtained that the condition for decoupling axial deformation from bending deformation for linear, elastic, and homogeneous material: the applied loads must pass through the centroid of the cross sections, and the centroids of all cross sections are on a straight line. However, the requirements for decoupling the axial from the bending problem may not be met for a number of reasons, some of which are given here:

- The column material may contain small holes, minute cracks, or other material inclusions. Hence the homogeneity requirement or the requirement that the centroids of all cross sections be on a straight line may not be met.
- The material processing may cause local strain hardening. Hence the condition of linear and elastic material behavior across the entire cross section may not be met.
- The theoretical design centroid and the actual centroid are offset due to manufacturing tolerances.
- Local conditions at the support cause the reaction force to be offset from the centroid.
- The transfer of loads from one member to another may not occur at the centroid.

The above partial list can be considered as imperfections in the column, which cause the application of axial loads to be offset from the centroid of the cross section. This offset loading is termed **eccentric loading** on columns. In this section we study the impact of eccentricity in loading on buckling.

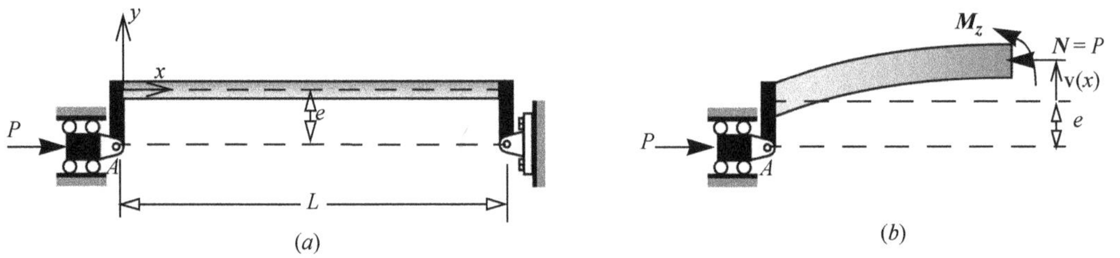


**Figure 4.9** Eccentrically loaded column.

Figure 4.9*a* shows a simply supported column on which an eccentric compressive axial load is applied at a distance *e* from the centroid of the cross section. Figure 4.9*b* shows the free-body diagram of the column segment. By balancing the moment at point *A* we obtain $\boldsymbol{M_z} + P(\mathrm{v} + e) = 0$. Substituting the moment–curvature relationship we obtain the differential equation

$$\frac{d^2\mathrm{v}}{dx^2} + \lambda^2\mathrm{v} = -\left(\frac{Pe}{EI}\right) \tag{4.11}$$

where $\lambda$ is given by Equation (4.1b). The boundary conditions are that displacements at $x = 0$ and $x = L$ are zero, as given by Equation (4.1c) and (4.1c). The homogeneous solution to Equation (4.11) is given by Equation (4.2). The particular solution to Equation (4.11) is $\mathrm{v}_P(x) = -e$. Thus the total solution is

$$\mathrm{v}(x) = \mathrm{v}_H + \mathrm{v}_P = A\cos\lambda x + B\sin\lambda x - e \tag{4.12}$$

From boundary conditions we obtain

$$\mathrm{v}(0) = A\cos(0) + B\sin(0) - e = 0 \quad \text{or} \quad A = e \tag{4.13}$$

$$\mathrm{v}(L) = A\cos\lambda L + B\sin\lambda L - e = 0 \quad \text{or} \quad B = \frac{e(1-\cos\lambda L)}{\sin\lambda L} = \frac{e2\sin^2(\lambda L/2)}{2\sin(\lambda L/2)\cos(\lambda L/2)} = e\tan(\lambda L/2) \tag{4.14}$$

Substituting for *A* and *B* in Equation (4.12), we obtain the deflection as

$$\mathrm{v}(x) = e[\cos\lambda x + \tan(\lambda L/2)\sin\lambda x - 1] \tag{4.15}$$

As $\lambda L/2 \to \pi/2$, the function $\tan(\lambda L/2) \to \infty$ and the displacement function $\mathrm{v}(x)$ becomes unbounded. Thus the critical load value can be found by substituting for $\lambda$ in the equation $\lambda L/2 = \pi/2$ to obtain the same critical value as given by Equation (4.6). In other words, *the buckling load value does not change with the eccentricity of the loading.* We will make use of this observation to extend our formulas to other types of support conditions.

In the eigenvalue approach discussed in Section 4.1, we were unable to determine the displacement function because we had an undetermined constant *B* in Equation (4.7). But here the displacement function is completely determined by Equation (4.15). The maximum deflection (by symmetry) will be at the midpoint. Substituting $x = L/2$ into Equation (4.15), we obtain

$$\mathrm{v}_{\max} = e[\cos(\lambda L/2) + \tan(\lambda L/2)\sin(\lambda L/2) - 1] = e[\sec(\lambda L/2) - 1] \tag{4.16a}$$

Substituting for $\lambda$ from Equation (4.1b), we obtain

$$\mathrm{v}_{max} = e\left[\sec\left(\frac{L}{2}\sqrt{\frac{P}{EI}}\right) - 1\right] = e\left[\sec\left(\frac{L}{2}\sqrt{\frac{PP_{\mathrm{cr}}}{P_{\mathrm{cr}}EI}}\right) - 1\right]$$

$$\boxed{\mathrm{v}_{max} = e\left[\sec\left(\frac{\pi}{2}\sqrt{\frac{P}{P_{\mathrm{cr}}}}\right) - 1\right]} \tag{4.17}$$

The maximum normal stress is the sum of compressive axial stress and maximum compressive bending stress:

$$\sigma_{max} = \frac{P}{A} + \frac{M_{max}y_{max}}{I} \tag{4.18a}$$

The maximum bending moment will be at the midpoint of the column, and its value is $M_{max} = P(e + v_{max})$. Substituting for $v_{max}$ we obtain

$$\sigma_{max} = \frac{P}{A} + \frac{Py_{max}}{I} e\left[\sec\left(\frac{L}{2}\sqrt{\frac{P}{EI}}\right)\right] \tag{4.18b}$$

Equation (4.18b) was derived for simply supported columns. We can extend the results to other supports by changing the length of the column to the effective length $L_{eff}$, as given in Table 4.1. We also substitute $y_{max} = c$, where $c$ represents the maximum distance from the buckling (bending) axis to a point on the cross section. Substituting $I = Ar^2$, where $A$ is the cross-sectional area and $r$ is the radius of gyration, we obtain

$$\boxed{\sigma_{max} = \frac{P}{A}\left[1 + \frac{ec}{r^2}\sec\left(\frac{L_{eff}}{2r}\sqrt{\frac{P}{EA}}\right)\right]} \tag{4.19}$$

Equation (4.19) is called the *secant formula.* The quantity $ec/r^2$ is called the *eccentricity ratio.*

By equating $\sigma_{max}$ to failure stress $\sigma_{fail}$ in Equation (4.19), we obtain the failure envelope for an imperfect column. The failure envelope equation can be written in nondimensional form as

$$\frac{P/A}{\sigma_{fail}}\left[1 + \frac{ec}{r^2}\sec\left(\frac{L_{eff}}{2r}\sqrt{\left(\frac{\sigma_{fail}}{E}\right)\frac{P/A}{\sigma_{fail}}}\right)\right] = 1 \tag{4.20}$$

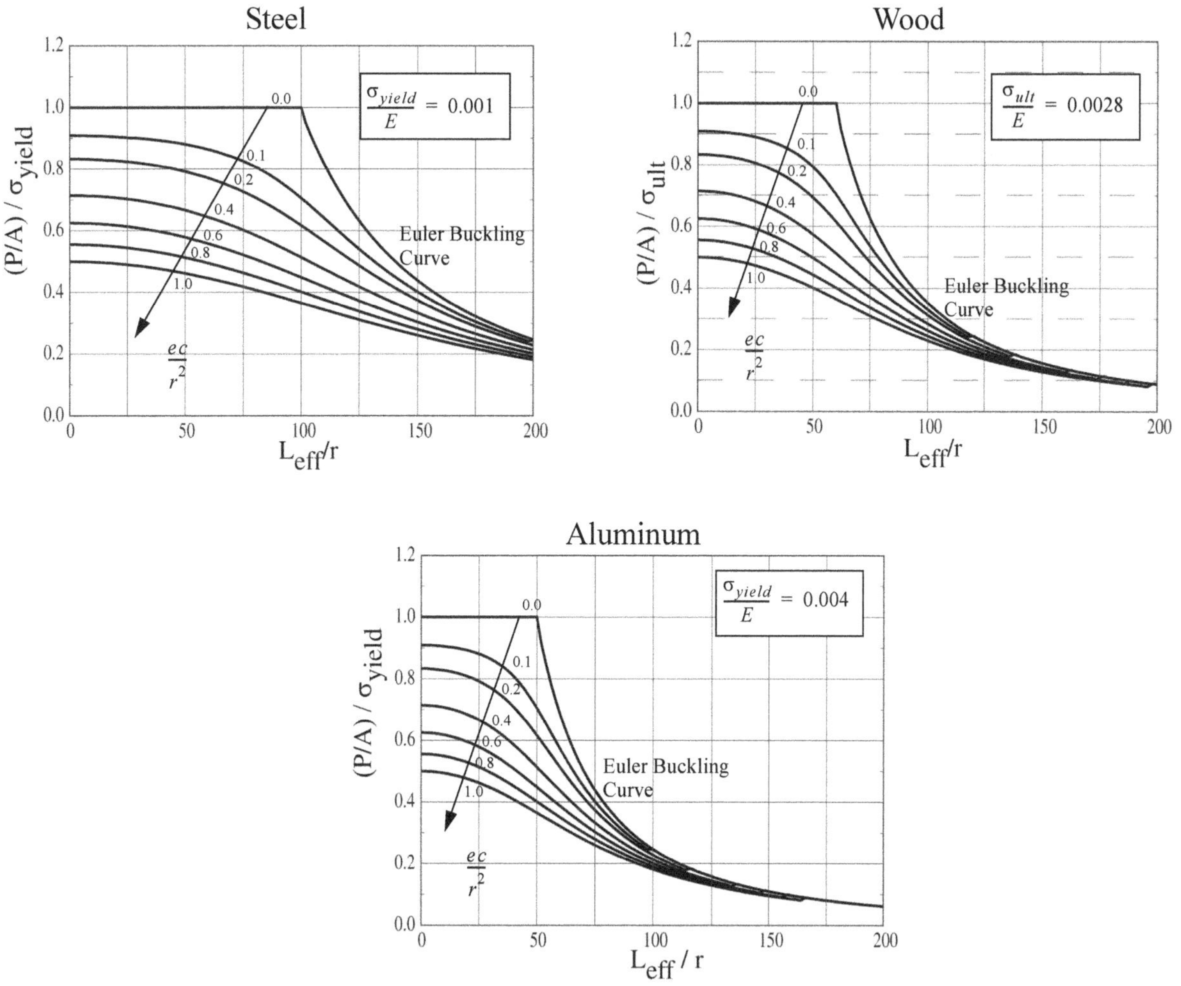


**Figure 4.10** Failure envelopes for imperfect columns.

Equation (4.20) can be plotted for different materials, as shown in Figure 4.10. These curves can be used for metric as well for U.S. customary units, since the variables used in creating the plots are nondimensional. The curves can be used for any material that has the same value for $\sigma_{yield}/E$. The failure stress in the cases of steel and aluminum would be the yield stress $\sigma_{yield}$, whereas for wood it would be the ultimate stress $\sigma_{ult}$. The curves can also be used for different end conditions by using the appropriate $L_{eff}$ as given in Table 4.1.

## EXAMPLE 4.4

A wooden box column ($E$ = 1800 ksi) is constructed by joining four pieces of lumber together, as shown in Figure 4.11. The load $P$ = 80 kips is applied at a distance of $e$ = 0.667 in. from the centroid of the cross section. (a) If the length is $L$ = 10 ft, what are the maximum stress and the maximum deflection? (b) If the allowable stress is 3 ksi, what is the maximum permissible length $L$ to the nearest inch?

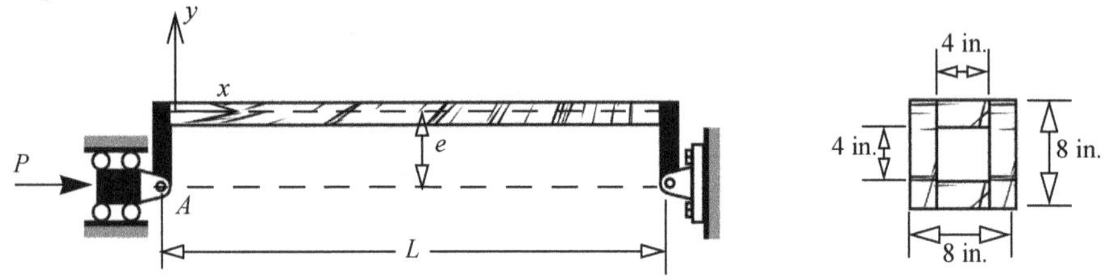


**Figure 4.11** Eccentrically loaded box column in Example 4.4.

**PLAN**

From cross section geometry we can find $A$, $I$, $r$, and $c$. (a) Substituting $L_{eff}$ = 120 in. and the values of the other variables into Equations (4.17) and (4.19), we can find the maximum stress and the maximum deflection. (b) Equating $\sigma_{max}$ in Equation (4.19) to 3 ksi and substituting the remaining variables, we find the length $L$.

**SOLUTION**

From the given cross section, the cross-sectional area $A$, the area moment of inertia $I$, and the radius of gyration $r$ can be found:

$$A = (8\text{ in.})(8\text{ in.}) - (4\text{ in.})(4\text{ in.}) = 48\text{ in.}^2 \qquad I = \frac{1}{12}[(8\text{ in.})^4 - (4\text{ in.})^4] = 320\text{ in.}^4 \qquad r = \sqrt{\frac{I}{A}} = 2.582\text{ in.} \tag{E1}$$

(a) Since the column is pinned at both ends, $L_{eff} = L$ = 10 ft = 120 in. Substituting $L_{eff}$, $I$, and $E$ = 1800 ksi into Equation (4.8) give the critical buckling load:

$$P_{cr} = [\pi^2(1800\text{ ksi})(320\text{ in.}^4)]/(120\text{ in.})^2 = 394.8\text{ kips} \tag{E2}$$

Substituting $e$ = 0.667in., $P$ = 80 kips, and Equation (E2) into Equation (4.17), we obtain the maximum deflection,

$$v_{max} = (0.667\text{ in.})\left[sec\left(\frac{\pi}{2}\sqrt{\frac{80\text{ kips}}{394.8\text{ kips}}}\right) - 1\right] = 0.2103\text{ in.} \tag{E3}$$

**ANS.** $v_{max}$ = 0.21 in.

Substituting $c$ = 4 in., $e$ = 0.667 in., $r$ = 2.582 in., $P$ = 80 kips, $E$ = 1800 ksi, and $A$ = 48 in.$^2$ into Equation (4.19), we obtain

$$\sigma_{max} = \frac{80\text{ kips}}{48\text{ in.}^2}\left[1 + \frac{(0.667\text{ in.})(4\text{ in.})}{(2.582\text{ in.})^2}sec\left(\frac{120\text{ in.}}{2(2.582\text{ in.})}\sqrt{\frac{80\text{ kips}}{(1800\text{ ksi})(48\text{ in.}^2)}}\right)\right] = 2.544\text{ ksi} \tag{E4}$$

**ANS.** $\sigma_{max}$ = 2.5 ksi (C)

(b) Substituting $\sigma_{max}$ = 3 ksi, $L_{eff}$ = $L$ in, and other variables into Equation (4.19), we obtain

$$3 = \frac{80\text{ kips}}{48\text{ in.}^2}\left[1 + \frac{(0.667\text{ in.})(4\text{ in.})}{(2.582\text{ in.})^2}sec\left\{\frac{L\text{ in.}}{2(2.582\text{ in.})}\sqrt{\frac{80\text{ kips}}{(1800\text{ ksi})(48\text{ in.}^2)}}\right\}\right] \tag{E5}$$

$$sec\{5.892(10^{-3})L\} = 2 \qquad \text{or} \qquad \cos(5.892 \times 10^{-3}L) = 0.5 \qquad \text{or} \qquad L = 177.7\text{ in.} \tag{E6}$$

Rounding downward, the maximum permissible length is: thus $L$ = 177 in.

**ANS.** $L$ = 177 in.

**COMMENTS**

1. The axial stress $P/A$ = (80 kips)/(48 in.$^2$) = 1.667 ksi, but the normal stress due to bending from eccentricity causes the normal stress to be significantly higher, as seen by the value of $\sigma_{max}$.
2. If the right end of the column shown in Figure 4.11 were built in rather than held by a pin, then from case 3 in Table 4.1, $L_{eff}$ = 0.7$L$ = 84 in. Using this value, we can find $P_{cr}$ = 805.7 kips, $v_{max}$ = 0.091 in., and $\sigma_{max}$ = 2.42 ksi.
3. In Equation (E6) we rounded downward, as shorter columns will result in a stress that is less than allowable.

## EXAMPLE 4.5

A wooden box column ($E$ = 1800 ksi) is constructed by joining four pieces of lumber together, as shown in Figure 4.11. The ultimate stress is 5 ksi. Determine the maximum load $P$ that can be applied.

**PLAN**

The eccentricity ratio and the slenderness ratio can be found using the values of the geometric quantities calculated in Example 4.4. Noting that $\sigma_{ult}/E$ = 0.0028, the failure envelopes for wood that are shown in Figure 4.10 can be used and $(P/A)/\sigma_{ult}$ can be found, from which the maximum load $P$ can be determined.

**SOLUTION**

From Equation (E1) in Example 4.4, $r = 2.582$ in. Thus the slenderness ratio $L_{eff}/r = (120\text{ in.})/(2.582\text{ in.}) = 46.48$. From Figure 4.11, $c = 4$ in. and $e = 0.667$ in. Thus the eccentricity ratio $ec/r^2 = 0.400$.

For a slenderness ratio of 46.48 and an eccentricity ratio of 0.4, we estimate the value of $(P/A)/\sigma_{ult} = 0.6$ from the failure envelope for wood in Figure 4.10. Substituting $\sigma_{ult} = 5$ ksi and $A = 48$ in.$^2$, we obtain the maximum load $P_{max} = (0.6)$ (5 ksi) (48in.$^2$).

**ANS.** $P_{max} = 144$ kips

**COMMENT**

1. If we let $x$ represent $(P/A)/\sigma_{ult}$ and substitute the remaining variables in Equation (4.20), we obtain the following nonlinear equation: $x[1 + 0.4\ sec(1.2297\sqrt{x})] = 1$. The root of the equation can be found using a numerical method. The value of the root to the third-place decimal is 0.593, which would yield a value of $P_{max} = 142.3$ kips, a difference of 1.18% from that reported in our example. The difference is small and an acceptable engineering approximation. Use of the plots in Figure 4.10 was a quick way of finding the load value with reasonable engineering approximation.

## 4.3 CLOSURE

Buckling of columns is affected by several factors. We saw that end conditions can be accounted for by defining effective column length which may be different from the actual column length. Elastic supports can be accounted by incorporating linear and torsional springs in the mathematical models. Tolerances, manufacturing flaws, support and loading misalignments can be incorporated by use of eccentricity factor. Stress levels beyond yield stress can be accounted for by use of tangent modulus. Temperature, material non-homogeneity, and various other complexities that we saw in Chapter 2 can be incorporated by changing the axial and / or bending formulas as we did in Chapter 2.

This chapter is stand alone in the book's current form and has no impact on subsequent chapters. However, the field of stability is very large and often there is an entire course devoted to the study of stability.

## 4.4 SYNOPSIS OF EQUATIONS

| | |
|---|---|
| | Euler Buckling |
| Buckling load | $P_{cr} = \dfrac{\pi^2 EI}{L^2};\ P_n = \dfrac{n^2\pi^2 EI}{L^2},\qquad v = B\ sin\left(n\pi\dfrac{x}{L}\right)$ |
| Column classification | $\sigma_{cr} = \dfrac{P_{cr}}{A} = \dfrac{\pi^2 E}{(L_{eff}/r)^2};\ P_{cr} = \dfrac{\pi^2 E_t I}{L_{eff}^2}$ |
| | Imperfect column |
| Displacement | $v_{max} = e\left[sec\left(\dfrac{\pi}{2}\sqrt{\dfrac{P}{P_{cr}}}\right) - 1\right]$ |
| Stress | $\sigma_{max} = \dfrac{P}{A}\left[1 + \dfrac{ec}{r^2}sec\left(\dfrac{L_{eff}}{2r}\sqrt{\dfrac{P}{EA}}\right)\right]$ |

## PROBLEMS

### Section 4.1

**4.1** Three column cross sections are shown in Figure P4.1. The area of each of the three cross sections is equal to $A$. Determine the ratios of critical loads $P_{cr1}$: $P_{cr2}$: $P_{cr3}$ assuming (a) the ends are simply supported; (b) the ends are built in. (c) How do you expect the ratios to change if the end conditions were as in cases 2 and 3 of Table 4.1?

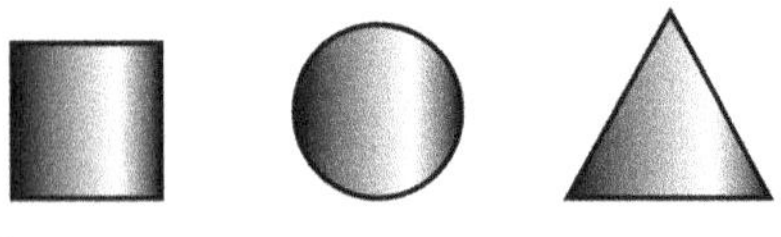

**Figure P4.1**

**4.2** Figure P4.2 shows two steel ($E$ = 30,000 ksi, $\sigma_{yield}$ = 30 ksi) bars of a diameter $d$ = 1/4 in. on which a force $F$ = 600 lb is applied. Bars $AP$ and $BP$ have lengths $L_{AP}$ = 7 in. and $L_{BP}$ = 10 in. Determine the factor of safety for the assembly.

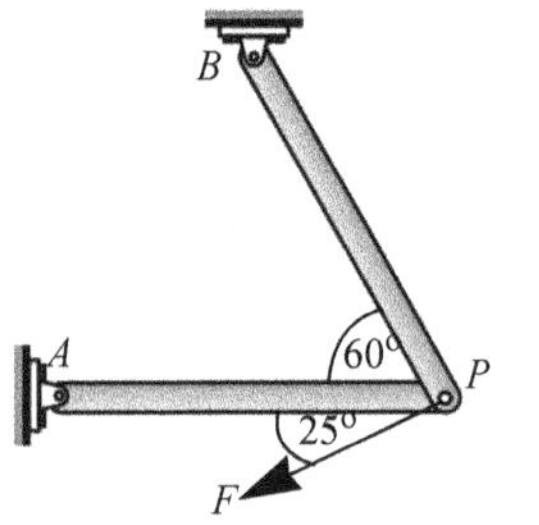


**Figure P4.2**

**4.3** Figure P4.3 shows two copper ($E$ = 15,000 ksi, $\sigma_{yield}$ = 12 ksi) bars of a diameter $d$ = 1/4 in. on which a force $F$ = 500 lb is applied. Bars $AP$ and $BP$ have lengths $L_{AP}$ = 7 in. and $L_{BP}$ = 9 in. Determine the factor of safety for the assembly.

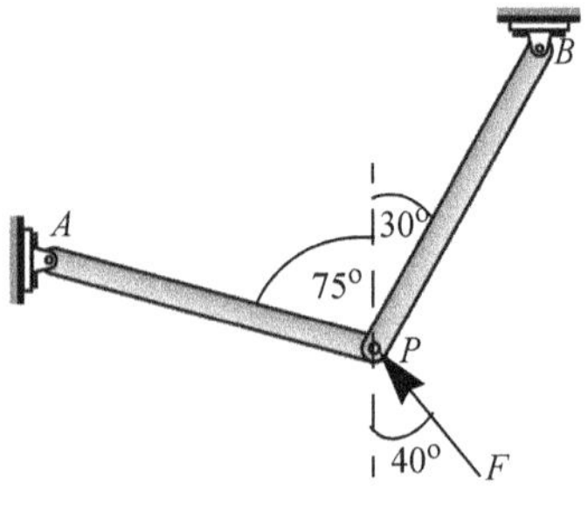


**Figure P4.3**

**4.4** (a) Solve the boundary-value problem for case 2 in Table 4.1 and obtain the critical load value $P_{cr}$ that is given in the table. (b) If buckling in mode 1 is prevented, then what would be the $P_{cr}$ value?

**4.5** (a) Solve the boundary-value problem for case 3 in Table 4.1 and obtain the critical load value $P_{cr}$ that is given in the table. (b) If buckling in mode 1 is prevented, then what would be the $P_{cr}$ value?

**4.6** (a) Solve the boundary-value problem for case 4 in Table 4.1 and obtain the critical load value $P_{cr}$ that is given in the table. (b) If buckling in mode 1 is prevented, then what would be the $P_{cr}$ value?

**4.7** A torsional spring with a spring constant $K$ is attached at one end of a column, as shown in Figure P4.7. Assume that bending about the $y$ axis is prevented. (a) Determine the characteristic equation for this buckling problem. (b) Show that for $K = 0$ and $K = \infty$ the critical load $P_{cr}$ is as given in Table 4.1 for cases 1 and 3, respectively.

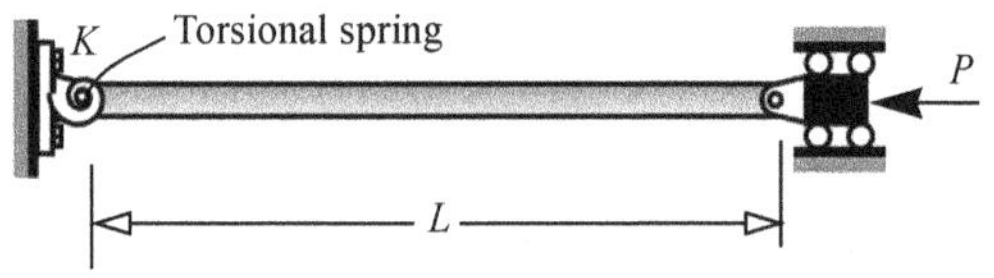


**Figure P4.7**

**4.8** A torsional spring with a spring constant $K$ is attached at one end of a column, as shown in Figure P4.8. Assume that bending about the $y$ axis is prevented. (a) Determine the characteristic equation for this buckling problem. (b) Show that for $K = 0$ the critical load $P_{cr}$ is as given for case 2 in Table 4.1. (c) For $K = \infty$ obtain the critical load $P_{cr}$.

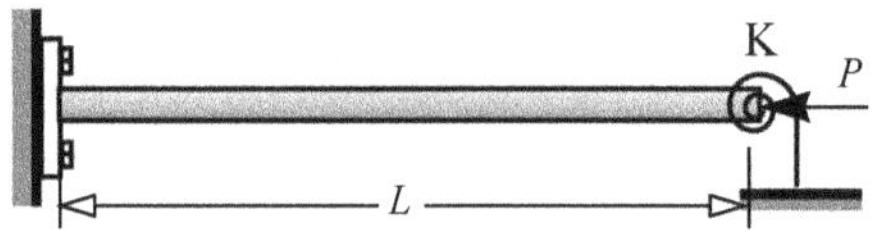


**Figure P4.8**

**4.9** Consider the column shown in Figure P4.9. (a) Determine the critical buckling in terms of $E$, $I$, $L$, and $\alpha$. (b) Show that when $\alpha = 0.5$, the critical load corresponds to mode 2, as shown in Figure 4.2.

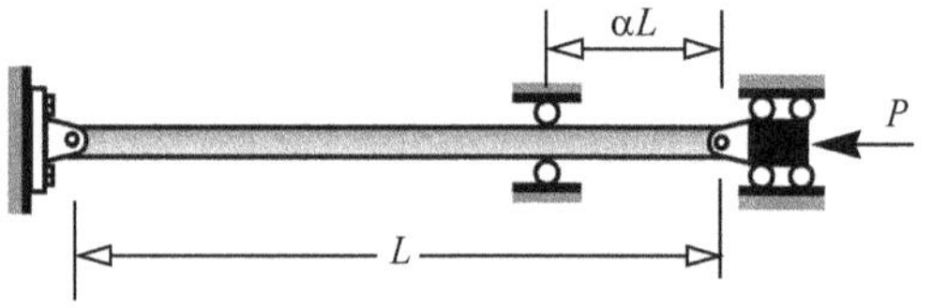


**Figure P4.9**

**4.10** For the column shown in Figure P4.10 determine (a) the deflection at $x = L$; (b) the critical load $P_{cr}$ in terms of the modulus of elasticity $E$, the column length $L$, the area moment of inertia $I$, and the force $P$.

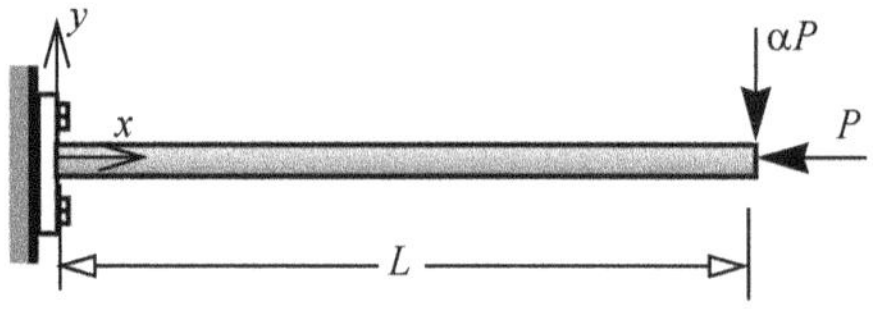


**Figure P4.10**

**4.11** For the column shown in Figure P4.11 determine (a) the deflection at $x = L$; (b) the critical load $P_{cr}$ in terms of the modulus of elasticity $E$, the column length $L$, the area moment of inertia $I$, and the force $P$.

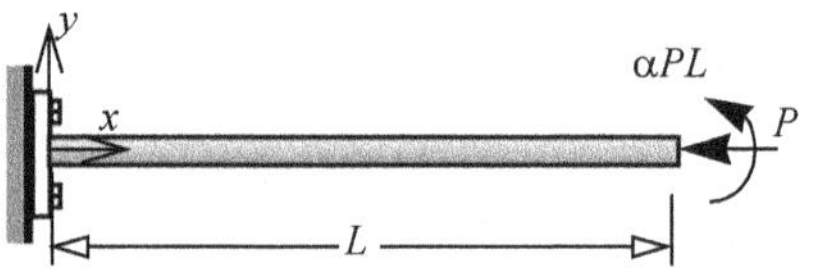


**Figure P4.11**

**4.12** The spreader shown in Figure P4.12 is to be made from an aluminum pipe ($E$ = 10,000 ksi) of 1 / 8 in thickness with an outer diameter of 2 in. The pipe lengths available for design start from 4 ft in 6-in. steps up to 8 ft. The allowable normal stress is 40 ksi. Develop a table for the lengths of pipe and the maximum force $F$ the spreader can support.

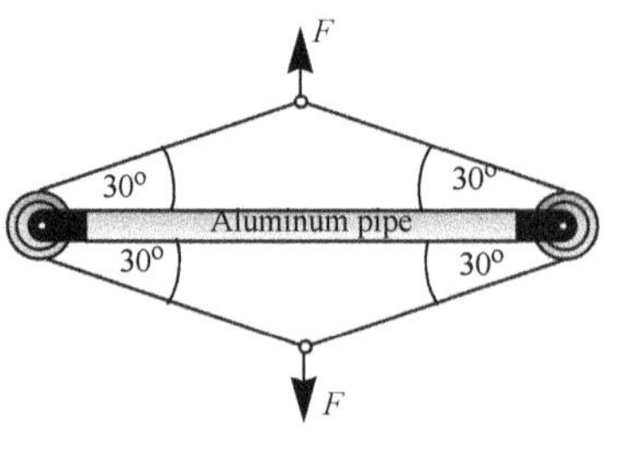


**Figure P4.12**

**4.13** A hoist is constructed using two wooden bars to lift a weight of 5 kips, as shown in Figure P4.13. The modulus of elasticity for wood $E = 1800$ ksi and the allowable normal stress is 3.0 ksi. Determine the maximum value of $L$ to the nearest inch that can be used in constructing the hoist.

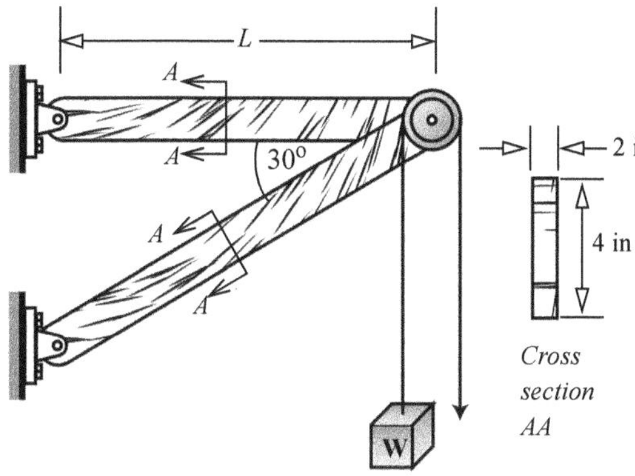


**Figure P4.13**

**4.14** Two 200-mm × 50-mm pieces of lumber ($E = 12.6$ GPa) form a part of a deck that is modeled as shown in Figure P4.14. The allowable stress for the lumber is 18 MPa. (a) Determine the maximum intensity of the distributed load $w$. (b) What is the factor of safety for column $BD$ corresponding to the answer in part (a)?

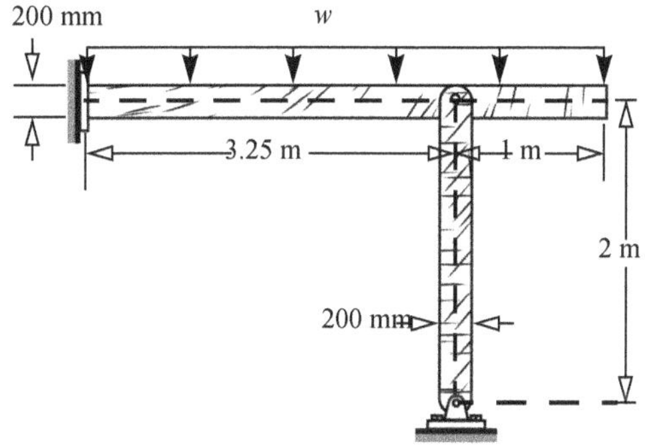


**Figure P4.14**

**4.15** Show that for a beam with a constant bending rigidity $EI$, the fourth-order differential equation for solving buckling problems is given by

$$EI\frac{d^4 v}{dx^4} + P\frac{d^2 v}{dx^2} = p_y \tag{4.21}$$

where $P$ is a compressive axial force and $p_y$ is the distributed force in the $y$ direction.

**4.16** A column with a constant bending rigidity $EI$ rests on an elastic foundation as shown in Figure P4.16. The foundation modulus is $k$, which exerts a spring force per unit length of $kv$. Show that the governing differential equation is given by Equation (4.22)

$$EI\frac{d^4 v}{dx^4} + P\frac{d^2 v}{dx^2} + kv = 0 \tag{4.22}$$

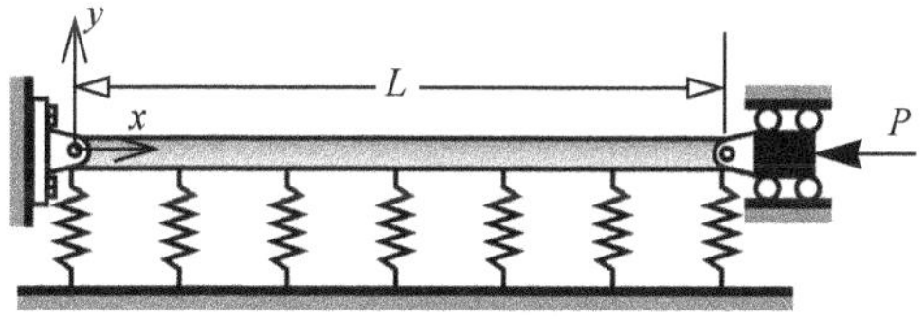


**Figure P4.16**

**4.17** Show that the buckling load for the column on an elastic foundation described in Problem 4.16 is given by the eigenvalues

$$P_n = \frac{\pi^2 EI}{L^2}\left[n^2 + \frac{1}{n^2}\left(\frac{kL^4}{\pi^4 EI}\right)\right], \qquad n = 1,2,3,\ldots \tag{4.23}$$

*Note:* For $n = 1$ and $k = 0$ Equation (4.23) gives the Euler buckling load.

**4.18** For a simply supported column with a symmetric composite cross section, show that the critical load $P_{cr}$ is given by

$$P_{cr} = \frac{\pi^2 \sum_{i=1}^{n} E_i I_i}{L_{eff}^2} \tag{4.24}$$

where $L_{eff}$ = the effective length of the column, $E_i$ is the modulus of elasticity for the i^th material, $I_i$ is the area moment of inertia about the buckling axis, and $n$ is the number of materials in the cross section.

**4.19** A composite column has the cross section shown in Figure P4.19. The modulus of elasticity of the outside material is twice that of the inside material. In terms of $E$, $d$, and $L$, determine the critical buckling load.

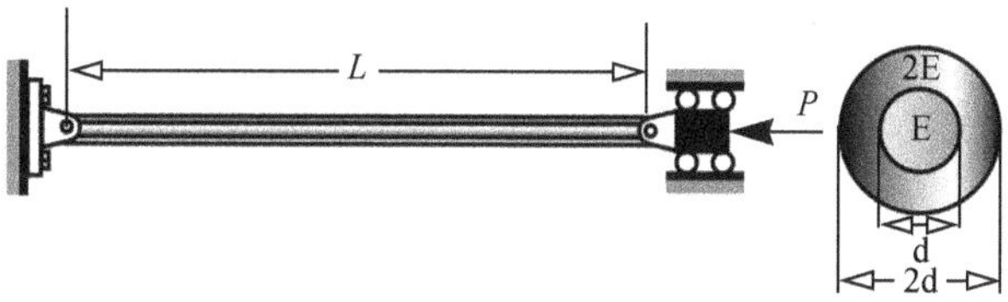


**Figure P4.19**

**4.20** Two strips of material of a modulus of elasticity of $2E$ are attached to a material with a modulus of elasticity $E$ to form a composite cross section of the column shown in Figure P4.20. In terms of $E$, $a$, and $L$, determine the critical buckling load. The column is free to buckle in any direction.

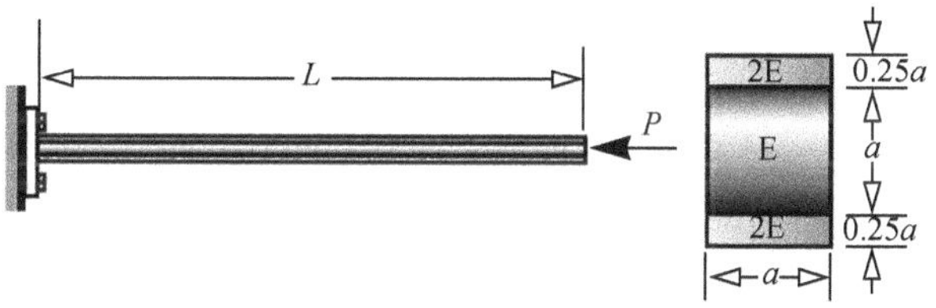


**Figure P4.20**

## Section 4.2

**4.21** A column built in on one end and free at the other end has a load that is eccentrically applied at a distance $e$ from the centroid, as shown in Figure P4.12. Show that the deflection curve is given by the equation $v(x) = e(1 - \cos \lambda x)/\cos \lambda L$., where $\lambda$ is as given by Equation (4.1b).

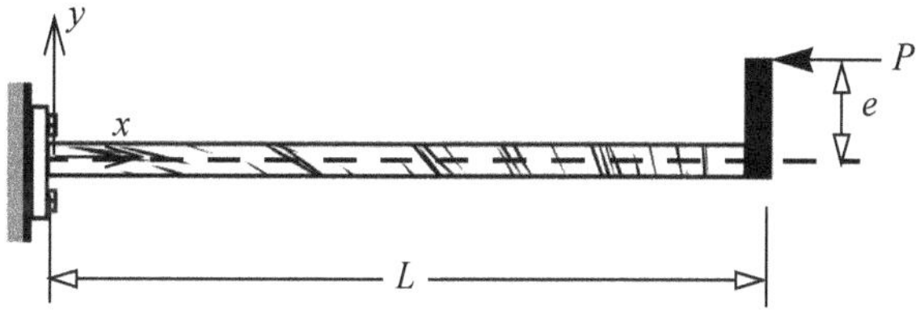


Figure 4.12

**4.22** On the cylinder shown in Figure P4.22 the applied load $P = 3$ kips, the length $L = 5$ ft, and the modulus of elasticity $E = 30{,}000$ ksi. What are the maximum stress and the maximum deflection?

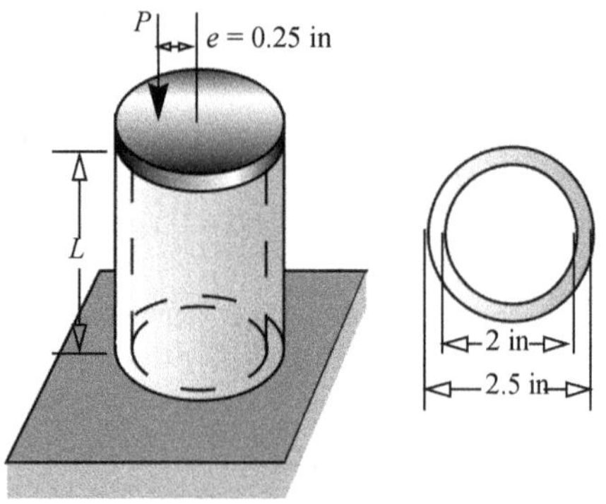


**Figure P4.22**

**4.23** On the cylinder shown in Figure P4.22 the applied load $P = 3$ kips and the modulus of elasticity $E = 30{,}000$ ksi. If the allowable normal stress is 8 ksi, what is the maximum permissible length $L$ of the cylinder?

**4.24** The length of the cylinder shown in Figure P4.22 is $L = 5$ ft. The yield stress of steel used in the cylinder is 30 ksi, and the modulus of elasticity $E = 30{,}000$ ksi. Determine the maximum load $P$ that can be applied. Use the plot for steel in Figure 4.10.

**4.25** On the column shown in Figure P4.25 the applied load $P = 100$ kN, the length $L = 2.0$ m, and the modulus of elasticity $E = 70$ GPa. What are the maximum stress and the maximum deflection?

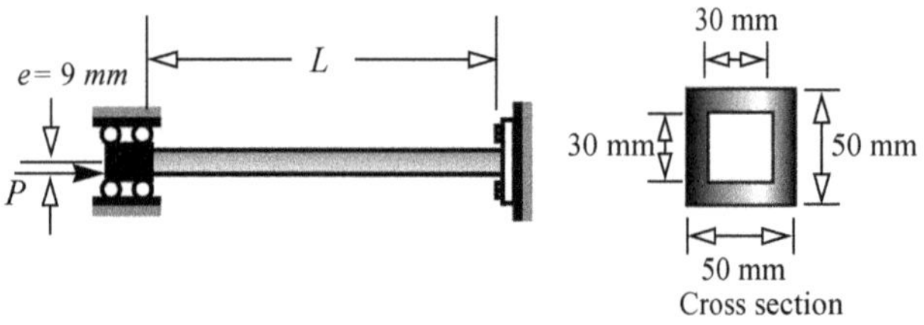


**Figure P4.25**

**4.26** On the column shown in Figure P4.25 the applied load $P = 100$ kN and the modulus of elasticity $E = 70$ GPa. If the allowable normal stress is 250 MPa, what is the maximum permissible length $L$ of the column?

**4.27** The length of the column shown in Figure P4.25 is $L = 2.0$ m. The yield stress of aluminum used in the column is 280 MPa, and the modulus of elasticity $E = 70$ GPa. Determine the maximum load $P$ that can be applied. Use the plot for aluminum in Figure 4.10.

**4.28** A simply supported 6-ft pipe has an outside diameter of 3 in. and a thickness of $1/8$ in. The pipe material has the stress–strain curve shown in Figure P4.28. Using Equation (4.10), determine the critical buckling load.

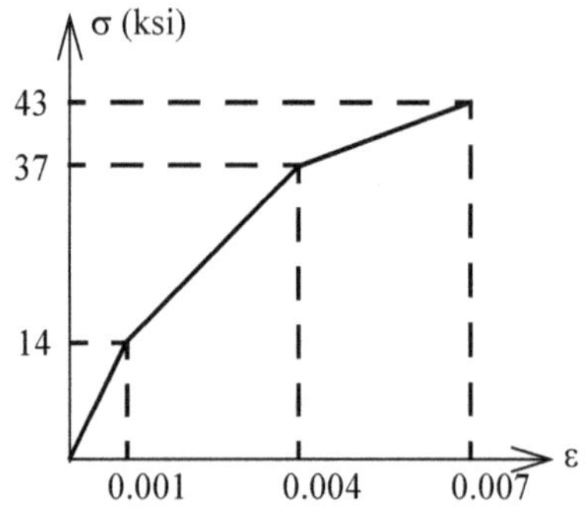


**Figure P4.28**

**4.29** A square box column is constructed from a sheet of 10-mm thickness. The outside dimensions of the square are 75 mm × 75 mm and the column has a length of 0.75 m. The material stress–strain curve is approximated as shown in Figure P4.29. Using Equation (4.10), determine the critical buckling load.

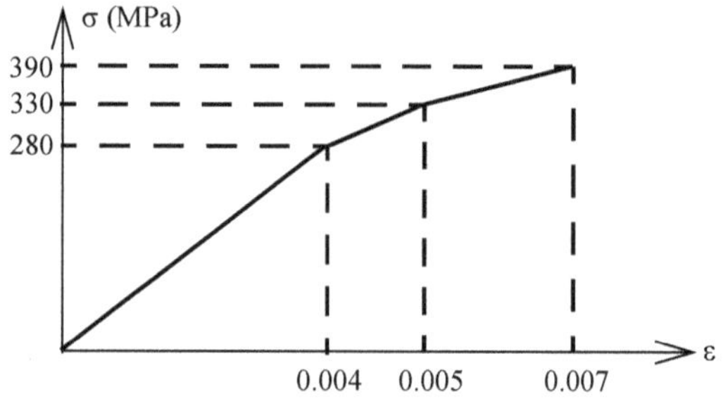


**Figure P4.29**

**4.30** A column that is pin held at its ends has a small initial curvature, which is approximated by the sine function shown in Figure P4.30. Show that the elastic curve of the column is given by the equation below.

$$v(x) = \frac{v_0}{1 - P/P_{cr}} \sin\frac{\pi x}{L}$$

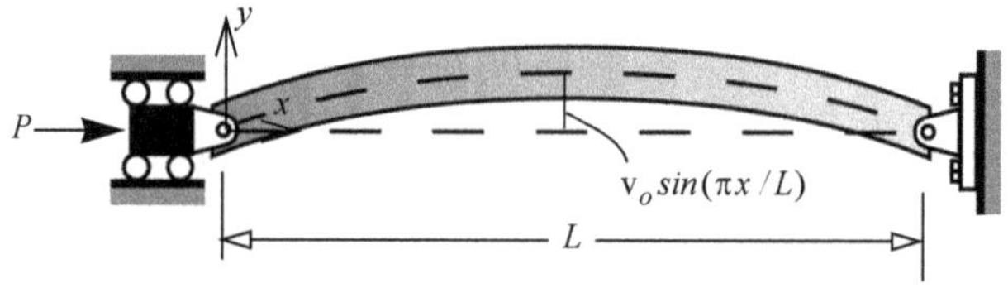


**Figure P4.30**

**4.31** In *double modulus theory,* also known as *reduced modulus theory* for intermediate columns, it is recognized that the bending action during buckling increases the compressive axial stress on the concave side of the beam but decreases the compressive stress on the convex side of the beam. Thus the use of the tangent modulus of elasticity $E_t$ is appropriate on the concave side, but on the convex side of the beam it may be better to use the original modulus of elasticity. Modeling the cross section material with the two moduli $E_t$ and $E$ and using Equation (4.10), show

$$P_{cr} = \frac{\pi^2 E_r I}{L_{eff}^2} \qquad E_r = E_t\frac{I_1}{I} + E\frac{I_2}{I}$$

where $E_r$ is the *reduced modulus of elasticity,* $I_1$ and $I_2$ are the moments of inertia of the areas on the concave and convex sides of the axis passing through the centroid, and $I$ is the moment of inertia of the entire cross section.

# 5 PLATES

## LEARNING OBJECTIVES

1. Understand the classical theory of bending of thin plates and its limitations.
2. Understand how complexities can be added to the classical plate theory.
3. Understand the solution techniques for rectangular and circular plates.

Floors, ceilings, window panes, disc brakes, ship decks, truck beds, are among the countless application of structural member called plate. A plate is a flat solid body whose thickness is small compared to the other dimensions and is subjected to bending loads. The theory of plate is a two-dimensional analog of beam theory. We approximate the variables in the thickness direction as the plate is thin. We integrate the stresses across the thickness to obtain internal forces and moments, which we then relate to external forces and moments by equilibrium.

We start by developing the simplest possible theory, that is, the classical thin plate theory (see Szilard [1974], Timoshenko and Woinowsky-Krieger [1959]). The limitations and assumptions that are made to get the simplest plate theory are identified with the intent of dropping them to include complexities in geometry, loading, and materials. The complexities that will be considered in examples and post-text problems are described in Section 5.2.9. Understanding methodology of adding complexities opens the door for many more possibilities that arise by including combinations of complexities.

The deflection of plates is governed by a fourth order partial differential equation called the bi-harmonic equation. In Sections 5.3 and 5.4 we study two solution techniques for rectangular plates called the Navier's Method and Nadai-Levy's method. In Section 5.5, we develop the equations for circular plates. To obtain solution for circular plates we simplify by assuming axi-symmetry in geometry, material properties, and loading.

## 5.1 LIMITATIONS ON PLATE.

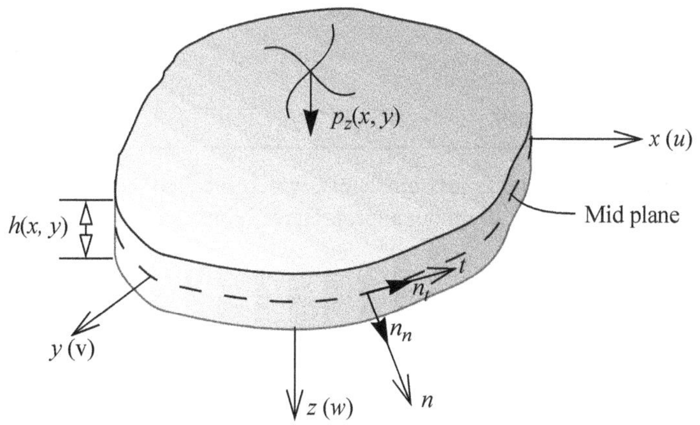


**Figure 5.1** Plate geometry

Figure 5.1 depicts plate geometry. We will develop the theory for thin plates for small deflections called the *Kirchhoff-Love theory*. The following limitations will apply.

1. The mid plane of the plate is flat. Shell theory accounts for curved mid plane.
2. The thickness of the plate $h(x, y)$ may vary *gradually* across the plan form but must be an order of magnitude smaller than the dimensions in the $x$ or $y$ direction.

3. The transverse loading $p_z(x, y)$ may vary *gradually* across the plan form
4. Sudden changes in $h(x, y)$ and $p_z(x, y)$ will result in stress concentration near the regions of rapid change but their impact will die out rapidly as per Saint Venant's principle. We will ignore these regions of stress concentration. Our theory will give the nominal values which are used in design after multiplying with appropriate stress concentration factors.
5. We will develop the initial theory for pure bending, that is, with no inplane forces. In post text problems we will relax this limitation by considering forces $n_n$ and $n_t$, but there will be no inplane forces on top and bottom surface of the plate.
6. Elasticity tells us that the normal stress $\sigma_{zz}$ on the surface should equal to the transverse load $p_z$. We will assume that the maximum $\sigma_{zz}$ will be an order of magnitude less than the maximum normal stresses $\sigma_{xx}$ and $\sigma_{yy}$, hence can be neglected.
7. The displacements in the $x$, $y$, and $z$, directions are given by $u$, v, and $w$, respectively. Our theory will be valid for transverse deflection in the range $0 \le w \le 0.2h$. This leads to a linear plate bending theory. Deflection range of $0.3h \le w \le h$ is called the Von-Karman range for plate bending and leads to non-linear theory. Deflection range of $w \ge 2h$ is called the Von-Karman range for membrane theory.

We once more will use the logic shown in Figure 5.2. We make assumptions that will let us develop the simplest possible theory. These assumptions will be points at which complexity can be added, as was done in beam theory.

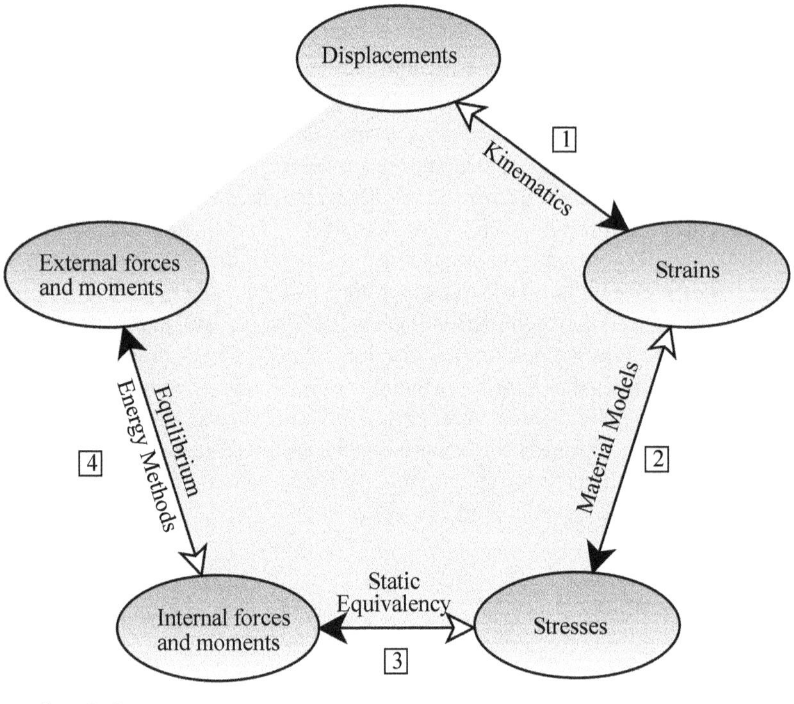


**Figure 5.2** Logic in structural analysis.

## 5.2 CLASSICAL PLATE THEORY

We will make assumptions to develop the simplest possible theory, which is called the classical plate theory. In Section 5.2.9, we will consider complexities that can be added by dropping appropriate assumption and then implementing the rest of the logic.

The theory objectives are:

1. To obtain equations relating the bending stresses $\sigma_{xx}$, $\sigma_{yy}$, and $\tau_{xy}$ to internal forces and moments.
2. To obtain the boundary value problem governing the transverse deflection $w$ of the plate.

### 5.2.1 Kinematics

We make assumptions similar to the beam theory and obtain an approximation for displacements and strain.

**Assumption 1** Deformations are not a functions of time.

The above assumption eliminates dynamic effects and make the theory applicable to static problems.

**Assumption 2** Squashing—dimensional changes in the $z$ direction, is significantly smaller than bending.

The above assumption implies that the normal strain in the $z$ direction is significantly smaller than the normal strains in $x$ and $y$ direction that will arise from bending, that is, $\varepsilon_{zz} = \partial w / \partial z \approx 0$. Thus, $w$ cannot be a function of $z$ as shown below.

$$u = u(x, y, z) \qquad \text{v} = \text{v}(x, y, z) \qquad w = w(x, y) \tag{5.1a}$$

**Assumption 3** Plane sections before deformation remain planes after deformation.

The above assumption implies that displacements $u$ and v are linear function of $z$. We will later show that if the plate is homogeneous in the thickness direction then $z$ is measured from the mid plane of the plate.

Figure 5.3 shows an exaggerated deformed shape of the plate as viewed from the $y$ axis and the $x$ axis. Line $ACB$ being part of the plane remains straight during deformation but it rotates by an angle $\psi_x$ and $\psi_y$ about $x$ and $y$ axis, respectively. $C$ is a point where $z = 0$ and it displaces by amount $u_o(x, y)$ and $v_o(x, y)$. Noting that the line rotation moves point $B$ in the negative $x$ and $y$ direction we can write the displacement field as

$$u \approx u_o(x, y) - z\sin\psi_y \qquad v \approx v_o(x, y) - z\sin\psi_x \qquad w \approx w(x, y) \tag{5.1b}$$

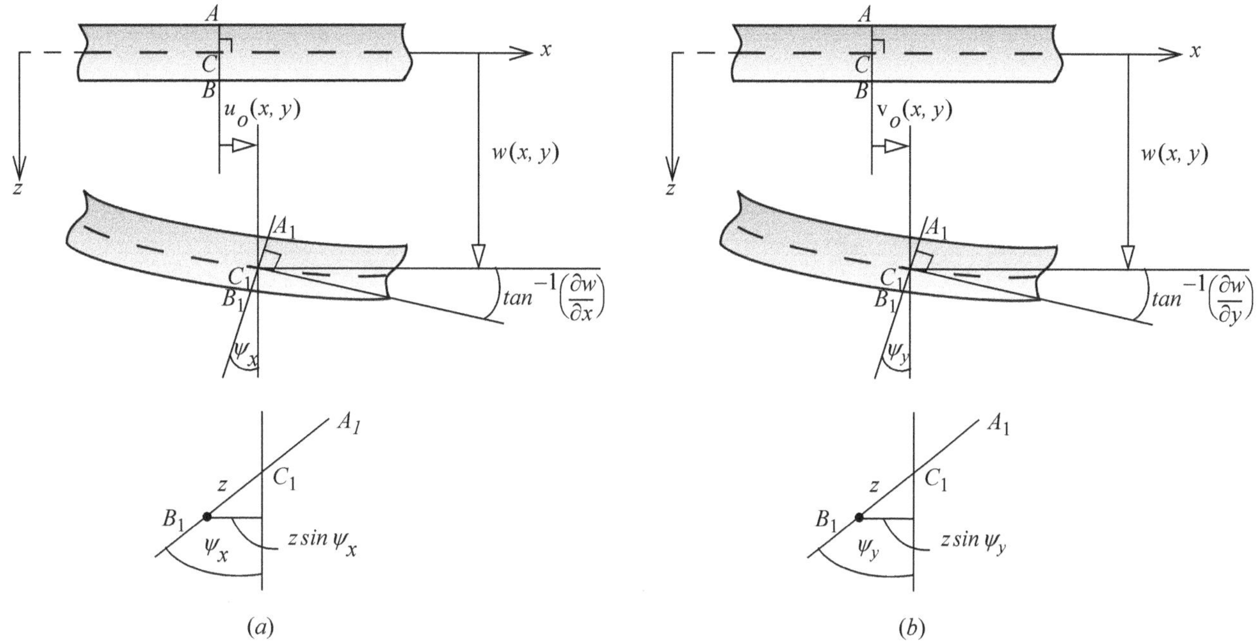


**Figure 5.3** Exaggerated deformed geometry. Rotation of line in thickness direction about (a) $y$ axis (b) $x$ axis.

Note our displacement field will be linear in $z$ only if the rotations $\psi_x$ and $\psi_y$ are functions of $x$ and $y$ but not of $z$. We have assumed there are no inplane forces and hence inplane displacements $u_o$ and $v_o$ will be zero as shown below.

$$u_o(x, y) = 0 \qquad v_o(x, y) = 0 \qquad \text{If there are no inplane forces.} \tag{5.1c}$$

**Assumption 4** Plane sections perpendicular to the plate mid surface remain *nearly* perpendicular after deformation.

The right angle at $C$ remains a right angle at $C_1$ according to the above assumption. This is only possible if the angles of rotation are equal to the angles of the tangent surface $w(x, y)$ makes with the $x$ and $y$ axis as shown in Figure 5.3, that is

$$\psi_x = tan^{-1}(\partial w/\partial x) \qquad \psi_y = tan^{-1}(\partial w/\partial y) \tag{5.1d}$$

**Assumption 5** Deflection and its derivatives are small.

With the above assumption we can approximate sine and tangent functions as shown below.

$$\sin\psi_x \approx \psi_x \qquad \sin\psi_y \approx \psi_y \qquad tan^{-1}(\partial w/\partial x) \approx (\partial w/\partial x) \qquad tan^{-1}(\partial w/\partial y) \approx (\partial w/\partial y) \tag{5.1e}$$

Substituting Equations (5.1d) and (5.1e) into Equation (5.1b) we obtain the displacement field

$$\boxed{u \approx -z(\partial w/\partial x) \qquad v \approx -z(\partial w/\partial y) \qquad w \approx w(x, y)} \tag{5.2}$$

From the above displacement field we find the following shear strains

$$\gamma_{xz} = \partial w/\partial x + \partial u/\partial z \approx 0 \qquad \gamma_{yz} = \partial w/\partial y + \partial v/\partial z \approx 0 \tag{5.3}$$

The zero shear strains are a consequence of Assumption 4. In symmetric bending of beams, we relaxed this type of assumption and developed the equations for *Timoshenko* beams. In a similar manner, relaxing Assumption 4 develops the equations for *Mindlin-Reissner* plate theory discussed in Example 5.3.

**Assumption 6** Strains are small.

We obtain the strains in plate bending as

$$\boxed{\varepsilon_{xx} = \frac{\partial u}{\partial x} = -z\frac{\partial^2 w}{\partial x^2} \qquad \varepsilon_{yy} = \frac{\partial v}{\partial y} = -z\frac{\partial^2 w}{\partial y^2} \qquad \gamma_{xy} = \frac{\partial u}{\partial y} + \frac{\partial v}{\partial x} = -2z\frac{\partial^2 w}{\partial x \partial y}} \tag{5.4}$$

Equation (5.4) shows that the strains vary linearly with $z$. The maximum strain values will be either on the top or bottom surface of the plate. The deformed plate forms a curved surface whose curvatures are given by $\partial^2 w/\partial x^2$, $\partial^2 w/\partial y^2$, and $\partial^2 w/\partial x\partial y$.

### 5.2.2 Material model

The simplest material model is a linear relationship between stresses and strains given by the Generalized Hooke's law. To use this simple model we make the following assumptions.

**Assumption 7** The material is isotropic.

**Assumption 8** The material is linearly elastic.

**Assumption 9** There are no thermal or other non-mechanical strains.

There are no inplane forces acting on top and bottom surfaces of the plate as per limitation 5. We further assumed that the stress $\sigma_{zz}$ can be neglected in comparison to the stresses $\sigma_{xx}$ and $\sigma_{yy}$. In other words, the top and bottom surfaces are free surfaces, that is the points on top and bottom surface are in plane stress. For thin plates we can thus say the *plate is in state of plane stress.* For plane stress we can write the following equations

$$\sigma_{xx} = \frac{E}{(1-\nu^2)}(\varepsilon_{xx} + \nu\varepsilon_{yy}) \qquad \sigma_{yy} = \frac{E}{(1-\nu^2)}(\varepsilon_{yy} + \nu\varepsilon_{xx}) \qquad \tau_{xy} = G\gamma_{xy} \tag{5.5}$$

where, $E$ is the modulus of elasticity, $\nu$ is the Poisson's ratio, and $G$ is the shear modulus of elasticity. For isotropic material $G = E/[2(1+\nu)]$. Substituting Equation (5.4) into Equation (5.5), we obtain

$$\sigma_{xx} = \frac{-Ez}{(1-\nu^2)}\left(\frac{\partial^2 w}{\partial x^2} + \nu\frac{\partial^2 w}{\partial y^2}\right) \qquad \sigma_{yy} = \frac{-Ez}{(1-\nu^2)}\left(\frac{\partial^2 w}{\partial y^2} + \nu\frac{\partial^2 w}{\partial x^2}\right) \qquad \tau_{xy} = -2Gz\frac{\partial^2 w}{\partial x \partial y} \tag{5.6}$$

### 5.2.3 Static equivalency

We replace the stresses by statically equivalent internal forces and moments (stress resultants) by integrating across the thickness $h$ as shown in the equations below. Figure 5.4 shows the positive directions of the stress resultants.

$$\boldsymbol{n}_{xx} = \int_h \sigma_{xx} dz \qquad \boldsymbol{n}_{yy} = \int_h \sigma_{yy} dz \qquad \boldsymbol{n}_{xy} = \boldsymbol{n}_{yx} = \int_h \tau_{xy} dz \tag{5.7}$$

$$\boldsymbol{m}_{xx} = \int_h z\sigma_{xx} dz \qquad \boldsymbol{m}_{yy} = \int_h z\sigma_{yy} dz \qquad \boldsymbol{m}_{xy} = \boldsymbol{m}_{yx} = \int_h z\tau_{xy} dz \tag{5.8}$$

$$\boldsymbol{q}_x = \int_h \tau_{xz} dz \qquad \boldsymbol{q}_y = \int_h \tau_{yz} dz \tag{5.9}$$

Note the following

- All *force stress resultants* will have the *units of force per unit length*
- All *moment stress resultants* will have the *units of moments per unit length.*

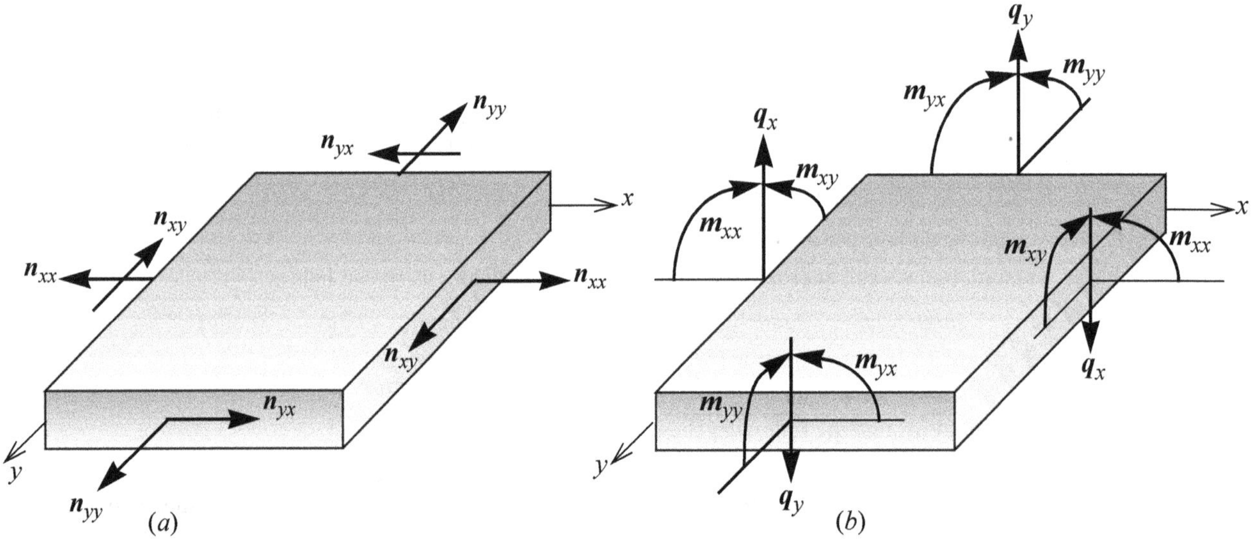


**Figure 5.4** Equivalent internal forces and moments (stress resultants) (a) In-plane. (b) Bending.

The internal shear forces $\boldsymbol{q}_x$ and $\boldsymbol{q}_y$ are necessary for force equilibrium in the $z$ direction, which implies that the stresses $\tau_{xz}$ and $\tau_{yz}$ cannot be zero. Equation (5.3) along with Hooke's law implies that the shear stress $\tau_{xz}$ and $\tau_{yz}$ must be small but not zero. Average values of these stresses can be found by dividing the internal shear forces $\boldsymbol{q}_x$ and $\boldsymbol{q}_y$ by the thickness. For our theory to be valid the maximum of these average values must be an order of magnitude smaller than the maximum inplane stresses.

$$\boxed{max|\tau_{xz}| \text{ and } max|\tau_{yz}| \ll max\{|\sigma_{xx}|, |\sigma_{yy}|, |\tau_{xy}|\}} \tag{5.10}$$

If our calculations show that the above equation is not valid, then the Assumption 4 is not valid and transverse shear must be accounted for in the theory.

### 5.2.4 Location of neutral surface

We have assumed there are no external inplane forces, hence the internal inplane forces must be zero. Substituting Equation (5.6) into Equation (5.7) and taking the curvatures outside the integrals as they are not function of $z$, we obtain

$$n_{xx} = \frac{\partial^2 w}{\partial x^2}\int_h \frac{Ez}{(1-\nu^2)}dz + \frac{\partial^2 w}{\partial y^2}\int_h \frac{E\nu z}{(1-\nu^2)}dz = 0 \qquad n_{yy} = \frac{\partial^2 w}{\partial y^2}\int_h \frac{Ez}{(1-\nu^2)}dz + \frac{\partial^2 w}{\partial x^2}\int_h \frac{E\nu z}{(1-\nu^2)}dz = 0$$

$$n_{xy} = -2\frac{\partial^2 w}{\partial x \partial y}\int_h Gzdz = 0 \tag{5.11a}$$

**Assumption 10** The material is homogeneous across the thickness of the plate.

The above assumption implies that the material constants can be taken outside the integrals and we obtain

$$n_{xx} = \left[\frac{\partial^2 w}{\partial x^2}\frac{E}{(1-\nu^2)} + \frac{\partial^2 w}{\partial y^2}\frac{E\nu}{(1-\nu^2)}\right]\int_h zdz = 0 \qquad n_{yy} = \left[\frac{\partial^2 w}{\partial y^2}\frac{E}{(1-\nu^2)} + \frac{\partial^2 w}{\partial x^2}\frac{E\nu}{(1-\nu^2)}\right]\int_h zdz = 0$$

$$n_{xy} = -2G\frac{\partial^2 w}{\partial x \partial y}\int_h zdz = 0 \tag{5.11b}$$

Deformation of plate implies that at least one of the curvature will not be zero. Hence, Equation (5.11b) implies

$$\boxed{\int_h zdz = 0} \tag{5.12}$$

The condition in Equation (5.12) will be met if the origin is at the mid surface. We record this observation for future use.

- The origin and the neutral surface must be at the mid surface for pure bending of thin plates.

### 5.2.5 Stress formulas

Substituting Equation (5.6) into Equation (5.8) we obtain

$$m_{xx} = -\left(\frac{\partial^2 w}{\partial x^2}\right)\int_h \frac{Ez^2}{(1-\nu^2)}dz - \left(\frac{\partial^2 w}{\partial y^2}\right)\int_h \frac{E\nu z^2}{(1-\nu^2)}dz \qquad m_{yy} = -\left(\frac{\partial^2 w}{\partial y^2}\right)\int_h \frac{Ez^2}{(1-\nu^2)}dz - \left(\frac{\partial^2 w}{\partial x^2}\right)\int_h \frac{E\nu z^2}{(1-\nu^2)}dz$$

$$m_{xy} = -2\left(\frac{\partial^2 w}{\partial x \partial y}\right)\int_h Gz^2 dz \tag{5.13a}$$

As per Assumption 10 of homogeneity across the thickness, we obtain

$$m_{xx} = -\left[\frac{E}{(1-\nu^2)}\right]\left[\left(\frac{\partial^2 w}{\partial x^2}\right) + \nu\left(\frac{\partial^2 w}{\partial y^2}\right)\right]\int_h z^2 dz \qquad m_{yy} = -\left[\frac{E}{(1-\nu^2)}\right]\left[\left(\frac{\partial^2 w}{\partial y^2}\right) + \nu\left(\frac{\partial^2 w}{\partial x^2}\right)\right]\int_h z^2 dz$$

$$m_{xy} = -2G\left(\frac{\partial^2 w}{\partial x \partial y}\right)\int_h z^2 dz \tag{5.13b}$$

Noting that $z$ is measured from the mid surface, we can write

$$\int_h z^2 dz = \int_{-h/2}^{h/2} z^2 dz = h^3/12 \tag{5.13c}$$

We obtain the final form of moment-curvature equations as

$$\boxed{m_{xx} = -D\left[\left(\frac{\partial^2 w}{\partial x^2}\right) + \nu\left(\frac{\partial^2 w}{\partial y^2}\right)\right] \qquad m_{yy} = -D\left[\left(\frac{\partial^2 w}{\partial y^2}\right) + \nu\left(\frac{\partial^2 w}{\partial x^2}\right)\right] \qquad m_{xy} = -D(1-\nu)\left(\frac{\partial^2 w}{\partial x \partial y}\right)} \tag{5.14}$$

where, the constant $D$ is called the *bending rigidity of the plate* and is given by

$$\boxed{D = \frac{Eh^3}{12(1-\nu^2)}} \tag{5.15}$$

Substituting Equation (5.14) into stress expressions Equation (5.6) we obtain the formulas for stresses

$$\boxed{\sigma_{xx} = \frac{m_{xx}z}{(h^3/12)} \qquad \sigma_{yy} = \frac{m_{yy}z}{(h^3/12)} \qquad \tau_{xy} = \frac{m_{xy}z}{(h^3/12)}} \tag{5.16}$$

- The bending stresses in homogeneous plates vary linearly through the thickness and are maximum at top and bottom surface of the plate $z = \pm h/2$.

### 5.2.6 Equilibrium

We consider a free body diagram of a differential element of the plate shown in Figure 5.5.

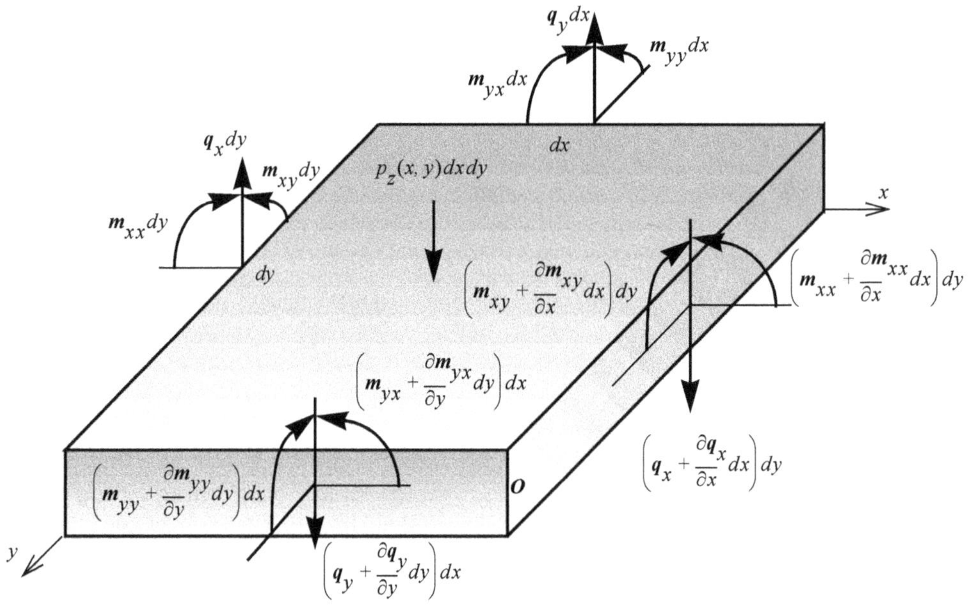


**Figure 5.5** Free body diagram of a differential plate element.

Note that all stress resultants are defined per unit length and hence must be multiplied by the differential lengths of the side on which they act. By equilibrium of forces in the $z$ direction we obtain

$$\left(\boldsymbol{q}_x + \frac{\partial \boldsymbol{q}_x}{\partial x}dx\right)dy - \boldsymbol{q}_x dy + \left(\boldsymbol{q}_y + \frac{\partial \boldsymbol{q}_y}{\partial y}dy\right)dx - \boldsymbol{q}_y dx + p_y(x,y)dxdy = 0$$

$$\boxed{\frac{\partial \boldsymbol{q}_x}{\partial x} + \frac{\partial \boldsymbol{q}_y}{\partial y} = -p_z(x,y)} \tag{5.17a}$$

We consider equilibrium of moment about the $y$ direction at point $O$. The moment from $p_z(x,y)$ and $q_y$ will have products of three differential while the rest of the terms will only have products of two differential, hence can be neglected. The remaining terms yield

$$\left(\boldsymbol{m}_{xx} + \frac{\partial \boldsymbol{m}_{xx}}{\partial x}dx\right)dy - \boldsymbol{m}_{xx}dy + \left(\boldsymbol{m}_{yx} + \frac{\partial \boldsymbol{m}_{yx}}{\partial y}dy\right)dx - \boldsymbol{m}_{yx}dx - (\boldsymbol{q}_x dy)dx = 0$$

$$\boxed{\boldsymbol{q}_x = \frac{\partial \boldsymbol{m}_{xx}}{\partial x} + \frac{\partial \boldsymbol{m}_{yx}}{\partial y}} \tag{5.17b}$$

Similarly by considering moment about the $x$ direction through point $O$ we obtain

$$\boxed{\boldsymbol{q}_y = \frac{\partial \boldsymbol{m}_{xy}}{\partial x} + \frac{\partial \boldsymbol{m}_{yy}}{\partial y}} \tag{5.17c}$$

### 5.2.7 Differential equation

Substituting Equation (5.14) into Equations (5.17b) and Equations (5.17c) we obtain

$$\boldsymbol{q}_x = -\frac{\partial}{\partial x}\left[D\left\{\frac{\partial^2 w}{\partial x^2} + \nu\frac{\partial^2 w}{\partial y^2}\right\}\right] - \frac{\partial}{\partial y}\left[\left\{D(1-\nu)\frac{\partial^2 w}{\partial x \partial y}\right\}\right] \tag{5.18a}$$

$$\boldsymbol{q}_y = -\frac{\partial}{\partial x}\left[\left\{D(1-\nu)\frac{\partial^2 w}{\partial x \partial y}\right\}\right] - \frac{\partial}{\partial y}\left[D\left\{\frac{\partial^2 w}{\partial y^2} + \nu\frac{\partial^2 w}{\partial x^2}\right\}\right] \tag{5.18b}$$

Substituting the above equations into Equation (5.17a) we obtain

$$\frac{\partial^2}{\partial x^2}\left[D\left\{\frac{\partial^2 w}{\partial x^2} + \nu\frac{\partial^2 w}{\partial y^2}\right\}\right] + 2\frac{\partial^2}{\partial x \partial y}\left[\left\{D(1-\nu)\frac{\partial^2 w}{\partial x \partial y}\right\}\right] + \frac{\partial^2}{\partial y^2}\left[D\left\{\frac{\partial^2 w}{\partial y^2} + \nu\frac{\partial^2 w}{\partial x^2}\right\}\right] = p_z(x,y) \tag{5.18c}$$

The above equation is the differential equation governing deflection of tapered plates. To get further simplification we make additional assumptions.

**Assumption 11** The plate is homogeneous in the $x$ and $y$ direction.

**Assumption 12** The plate is of uniform thickness.

From above assumptions $h$, $E$, and $\nu$ are constant, hence $D$ is a constant. Equations (5.18a), and (5.18b) yield the following

$$q_x = -D\frac{\partial}{\partial x}\left[\frac{\partial^2 w}{\partial x^2} + \frac{\partial^2 w}{\partial y^2}\right] = -D\frac{\partial}{\partial x}(\nabla^2 w) \qquad q_y = -D\frac{\partial}{\partial y}\left[\frac{\partial^2 w}{\partial x^2} + \frac{\partial^2 w}{\partial y^2}\right] = -D\frac{\partial}{\partial y}(\nabla^2 w) \tag{5.19}$$

where, $\nabla^2 w = \partial^2 w/\partial x^2 + \partial^2 w/\partial y^2$ is the Laplace or harmonic operator. From Equation (5.18c) we obtain

$$\text{Differential Equation:}\quad D\nabla^4 w = D\left[\frac{\partial^4 w}{\partial x^4} + 2\frac{\partial^4 w}{\partial x^2 \partial y^2} + \frac{\partial^4 w}{\partial y^4}\right] = p_z(x, y) \tag{5.20}$$

where, $\nabla^4 w = \nabla^2(\nabla^2 w)$ is called the *bi-harmonic* operator.

## 5.2.8 Boundary conditions

A fourth order partial differential equation governs the plate deflection as seen from Equation (5.20). So like beam deflection which is also governed by fourth order differential equation, we need to specify two conditions at each boundary point. In beams one condition was on deflection or internal shear force and the second condition was on slope or internal moment. But in plates there is a shear force and two moments at the boundary. Which two do we use? The answer was provided by *Kirchhoff*, who showed that the variation of the twisting moment $\boldsymbol{m}_{xy}$ (or $\boldsymbol{m}_{yx}$) results in a shear force that adds to the existing shear force.

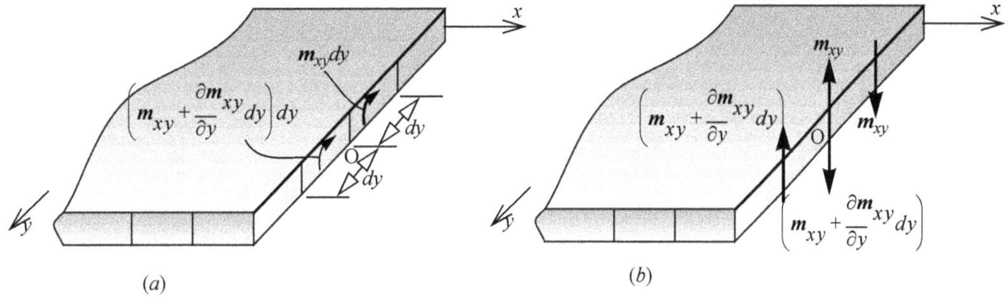


**Figure 5.6** Shear force generated from twisting moment.

Consider a boundary $x$ = constant. We can interpret the moment $\boldsymbol{m}_{xy}dy$ as a couple of two forces acting in opposite directions that are separated by a distance of $dy$ as shown in Figure 5.6. In a similar manner the variation of twisting moment given by $[\boldsymbol{m}_{xy} + (\partial \boldsymbol{m}_{xy}/\partial y)dy]dy$ can also be interpreted as a couple of opposite forces. The resultant of the two forces at point O thus produces a shear force of $\partial \boldsymbol{m}_{xy}/\partial y$ in the direction of the shear force $\boldsymbol{q}_x$. Thus, the total boundary shear force is $V_x = \boldsymbol{q}_x + \partial \boldsymbol{m}_{xy}/\partial y$. Substituting for $\boldsymbol{q}_x$ and $\boldsymbol{m}_{xy}$ we obtain

$$V_x = -D\left[\frac{\partial^3 w}{\partial x^3} + (2-\nu)\frac{\partial^3 w}{\partial x \partial y^2}\right] \tag{5.21a}$$

In a similar manner on a boundary $y$ = constant we obtain:

$$V_y = -D\left[\frac{\partial^3 w}{\partial y^3} + (2-\nu)\frac{\partial^3 w}{\partial y \partial x^2}\right] \tag{5.21b}$$

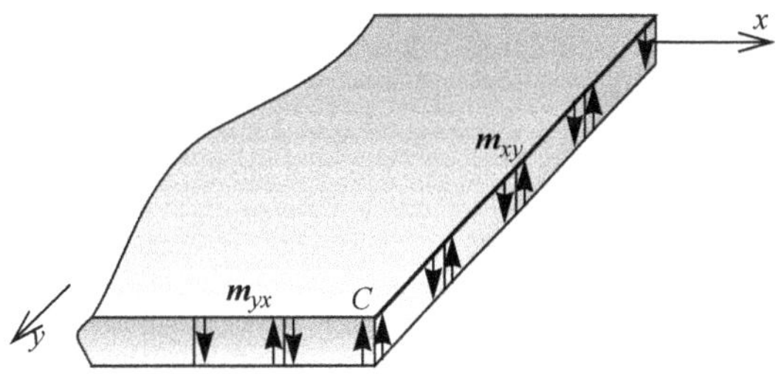


**Figure 5.7** Explanation of corner force.

The twisting moment produces another feature at a corner. The interpretation of the twisting moment as a couple of equal and opposite forces distributed on two edges is shown in Figure 5.7. In the middle of an edge the forces from two successive couples cancel out. But at the corner a force from couple $\boldsymbol{m}_{xy}$ and force from $\boldsymbol{m}_{yx}$ are in the same direction giving a resultant force. This force is called the corner force and is given by the equation below. If you see cracks near the corner of floor or ceiling in old buildings, then it is likely due to the corner force.

$$R_{corner} = \boldsymbol{m}_{xy} + \boldsymbol{m}_{yx} = 2\boldsymbol{m}_{xy} = -2D(1-\nu)(\partial^2 w/\partial x \partial y) \tag{5.22}$$

We can now state all possible boundary conditions.

$$\text{Boundary Conditions:}\quad \boxed{\begin{array}{ll} \text{On } x = \text{constant} & \text{Specify } [V_x \text{ or } w] \text{ and } [m_{xx} \text{ or } \partial w/\partial x] \\ \text{On } y = \text{constant} & \text{Specify } [V_y \text{ or } w] \text{ and } [m_{yy} \text{ or } \partial w/\partial y] \\ \text{On a corner} & \text{Specify } [R_{corner} \text{ or } w] \end{array}} \qquad \textbf{(5.23)}$$

**Table 5.1** Boundary conditions at edge $x = a$

| Type of support | | Boundary Conditions: Specify $[V_x \text{ or } w]$ | Boundary Conditions: Specify $[m_{xx} \text{ or } \partial w/\partial y]$ |
|---|---|---|---|
| Clamped | 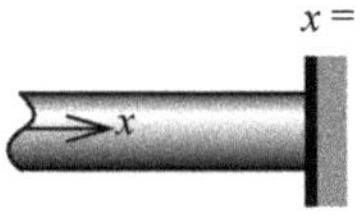 | $w(a,y) = 0$ | $\dfrac{\partial w(a,y)}{\partial x} = 0$ |
| Simply Supported | 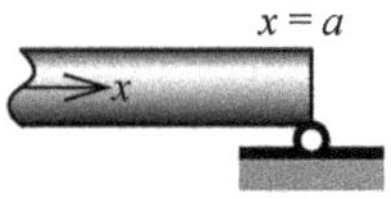 | $w(a,y) = 0$ | $m_{xx}(a,y) = 0$ or $\dfrac{\partial^2 w(a,y)}{\partial x^2} = 0$ |
| Free | 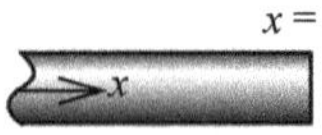 | $V_x(a,y) = 0$ or $\left[\dfrac{\partial^3 w}{\partial x^3} + (2-\nu)\dfrac{\partial^3 w}{\partial x \partial y^2}\right]\Bigg\vert_{x=a} = 0$ | $m_{xx}(a,y) = 0$ or $\left[\dfrac{\partial^2 w}{\partial x^2} + \nu\dfrac{\partial^2 w}{\partial y^2}\right]\Bigg\vert_{x=a} = 0$ |
| Roller | 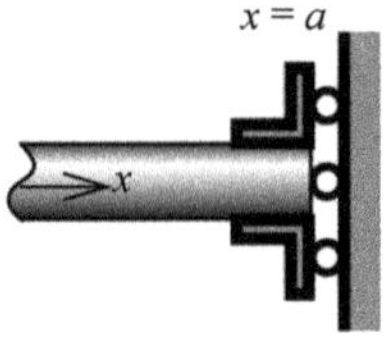 | $V_x(a,y) = 0$ or $\left[\dfrac{\partial^3 w}{\partial x^3} + (2-\nu)\dfrac{\partial^3 w}{\partial x \partial y^2}\right]\Bigg\vert_{x=a} = 0$ | $\dfrac{\partial w(a,y)}{\partial x} = 0$ |
| Elastic | 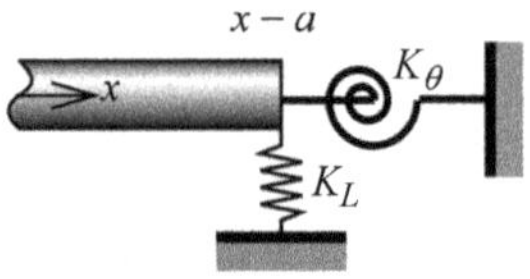 | $V_x(a,y) = -K_L w(a,y)$ or $\left[\dfrac{\partial^3 w}{\partial x^3} + (2-\nu)\dfrac{\partial^3 w}{\partial x \partial y^2}\right]\Bigg\vert_{x=a} = \left(\dfrac{K_L}{D}\right) w(a,y)$ | $m_{xx}(a,y) = K_\theta \dfrac{\partial w(a,y)}{\partial x}$ or $\left[\dfrac{\partial^2 w}{\partial x^2} + \nu\dfrac{\partial^2 w}{\partial y^2}\right]\Bigg\vert_{x=a} = -\left(\dfrac{K_\theta}{D}\right)\dfrac{\partial w(a,y)}{\partial x}$ |

We will consider the boundary at $x = a$ and the extension to $y = b$ is left to the reader. Table 5.1 shows several types of support and list the associated boundary conditions obtained using Equation (5.23).

The simply supported boundary condition needs some explanation. With $w(a, y) = 0$, the $y$-derivatives will also be zero. Setting the second $y$ derivative of $w$ to zero in the zero moment condition results in the second $x$ derivative of $w$ to be zero as shown in Table 5.1.

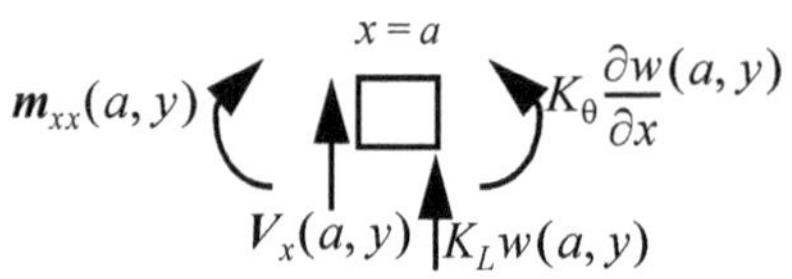


**Figure 5.8 Free body diagram of an infinitesimal element at $x = a$ boundary.**

The boundary condition for elastic support were obtained by drawing a free body diagram of an infinitesimal element at the boundary end as shown in Figure 5.8. It also highlights that such free body diagrams must be drawn to get the right signs for internal moment and internal boundary shear forces whenever an external force or moment acts on the edge.

### 5.2.9 Complexity map

In Section 2.3.3, in context of one-dimensional structural members, two observations were made which are analogous to the ones given below.

- The formulas for stresses given by Equation (5.16) or their equivalent form are obtained by substituting Equation (5.14) or its equivalent forms into Equation (5.6) or its equivalent forms, irrespective of the complexity included.
- The differential equation(s) given by Equation (5.20) or their equivalent form are obtained from by substituting Equation (5.14) or its equivalent form into Equations (5.17a), (5.17b), and (5.17c) or their equivalent forms, irrespective of the complexity included.

Below are listed the complexities that are added in this chapter to the classical theory of plates.

1. Dropping of Assumption 1 implies that inertial forces must be included in equilibrium equations. In Problem 5.6 we consider the vibration of plates.

2. We will not be dropping Assumptions 2 or 3 in this book. However, in Problems 5.1 and 5.2, we will not set the inplane displacements to be zero as we did in Equation (5.1c). In other words, we will develop the theory with inplane and bending displacements simultaneously.
3. Dropping Assumption 4 permits us to incorporate out of plane shear and develop the *Mindlin-Reissner* plate theory as shown in Example 5.3. This is analogous to the Timoshenko beam theory discussed in Example 2.6.
4. Dropping of Assumptions 5 or 6 permits us incorporates geometric nonlinearity and develop the theory for Von-Karman region discussed in Problem 5.7. In Example 8.6 we derive the boundary value problem after incorporating large strains.
5. Assumption 7 is dropped in Example 5.2 to develop the theory for orthotropic plates.
6. Assumption 9 is dropped in Problem 5.3 to incorporate effect of temperature in plate theory.
7. Assumption 10 of material on material homogeneity across the thickness is dropped in Example 5.1 and Problems 5.4 and 5.5 to develop theories for laminated plates.
8. Assumption 12 of untapered plate is dropped in Problems 5.10 and 5.11 to appreciate the impact on boundary conditions.

In Problem 5.8 the differential equation for a plate on elastic foundation is considered. More than one complexity may be added simultaneously and the possible combinations of complexities is very large, but hopefully the reader by end of this chapter will develop a sense of how these complexities can be added to classical theory of plates.

Addition of complexities leads to more general theories. *An important check on these complex theories is that we should be able to reproduce results of classical plate theory by appropriate substitution of variables.* This check will be performed in the examples presented next.

## EXAMPLE 5.1

The cross section of laminated plate made from two materials is shown in Figure 5.9. Both materials have the same Poisson's ratio but different modulus of elasticity. The displacement field is given

$$u \approx -(z-z_o)(\partial w/\partial x) \qquad v \approx -(z-z_o)(\partial w/\partial y) \qquad w \approx w(x,y)$$

where, $z$ is measured from the mid surface and $z_o$ is the location of the neutral surface. (*a*) Determine the value of $z_o$ assuming all assumptions in Section 5.2 except for material homogeneity across the thickness are valid. (*b*) Obtain stress formulas and the differential equation governing the deflection of the laminated plate. (*c*) By substituting $E_1 = E_2$, show that the results for parts *a* and *b* give the same results as classical plate theory for homogeneous material.

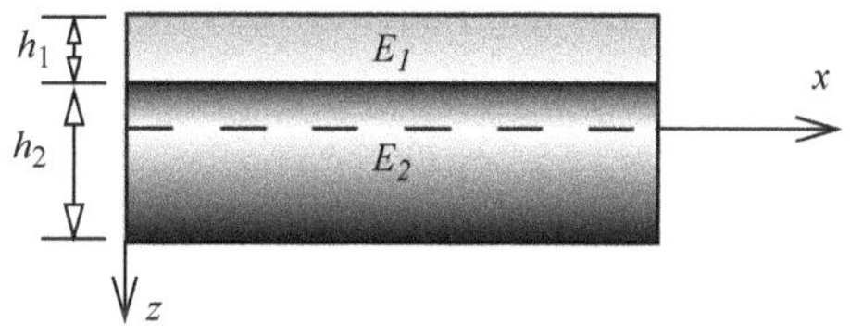


**Figure 5.9** Edge view of laminated plate in Example 5.1.

**PLAN**

We follow the logic of Figure 5.2 in the same manner as the sequence of steps in Section 5.2 to obtain the results.

**SOLUTION**

**Strains:** For small strain we obtain the following from the displacement function.

$$\varepsilon_{xx} = \frac{\partial u}{\partial x} = -(z-z_o)(\partial^2 w/\partial x^2) \qquad \varepsilon_{yy} = \frac{\partial v}{\partial y} = -(z-z_o)(\partial^2 w/\partial y^2) \qquad \gamma_{xy} = \frac{\partial u}{\partial y}+\frac{\partial v}{\partial x} = -2(z-z_o)(\partial^2 w/\partial x\partial y) \quad \text{(E1)}$$

**Stresses:** If Assumptions 7 through 9 are valid then from Equation (5.5) we have

$$\sigma_{xx} = \frac{E}{(1-\nu^2)}(\varepsilon_{xx}+\nu\varepsilon_{yy}) \qquad \sigma_{yy} = \frac{E}{(1-\nu^2)}(\varepsilon_{yy}+\nu\varepsilon_{xx}) \qquad \tau_{xy} = G\gamma_{xy} \quad \text{(E2)}$$

where $E$ could be either $E_1$ or $E_2$ depending upon the point at which stress is being evaluated. Substituting the strains we obtain

$$\sigma_{xx} = \frac{-E(z-z_o)}{(1-\nu^2)}\left(\frac{\partial^2 w}{\partial x^2}+\nu\frac{\partial^2 w}{\partial y^2}\right) \qquad \sigma_{yy} = \frac{-E(z-z_o)}{(1-\nu^2)}\left(\frac{\partial^2 w}{\partial y^2}+\nu\frac{\partial^2 w}{\partial x^2}\right) \qquad \tau_{xy} = -2G(z-z_o)\frac{\partial^2 w}{\partial x\partial y} \quad \text{(E3)}$$

**Stress Resultants:** From Equation (5.7) and the fact there are no inplane forces we have

$$\boldsymbol{n}_{xx} = \int_h \sigma_{xx}dz = 0 \qquad \boldsymbol{n}_{yy} = \int_h \sigma_{yy}dz = 0 \qquad \boldsymbol{n}_{xy} = \boldsymbol{n}_{yx} = \int_h \tau_{xy}dz = 0 \quad \text{(E4)}$$

As the axis neutral passes through $z_o$, we define the moment resultants at the neutral axis as

$$\boldsymbol{m}_{xx} = \int_h (z-z_o)\sigma_{xx}dz \qquad \boldsymbol{m}_{yy} = \int_h (z-z_o)\sigma_{yy}dz \qquad \boldsymbol{m}_{xy} = \boldsymbol{m}_{yx} = \int_h (z-z_o)\tau_{xy}dz \quad \text{(E5)}$$

We consider the evaluation of $\boldsymbol{n}_{xx}$ and $\boldsymbol{m}_{xx}$. Substituting the stress $\sigma_{xx}$ and noting that curvatures are not function of z and $\nu$ is same for all materials, we obtain

$$\boldsymbol{n}_{xx} = -\frac{1}{(1-\nu^2)}\left(\frac{\partial^2 w}{\partial x^2}+\nu\frac{\partial^2 w}{\partial y^2}\right)\int_h E(z-z_o)dz = 0 \quad \text{or} \quad \int_h E(z-z_o)dz = 0 \quad \text{or} \quad z_o = \left(\int_h Ez\,dz\right)\Big/\left(\int_h E\,dz\right) \text{(E6)}$$

$$m_{xx} = -\frac{1}{(1-\nu^2)}\left(\frac{\partial^2 w}{\partial x^2}+\nu\frac{\partial^2 w}{\partial y^2}\right)\int_h E(z-z_o)^2 dz = -D_{eq}\left(\frac{\partial^2 w}{\partial x^2}+\nu\frac{\partial^2 w}{\partial y^2}\right) \quad \text{(E7)}$$

$$\text{where, } D_{eq} = \frac{I}{(1-\nu^2)} \qquad I = \left[\int_h E(z-z_o)^2 dz\right] \quad \text{(E8)}$$

The zero values of $n_{yy}$ and $n_{xy}$ will lead to the same values of $z_o$ as given in Equation (E6). In a similar manner the moment expressions can be written as

$$m_{yy} = -D_{eq}\left(\frac{\partial^2 w}{\partial y^2}+\nu\frac{\partial^2 w}{\partial x^2}\right) \qquad m_{xy} = -D_{eq}(1-\nu)\left(\frac{\partial^2 w}{\partial x \partial y}\right) \quad \text{(E9)}$$

**Formulas:** To evaluate the integrals, we write each integral as the sum of integrals over each material. As $z$ is measured from the mid surface, the limits of integration for:

material 1 are from $-(h_1+h_2)/2$ to $-(h_1+h_2)/2+h_1 = (h_1-h_2)/2$

material 2 are $(h_1-h_2)/2$ to $(h_1+h_2)/2$.

(*a*) We evaluate $z_o$ as shown below

$$z_o = \frac{\int_{-(h_1+h_2)/2}^{(h_1-h_2)/2} E_1 z dz + \int_{(h_1-h_2)/2}^{(h_1-h_2)/2} E_2 z dz}{\int_{-(h_1+h_2)/2}^{(h_1-h_2)/2} E_1 dz + \int_{(h_1-h_2)/2}^{(h_1-h_2)/2} E_2 dz} = \frac{E_1 z^2/2\Big|_{-(h_1+h_2)/2}^{(h_1-h_2)/2} + E_2 z^2/2\Big|_{(h_1-h_2)/2}^{(h_1+h_2)/2}}{E_1 z\Big|_{-(h_1+h_2)/2}^{(h_1-h_2)/2} + E_2 z\Big|_{(h_1-h_2)/2}^{(h_1+h_2)/2}}$$

$$z_o = \frac{E_1[(h_1-h_2)^2-(h_1+h_2)^2]+E_2[(h_1+h_2)^2-(h_1-h_2)^2]}{8(E_1h_1+E_2h_2)} = \frac{E_1[-4h_1h_2]+E_2[4h_1h_2]}{8(E_1h_1+E_2h_2)} = \frac{(E_2-E_1)h_1h_2}{2(E_1h_1+E_2h_2)} \quad \text{(E10)}$$

**ANS.** $z_o = \dfrac{(E_2-E_1)h_1h_2}{2(E_1h_1+E_2h_2)}$

(*b*) Substituting the curvatures from Equation (E7) into Equation (E3) we obtain $\sigma_{xx}$ and other stress components.

$$\sigma_{xx} = \frac{E(z-z_o)m_{xx}}{(1-\nu^2)D_{eq}} = E(z-z_o)\left(\frac{m_{xx}}{I}\right) \quad \text{(E11)}$$

where $E$ could be either $E_1$ or $E_2$ depending upon the point at which stress is being evaluated.

**ANS.** $\sigma_{xx} = E(z-z_o)(m_{xx}/I) \qquad \sigma_{yy} = E(z-z_o)(m_{yy}/I) \qquad \tau_{xy} = E(z-z_o)(m_{xy}/I)$

The equilibrium equations Equations (5.17a), (5.17b), and (5.17c) are unchanged. Since the only difference in moment expression for laminated plate and those for homogeneous plate is $D_{eq}$ instead of $D$, we obtain the differential equation shown below.

**ANS.** $D_{eq}\nabla^4 w = p_z(x, y)$

(*c*) Substituting $E_1 = E_2$ into Equation (E10) we obtain $z_o = 0$, implying the neutral axis is at the mid-surface. With $E_1 = E_2 = E$ and $z_o = 0$, the value of the integral in Equation (E7) is $I = (Eh^3)/12$, where $h = h_1+h_2$. Substituting $z_0$ and $I$ into Equation (E11) we obtain the stress expressions for homogeneous plate given by Equation (5.16).Substituting the value of $I$ in Equation (E7) we obtain $D_{eq} = D$ and we obtain the differential equation given by Equation (5.20). Thus, our results reduce to those of classical plate theory for homogeneous plate.

**COMMENTS.**

1. The example emphasizes the modular character of logic. Though the algebra changed, the flow of ideas did not. The kinematics, the material model, static equivalency, and equilibrium equations remained unchanged.
2. The example also emphasizes the two observations made Section 5.2.9 with regard to obtaining stress formulas and differential equations.

## EXAMPLE 5.2

For an orthotropic plate the constitutive equations are

$$\sigma_{xx} = S_{11}\varepsilon_{xx}+S_{12}\varepsilon_{yy} \qquad \sigma_{yy} = S_{21}\varepsilon_{yy}+S_{22}\varepsilon_{xx} \qquad \tau_{xy} = S_{33}\gamma_{xy} \qquad S_{21} = S_{12} \quad \textbf{(5.24)}$$

(a) Starting with the displacement field of Equation (5.2) and assuming $z$ is measured from the mid-surface and all assumptions except Assumption 7 are valid, obtain the stress formulas and differential equation governing the deflection of orthotropic plate. (b) Show that the results reduce to those for isotropic material by appropriate substitution of material constants.

**PLAN**

We follow the logic of Figure 5.2 in the same manner as the sequence of steps in Section 5.2 to obtain the differential equations.

**SOLUTION**

(a) **Displacements**: The displacement in Equation (5.2) are:

$$u \approx -z(\partial w/\partial x) \qquad v \approx -z(\partial w/\partial y) \qquad w \approx w(x, y) \quad \text{(E1)}$$

**Strains:** The strain are

$$\varepsilon_{xx} = \frac{\partial u}{\partial x} = -z(\partial^2 w/\partial x^2) \qquad \varepsilon_{yy} = \frac{\partial v}{\partial y} = -z(\partial^2 w/\partial y^2) \qquad \gamma_{xy} = \frac{\partial u}{\partial y} + \frac{\partial v}{\partial x} = -2z(\partial^2 w/\partial x \partial y) \tag{E2}$$

**Stresses:** Substituting Equation (E2) into Equation (5.24) we obtain the stresses as

$$\sigma_{xx} = -z\left(S_{11}\frac{\partial^2 w}{\partial x^2} + S_{12}\frac{\partial^2 w}{\partial y^2}\right) \qquad \sigma_{yy} = -z\left(S_{21}\frac{\partial^2 w}{\partial y^2} + S_{22}\frac{\partial^2 w}{\partial x^2}\right) \qquad \tau_{xy} = -2zS_{33}\frac{\partial^2 w}{\partial x \partial y} \tag{E3}$$

**Stress Resultants:** The curvatures are functions of $x$, $y$ and integration is with respect to $z$. Hence, the curvatures can be taken outside the integration. Furthermore, we assume material is homogeneous in the z direction and the constants $S_{11}$, $S_{12}$, $S_{22}$, and $S_{33}$ can be taken outside the integrals. As $z$ is measured from the mid-plane we have the following:

$$\int_h z\,dz = 0 \qquad \text{and} \qquad \int_h z^2\,dz = h^3/12 \tag{E4}$$

We define the following:

$$D_{11} = S_{11}(h^3/12) \qquad D_{22} = S_{22}(h^3/12) \qquad D_{33} = S_{33}(h^3/12) \qquad D_{12} = S_{12}(h^3/12) = D_{21} \tag{E5}$$

The stress resultants are as shown below.

$$\boldsymbol{n}_{xx} = -\left(S_{11}\frac{\partial^2 w}{\partial x^2} + S_{12}\frac{\partial^2 w}{\partial y^2}\right)\int_h z\,dz = 0 \qquad \boldsymbol{n}_{yy} = -\left(S_{21}\frac{\partial^2 w}{\partial y^2} + S_{22}\frac{\partial^2 w}{\partial x^2}\right)\int_h z\,dz = 0 \qquad \boldsymbol{n}_{xy} = -2S_{33}\frac{\partial^2 w}{\partial x \partial y}\int_h z\,dz = 0 \tag{E6}$$

$$\boldsymbol{m}_{xx} = \int_h z\sigma_{xx}\,dz = -\left(S_{11}\frac{\partial^2 w}{\partial x^2} + S_{12}\frac{\partial^2 w}{\partial y^2}\right)\int_h z^2\,dz = -\left(D_{11}\frac{\partial^2 w}{\partial x^2} + D_{12}\frac{\partial^2 w}{\partial y^2}\right) \tag{E7}$$

$$\boldsymbol{m}_{yy} = \int_h z\sigma_{yy}\,dz = -\left(S_{21}\frac{\partial^2 w}{\partial y^2} + S_{22}\frac{\partial^2 w}{\partial x^2}\right)\int_h z^2\,dz = -\left(D_{21}\frac{\partial^2 w}{\partial x^2} + D_{22}\frac{\partial^2 w}{\partial y^2}\right) \tag{E8}$$

$$\boldsymbol{m}_{xy} = \boldsymbol{m}_{yx} = \int_h z\tau_{xy}\,dz = -2S_{33}\frac{\partial^2 w}{\partial x \partial y}\int_h z^2\,dz = -2D_{33}\frac{\partial^2 w}{\partial x \partial y} \tag{E9}$$

**Stress Formulas**: Using Cramer's rule, we can solve Equations (E7) and (E8) to obtain the following equations for curvatures

$$\frac{\partial^2 w}{\partial x^2} = -\left[\frac{D_{22}\boldsymbol{m}_{xx} - D_{12}\boldsymbol{m}_{yy}}{D_{22}D_{11} - D_{12}D_{21}}\right] \qquad \frac{\partial^2 w}{\partial y^2} = -\left[\frac{D_{11}\boldsymbol{m}_{yy} - D_{21}\boldsymbol{m}_{xx}}{D_{22}D_{11} - D_{12}D_{21}}\right] \qquad \frac{\partial^2 w}{\partial x \partial y} = -\left[\frac{\boldsymbol{m}_{xy}}{2D_{33}}\right] \tag{E10}$$

Substituting Equation (E10) into Equation (E3) we obtain

$$\sigma_{xx} = z\left(S_{11}\frac{D_{22}\boldsymbol{m}_{xx} - D_{12}\boldsymbol{m}_{yy}}{D_{22}D_{11} - D_{12}D_{21}} + S_{12}\frac{D_{11}\boldsymbol{m}_{yy} - D_{21}\boldsymbol{m}_{xx}}{D_{22}D_{11} - D_{12}D_{21}}\right) = \frac{D_{22}S_{11} - S_{12}D_{21}}{D_{22}D_{11} - D_{12}D_{21}}z\boldsymbol{m}_{xx} + \frac{D_{11}S_{12} - S_{11}D_{12}}{D_{22}D_{11} - D_{12}D_{21}}z\boldsymbol{m}_{yy} \tag{E11}$$

Substituting Equation (E5) we obtain

$$\sigma_{xx} = \frac{h^3/12}{(h^3/12)^2}\frac{S_{22}S_{11} - S_{12}S_{21}}{S_{22}S_{11} - S_{12}S_{21}}z\boldsymbol{m}_{xx} + \frac{h^3/12}{(h^3/12)^2}\frac{S_{11}S_{12} - S_{11}S_{12}}{S_{22}S_{11} - S_{12}S_{21}}z\boldsymbol{m}_{yy} = \frac{z\boldsymbol{m}_{xx}}{(h^3/12)} \tag{E12}$$

In a similar manner we obtain the other stress expressions as

$$\textbf{ANS.} \quad \sigma_{xx} = \frac{\boldsymbol{m}_{xx}z}{(h^3/12)} \qquad \sigma_{yy} = \frac{\boldsymbol{m}_{yy}z}{(h^3/12)} \qquad \tau_{xy} = \frac{\boldsymbol{m}_{xy}z}{(h^3/12)}$$

**Equilibrium equations:** From Equations (5.17a) through (5.17c) we have the following equilibrium equations.

$$\frac{\partial \boldsymbol{q}_x}{\partial x} + \frac{\partial \boldsymbol{q}_y}{\partial y} = -p_z(x,y) \qquad \boldsymbol{q}_x = \frac{\partial \boldsymbol{m}_{xx}}{\partial x} + \frac{\partial \boldsymbol{m}_{yx}}{\partial y} \qquad \boldsymbol{q}_y = \frac{\partial \boldsymbol{m}_{xy}}{\partial x} + \frac{\partial \boldsymbol{m}_{yy}}{\partial y} \tag{E13}$$

**Differential equations:** Substituting the moment equations into the equilibrium equations we obtain the shear forces as

$$\boldsymbol{q}_x = -\frac{\partial}{\partial x}\left(D_{11}\frac{\partial^2 w}{\partial x^2} + D_{12}\frac{\partial^2 w}{\partial y^2}\right) - \frac{\partial}{\partial y}\left(2D_{33}\frac{\partial^2 w}{\partial x \partial y}\right) = -D_{11}\frac{\partial^3 w}{\partial x^3} - (D_{12} + 2D_{33})\frac{\partial^3 w}{\partial x \partial y^2} \tag{E14}$$

$$\boldsymbol{q}_y = -\frac{\partial}{\partial x}\left(2D_{33}\frac{\partial^2 w}{\partial x \partial y}\right) - \frac{\partial}{\partial y}\left(D_{21}\frac{\partial^2 w}{\partial y^2} + D_{22}\frac{\partial^2 w}{\partial x^2}\right) = -(D_{12} + 2D_{33})\frac{\partial^3 w}{\partial x^2 \partial y} - D_{22}\frac{\partial^3 w}{\partial y^3} \tag{E15}$$

Substituting the above equation into the remaining equilibrium equation we obtain the differential equation as shown below.

$$-D_{11}\frac{\partial^4 w}{\partial x^4} - (D_{12} + 2D_{33})\frac{\partial^4 w}{\partial x^2 \partial y^2} - (D_{12} + 2D_{33})\frac{\partial^4 w}{\partial x^2 \partial y^2} - D_{22}\frac{\partial^4 w}{\partial y^4} = -p_z(x,y) \tag{E16}$$

$$\textbf{ANS. } D_{11}\frac{\partial^4 w}{\partial x^4} + 2(D_{12} + 2D_{33})\frac{\partial^4 w}{\partial x^2 \partial y^2} + D_{22}\frac{\partial^4 w}{\partial y^4} = p_z(x,y) \qquad \text{(E17)}$$

(b) Comparing orthotropic constitutive Equation (5.24) with that the isotropic Equation (5.5) we obtain

$$S_{11} = S_{22} = \frac{E}{(1-\nu^2)} \qquad S_{33} = G = \frac{E}{2(1+\nu)} = \frac{(1-\nu)E}{2(1-\nu^2)} \qquad S_{12} = \frac{\nu E}{(1-\nu^2)} \qquad \text{(E18)}$$

Substituting the above equation into Equation (E5) we obtain

$$D_{11} = \frac{S_{11}h^3}{12} = \frac{Eh^3}{12(1-\nu^2)} = D_{22} = D \qquad D_{12} + 2D_{33} = \frac{(S_{12} + 2S_{33})h^3}{12} = \frac{Eh^3}{12(1-\nu^2)} = D \qquad \text{(E19)}$$

Substituting the above equation into Equation (E17) we obtain the bi-harmonic equation of the classical plate theory. The formulas for stresses $\sigma_{xx}$, $\sigma_{yy}$, and $\tau_{xy}$ are the same as for classical plate theory.

**COMMENTS**

1. This example highlights the modular character of the logic in Figure 5.2. The fundamental equations did not change, i.e., the strain-displacement equations, the stress resultant equations, and the equilibrium equations were all unchanged.
2. The example also emphasizes the two observations made Section 5.2.9 with regard to obtaining stress formulas and differential equations.

## EXAMPLE 5.3

In *Mindlin-Reissner* plate theory Assumption 4 of planes sections perpendicular to the plate mid surface remain nearly perpendicular after deformation is dropped to account for shear. (*a*) Starting with the displacement field below obtain the stress formulas and differential equations if all assumptions except Assumption 4 are valid.

$$u \approx -z\psi_x(x,y) \qquad \mathrm{v} \approx -z\psi_y(x,y) \qquad w \approx w(x,y) \qquad \textbf{(5.25)}$$

(*b*) Show the results reduce to those of classical plate theory by substituting $\psi_x = \partial w/\partial x$ and $\psi_y = \partial w/\partial y$.

**PLAN**

We follow the logic Figure 5.2 in the same manner as the sequence of steps in Section 5.2 to obtain the results.

**SOLUTION**

**Strains:** We obtain the strains from the given displacement field as follows.

$$\varepsilon_{xx} = \partial u/\partial x = -z\partial\psi_x/\partial x \qquad \varepsilon_{yy} = \partial \mathrm{v}/\partial y = -z(\partial\psi_y/\partial y) \qquad \gamma_{xy} = \partial u/\partial y + \partial \mathrm{v}/\partial x = -z(\partial\psi_x/\partial y + \partial\psi_y/\partial x) \qquad \text{(E1)}$$

$$\gamma_{xz} = \partial w/\partial x + \partial u/\partial z = -\psi_x + \partial w/\partial x \qquad \gamma_{yz} = \partial w/\partial y + \partial \mathrm{v}/\partial z = -\psi_y + \partial w/\partial y \qquad \text{(E2)}$$

**Stresses:** for plane stress, we obtain the stresses as follows.

$$\sigma_{xx} = \frac{E}{(1-\nu^2)}(\varepsilon_{xx} + \nu\varepsilon_{yy}) = \frac{-Ez}{(1-\nu^2)}\left(\frac{\partial\psi_x}{\partial x} + \nu\frac{\partial\psi_y}{\partial y}\right) \qquad \sigma_{yy} = \frac{E}{(1-\nu^2)}(\varepsilon_{yy} + \nu\varepsilon_{xx}) = \frac{-Ez}{(1-\nu^2)}\left(\frac{\partial\psi_y}{\partial y} + \nu\frac{\partial\psi_x}{\partial x}\right) \qquad \text{(E3)}$$

$$\tau_{xy} = G\gamma_{xy} = -Gz(\partial\psi_x/\partial y + \partial\psi_y/\partial x) \qquad \tau_{xz} = G\gamma_{xz} = G(-\psi_x + \partial w/\partial x) \qquad \tau_{yz} = G(-\psi_y + \partial w/\partial y) \qquad \text{(E4)}$$

**Stress Resultants:** The inplane stress resultants are zero as z is measured from the mid-plane. The bending stress resultants are obtained as follows assuming material is homogeneous across the thickness.

$$\boldsymbol{m}_{xx} = \int_h z\sigma_{xx}dz = \frac{-E}{(1-\nu^2)}\left(\frac{\partial\psi_x}{\partial x} + \nu\frac{\partial\psi_y}{\partial y}\right)\int_h z^2 dz = -D\left(\frac{\partial\psi_x}{\partial x} + \nu\frac{\partial\psi_y}{\partial y}\right) \qquad \text{(E5)}$$

$$\boldsymbol{m}_{yy} = \int_h z\sigma_{yy}dz = \frac{-E}{(1-\nu^2)}\left(\frac{\partial\psi_y}{\partial y} + \nu\frac{\partial\psi_x}{\partial x}\right)\int_h z^2 dz = -D\left(\frac{\partial\psi_y}{\partial y} + \nu\frac{\partial\psi_x}{\partial x}\right) \qquad \text{(E6)}$$

$$\boldsymbol{m}_{xy} = \boldsymbol{m}_{yx} = \int_h z\tau_{xy}dz = -G\left(\frac{\partial\psi_x}{\partial y} + \frac{\partial\psi_y}{\partial x}\right)\int_h z^2 dz = -\frac{D}{2}(1-\nu)\left(\frac{\partial\psi_x}{\partial y} + \frac{\partial\psi_y}{\partial x}\right) \qquad \text{(E7)}$$

$$\boldsymbol{q}_x = \int_h \tau_{xz}dz = G\left(-\psi_x + \frac{\partial w}{\partial x}\right)\int_h dz = Gh\left(-\psi_x + \frac{\partial w}{\partial x}\right) \qquad \text{(E8)}$$

$$\boldsymbol{q}_y = \int_h \tau_{yz}dz = G\left(-\psi_y + \frac{\partial w}{\partial y}\right)\int_h dz = Gh\left(-\psi_y + \frac{\partial w}{\partial y}\right) \qquad \text{(E9)}$$

(*a*) **Stress Formulas:** Substituting Equations (E5) through (E9) into Equations (E3) and (E4) we obtain

$$\textbf{ANS. } \sigma_{xx} = \frac{\boldsymbol{m}_{xx}z}{(h^3/12)} \qquad \sigma_{yy} = \frac{\boldsymbol{m}_{yy}z}{(h^3/12)} \qquad \tau_{xy} = \frac{\boldsymbol{m}_{xy}z}{(h^3/12)} \qquad \tau_{xz} = \boldsymbol{q}_x\frac{z}{h} \qquad \tau_{yz} = \boldsymbol{q}_y\frac{z}{h} \qquad \textbf{(5.26a)}$$

**Equilibrium**: Equilibrium equations are as before. Substituting the above stress resultants into Equation (5.17a) we obtain

$$\frac{\partial \boldsymbol{q}_x}{\partial x} + \frac{\partial \boldsymbol{q}_y}{\partial y} = -p_z(x,y) \quad \text{or} \quad \frac{\partial}{\partial x}\left(-\psi_x + \frac{\partial w}{\partial x}\right) + \frac{\partial}{\partial y}\left(-\psi_y + \frac{\partial w}{\partial y}\right) = -\frac{p_z(x,y)}{Gh} \quad \text{or} \qquad \text{(E10)}$$

$$\frac{\partial^2 w}{\partial x^2}+\frac{\partial^2 w}{\partial y^2}-\left(\frac{\partial\psi_x}{\partial x}+\frac{\partial\psi_y}{\partial y}\right) = -\frac{p_z(x,y)}{Gh} \qquad \text{(E11)}$$

$$\textbf{ANS. } \nabla^2 w-\left(\frac{\partial\psi_x}{\partial x}+\frac{\partial\psi_y}{\partial y}\right) = -\frac{p_z(x,y)}{Gh} \qquad \textbf{(5.26b)}$$

Substituting the stress resultants above into Equations (5.17b) and (5.17c) we obtain

$$\boldsymbol{q}_x = \frac{\partial \boldsymbol{m}_{xx}}{\partial x}+\frac{\partial \boldsymbol{m}_{yx}}{\partial y} = -D\left[\frac{\partial^2\psi_x}{\partial x^2}+\nu\frac{\partial^2\psi_y}{\partial x\partial y}\right]-\frac{D}{2}(1-\nu)\left(\frac{\partial^2\psi_x}{\partial y^2}+\frac{\partial^2\psi_y}{\partial x\partial y}\right) = Gh\left(-\psi_x+\frac{\partial w}{\partial x}\right) \qquad \text{(E12)}$$

$$\boldsymbol{q}_y = \frac{\partial \boldsymbol{m}_{xy}}{\partial x}+\frac{\partial \boldsymbol{m}_{yy}}{\partial y} = -\frac{D}{2}(1-\nu)\left(\frac{\partial^2\psi_x}{\partial x\partial y}+\frac{\partial^2\psi_y}{\partial x^2}\right)-D\left[\frac{\partial^2\psi_y}{\partial y^2}+\nu\frac{\partial^2\psi_x}{\partial y\partial x}\right] = Gh\left(-\psi_y+\frac{\partial w}{\partial y}\right) \qquad \text{(E13)}$$

Substituting the above two equations into Equation (E10) we obtain

$$-D\left[\frac{\partial^3\psi_x}{\partial x^3}+\nu\frac{\partial^3\psi_y}{\partial x^2\partial y}\right]-\frac{D}{2}(1-\nu)\left(\frac{\partial^3\psi_x}{\partial y^2\partial x}+\frac{\partial^3\psi_y}{\partial x^2\partial y}\right)-\frac{D}{2}(1-\nu)\left(\frac{\partial^3\psi_x}{\partial x\partial y^2}+\frac{\partial^2\psi_y}{\partial x^2\partial y}\right)-D\left[\frac{\partial^3\psi_y}{\partial y^3}+\nu\frac{\partial^3\psi_x}{\partial y^2\partial x}\right] = -p_z(x,y)$$

$$\frac{\partial}{\partial x}\left[\frac{\partial^2\psi_x}{\partial x^2}+\frac{\partial^2\psi_x}{\partial y^2}\right]+\frac{\partial}{\partial y}\left[\frac{\partial^2\psi_y}{\partial x^2}+\frac{\partial^2\psi_y}{\partial y^2}\right] = \frac{p_z(x,y)}{D} \qquad \text{(E14)}$$

$$\textbf{ANS. } \nabla^2\left(\frac{\partial\psi_x}{\partial x}+\frac{\partial\psi_y}{\partial y}\right) = \frac{p_z(x,y)}{D} \qquad \textbf{(5.26c)}$$

We eliminate $w$ from Equations (E12) and (E13) to obtain the following.

$$\frac{\partial \boldsymbol{q}_x}{\partial y}-\frac{\partial \boldsymbol{q}_y}{\partial x} = Gh\left[-\frac{\partial\psi_x}{\partial y}+\frac{\partial\psi_y}{\partial x}\right]$$

$$= -D\left[\frac{\partial^3\psi_x}{\partial x^2\partial y}+\nu\frac{\partial^3\psi_y}{\partial x\partial y^2}\right]-\frac{D}{2}(1-\nu)\left[\frac{\partial^3\psi_x}{\partial y^3}+\frac{\partial^3\psi_y}{\partial y^2\partial x}\right]+\frac{D}{2}(1-\nu)\left[\frac{\partial^3\psi_x}{\partial x^2\partial y}+\frac{\partial^3\psi_y}{\partial x^3}\right]+D\left[\frac{\partial^3\psi_y}{\partial x\partial y^2}+\nu\frac{\partial^3\psi_x}{\partial x^2\partial y}\right]$$

$$\frac{Gh}{D}\left[-\frac{\partial\psi_x}{\partial y}+\frac{\partial\psi_y}{\partial x}\right] = \frac{(1-\nu)}{2}\left[\frac{\partial^3\psi_y}{\partial x^3}-\frac{\partial^3\psi_x}{\partial y^3}\right]+\frac{\partial^3\psi_x}{\partial x^2\partial y}\left[-1+\frac{(1-\nu)}{2}+\nu\right]+\frac{\partial^3\psi_y}{\partial y^2\partial x}\left[-\nu-\frac{(1-\nu)}{2}+1\right]$$

$$\frac{2Gh}{(1-\nu)D}\left[-\frac{\partial\psi_x}{\partial y}+\frac{\partial\psi_y}{\partial x}\right] = \left[\frac{\partial^3\psi_y}{\partial x^3}-\frac{\partial^3\psi_x}{\partial y^3}\right]-\frac{\partial^3\psi_x}{\partial x^2\partial y}+\frac{\partial^3\psi_y}{\partial y^2\partial x} = \frac{\partial}{\partial x}\left[\frac{\partial^2\psi_y}{\partial x^2}+\frac{\partial^2\psi_y}{\partial y^2}\right]-\frac{\partial}{\partial y}\left[\frac{\partial^2\psi_x}{\partial x^2}+\frac{\partial^2\psi_x}{\partial y^2}\right] \qquad \text{(E15)}$$

$$\textbf{ANS. } \nabla^2\left(-\frac{\partial\psi_x}{\partial y}+\frac{\partial\psi_y}{\partial x}\right) = \frac{2Gh}{D(1-\nu)}\left(-\frac{\partial\psi_x}{\partial y}+\frac{\partial\psi_y}{\partial x}\right) \qquad \textbf{(5.26d)}$$

(*b*) We substitute $\psi_x = \partial w/\partial x$ and $\psi_y = \partial w/\partial y$, a consequence of Assumption 4, then Equation (E2) shows $\gamma_{xz} = 0$ and $\gamma_{yz} = 0$. Furthermore, Equations (E3) through Equations (E8) reduce to the same corresponding equations in Section 5.2. Equation (5.26c) also reduced to the differential equation of classical plate theory. There are no equations corresponding to Equation (5.26b) and Equation (5.26d) in classical plate theory. Because of Assumption 4 and $\psi_x = \partial w/\partial x$ and $\psi_y = \partial w/\partial y$, we have only one independent variable $w$. Thus, one may look at the motivation of Assumption 4, as a simplifying assumption from three variables to one.

The formulas for stresses $\sigma_{xx}$, $\sigma_{yy}$, and $\tau_{xy}$ are the same as for classical plate theory. Thus, our results reduce to those of classical plate theory. However, we get additional information with *Mindlin-Reissner* plate theory.

**COMMENTS**

1. Once more, the fundamental equations did not change, i.e., the strain-displacement equations, the stress-strain equations, the resultant-stress equations, and the equilibrium-resultant equations were all unchanged. What changed was what we substituted in them as we carried the new displacement field through the logic.
2. We note that the three differential equation in this example are second order differential equations, while in the classical plate theory the differential equation is of fourth order. In finite element method (*FEM, see* Section 7.13) we approximate the primary variables ($w$, $\psi_x$, and $\psi_y$) by polynomials over small elements. The second order differential equation implies we need to only ensure continuity of the variables across the boundary. The fourth order differential equations however requires higher order continuity which leads to a variety of difficulties. Thus, *Mindlin-Reissner* formulation is a better alternative in *FEM* over the classical plate theory formulation.
3. Incorporating other complexities in the formulation are given in several post-text problems. It should be however emphasized understanding the implications of these complexities require studying the solution, which can at times be a chapter in itself.

## 5.3 NAVIER'S SOLUTION FOR RECTANGULAR PLATES

In 1820, Navier found a solution for a rectangular plate that is simply supported on all its boundary as shown in Figure 5.10. In this section we discuss Navier's approach.

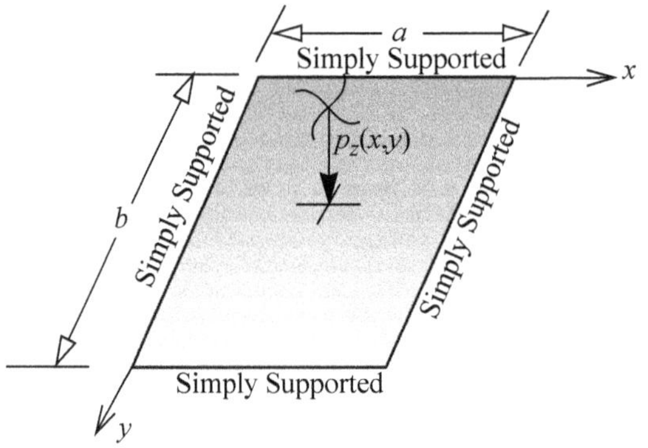


**Figure 5.10** Simply supported rectangular plate.

The boundary value problem for the plate shown in Figure 5.10 can be written as

$$\frac{\partial^4 w}{\partial x^4} + 2\frac{\partial^4 w}{\partial x^2 \partial y^2} + \frac{\partial^4 w}{\partial y^4} = \frac{p_z(x,y)}{D} \quad \textbf{(5.27)}$$

$$\begin{array}{llllll} \text{On } x = 0 & w(0,y) = 0 & \dfrac{\partial^2 w}{\partial x^2}(0,y) = 0 & \text{On } y = 0 & w(x,0) = 0 & \dfrac{\partial^2 w(x,0)}{\partial y^2} = 0 \\ \text{On } x = a & w(a,y) = 0 & \dfrac{\partial^2 w}{\partial x^2}(a,y) = 0 & \text{On } y = b & w(x,b) = 0 & \dfrac{\partial^2 w(x,b)}{\partial y^2} = 0 \end{array} \quad \textbf{(5.27a)}$$

The functions $sin(m\pi x/a)$ and $sin(m\pi y/b)$ have two properties that can be used for finding solutions. (i) Its even derivatives result in the same function multiplied by a constant and we have only even derivatives in our boundary value problem. (ii) The product of these two functions satisfy all boundary conditions. So we define

$$\boxed{w(x,y) = \sum_{m=1}^{\infty}\sum_{n=1}^{\infty} W_{mn} sin(m\pi x/a) sin(n\pi y/b) \qquad p_z(x,y) = \sum_{m=1}^{\infty}\sum_{n=1}^{\infty} P_{mn} sin(m\pi x/a) sin(n\pi y/b)} \quad \textbf{(5.28)}$$

where, $W_{mn}$ and $P_{mn}$ are constant coefficients to be determined. It should be noted that the functions $sin(m\pi x/a)$ and $sin(n\pi y/b)$ are independent functions and the infinite series is complete, hence the displacement field will be correct irrespective of the loading. However, depending upon the loading the solution may have poor convergence rate, that is, it requires too many terms to produce acceptable accuracy.

Substituting Equation (5.28) into Equation (5.27) we obtain

$$\sum_{m=1}^{\infty}\sum_{n=1}^{\infty} W_{mn}\left[\left(\frac{m\pi}{a}\right)^4 + 2\left(\frac{m\pi}{a}\right)^2\left(\frac{n\pi}{b}\right)^2 + \left(\frac{n\pi}{b}\right)^4\right] sin\left(m\pi\frac{x}{a}\right) sin\left(n\pi\frac{y}{b}\right) = \frac{1}{D}\sum_{m=1}^{\infty}\sum_{n=1}^{\infty} P_{mn} sin\left(m\pi\frac{x}{a}\right) sin\left(n\pi\frac{y}{b}\right) \quad \textbf{(5.29)}$$

As the Sine functions form an independent set, the above equation implies the coefficients of the series on the left must equal to coefficients of the series on the right, yielding the following

$$W_{mn} = \frac{1}{D}\frac{P_{mn}}{[(m\pi/a)^4 + 2(m\pi/a)^2(n\pi/b)^2 + (n\pi/b)^4]}$$

$$\boxed{W_{mn} = \frac{1}{D\pi^4}\frac{P_{mn}}{[(m/a)^2 + (n/b)^2]^2}} \quad \textbf{(5.30)}$$

To determine the constants $P_{mn}$ we use the orthogonality conditions of Sine functions given below.

$$\int_0^{\pi} sin(m\theta) sin(n\theta) d\theta = \begin{cases} 0 & m \neq n \\ \pi/2 & m = n \end{cases} \quad \textbf{(5.31a)}$$

Multiplying Equation (5.28) by $sin(i\pi x/a) sin(j\pi y/b)$ and integrating we obtain

$$\sum_{m=1}^{\infty}\sum_{n=1}^{\infty} P_{mn}\left[\int_0^a sin\left(i\pi\frac{x}{a}\right) sin\left(m\pi\frac{x}{a}\right) dx\right]\left[\int_0^b sin\left(j\pi\frac{y}{b}\right) sin\left(n\pi\frac{y}{b}\right) dy\right] = \int_0^b\int_0^a p_z(x,y) sin\left(i\pi\frac{x}{a}\right) sin\left(j\pi\frac{y}{b}\right) dx\,dy \quad \textbf{(5.32)}$$

Substituting $\theta_1 = \pi x/a$ and $\theta_2 = \pi y/b$ into Equation (5.32) and using Equation (5.31a) we obtain Equation (5.33).

$$\sum_{m=1}^{\infty}\sum_{n=1}^{\infty} P_{mn}\left[\frac{a}{\pi}\int_0^{\pi} sin(i\theta_1) sin(m\theta_1) d\theta_1\right]\left[\frac{b}{\pi}\int_0^{\pi} sin(j\theta_2) sin(m\theta_2) d\theta_2\right] = \int_0^b\int_0^a p_z(x,y) sin\left(i\pi\frac{x}{a}\right) sin\left(j\pi\frac{y}{b}\right) dx\,dy$$

$$\boxed{P_{mn} = \frac{4}{ab}\int_0^b\int_0^a p_z(x,y) sin(m\pi x/a) sin(n\pi y/b) dx\,dy} \quad \textbf{(5.33)}$$

The right hand side of Equation (5.33) can be evaluated once $p_z(x, y)$ is known, hence we know all $P_{mn}$. Now from Equation (5.30) we know all $W_{mn}$, hence from Equation (5.28) we know $w(x, y)$ at all points on the plate. If the displacement field is known, then all mechanics variables can be calculated. Table 5.2 lists all the derivatives of $w$ that we will need in the calculations of the mechanics variables.

**Table 5.2** Derivatives of plate deflection.

$$\frac{\partial w}{\partial x} = \left(\frac{\pi}{a}\right)\sum_{m=1}^{\infty}\sum_{n=1}^{\infty} W_{mn} m \cos\left(m\pi\frac{x}{a}\right)\sin\left(n\pi\frac{y}{b}\right) \qquad \textbf{(5.34a)}$$

$$\frac{\partial w}{\partial y} = \left(\frac{\pi}{b}\right)\sum_{m=1}^{\infty}\sum_{n=1}^{\infty} W_{mn} n \sin\left(m\pi\frac{x}{a}\right)\cos\left(n\pi\frac{y}{b}\right) \qquad \textbf{(5.34b)}$$

$$\frac{\partial^2 w}{\partial x^2} = -\left(\frac{\pi}{a}\right)^2\sum_{m=1}^{\infty}\sum_{n=1}^{\infty} W_{mn} m^2 \sin\left(m\pi\frac{x}{a}\right)\sin\left(n\pi\frac{y}{b}\right) \qquad \textbf{(5.34c)}$$

$$\frac{\partial^2 w}{\partial y^2} = -\left(\frac{\pi}{b}\right)^2\sum_{m=1}^{\infty}\sum_{n=1}^{\infty} W_{mn} n^2 \sin\left(m\pi\frac{x}{a}\right)\sin\left(n\pi\frac{y}{b}\right) \qquad \textbf{(5.34d)}$$

$$\frac{\partial^2 w}{\partial x \partial y} = \left(\frac{\pi^2}{ab}\right)\sum_{m=1}^{\infty}\sum_{n=1}^{\infty} W_{mn} m n \cos\left(m\pi\frac{x}{a}\right)\cos\left(n\pi\frac{y}{b}\right) \qquad \textbf{(5.34e)}$$

$$\frac{\partial^3 w}{\partial x^2 \partial y} = -\left(\frac{\pi^3}{a^2 b}\right)\sum_{m=1}^{\infty}\sum_{n=1}^{\infty} W_{mn} m^2 n \sin\left(m\pi\frac{x}{a}\right)\cos\left(n\pi\frac{y}{b}\right) \qquad \textbf{(5.34f)}$$

$$\frac{\partial^3 w}{\partial x^3} = -\left(\frac{\pi}{a}\right)^3\sum_{m=1}^{\infty}\sum_{n=1}^{\infty} W_{mn} m^3 \cos\left(m\pi\frac{x}{a}\right)\sin\left(n\pi\frac{y}{b}\right) \qquad \textbf{(5.34g)}$$

$$\frac{\partial^3 w}{\partial x \partial y^2} = -\left(\frac{\pi^3}{ab^2}\right)\sum_{m=1}^{\infty}\sum_{n=1}^{\infty} W_{mn} m n^2 \cos\left(m\pi\frac{x}{a}\right)\sin\left(n\pi\frac{y}{b}\right) \qquad \textbf{(5.34h)}$$

$$\frac{\partial^3 w}{\partial y^3} = -\left(\frac{\pi}{b}\right)^3\sum_{m=1}^{\infty}\sum_{n=1}^{\infty} W_{mn} n^3 \sin\left(m\pi\frac{x}{a}\right)\cos\left(n\pi\frac{y}{b}\right) \qquad \textbf{(5.34i)}$$

## EXAMPLE 5.4

The simply supported plate in Figure 5.10 is subject to a uniform distributed force of $p_o$. (a) Obtain a series solution for deflection $w$ and bending moment $m_{xx}$. (b) Assume the plate is a square plate $b = a$ and Poisson's ratio is $\nu = 1/3$, determine the maximum deflection $w$ and bending moment $m_{xx}$ and comment on the convergence of the two series.

**PLAN**

(a) Substituting $p_z(x, y) = p_o$ in Equation (5.33), we obtain $P_{mn}$ by integration. We then obtain $W_{mn}$ from Equation (5.30), which gives us the solution. (b) The maximum deflection $w$ and bending moment $m_{xx}$ will be at the center of the plate, which we can find from the solution in part a. We plot the solution vs. number of terms in the series and discuss the results.

**SOLUTION**

Substituting $p_z(x, y) = p_o$ into Equation (5.33) we obtain

$$P_{mn} = \frac{4p_o}{ab}\int_0^b\int_0^a \sin\left(m\pi\frac{x}{a}\right)\sin\left(n\pi\frac{y}{b}\right)dx\,dy = \left(\frac{4p_o}{ab}\right)\left[-\frac{a}{m\pi}\cos m\pi\frac{x}{a}\right]\Bigg|_{x=0}^{x=a}\left[-\frac{b}{n\pi}\cos m\pi\frac{y}{b}\right]\Bigg|_{y=0}^{y=b}$$

$$P_{mn} = \left(\frac{4p_o}{ab}\right)\left[\frac{a}{m\pi}(1-\cos m\pi)\right]\left[\frac{b}{n\pi}(1-\cos n\pi)\right] = \left(\frac{4p_o}{\pi^2 mn}\right)(1-\cos m\pi)(1-\cos n\pi) = \left(\frac{4p_o}{\pi^2 mn}\right)[1-(-1)^m][1-(-1)^n]$$

$$P_{mn} = \begin{cases} 0 & m, n = 2, 4, 6.\cdots \\ (16p_o)/(\pi^2 mn) & m, n = 1, 3, 5\cdots \end{cases} \qquad \text{(E1)}$$

From Equation (5.30) we obtain

$$W_{mn} = \begin{cases} 0 & m, n = 2, 4, 6.\cdots \\ \dfrac{16p_o}{D\pi^6 mn[(m/a)^2+(n/b)^2]^2} & m, n = 1, 3, 5\cdots \end{cases} \qquad \text{(E2)}$$

Substituting the above in Equation (5.28) we obtain

$$w(x, y) = \frac{16p_o}{D\pi^6}\sum_{m=1,3,5\cdots}^{\infty}\sum_{n=1,3,5\cdots}^{\infty}\frac{1}{mn[(m/a)^2+(n/b)^2]^2}\sin\left(m\pi\frac{x}{a}\right)\sin\left(n\pi\frac{y}{b}\right) \qquad \text{(E3)}$$

To calculate bending moment $m_{xx}$ we substitute Equations (5.34c) and (5.34d) into (5.14) to obtain

$$m_{xx}(x, y) = \left(\frac{16p_o}{\pi^4}\right)\sum_{m=1,3,5\cdots}^{\infty}\sum_{n=1,3,5\cdots}^{\infty}\frac{[(m/a)^2+\nu(n/b)^2]}{mn[(m/a)^2+(n/b)^2]^2}\sin\left(m\pi\frac{x}{a}\right)\sin\left(n\pi\frac{y}{b}\right) \qquad \text{(E4)}$$

The maximum values of $w$ and $m_{xx}$ will be where the Sine function will be maximum, which is at the center of the plate, i.e., at $x = a/2$ and $y = b/2$. To study the convergence of the series consider a square plate ($b = a$) with a Poisson's ratio of $\nu = 1/3$. We obtain the following maximum values

$$w_{max} = \frac{16p_o a^4}{D\pi^6}\sum_{m=1,3,5\cdots}^{\infty}\sum_{n=1,3,5\cdots}^{\infty}\frac{\sin(m\pi/2)\sin(n\pi/2)}{mn[m^2+n^2]^2} = \frac{16p_o a^4}{D\pi^6}\sum_{m=1,3,5\cdots}^{\infty}\sum_{n=1,3,5\cdots}^{\infty}\frac{(-1)^{(m+n-2)/2}}{mn[m^2+n^2]^2} \qquad \text{(E5)}$$

$$(\boldsymbol{m}_{xx})_{max} = \left(\frac{16p_o a^2}{\pi^4}\right)\sum_{m=1,3,5\cdots}^{\infty}\sum_{n=1,3,5\cdots}^{\infty}\frac{[m^2+\nu n^2]}{mn[m^2+n^2]^2}sin\left(\frac{m\pi}{2}\right)sin\left(\frac{n\pi}{2}\right)$$

$$(\boldsymbol{m}_{xx})_{max} = \left(\frac{16p_o a^2}{\pi^4}\right)\sum_{m=1,3,5\cdots}^{\infty}\sum_{n=1,3,5\cdots}^{\infty}\frac{[m^2+\nu n^2](-1)^{(m+n-2)/2}}{mn[m^2+n^2]^2} \qquad \text{(E6)}$$

Table 5.3 shows the evaluation of maximum deflection and moment for 16 terms in the series and Figure 5.11 shows the variation in the values of these quantities as more terms are included. The column % shows the percentage difference of the value compared to the sixteenth term calculated as follows.

$$\% \text{ difference} = [(T_n - T_{16})/T_{16}] \times 100 \text{ where } T_n \text{ is the value after } n \text{ terms.} \qquad \text{(E7)}$$

**Table 5.3** Series convergence in Example 5.4.

| No. of terms | m | n | $\frac{Dw_{max}(10^{-3})}{(p_o a^4)}$ Value | % | $\frac{(\boldsymbol{m}_{xx})_{max}(10^{-3})}{p_o a^2}$ Value | % |
|---|---|---|---|---|---|---|
| 1 | 1 | 1 | 4.1606 | -2.4278 | 54.7519 | -11.8235 |
| 2 | 3 | 1 | 4.1052 | -1.0621 | 49.6417 | -1.3867 |
| 3 | 1 | 3 | 4.0497 | 0.3036 | 47.4517 | 3.0863 |
| 4 | 3 | 3 | 4.0554 | 0.1631 | 48.1276 | 1.7057 |
| 5 | 5 | 1 | 4.0603 | 0.0419 | 49.3587 | -0.8087 |
| 6 | 1 | 5 | 4.0653 | -0.0793 | 49.8123 | -1.7350 |
| 7 | 5 | 3 | 4.0643 | -0.0557 | 49.5470 | -1.1933 |
| 8 | 3 | 5 | 4.0633 | -0.0321 | 49.3828 | -0.8580 |
| 9 | 5 | 5 | 4.0636 | -0.0386 | 49.4705 | -1.0369 |
| 10 | 7 | 1 | 4.0626 | -0.0152 | 49.0074 | -0.0912 |
| 11 | 1 | 7 | 4.0617 | 0.0082 | 48.8447 | 0.2411 |
| 12 | 7 | 3 | 4.0619 | 0.0024 | 48.9656 | -0.0058 |
| 13 | 3 | 7 | 4.0622 | -0.0034 | 49.0245 | -0.1261 |
| 14 | 7 | 5 | 4.0621 | -0.0013 | 48.9754 | -0.0258 |
| 15 | 5 | 7 | 4.0620 | 0.0009 | 48.9400 | 0.0466 |
| 16 | 7 | 7 | 4.0620 | 0.0000 | 48.9628 | 0.0000 |

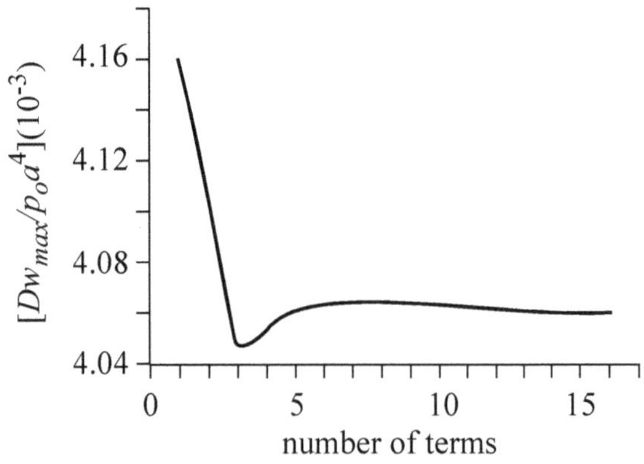


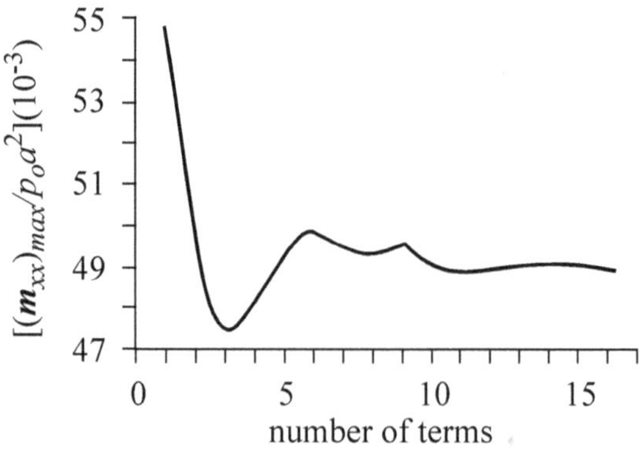


**Figure 5.11** Convergence of series for maximum deflection and moment.

**COMMENTS**

1. The maximum deflection converges quickly as seen in Figure 5.11. If we compare the result after 16 terms to those after one term, the difference is 2.5%, while after 3 terms the difference is 0.3%. The maximum moment converges more slowly. Comparing moment results after 16 terms to those after one term the difference is 11.8%, while after 3 terms the difference is 3.1%. This behavior is consistent with the maxim in approximation, namely, *differentiation increases and integration decreases the error of approximation.*

## EXAMPLE 5.5

The rectangular plate Figure 5.10 is subjected to a transverse load that varies as $p_z = p_o(x/a)(y/b)$. Determine (*i*) the deflection $w$ and moment $\boldsymbol{m}_{xx}$ field on the entire plate. (*ii*) the deflection $w$ and moment $\boldsymbol{m}_{xx}$ at the center of the plate using at least 4 terms in the series assuming the plate is square i.e., $b = a$. Use $\nu = 1/3$.

**SOLUTION**

(*i*) We substitute the given load into Equation (5.33) to obtain

$$P_{mn} = \frac{4p_o}{ab}\int_0^b\int_0^a (x/a)(y/b)sin(m\pi x/a)sin(n\pi y/b)dxdy \qquad \text{(E1)}$$

We substitute $\xi = \pi x/a$ and $\eta = \pi y/b$ to obtain

$$P_{mn} = \frac{4p_o}{\pi^4}\int_0^{\pi}\int_0^{\pi}\xi\eta\, sin(m\xi)sin(n\eta)d\xi d\eta = \frac{4p_o}{\pi^4}\left[\int_0^{\pi}\eta\, sin(n\eta)d\eta\right]\left[\int_0^{\pi}\xi\, sin(m\xi)d\xi\right] \quad \text{(E2)}$$

We consider the evaluation of the following integral by integration by parts.

$$I_1 = \int_0^{\pi}\eta\, sin(n\eta)d\eta = -(\eta/n)cos(n\eta)\Big|_0^{\pi} + (1/n)\int_0^{\pi}cos(n\eta)d\eta = [-(\eta/n)cos(n\eta) + sin(n\eta)/n^2]\Big|_0^{\pi} \text{ or}$$

$$I_1 = [-(\pi/n)cos(n\pi) + sin(n\pi)/n^2] = -(\pi/n)(-1)^n \quad \text{(E3)}$$

The value of other integral in Equation (E2) can be obtained by replacing *j* with *i*. Substituting the values of the integrals in Equation (E2) we obtain

$$P_{mn} = [4p_o/(\pi^2 mn)](-1)^{m+n} \quad \text{(E4)}$$

Substituting Equation (E4) into Equation (5.29) we obtain

$$W_{mn} = \frac{1}{D\pi^6}\frac{4p_o(-1)^{m+n}}{mn[(m/a)^2 + (n/b)^2]^2} \quad \text{(E5)}$$

We substitute the above equation into Equation (5.28) to obtain the deflection field as:

$$\text{ANS: } w(x,y) = \frac{4p_o}{D\pi^6}\sum_{m=1}^{\infty}\sum_{n=1}^{\infty}\frac{(-1)^{m+n}}{mn[(m/a)^2 + (n/b)^2]^2}sin\left(m\pi\frac{x}{a}\right)sin\left(n\pi\frac{y}{b}\right) \quad \text{(E6)}$$

We substitute Equations (5.34c) and (5.34d) into (5.14) to obtain the bending moment $\boldsymbol{m}_{xx}$

$$\text{ANS: } \boldsymbol{m}_{xx}(x,y) = \left(\frac{4p_o}{\pi^4}\right)\sum_{m=1}^{\infty}\sum_{n=1}^{\infty}\frac{(-1)^{m+n}[(m/a)^2 + \nu(n/b)^2]}{mn[(m/a)^2 + (n/b)^2]^2}sin\left(m\pi\frac{x}{a}\right)sin\left(n\pi\frac{y}{b}\right) \quad \text{(E7)}$$

(*ii*) For a square plate at the center we substitute $b = a$, $x = a/2$, and $y = a/2$ in Equations (E6) and (E7) to obtain

$$w\left(\frac{a}{2},\frac{a}{2}\right) = \frac{4p_oa^4}{D\pi^6}\sum_{m=1}^{\infty}\sum_{n=1}^{\infty}\frac{(-1)^{m+n}}{mn[m^2+n^2]^2}sin\left(m\frac{\pi}{2}\right)sin\left(m\frac{\pi}{2}\right) = \frac{4p_oa^4}{D\pi^6}\sum_{m=1,3,5,\ldots}^{\infty}\sum_{n=1,3,5,\ldots}^{\infty}\frac{(-1)^{(3m+3n-2)/2}}{mn[m^2+n^2]^2} \quad \text{(E8)}$$

$$\boldsymbol{m}_{xx}\left(\frac{a}{2},\frac{a}{2}\right) = \left(\frac{4p_oa^2}{\pi^4}\right)\sum_{m=1}^{\infty}\sum_{n=1}^{\infty}\frac{(-1)^{m+n}[m^2+\nu n^2]}{mn[m^2+n^2]^2}sin\left(m\frac{\pi}{2}\right)sin\left(m\frac{\pi}{2}\right)$$

$$\boldsymbol{m}_{xx}\left(\frac{a}{2},\frac{a}{2}\right) = \left(\frac{4p_oa^2}{\pi^4}\right)\sum_{m=1,3,5,\ldots}^{\infty}\sum_{n=1,3,5,\ldots}^{\infty}\frac{(-1)^{(3m+3n-2)/2}[m^2+\nu n^2]}{mn[m^2+n^2]^2} \quad \text{(E9)}$$

Table 5.4 shows the evaluation of deflection and moment in the center of the plate for 16 terms.

**Table 5.4** Series convergence in Example 5.5

| No. of terms | m | n | $\frac{Dw_{mid}(10^{-3})}{p_oa^4}$ | $\frac{(\boldsymbol{m}_{xx})_{mid}(10^{-3})}{p_oa^2}$ | No. of terms | m | n | $\frac{Dw_{mid}(10^{-3})}{p_oa^4}$ | $\frac{(\boldsymbol{m}_{xx})_{mid}(10^{-3})}{p_oa^2}$ |
|---|---|---|---|---|---|---|---|---|---|
| 1 | 1 | 1 | 1.0402 | 13.6880 | 9 | 5 | 5 | 1.0159 | 12.3676 |
| 2 | 3 | 1 | 1.0263 | 12.4104 | 10 | 7 | 1 | 1.0157 | 12.2519 |
| 3 | 1 | 3 | 1.0124 | 11.8629 | 11 | 1 | 7 | 1.0154 | 12.2112 |
| 4 | 3 | 3 | 1.0139 | 12.0319 | 12 | 7 | 3 | 1.0155 | 12.2414 |
| 5 | 5 | 1 | 1.0151 | 12.3397 | 13 | 3 | 7 | 1.0155 | 12.2561 |
| 6 | 1 | 5 | 1.0163 | 12.4531 | 14 | 7 | 5 | 1.0155 | 12.2438 |
| 7 | 5 | 3 | 1.0161 | 12.3868 | 15 | 5 | 7 | 1.0155 | 12.2350 |
| 8 | 3 | 5 | 1.0158 | 12.3457 | 16 | 7 | 7 | 1.0155 | 12.2407 |

After four term the values are as given below.

**ANS.** $w(a/2, a/2) = 1.0139(10^{-3})(p_oa^4/D)$; $\boldsymbol{m}_{xx}(a/2, a/2) = 12.032(10^{-3})(p_oa^2)$

## COMMENTS

1. Comparing the results for deflection after 16 terms to those after one term the difference is 2.42% while after 4 terms the difference is 0.16%. Comparing moment results after 16 terms to those after one term the difference is 11.8% while after 4 terms the difference is 1.7%.
2. Figure 5.12 shows the change in values of deflection and moment with the number of terms. For deflection, after 6 terms there is very little variation. However, the moment values show greater changes till about the 12th term.

3. The two comments above once more emphasizes that deflection converges faster than moment. Alternatively, derivatives increase the error in approximations.

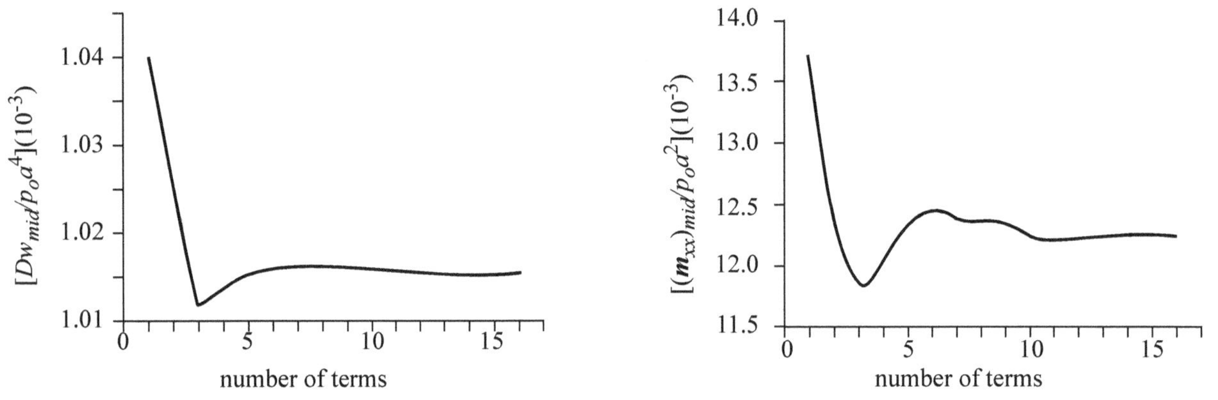


**Figure 5.12** Convergence of series in Example 5.5.

## 5.4 NADAI-LEVY SOLUTION FOR RECTANGULAR PLATES

The Navier's solution is limited to plates that are simply supported on all sides. Furthermore, the double summation can lead to slow convergence of the series for complex loading, particularly for moments as we have seen in the previous section. Nadai-Levy overcomes these difficulty by imposing a different set of limitations. The limitations are:

1. Plate is simply supported on two opposite sides as shown in Figure 5.13. The other two sides can have any type of support or boundary conditions. The boundary conditions at $x = 0$ and $x = a$ can be written as

$$w(0, y) = 0 \qquad \frac{\partial^2 w}{\partial x^2}(0, y) = 0 \qquad w(a, y) = 0 \qquad \frac{\partial^2 w}{\partial x^2}(a, y) = 0 \tag{5.35a}$$

2. The loading $p_z$ is only dependent on $x$, that is, $p_z(x, y) = p_z(x)$.

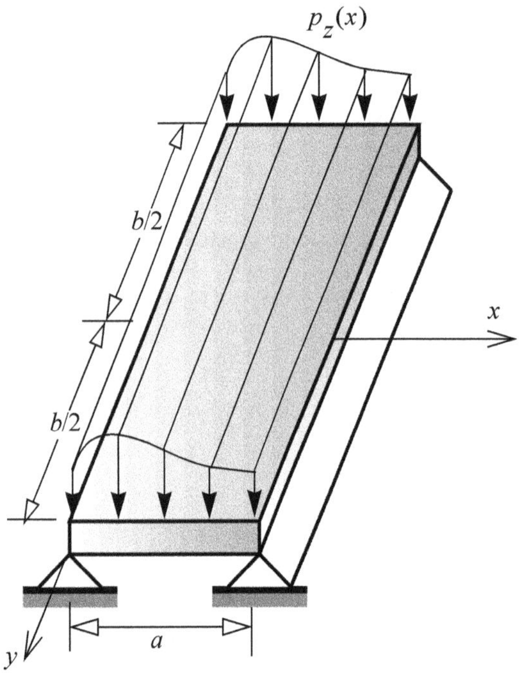


**Figure 5.13** Plate geometry and loading for Nadai-Levy's method

We consider the deflection solution in two parts as shown below.

$$w(x, y) = w_p(x) + w_h(x, y) \tag{5.36}$$

Substituting the above into the bi-harmonic equation (5.20) we obtain:

$$\nabla^4 w = \frac{d^4 w_p}{dx^4} + \left[\frac{\partial^4 w_h}{\partial x^4} + 2\frac{\partial^4 w_h}{\partial x^2 \partial y^2} + \frac{\partial^4 w_h}{\partial y^4}\right] = \frac{p_z(x)}{D}$$

Writing the above equation as a set of two equations shown below.

$$\frac{d^4 w_p}{dx^4} = \frac{p_z(x)}{D} \tag{5.37a}$$

$$\frac{\partial^4 w_h}{\partial x^4} + 2\frac{\partial^4 w_h}{\partial x^2 \partial y^2} + \frac{\partial^4 w_h}{\partial y^4} = 0 \tag{5.37b}$$

### 5.4.1 Homogeneous solution

Noting the fact that the $sin(m\pi x/a)$, satisfy the boundary conditions in Equation (5.35a), we choose the following for the homogeneous solution

$$w_h(x, y) = \sum_{m=1}^{\infty} H_m(y) sin(m\pi x/a) \tag{5.37c}$$

Substituting the above into Equation (5.37b) we obtain

$$\sum_{m=1}^{\infty} \left[\left(\frac{m\pi}{a}\right)^4 H_m - 2\left(\frac{m\pi}{a}\right)^2 \frac{d^2 H_m}{dy^2} + \frac{d^4 H_m}{dy^4}\right] sin(m\pi x/a) = 0 \tag{5.37d}$$

For the above equation to be valid at all points on the plate the following equation must be satisfied.

$$\left(\frac{m\pi}{a}\right)^4 H_m - 2\left(\frac{m\pi}{a}\right)^2 \frac{d^2 H_m}{dy^2} + \frac{d^4 H_m}{dy^4} = 0 \tag{5.37e}$$

The above equation is an ordinary differential equation with constant coefficients, whose solution can be found by substituting $H_m = e^{\lambda y}$ to obtain the characteristic equation below.

$$(m\pi/a)^4 - 2(m\pi/a)^2\lambda^2 + \lambda^4 = 0 \quad \text{or} \quad [\lambda^2 - (m\pi/a)^2]^2 = 0 \quad \text{or} \quad \lambda = \pm(m\pi/a); \pm(m\pi/a) \tag{5.37f}$$

The repeated roots imply the following solution

$$H_m = C_{1m}e^{(m\pi/a)y} + C_{2m}ye^{(m\pi/a)y} + C_{3m}e^{-(m\pi/a)y} + C_{4m}ye^{-(m\pi/a)y} \tag{5.37g}$$

The above equation can also be expressed in terms of hyperbolic functions as shown below.

$$H_m = A_m cosh(m\pi y/a) + B_m(m\pi y/a) sinh(m\pi y/a) + C_m sinh(m\pi y/a) + D_m(m\pi y/a) cosh(m\pi y/a) \quad \text{or}$$

$$H_m = [A_m + D_m(m\pi y/a)] cosh(m\pi y/a) + [C_m + B_m(m\pi y/ay)] sinh(m\pi y/a) \tag{5.37h}$$

The homogeneous solution is

$$\boxed{w_h(x, y) = \sum_{m=1}^{\infty} \{[A_m + D_m(m\pi y/a)] cosh(m\pi y/a) + [C_m + B_m(m\pi y/a)] sinh(m\pi y/a)\} sin(m\pi y/a)} \tag{5.38}$$

### 5.4.2 Particular solution

There are two ways to find the particular solution. In Method I we find a series solution and in Method II we use direct integration. Method I is useful in relating the constants $A_m$, $B_m$, $C_m$, and $D_m$ to the loading parameters in $p_z$. Method II can give closed form solutions when $p_z$ is a simple function of $x$, such as a polynomial. We can use a combination of the two approaches to obtain the complete solution.

**Method I**: We use series to represent the deflection and distributed loads as shown below.

$$w_p(x) = \sum_{m=1}^{\infty} W_m sin(m\pi x/a) \tag{5.39a}$$

$$p_z(x) = \sum_{m=1}^{\infty} P_m sin(m\pi x/a) \tag{5.39b}$$

$P_m$ can be found using the orthogonality condition Equation (5.31a) as shown below

$$\sum_{m=1}^{\infty} P_m \int_0^a sin(m\pi x/a) sin(n\pi x/a) dx = \int_0^a p_z(x) sin(n\pi x/a)(dx) dx$$

$$\boxed{P_n = \frac{2}{a}\int_0^a p_z(x) sin(n\pi x/a) dx} \tag{5.40}$$

Substituting Equations (5.39a) and Equations (5.39b) into Equation (5.37a) we obtain

$$\sum_{m=1}^{\infty} W_m \left(\frac{m\pi}{a}\right)^4 sin\left(m\pi\frac{x}{a}\right) = \frac{1}{D}\sum_{m=1}^{\infty} P_m sin\left(m\pi\frac{x}{a}\right)$$

For the equality to hold for each term the following equations must be satisfied

$$W_m = \left(\frac{a}{m\pi}\right)^4 \frac{P_m}{D} \tag{5.41a}$$

The particular solution can be written as

$$w_p(x) = \sum_{m=1}^{\infty}\left\{\left(\frac{a}{m\pi}\right)^4 \frac{P_m}{D} sin(m\pi x/a)\right\} \tag{5.42a}$$

Substituting Equations (5.37h), (5.39a), and (5.41a) into Equation (5.36) we obtain the complete solution as

$$w(x,y) = \sum_{m=1}^{\infty}\left[\left\{\left(\frac{a}{m\pi}\right)^4 \frac{P_m}{D} + \left[A_m + D_m\left(\frac{m\pi}{a}y\right)\right]cosh\left(\frac{m\pi}{a}y\right) + \left[C_m + B_m\left(\frac{m\pi}{a}y\right)\right]sinh\left(\frac{m\pi}{a}y\right)\right\}sin\left(m\pi\frac{x}{a}\right)\right] \tag{5.43}$$

The constant $P_m$ is determined from Equation (5.40). The constants $A_m$, $B_m$, $C_m$, and $D_m$ are determined from the boundary conditions on $y = \pm b/2$.

**Method II**: $w_h$ satisfies the boundary condition at $x = 0$ and $x = a$. Thus the particular solution must satisfy the boundary conditions of zero displacement and moment at $x = 0$ and $x = a$ *given by* Equation (5.35a). We consider a uniform load $p_z = p_0$. By integrating Equation (5.37a) four times we obtain

$$w_p = \frac{p_0}{D}\left[\frac{x^4}{24} + C_1x^3 + C_2x^2 + C_3x + C_4\right] \tag{5.43a}$$

From the boundary conditions in Equation (5.35a) we obtain the constants and hence the solution as in

$$w_p = \frac{p_0 a^4}{24D}\left[\left(\frac{x}{a}\right)^4 - 2\left(\frac{x}{a}\right)^3 + \left(\frac{x}{a}\right)\right] \tag{5.43b}$$

For other loadings, we can find the particular solution in a similar manner.

## EXAMPLE 5.6

The simply supported plate in Figure 5.14 is subject to a uniform distributed force of $p_o$. (a) Obtain a series solution for deflection $w$ using Nadai-Levy Method. (b) For maximum deflection $w$ and bending moment $m_{xx}$, compare the convergence of Nadai-Levy Method with Navier's method for a square plate and Poisson's ratio of $\nu = 1/3$.

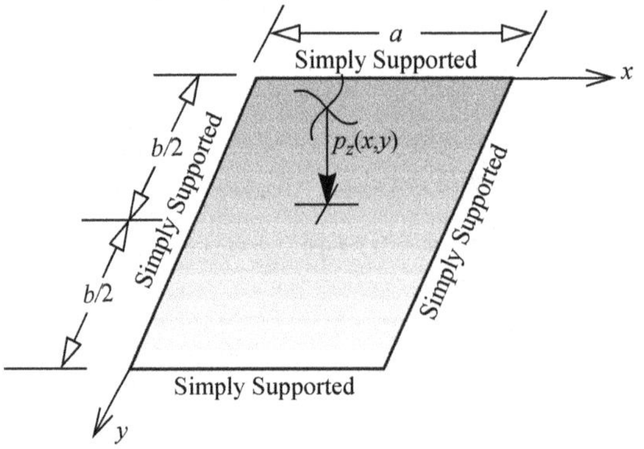


**Figure 5.14** Simply supported rectangular plate.

**PLAN**

The coefficients $P_m$ can be found from Equation (5.40). The loading and boundary conditions are symmetric in the $y$ direction, hence the deflection has to be symmetric, i.e., $w(x,-y) = w(x,y)$. Thus, the coefficients of odd functions of $y$ must be zero, i.e., $C_m = 0$ and $D_m = 0$. We can determine $A_m$ and $B_m$ from the zero deflection and moment conditions at $y = b/2$. Once the constants are determined, the deflection is known and moments can be found and compared with Navier's solution.

**SOLUTION**

Substituting $p_z = p_0$ into Equation (5.40) and integrating we obtain the value of $P_m$ as shown below.

$$P_m = \frac{2p_0}{a}\int_0^a sin\left(m\pi\frac{x}{a}\right)dx = -\left(\frac{2p_0}{m\pi}\right)cos\, m\pi\frac{x}{a}\Bigg|_0^a = \left(\frac{2p_0}{m\pi}\right)(1 - cos\, m\pi) = \begin{cases} 4p_0/(m\pi) & \text{odd values of } m \\ 0 & \text{even values of } m \end{cases} \tag{E1}$$

Because of symmetry the deflection solution is given by the following

$$w(x,y) = \sum_{m=1}^{\infty}\left[\left\{\left(\frac{a}{m\pi}\right)^4 \frac{P_m}{D} + A_m cosh\left(\frac{m\pi}{a}y\right) + B_m\left(\frac{m\pi}{a}y\right)sinh\left(\frac{m\pi}{a}y\right)\right\}sin\left(m\pi\frac{x}{a}\right)\right]$$

$$w(x,y) = \sum_{m=1}^{\infty}\left[\left\{\left(\frac{1}{\alpha_m}\right)^4 \frac{P_m}{D} + A_m cosh\alpha_m y + B_m\alpha_m y\, sinh\alpha_m y\right\}sin(\alpha_m x)\right] \quad \text{where} \quad \alpha_m = \frac{m\pi}{a} \tag{E2}$$

The zero deflection and zero moment boundary conditions yield the following

$$w(x, b/2) = 0 \qquad \frac{\partial^2 w}{\partial y^2}(x, b/2) = 0 \tag{E3}$$

We can find the derivatives Equation (E2) as follows

$$\frac{\partial w}{\partial y} = \sum_{m=1}^{\infty} [\{A_m \alpha_m sinh\alpha_m y + B_m \alpha_m sinh\alpha_m y + B_m \alpha_m^2 y cosh\alpha_m y\} sin(\alpha_m x/a)] \quad \text{(E4)}$$

$$\frac{\partial^2 w}{\partial y^2} = \sum_{m=1}^{\infty} [\{A_m \alpha_m^2 cosh\alpha_m y + 2B_m \alpha_m^2 cosh\alpha_m y + B_m y \alpha_m^3 sinh\alpha_m y\} sin(\alpha_m x/a)] \quad \text{(E5)}$$

Substituting Equations (E2) and Equations (E5) into Equation (E3) and noting the condition must be satisfied at all $x$, we obtain

$$\left(\frac{1}{\alpha_m}\right)^4 \frac{P_m}{D} + A_m cosh(\alpha_m b/2) + B_m(\alpha_m b/2) sinh(\alpha_m b/2) = 0 \quad \text{(E6)}$$

$$A_m \alpha_m^2 cosh(\alpha_m b/2) + B_m(2\alpha_m^2 cosh(\alpha_m b/2) + (b/2)\alpha_m^3 sinh(\alpha_m b/2)) = 0 \quad \text{(E7)}$$

Solving the above two equations we obtain

$$A_m = -\left(\frac{P_m}{4D\alpha_m^4}\right)\frac{[\alpha_m b\, tanh(\alpha_m b/2) + 4]}{cosh(\alpha_m b/2)} = \begin{cases} -\left(\dfrac{p_0 a^4}{Dm^5\pi^5}\right)\dfrac{[\alpha_m b\, tanh(\alpha_m b/2) + 4]}{cosh(\alpha_m b/2)} & \text{odd values of } m \\ 0 & \text{even values of } m \end{cases} \quad \text{(E8)}$$

$$B_m = \frac{P_m}{2D\alpha_m^4}\frac{1}{cosh(\alpha_m b/2)} = \begin{cases} \left(\dfrac{2p_0 a^4}{Dm^5\pi^5}\right)\dfrac{1}{cosh(\alpha_m b/2)} & \text{odd values of } m \\ 0 & \text{even values of } m \end{cases} \quad \text{(E9)}$$

Using Equation (5.43b) for particular solution we can write the complete solutions as shown below.

$$w(x,y) = \frac{p_0 a^4}{24D}\left[\left(\frac{x}{a}\right)^4 - 2\left(\frac{x}{a}\right)^3 + \left(\frac{x}{a}\right)\right] + \sum_{m=1,3,5,\ldots} [\{A_m cosh(\alpha_m y) + B_m(\alpha_m y) sinh(\alpha_m y)\} sin(\alpha_m x)] \quad \text{(E10)}$$

(b) For a square plate $b = a$, thus $\alpha_m b = m\pi$. From Equations (E8) and (E9) we obtain the constants as shown below.

$$\frac{A_m}{(p_0 a^4/D)} = \begin{cases} -\left[\dfrac{m\pi\, tanh(m\pi/2) + 4}{m^5\pi^5 cosh(m\pi/2)}\right] & \text{odd values of } m \\ 0 & \text{even values of } m \end{cases} \quad \text{(E11)}$$

$$\frac{B_m}{(p_0 a^4/D)} = \begin{cases} \left[\dfrac{2}{m^5\pi^5 cosh(m\pi/2)}\right] & \text{odd values of } m \\ 0 & \text{even values of } m \end{cases} \quad \text{(E12)}$$

The maximum deflection will be at the center of the plate where $x = a/2$ and $y = 0$.

$$w_{max} = w(a/2, 0) = \frac{p_0 a^4}{24D}[(1/2)^4 - 2(1/2)^3 + (1/2)] + \sum_{m=1,3,5,\ldots}^{\infty} A_m sin(m\pi/2)$$

$$\frac{w_{max}}{(p_0 a^4/D)} = 13.02(10^{-3}) + \sum_{m=1,3,5,\ldots}^{\infty} \frac{A_m}{(p_0 a^4/D)}(-1)^{(m-1)/2} \quad \text{(E13)}$$

The maximum moment $\boldsymbol{m_{xx}}$ will be at the center of the plate where $x = a/2$ and $y = 0$. We find the values as shown below. To find the moment expression, we find the $x$ derivatives as shown below and evaluate it at the center of the plate.

$$\frac{\partial^2 w}{\partial x^2} = \frac{p_0 a^2}{2D}\left[\left(\frac{x}{a}\right)^2 - \left(\frac{x}{a}\right)\right] + \sum_{m=1,3,5,\ldots} [-\alpha_m^2\{A_m cosh(\alpha_m y) + B_m(\alpha_m y) sinh(\alpha_m y)\} sin(\alpha_m x)] \quad \text{(E14)}$$

$$\frac{\partial^2 w}{\partial x^2}\left(\frac{a}{2}, 0\right) = \frac{p_0 a^2}{2D}\left[\left(\frac{1}{2}\right)^2 - \left(\frac{1}{2}\right)\right] - \sum_{m=1,3,5,\ldots} \left[\left(\frac{m\pi}{a}\right)^2 A_m sin\left(\frac{m\pi}{2}\right)\right]$$

$$\frac{\partial^2 w}{\partial x^2}\left(\frac{a}{2}, 0\right) = -\left(\frac{p_0 a^2}{D}\right)\left[0.125 + \sum_{m=1,3,5,\ldots} \left[(m\pi)^2 \frac{A_m}{(p_0 a^4/D)}(-1)^{(m-1)/2}\right]\right] \quad \text{(E15)}$$

From Equation (E5) we obtain

$$\frac{\partial^2 w}{\partial y^2}\left(\frac{a}{2}, 0\right) = \sum_{m=1,3,5,\ldots}^{\infty} \left[\{A_m \alpha_m^2 + 2B_m \alpha_m^2\} sin\left(\frac{m\pi}{2}\right)\right] = \sum_{m=1,3,5,\ldots} \left[\left(\frac{m\pi}{a}\right)^2 \{A_m + 2B_m\} sin\left(\frac{m\pi}{2}\right)\right]$$

$$\frac{\partial^2 w}{\partial y^2}\left(\frac{a}{2}, 0\right) = \frac{p_0 a^2}{D}\left[\sum_{m=1,3,5,\ldots}\left[(m\pi)^2\left\{\frac{A_m + 2B_m}{(p_0 a^4/D)}\right\}(-1)^{(m-1)/2}\right]\right] \quad \text{(E16)}$$

The maximum moment $m_{xx}$ for $\nu = 1/3$ is as follows.

$$M_{max} = m_{xx}\left(\frac{a}{2}, 0\right) = p_0 a^2\left[0.125 + \sum_{m=1,3,5,\ldots}\left\{(m\pi)^2\frac{[A_m - (A_m + 2B_m)/3]}{(p_0 a^4/D)}(-1)^{\frac{(m-1)}{2}}\right\}\right]$$

$$\frac{M_{max}}{p_0 a^2} = \left[0.125 + \sum_{m=1,3,5,\ldots}\left[\frac{2}{3}(m\pi)^2\frac{[A_m - B_m]}{(p_0 a^4/D)}(-1)^{(m-1)/2}\right]\right] \quad \text{(E17)}$$

The constants can be evaluated from Equations (E8) and (E9) and the maximum deflection and moments from Equations (E13) and (E17) as shown in Table 5.4.

**Table 5.4** Series convergence in Example 5.6.

| m | $\frac{A_m}{(p_0 a^4/D)}$ | $\frac{B_m}{(p_0 a^4/D)}$ | $\frac{w_{max}(10^{-3})}{(p_o a^4)/D}$ | $\frac{(m_{xx})_{max}(10^{-3})}{p_o a^2}$ | m | $\frac{A_m}{(p_0 a^4/D)}$ | $\frac{B_m}{(p_0 a^4/D)}$ | $\frac{w_{max}(10^{-3})}{(p_o a^4)/D}$ | $\frac{(m_{xx})_{max}(10^{-3})}{p_o a^2}$ |
|---|---|---|---|---|---|---|---|---|---|
| 1 | -8.962E-03 | 2.605E-03 | 4.0591 | 48.8964 | 9 | -2.590E-12 | 1.605E-13 | 4.0624 | 49.1142 |
| 3 | -3.243E-06 | 4.832E-07 | 4.0624 | 49.1171 | 11 | -4.902E-14 | 2.543E-15 | 4.0624 | 49.1142 |
| 5 | -1.600E-08 | 1.624E-09 | 4.0624 | 49.1142 | 13 | -1.069E-15 | 4.766E-17 | 4.0624 | 49.1142 |
| 7 | -1.696E-10 | 1.305E-11 | 4.0624 | 49.1142 | 15 | -2.574E-17 | 1.007E-18 | 4.0624 | 49.1142 |

**COMMENTS**

1. Up to five significant figures, the deflection value converged in two terms and moment values in 3 terms. In contrast, Navier's solution took 16 terms for deflection convergence and moment values showed oscillations in the fifth significant figure even for 16 terms.
2. The fast convergence for Nadai-Levy method in contrast to Navier's solution is for two reasons. Nadai-Levy's solution has a single series, while Navier's solution is a double series. The second reason for the fast convergence of Nadai-Levy method are the fast decrease in the values of hyperbolic functions in the definitions of constants $A_m$ and $B_m$. The values of these constants decrease by orders of magnitude with the increase in the value of $m$.
3. As $b/a \to \infty$ the functions $tanh(\alpha_m b/2) \to 1$ and $cosh(\alpha_m b/2) \to \infty$, which from Equations (E8) and (E9) imply that the constants $A_m$ and $B_m$ tend to zero and the effect of series terms die out.

## 5.5 CIRCULAR PLATES

The equations of bending of circular plates are derived in this section. We will use cylindrical coordinates shown in Figure 5.15 and transform the equations in Section 5.2 into cylindrical coordinates.

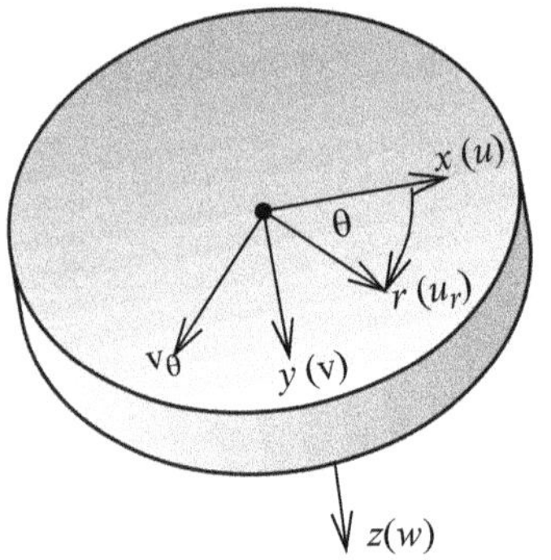


**Figure 5.15** Circular plate.

We can relate the coordinates in the cartesian $(x, y, z)$ and cylindrical coordinates $(r, \theta, z)$ as shown below.

$$x = r\cos\theta \qquad y = r\sin\theta \qquad r^2 = x^2 + y^2 \qquad \theta = \tan^{-1}(y/x) \quad \textbf{(5.44a)}$$

We can write

$$\frac{\partial r}{\partial x} = \frac{x}{r} = \cos\theta \qquad \frac{\partial r}{\partial y} = \frac{y}{r} = \sin\theta \qquad \frac{\partial\theta}{\partial x} = \frac{(-y/x^2)}{1 + (y/x)^2} = -\frac{y}{r^2} = \frac{-\sin\theta}{r} \qquad \frac{\partial\theta}{\partial y} = \frac{(1/x)}{1 + (y/x)^2} = \frac{x}{r^2} = \frac{\cos\theta}{r} \quad \textbf{(5.44b)}$$

We can relate the first and derivatives in cartesian to cylindrical as shown below.

$$\frac{\partial}{\partial x} = \frac{\partial r}{\partial x}\frac{\partial}{\partial r} + \frac{\partial\theta}{\partial x}\frac{\partial}{\partial\theta} = \cos\theta\frac{\partial}{\partial r} - \frac{\sin\theta}{r}\frac{\partial}{\partial\theta} \qquad \frac{\partial}{\partial y} = \frac{\partial r}{\partial y}\frac{\partial}{\partial r} + \frac{\partial\theta}{\partial y}\frac{\partial}{\partial\theta} = \sin\theta\frac{\partial}{\partial r} + \frac{\cos\theta}{r}\frac{\partial}{\partial\theta} \quad \textbf{(5.44c)}$$

$$\frac{\partial^2}{\partial x^2} = \frac{\partial r}{\partial x}\frac{\partial}{\partial r}\left(cos\theta\frac{\partial}{\partial r} - \frac{sin\theta}{r}\frac{\partial}{\partial \theta}\right) + \frac{\partial \theta}{\partial x}\frac{\partial}{\partial \theta}\left(cos\theta\frac{\partial}{\partial r} - \frac{sin\theta}{r}\frac{\partial}{\partial \theta}\right)$$

$$\frac{\partial^2}{\partial x^2} = cos\theta\left(cos\theta\frac{\partial^2}{\partial r^2} + \frac{sin\theta}{r^2}\frac{\partial}{\partial \theta}\right) - \frac{sin\theta}{r}\left(-sin\theta\frac{\partial}{\partial r} - \frac{sin\theta}{r}\frac{\partial^2}{\partial \theta^2} - \frac{cos\theta}{r}\frac{\partial}{\partial \theta}\right) \tag{5.44d}$$

$$\frac{\partial^2}{\partial y^2} = \frac{\partial r}{\partial y}\frac{\partial}{\partial r}\left(sin\theta\frac{\partial}{\partial r} + \frac{cos\theta}{r}\frac{\partial}{\partial \theta}\right) + \frac{\partial \theta}{\partial y}\frac{\partial}{\partial \theta}\left(sin\theta\frac{\partial}{\partial r} + \frac{cos\theta}{r}\frac{\partial}{\partial \theta}\right)$$

$$\frac{\partial^2}{\partial y^2} = sin\theta\left(sin\theta\frac{\partial^2}{\partial r^2} - \frac{cos\theta}{r^2}\frac{\partial}{\partial \theta}\right) + \frac{cos\theta}{r}\left(cos\theta\frac{\partial}{\partial r} + \frac{cos\theta}{r}\frac{\partial^2}{\partial \theta^2} - \frac{sin\theta}{r}\frac{\partial}{\partial \theta}\right) \tag{5.44e}$$

Adding Equations (5.44d) and (5.44e) we obtain:

$$\boxed{\nabla^2 = \frac{\partial^2}{\partial x^2} + \frac{\partial^2}{\partial y^2} = \frac{\partial^2}{\partial r^2} + \frac{1}{r}\frac{\partial}{\partial r} + \frac{1}{r^2}\frac{\partial^2}{\partial \theta^2} = \frac{1}{r}\frac{\partial}{\partial r}\left[r\frac{\partial}{\partial r}\right] + \frac{1}{r^2}\frac{\partial^2}{\partial \theta^2}} \tag{5.44f}$$

We can relate the displacements in the cartesian ($u$, v, $w$) and cylindrical coordinates ($u_r$, $v_\theta$, $w$)as shown below.

$$u_r = u\,cos\theta + v\,sin\theta \qquad v_\theta = -u\,sin\theta + v\,cos\theta \tag{5.44g}$$

We derive the circular plate equations assuming all assumptions of Section 5.2 are applicable. Some of the algebraic details are skipped as these details are same as in Section 5.2.

**1. Displacements**

Substituting Equation (5.2) into Equation (5.44g) we obtain

$$u_r = \left(-z\frac{\partial w}{\partial x}\right)cos\theta + \left(-z\frac{\partial w}{\partial y}\right)sin\theta \qquad v_\theta = -\left(-z\frac{\partial w}{\partial x}\right)sin\theta + \left(-z\frac{\partial w}{\partial y}\right)cos\theta \qquad w \approx w(r,\theta)$$

Substituting Equation (5.44c) into the above equation we obtain

$$u_r = -z\left(\frac{\partial w}{\partial r}\right) \qquad v_\theta = -\frac{z}{r}\left(\frac{\partial w}{\partial \theta}\right) \qquad w \approx w(r,\theta) \tag{5.45}$$

**2. Strains**

Substituting Equation (5.45) into Equations (1.31a) through (1.31c) we obtain the strains as

$$\varepsilon_{rr} = \frac{\partial u_r}{\partial r} = -z\left(\frac{\partial^2 w}{\partial r^2}\right) \qquad \varepsilon_{\theta\theta} = \frac{u_r}{r} + \frac{1}{r}\frac{\partial v_\theta}{\partial \theta} = -z\left(\frac{1}{r}\frac{\partial w}{\partial r} + \frac{1}{r^2}\frac{\partial^2 w}{\partial \theta^2}\right) \qquad \gamma_{r\theta} = \frac{1}{r}\frac{\partial u_r}{\partial \theta} + \frac{\partial v_\theta}{\partial r} - \frac{v_\theta}{r} = -2z\left(\frac{1}{r}\frac{\partial^2 w}{\partial r\partial \theta} - \frac{1}{r^2}\frac{\partial w}{\partial \theta}\right) \tag{5.46}$$

**3. Stresses**

For plane stress, the stresses can be written as

$$\sigma_{rr} = \frac{E}{(1-\nu^2)}(\varepsilon_{rr} + \nu\varepsilon_{\theta\theta}) \qquad \sigma_{\theta\theta} = \frac{E}{(1-\nu^2)}(\varepsilon_{\theta\theta} + \nu\varepsilon_{rr}) \qquad \tau_{r\theta} = G\gamma_{r\theta}$$

$$\sigma_{rr} = \frac{-zE}{(1-\nu^2)}\left(\frac{\partial^2 w}{\partial r^2} + \frac{\nu}{r}\frac{\partial w}{\partial r} + \frac{\nu}{r^2}\frac{\partial^2 w}{\partial \theta^2}\right) \qquad \sigma_{\theta\theta} = \frac{-zE}{(1-\nu^2)}\left(\frac{1}{r}\frac{\partial w}{\partial r} + \frac{1}{r^2}\frac{\partial^2 w}{\partial \theta^2} + \nu\frac{\partial^2 w}{\partial r^2}\right) \qquad \tau_{r\theta} = -2Gz\left(\frac{1}{r}\frac{\partial^2 w}{\partial r\partial \theta} - \frac{1}{r^2}\frac{\partial w}{\partial \theta}\right) \tag{5.47}$$

**4. Stress Resultants**

We can replace the stresses with equivalent forces and moments as shown in Equation (5.48). Figure 5.16 shows the positive directions of the stress resultants.

$$\boldsymbol{m}_{rr} = \int_h \sigma_{rr}dz \qquad \boldsymbol{m}_{\theta\theta} = \int_h \sigma_{\theta\theta}dz \qquad \boldsymbol{m}_{r\theta} = \int_h \tau_{r\theta}dz \qquad \boldsymbol{m}_{\theta r} = \int_h \tau_{\theta r}dz \qquad \boldsymbol{q}_r = \int_h \tau_{rz}dz \qquad \boldsymbol{q}_\theta = \int_h \tau_{\theta z}dz \tag{5.48}$$

Substituting Equation (5.47) into Equation (5.48) we obtain the internal moments as shown below.

$$\boldsymbol{m}_{rr} = -D\left(\frac{\partial^2 w}{\partial r^2} + \frac{\nu}{r}\frac{\partial w}{\partial r} + \frac{\nu}{r^2}\frac{\partial^2 w}{\partial \theta^2}\right) \qquad \boldsymbol{m}_{\theta\theta} = -D\left(\frac{1}{r}\frac{\partial w}{\partial r} + \frac{1}{r^2}\frac{\partial^2 w}{\partial \theta^2} + \nu\frac{\partial^2 w}{\partial r^2}\right) \qquad \boldsymbol{m}_{r\theta} = -D(1-\nu)\left(\frac{1}{r}\frac{\partial^2 w}{\partial r\partial \theta} - \frac{1}{r^2}\frac{\partial w}{\partial \theta}\right) \tag{5.49}$$

Substituting Equation (5.49) into Equation (5.47) we obtain the stresses as shown below.

$$\sigma_{rr} = \frac{\boldsymbol{m}_{rr}z}{(h^3/12)} \qquad \sigma_{\theta\theta} = \frac{\boldsymbol{m}_{\theta\theta}z}{(h^3/12)} \qquad \tau_{r\theta} = \frac{\boldsymbol{m}_{r\theta}z}{(h^3/12)} \tag{5.50}$$

**5. Equilibrium and differential equation**

We could consider a free body diagram of a differential element and write the equilibrium equations as we did in Section 5.2. However we will take a simpler alternative route. We note that $\boldsymbol{q}_x$, $\boldsymbol{q}_y$ and $\boldsymbol{q}_r$, $\boldsymbol{q}_\theta$ are vectors in Cartesian and cylindrical coordinate system related as were the displacements as in Equation (5.44g).

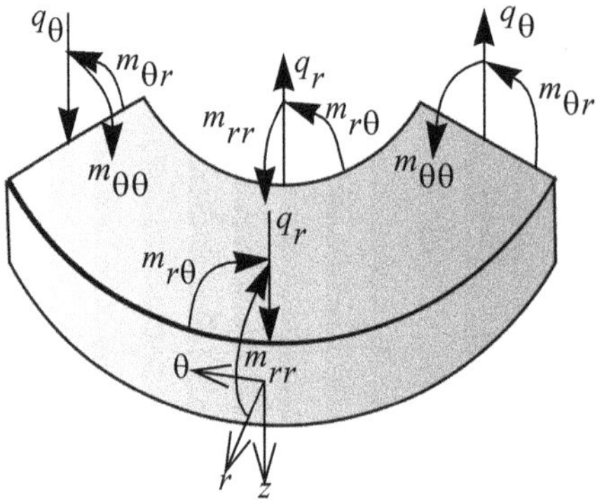


**Figure 5.16** Positive stress resultants.

We also can substitute for $q_x$, $q_y$ given by Equation (5.19) and obtain the results for $q_r$, $q_\theta$ as below.

$$q_r = q_x cos\theta + q_y sin\theta = -D\frac{\partial}{\partial x}(\nabla^2 w)cos\theta - D\frac{\partial}{\partial y}(\nabla^2 w)sin\theta$$

$$q_\theta = -q_x sin\theta + q_y cos\theta = D\frac{\partial}{\partial x}(\nabla^2 w)sin\theta - D\frac{\partial}{\partial y}(\nabla^2 w)cos\theta$$

Substituting Equation (5.44c) into the above equations we obtain

$$q_r = -D\frac{\partial}{\partial r}(\nabla^2 w) \qquad q_\theta = -\frac{D}{r}\frac{\partial}{\partial\theta}(\nabla^2 w) \tag{5.51}$$

In a similar manner the resultant shear force on the boundary is

$$V_r = q_r + \frac{1}{r}\frac{\partial m_{r\theta}}{\partial\theta} \qquad V_\theta = q_\theta + \frac{\partial m_{r\theta}}{\partial r} \tag{5.52}$$

The differential equation is the bi harmonic equations written as shown below.

$$\nabla^2\nabla^2 w = p_z(r, \theta) \qquad \text{where} \qquad \nabla^2 = \frac{1}{r}\frac{\partial}{\partial r}\left[r\frac{\partial}{\partial r}\right] + \frac{1}{r^2}\frac{\partial^2}{\partial\theta^2} \tag{5.53}$$

**6. Boundary conditions**

The boundary conditions on an edge follows the basic rule that either resultant shear force $V_r$ is specified or the deflection $w$, *and* either the normal moment $m_{rr}$ is specified or the normal slope $\partial w/\partial r$ as shown below.

$$\text{At each boundary point specify: } [V_r \quad \text{or} \quad w] \quad \text{and} \quad \left[m_{rr} \quad \text{or} \quad \frac{\partial w}{\partial r}\right] \tag{5.54}$$

The one exception to the above is a solid circular plate, which has only one boundary. We need two more conditions. These two conditions are that deflection and slope are bounded in the center. We will see this type of boundary condition in the next section.

## 5.6 AXISYMMETRIC PLATES

In this section we consider a simplified set of equations from the general equations of circular plates. We assume loading, material property, and geometry are independent of the angular coordinate $\theta$. This is an axisymmetric problem in which solution is independent of angular coordinate $\theta$, that is, $w = w(r)$. Hence $\partial w/\partial\theta = 0$ and $\partial w/\partial r = dw/dr$. Substituting this result in the equations of Section 5.5 we obtain the equations given below.

Displacements
$$u_r = -z(dw/dr) \qquad v_\theta = 0 \qquad w \approx w(r) \tag{5.55a}$$

Strain
$$\varepsilon_{rr} = -z(d^2w/dr^2) \qquad \varepsilon_{\theta\theta} = -\frac{z}{r}(dw/dr) \tag{5.55b}$$

Moments
$$m_{rr} = -D\left(\frac{d^2w}{dr^2} + \frac{\nu}{r}\frac{\partial w}{\partial r}\right) \qquad m_{\theta\theta} = -D\left(\frac{1}{r}\frac{dw}{dr} + \nu\frac{d^2w}{dr^2}\right) \qquad m_{r\theta} = 0 \tag{5.55c}$$

Laplace Operator
$$\nabla^2 = \frac{1}{r}\frac{d}{dr}\left(r\frac{d}{dr}\right) \tag{5.55d}$$

Stresses
$$\sigma_{rr} = \frac{m_{rr}z}{(h^3/12)} \qquad \sigma_{\theta\theta} = \frac{m_{\theta\theta}z}{(h^3/12)} \qquad \tau_{r\theta} = 0 \tag{5.55e}$$

Shear forces
$$q_r = -D\frac{d}{dr}\left\{\frac{1}{r}\frac{d}{dr}\left(r\frac{dw}{dr}\right)\right\} \qquad q_\theta = 0 \tag{5.55f}$$

Resultant shear forces $\quad V_r = q_r \qquad V_\theta = 0$ **(5.55g)**

Differential equation $\quad \frac{D}{r}\frac{d}{dr}\left[r\frac{d}{dr}\left\{\frac{1}{r}\frac{d}{dr}\left(r\frac{dw}{dr}\right)\right\}\right] = p_z(r)$ **(5.55h)**

Equation (5.55h) is an ordinary differential equation that can be integrated four times to obtain the solution as shown below.

$$D\left[r\frac{d}{dr}\left\{\frac{1}{r}\frac{d}{dr}\left(r\frac{dw}{dr}\right)\right\}\right] = I_1(r) + C_1 \qquad \text{where} \qquad I_1(r) = \int rp_z(r)dr \tag{5.56a}$$

$$D\frac{1}{r}\frac{d}{dr}\left(r\frac{dw}{dr}\right) = I_2(r) + C_1 lnr + C_2 \qquad \text{where} \qquad I_2(r) = \int \frac{I_1(r)}{r}dr \tag{5.56b}$$

$$D\left(r\frac{dw}{dr}\right) = I_3(r) + C_1\left(\frac{r^2}{2}lnr - \frac{r^2}{4}\right) + C_2\frac{r^2}{2} + C_3 \qquad \text{where} \qquad I_3(r) = \int rI_2(r)dr \tag{5.56c}$$

$$Dw(r) = I_4(r) + \frac{C_1}{4}(r^2 lnr - r^2) + C_2\frac{r^2}{4} + C_3 lnr + C_4 \qquad \text{where} \qquad I_4(r) = \int \frac{I_3(r)}{r}dr \tag{5.56d}$$

The indefinite integrals are called the **loading integrals** and can be found as $p_z(r)$ is known; $C_1$, $C_2$, $C_3$, $C_4$ are constants to be determined from the boundary conditions. The particular solution is $I_4(r)/D$ and the homogeneous solution is $[(C_1/4)(r^2 lnr - r^2) + C_2(r^2/2) + C_3 lnr + C_4]/D$. By substituting Equation (5.56d) into Equations (5.55c) and (5.55f) we obtain the solutions for moments and shear force as shown below.

$$m_{rr} = -[I_2 - (1-\nu)(I_3/r^2)] - (C_1/2)[(1+\nu)lnr + (1-\nu)/2] - C_2(1+\nu)/2 + C_3(1-\nu)/r^2 \tag{5.56e}$$

$$m_{\theta\theta} = -[(1-\nu)(I_3/r^2) + \nu I_2] - (C_1/2)[(1+\nu)lnr - (1-\nu)/2] - C_2(1+\nu)/2 - C_3(1-\nu)/r^2 \tag{5.56f}$$

$$q_r = -(I_1 + C_1)/r \tag{5.56g}$$

## 5.6.1 Uniform load

Uniform load, i.e, $p_z(r) = p_o$ is an approximation used in many practical applications. We assume the load acts on the entire plate. We obtain the loading integrals by successive integration as shown below.

$$I_1(r) = p_o\int rdr = \frac{p_o r^2}{2} \qquad I_2(r) = \int\frac{I_1}{r}dr = \frac{p_o r^2}{4} \qquad I_3(r) = \int rI_2 dr = \frac{p_o r^4}{16} \qquad I_4(r) = \int\frac{I_3}{r}dr = \frac{p_o r^4}{64} \tag{5.57}$$

In the above equations no integration constants are included as these are accounted in the homogenous solution. If the loading is only on part of the plate, then the loading integrals will have to be re-evaluated. Numerical integration could be used for more complex loadings.

## 5.6.2 Solid circular plates

Solid circular plates shown in Figure 5.15 has only one boundary, hence we have only two boundary conditions for four unknown constants. We overcome this problem by requiring that displacement $w$ and slope $dw/dr$ must be bounded at $r = 0$. This requirement of boundedness for deflection and slope are called **regularity conditions**. The function $lnr$ would tend to infinity as $r$ tend to zero, thus if deflection and slope are to remain bounded then the constants multiplying $lnr$ must be zero. Thus, from Equations (5.56c) and (5.56d) we obtain:

$$C_1 = 0 \qquad \text{and} \qquad C_3 = 0 \tag{5.58}$$

The remaining two constants can be obtained from the two boundary conditions at the outer boundary as demonstrated in Example 5.7.

### EXAMPLE 5.7

The edge view of a solid circular plate that is simply supported and loaded by a moment is shown in Figure 5.17. Determine the deflection and bending moments $m_{rr}$ and $m_{\theta\theta}$ in terms of $D$, $\nu$, $r$, $M_o$, and $R_o$.

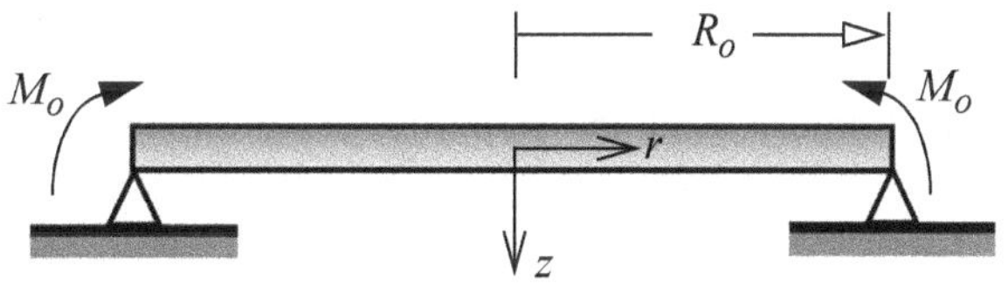


**Figure 5.17** Simply supported solid circular plate in Example 5.7.

**PLAN**

The loading integrals are all zero as the distributed load is zero. The boundary conditions at $r = R_o$ are $w = 0$ and $m_{rr} = M_o$. Using these conditions the two constants $C_2$ and $C_4$ can be found and the desired solution obtained.

**SOLUTION**

Substituting zero for all loading integrals and Equation (5.58) into Equations (5.56d), (5.56e), and (5.56f) we obtain

$$Dw(r) = C_2\frac{r^2}{4} + C_4 \qquad m_{rr} = -\frac{C_2(1+\nu)}{2} \qquad m_{\theta\theta} = -\frac{C_2(1+\nu)}{2} \tag{E1}$$

From the boundary conditions we obtain the following:

$$m_{rr}(r = R_o) = -C_2(1+\nu)/2 = M_o \qquad \text{or} \qquad C_2 = -2M_o/(1+\nu) \tag{E2}$$

$$w(r = R_o) = C_2\frac{R_o^2}{4} + C_4 = 0 \qquad \text{or} \qquad C_4 = \frac{M_o R_o^2}{2(1+\nu)} \tag{E3}$$

Substituting Equations (E2) and (E3) into Equation (E1), we obtain the solution as given below.

$$\textbf{ANS.} \quad w(r) = \frac{M_o(R_o^2 - r^2)}{2D(1+\nu)} \qquad m_{rr} = m_{\theta\theta} = -M_o$$

**COMMENTS**

1. The deflection solution shows a parabolic dishing of the plate, with the maximum value at the center ($r = 0$).
2. The radial and tangential bending moments are constants over the entire plate with a magnitude of the applied moment. This is pure bending of the plate.
3. The applied moment $M_o$ is moment per unit length, which is consistent with the units of moment resultants. A further check can be done with the dimensions of deflection on the left and right hand side of the answer.

## 5.6.3 Annular plates

The geometry of an annular plate is shown in Figure 5.18. Two boundary conditions will be imposed on each boundary, resulting in four conditions that can be used to evaluate the four constants in Equation (5.56d). When the distributed load is zero, the loading integrals are zero and the algebra is straight forward. When the distributed force is non-zero the resulting algebra becomes tedious. Simplification of algebra can be done by non-dimensionalizing the variables and equations for annular plate as discussed in the next section.

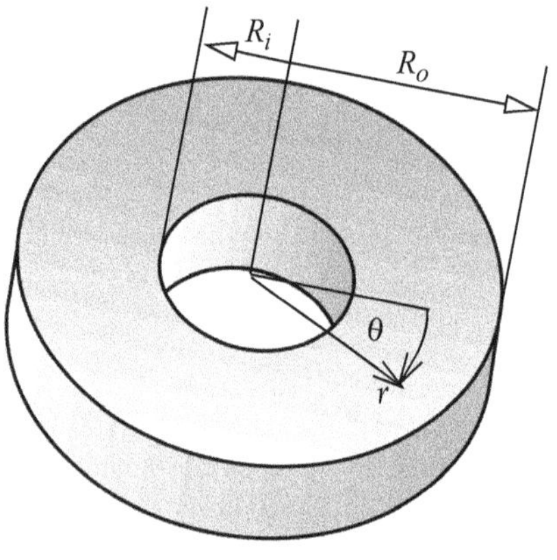


**Figure 5.18** Annular plate.

## 5.6.4 Non-dimensional equations in annular plates

The algebra for evaluating the constants in annular plates with distributed loads can be tedious. Simplification in the algebra can be achieved by non-dimensionalizing the variables. We start with non-dimensionalizing the radial coordinate $r$ with respect to the inner radius $R_i$ and the distributed load $p_z$ with respect to some characteristic value $p_o$ as shown below.

$$\frown{r} = r/R_i \qquad \frown{p_o} = p_z/p_o \tag{5.59a}$$

Variables with curved bars will refer to the non-dimensional variables. *When we impose boundary conditions on the inner boundary ($\frown{r} = 1$) the logarithm function will become zero and the equations will be simplified.* As was discussed in Section 3.1.5, we can define other non-dimensional variables using the definitions of the variables. The variables and the equations we need are given below in non-dimensional form. The loading integrals are given for uniform load.

$$R = \frac{R_0}{R_i} \qquad \frown{w} = \frac{Dw}{p_o R_i^4} \qquad \frac{d\frown{w}}{d\frown{r}} = \frac{D(dw/dr)}{p_o R_i^3} \qquad \frown{m_{rr}} = \frac{m_{rr}}{p_o R_i^2} \qquad \frown{m_{\theta\theta}} = \frac{m_{\theta\theta}}{p_o R_i^2} \qquad \frown{q_r} = \frac{q_r}{p_o R_i} \tag{5.59b}$$

$$\frown{I_1}(\frown{r}) = \frown{r}^2/2 \qquad \frown{I_2}(\frown{r}) = \frown{r}^2/4 \qquad \frown{I_3}(\frown{r}) = \frown{r}^4/16 \qquad \frown{I_4}(\frown{r}) = \frown{r}^4/64 \tag{5.59c}$$

$$\widehat{w}(\widehat{r}) = \widehat{I_4}(\widehat{r}) + A_1(\widehat{r}^2 ln\widehat{r} - \widehat{r}^2)/4 + A_2\widehat{r}^2/4 + A_3 ln\widehat{r} + A_4 \quad \textbf{(5.59d)}$$

$$\widehat{r}\frac{d\widehat{w}}{d\widehat{r}} = \widehat{I_3}(\widehat{r}) + A_1\left(\frac{\widehat{r}^2}{2}ln\widehat{r} - \frac{\widehat{r}^2}{4}\right) + A_2\frac{\widehat{r}^2}{2} + A_3 \quad \textbf{(5.59e)}$$

$$\widehat{\boldsymbol{m}}_{rr} = -[\widehat{I_2} - (\widehat{I_3}/\widehat{r}^2)(1-\nu)] - A_1[(1+\nu)ln\widehat{r} + (1-\nu)/2]/2 - \frac{A_2(1+\nu)}{2} + A_3(1-\nu)/\widehat{r}^2 \quad \textbf{(5.59f)}$$

$$\widehat{\boldsymbol{m}}_{\theta\theta} = -[(\widehat{I_3}/\widehat{r}^2)(1-\nu) + \nu\widehat{I_2}] - A_1[(1+\nu)ln\widehat{r} - (1-\nu)/2]/2 - A_2(1+\nu)/2 - A_3(1-\nu)/\widehat{r}^2 \quad \textbf{(5.59g)}$$

$$\widehat{\boldsymbol{q}_r} = -(\widehat{I_1} + A_1)/\widehat{r} \quad \textbf{(5.59h)}$$

where, $A_1, A_2, A_3,$ and $A_4$ are constants to be determine from boundary conditions. The above equations are valid for $1 \le \widehat{r} \le R$.

## EXAMPLE 5.8

The edge of an annular plate that is simply supported at the outer boundary and loaded with moments $M_i$ and $M_o$ is shown in Figure 5.19. Determine the deflection of the plate in terms of $r$, $D$, $R_i$, $R_o$, $M_i$, and $M_o$.

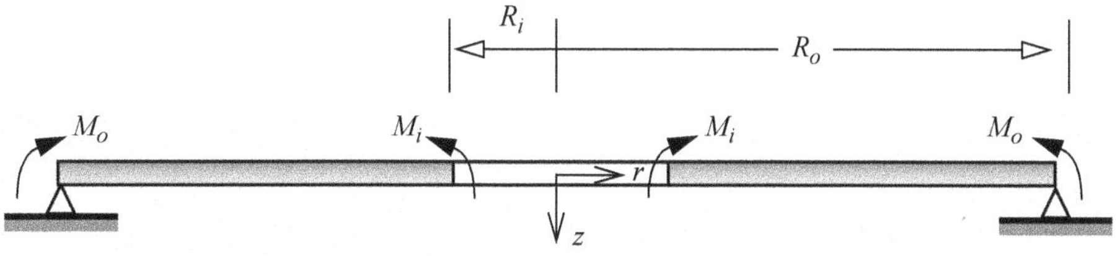


**Figure 5.19** A simply supported annular plate in Example 5.8.

**PLAN**

All loading integrals are zero as the distributed load is zero. The four boundary conditions are: shear force $\boldsymbol{q}_r = 0$ and moment $\boldsymbol{m}_{rr} = M_i$ are zero on the inner boundary and deflection $w = 0$ and moment $\boldsymbol{m}_{rr} = M_o$ are zero on the outer boundary. Using these four conditions the four constants in Equation (5.56d) can be found to determine the deflection solution.

**SOLUTION**

All loading integrals are zero. From Equation (5.56g) at the inner boundary we obtain the following.

$$\boldsymbol{q}_r = -C_1/R_i = 0 \quad \text{or} \quad C_1 = 0 \quad \text{(E1)}$$

From Equation (5.56e) on the inner and outer boundary we obtain

$$\boldsymbol{m}_{rr}(R_i) = -\frac{C_2(1+\nu)}{2} + \frac{C_3(1-\nu)}{R_i^2} = M_i \qquad \boldsymbol{m}_{rr}(R_o) = -\frac{C_2(1+\nu)}{2} + \frac{C_3(1-\nu)}{R_o^2} = M_o \quad \text{(E2)}$$

Solving Equations (E2) and (E2) we obtain

$$C_2 = \frac{2}{1+\nu}\left(\frac{R_o^2M_o - R_i^2M_i}{R_o^2 - R_i^2}\right) \qquad C_3 = \frac{R_i^2R_o^2}{1-\nu}\left(\frac{M_i - M_o}{R_o^2 - R_i^2}\right) \quad \text{(E3)}$$

From Equation (5.56d) om outer boundary we obtain

$$Dw(R_o) = C_2R_o^2/4 + C_3 lnR_o + C_4 = 0 \quad \text{or} \quad C_4 = -(C_2R_o^2/4 + C_3 lnR_o) \quad \text{(E4)}$$

Substituting Equations (E1), (E3) and (E4) into Equation (5.56d) we obtain the solution as shown below.

$$Dw(r) = C_2\frac{r^2}{4} + C_3 lnr - \left(C_2\frac{R_o^2}{4} + C_3 lnR_o\right) = C_2\frac{r^2 - R_o^2}{4} + C_3 ln\frac{r}{R_o} \quad \text{(E5)}$$

Substituting Equations (E3) and (E4) into Equation (E5), we obtain the solution below.

$$\textbf{ANS: } w(r) = -\left[\frac{R_o^2M_o - R_i^2M_i}{2D(1+\nu)}\right]\frac{(R_o^2 - r^2)}{R_o^2 - R_i^2} + \frac{M_i - M_o}{D(1-\nu)}\frac{R_i^2R_o^2}{R_o^2 - R_i^2}ln\frac{r}{R_o} \quad \text{(E6)}$$

**COMMENT**

1. If the simple support was moved from outer boundary to the inner boundary then the only constant that will change its value would be $C_4$, which corresponds to rigid body translation.

## EXAMPLE 5.9

An edge view of an annular plate that is attached to a rigid shaft and subjected to a uniform distributed load $p_o$ is shown in Figure 5.20. Determine the maximum deflection $w$ and maximum bending moments $m_{rr}$ and $m_{\theta\theta}$ in terms of $D$, $p_o$, and $R_o$. Use $\nu = 1/3$ and $R = 2$

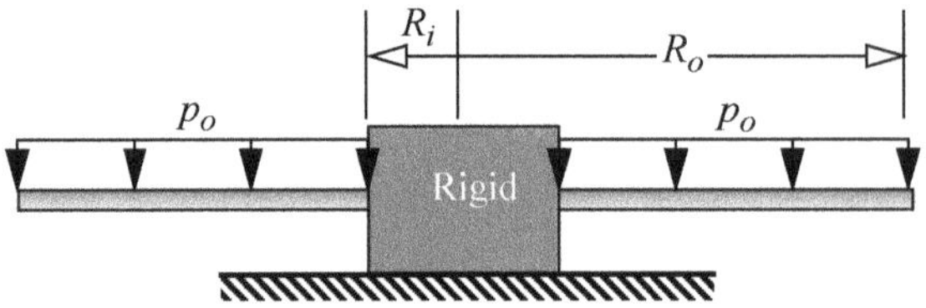


**Figure 5.20** Annular plate attached to a rigid shaft in Example 5.9

**PLAN**

The deflection and slope are zero at the inner boundary and radial shear force and radial moment are zero at the outer boundary. Using these conditions the four constants in Equation (5.59d) can be determined. The maximum deflection will be at the outer boundary and maximum moments will be on the inner boundary and can be determined once the constants of integration are known.

**SOLUTION**

The loading integrals can be determined at $\hat{r} = 1$ and $\hat{r} = R = 2$ from Equation (5.59c) as shown below.

$$\hat{I}_1(1) = 1/2 \qquad \hat{I}_2(1) = 1/4 \qquad \hat{I}_3(1) = 1/16 \qquad \hat{I}_4(1) = 1/64 \tag{E1}$$

$$\hat{I}_1(R) = R^2/2 = 2 \qquad \hat{I}_2(R) = R^2/4 = 1 \qquad \hat{I}_3(R) = R^4/16 = 1/4 \qquad \hat{I}_4(R) = R^4/64 = 1/4 \tag{E2}$$

Equating the radial shear force in Equation (5.59h) to zero at $\hat{r} = R = 2$ we obtain

$$\hat{q}_r(\hat{r} = R) = -(\hat{I}_1(R) + A_1)/R \qquad \text{or} \qquad A_1 = -2 \tag{E3}$$

Equating the radial bending moment in Equation (5.59f) to zero at $\hat{r} = R = 2$ we obtain

$$\hat{m}_{rr}(\hat{r} = R) = -\left[\hat{I}_2(R) - \frac{\hat{I}_3(R)(1-\nu)}{R^2}\right] - \frac{A_1}{2}\left[(1+\nu)\ln R + \frac{(1-\nu)}{2}\right] - \frac{A_2(1+\nu)}{2} + \frac{A_3(1-\nu)}{R^2} = 0 \tag{E4}$$

Substituting $\nu = 1/3$, $R = 2$, and $A_1 = -2$ we obtain

$$2A_2 - A_3/2 = 1.2726 \tag{E5}$$

Equating the deflection in Equation (5.59d) to zero at $\hat{r} = 1$ we obtain

$$\hat{w}(1) = \hat{I}_4(1) + \frac{A_1}{4}(-1^2) + A_2\frac{1}{4} + A_4 = 0 \qquad \text{or} \qquad A_2 + 4A_4 = -2.0625 \tag{E6}$$

Equating the slope in Equation (5.59e) to zero at $\hat{r} = 1$ we obtain

$$\frac{d\hat{w}}{d\hat{r}}(\hat{r} = 1) = \hat{I}_3(1) + A_1\left(-\frac{1}{4}\right) + A_2\frac{1}{2} + A_3 = 0 \qquad \text{or} \qquad A_2 + 2A_3 = -1.125 \tag{E7}$$

Solving Equations (E5) and (E7), we obtain

$$A_2 = 0.4406 \qquad A_3 = -0.7828 \tag{E8}$$

Substituting the above constants into Equation (E6), we obtain

$$A_4 = -0.6258 \tag{E9}$$

Substituting Equations (E8) and (E9) into Equation (5.59d) and evaluating it at $\hat{r} = R = 2$ we obtain the maximum non-dimensional deflection as shown below.

$$\hat{w}(\hat{r} = 2) = \hat{I}_4(2) + \frac{A_1}{4}(2^2\ln 2 - 2^2) + A_2\frac{2^2}{4} + A_3\ln 2 + A_4 = 0.1359 \tag{E10}$$

Substituting Equations (E8) and (E9) into Equations (5.59f) and (5.59g) and evaluating it at $\hat{r} = 1$ we obtain the maximum non-dimensional moments as shown below.

$$\hat{m}_{rr}(\hat{r} = 1) = -[\hat{I}_2(1) - (1 - 1/3)\hat{I}_3(1)] - \frac{A_1}{2}\left[\frac{(1 - 1/3)}{2}\right] - \frac{A_2(1 + 1/3)}{2} + A_3(1 - 1/3) = -0.6906 \tag{E11}$$

$$\hat{m}_{\theta\theta}(\hat{r} = 1) = -[(1 - 1/3)\hat{I}_3(1) + (1/3)\hat{I}_2] - \frac{A_1}{2}\left[-\frac{(1 - 1/3)}{2}\right] - \frac{A_2(1 + 1/3)}{2} - A_3(1 - 1/3) = -0.2302 \tag{E12}$$

The above results can be re-dimensionalized to obtain the maximum deflection and moments as shown below.

$$w_{max} = \left(\frac{p_o R_i^4}{D}\right)\hat{w}(\hat{r} = 2) = \left(\frac{p_o R_o^4}{D}\right)\left(\frac{R_i}{R_o}\right)^4(0.1359) = \left(\frac{p_o R_o^4}{D}\right)\left(\frac{1}{2}\right)^4(0.1359) = 8.496(10^{-3})\frac{p_o R_o^4}{D} \tag{E13}$$

$$(\boldsymbol{m}_{rr})_{max} = (p_o R_i^2)\widehat{\boldsymbol{m}}_{rr}(\widehat{r} = 1) = (p_o R_o^2)\left(\frac{R_i}{R_o}\right)^2(-0.6906) = (p_o R_o^2)\left(\frac{1}{2}\right)^2(-0.6906) = -0.1726 p_o R_o^2 \quad \text{(E14)}$$

$$(\boldsymbol{m}_{\theta\theta})_{max} = (p_o R_i^2)\widehat{\boldsymbol{m}}_{\theta\theta}(\widehat{r} = 1) = (p_o R_o^2)\left(\frac{R_i}{R_o}\right)^2(-0.2302) = (p_o R_o^2)\left(\frac{1}{2}\right)^2(-0.2302) = -0.05755 p_o R_o^2 \quad \text{(E15)}$$

**ANS.** $w_{max} = 8.496(10^{-3})\dfrac{p_o R_o^4}{D}$ $\quad (\boldsymbol{m}_{rr})_{max} = -0.1726 p_o R_o^2$ $\quad (\boldsymbol{m}_{\theta\theta})_{max} = -0.05755 p_o R_o^2$

**COMMENTS**

1. In this example we were able to determine the location of maximum deflection and moment by inspection. For more complex situations, we would plot $w$, $\boldsymbol{m}_{rr}$ and $\boldsymbol{m}_{\theta\theta}$ vs. $\widehat{r}$ from 1 to $R$ and determine the maximum value from the plot.
2. Timoshenko and Woinowsky-Krieger (1959) present a table for several annular plate for various values of $R$. The table is based on the work of Wahl and Lobo (1930). The table uses Poisson's ratio of $\nu = 0.3$. Another difference is they use modulus of elasticity $E$ rather than plate rigidity $D$ for nondimensionalizing. Once these differences are accounted for, the results of this example are same as the one they report.
3. Note the simplification in Equations (E6) and (E7) because of $ln(\widehat{r})$ being zero at the inner boundary.

## 5.7 CLOSURE

In this chapter we once more saw the modularity of logic shown in Figures 2.7. We used the logic to develop the simplest possible theory (classical theory) and identified the assumptions needed for simplification. We then incorporated complexities in examples and problems to develop new formulas and checked the validity of these formulas by appropriate substitution of variables to reproduce results of the classical plate theory.

For classical plate theories we also developed solution techniques by imposing limitations on geometry, loading, and boundary conditions. Navier's solution is restricted to rectangular plates with simply supported edges on all sides. Nadai-Levy solution is restricted to long rectangular plates with two opposite boundaries on the short side to be simply supported. By restricting loading, geometry, and material properties to be axisymmetric we were able to find solution for circular plates. By nondimensionalizing annular plate equations we were able to reduce algebra in finding solutions with variety of boundary conditions.

We will see derivations of plate equations with complexities in Chapter 7 from energy perspective. We will do the derivation using variational calculus, which will let us develop equations for plates with arbitrary shapes. By combining variational calculus and indicial notation in Chapter 8 we will see a very compact and elegant way of obtaining plate equations. It will let us obtain equations of bending of plate with large strains.

## 5.8 SYNOPSIS OF EQUATIONS

| | Classical plate theory |
|---|---|
| Displacements | $u \approx -z(\partial w/\partial x)$ $\quad v \approx -z(\partial w/\partial y)$ $\quad w \approx w(x, y)$ |
| Strains | $\varepsilon_{xx} = -z(\partial^2 w/\partial x^2)$ $\quad \varepsilon_{xx} = -z(\partial^2 w/\partial y^2)$ $\quad \gamma_{xy} = -2z(\partial^2 w/\partial x \partial y)$ |
| Stresses | $\sigma_{xx} = \dfrac{-Ez}{(1-\nu^2)}(\partial^2 w/\partial x^2 + \nu\partial^2 w/\partial y^2)$ $\quad \sigma_{yy} = \dfrac{-Ez}{(1-\nu^2)}(\partial^2 w/\partial y^2 + \nu\partial^2 w/\partial x^2)$ $\quad \tau_{xy} = -2Gz(\partial^2 w/\partial x \partial y)$ |
| Stress Resultants | $\boldsymbol{n}_{xx} = \int_h \sigma_{xx} dz = 0$ $\quad \boldsymbol{n}_{yy} = \int_h \sigma_{yy} dz$ $\quad \boldsymbol{n}_{xy} = \boldsymbol{n}_{yx} = \int_h \tau_{xy} dz = 0$<br>$\boldsymbol{m}_{xx} = \int_h z\sigma_{xx} dz = -D[\partial^2 w/\partial x^2 + \nu\partial^2 w/\partial y^2]$ $\quad \boldsymbol{m}_{yy} = \int_h z\sigma_{yy} dz = -D[\partial^2 w/\partial y^2 + \nu\partial^2 w/\partial x^2]$<br>$\boldsymbol{m}_{xy} = \boldsymbol{m}_{yx} = \int_h z\tau_{xy} dz = -D(1-\nu)(\partial^2 w/\partial x \partial y)$ $\quad D = \dfrac{Eh^3}{12(1-\nu^2)}$ $\quad \boldsymbol{q}_x = \int_h \tau_{xz} dz$ $\quad \boldsymbol{q}_y = \int_h \tau_{yz} dz$ |
| Stress Formulas | $\sigma_{xx} = \dfrac{\boldsymbol{m}_{xx} z}{(h^3/12)}$ $\quad \sigma_{yy} = \dfrac{\boldsymbol{m}_{yy} z}{(h^3/12)}$ $\quad \tau_{xy} = \dfrac{\boldsymbol{m}_{xy} z}{(h^3/12)}$ |
| Equilibrium Equations | $\dfrac{\partial \boldsymbol{q}_x}{\partial x} + \dfrac{\partial \boldsymbol{q}_y}{\partial y} = -p_z(x, y)$ $\quad \boldsymbol{q}_x = \dfrac{\partial \boldsymbol{m}_{xx}}{\partial x} + \dfrac{\partial \boldsymbol{m}_{yx}}{\partial y} = -D\dfrac{\partial}{\partial x}(\nabla^2 w)$ $\quad \boldsymbol{q}_y = \dfrac{\partial \boldsymbol{m}_{xy}}{\partial x} + \dfrac{\partial \boldsymbol{m}_{yy}}{\partial y} = -D\dfrac{\partial}{\partial y}(\nabla^2 w)$ |
| Differential Equation. | $D\nabla^4 w = D\left[\dfrac{\partial^4 w}{\partial x^4} + 2\dfrac{\partial^4 w}{\partial x^2 \partial y^2} + \dfrac{\partial^4 w}{\partial y^4}\right] = p_z(x, y)$ |
| Boundary Conditions | On $x$ = constant: Specify $[V_x$ or $w]$ and $\left[\boldsymbol{m}_{xx}\right.$ or $\left.\dfrac{\partial w}{\partial x}\right]$ $\quad V_x = \boldsymbol{q}_x + \dfrac{\partial \boldsymbol{m}_{xy}}{\partial y} = -D\left[\dfrac{\partial^3 w}{\partial x^3} + (2-\nu)\dfrac{\partial^3 w}{\partial x \partial y^2}\right]$<br>On $y$ = constant: Specify $[V_y$ or $w]$ and $\left[\boldsymbol{m}_{yy}\right.$ or $\left.\dfrac{\partial w}{\partial y}\right]$ $\quad V_y = \boldsymbol{q}_y + \dfrac{\partial \boldsymbol{m}_{yx}}{\partial x} = -D\left[\dfrac{\partial^3 w}{\partial y^3} + (2-\nu)\dfrac{\partial^3 w}{\partial y \partial x^2}\right]$<br>On a corner: Specify $[R_{corner}$ or $w]$ $\quad R_{corner} = \boldsymbol{m}_{xy} + \boldsymbol{m}_{yx} = 2\boldsymbol{m}_{xy} = -2D(1-\nu)\dfrac{\partial^2 w}{\partial x \partial y}$ |
| Orthogonality Conditions | $\int_0^\pi sin(m\theta)\,sin(n\theta)d\theta = \begin{cases} 0 & m \neq n \\ \pi/2 & m = n \end{cases}$ $\quad \int_0^\pi cos(m\theta)\,cos(n\theta)d\theta = \begin{cases} 0 & m \neq n \\ \pi/2 & m = n \end{cases}$ |
| | **Rectangular plates** |
| Navier's solution | $w(x, y) = \sum_{m=1}^{\infty}\sum_{n=1}^{\infty} W_{mn}\,sin(m\pi x/a)\,sin(n\pi y/b)$<br>$W_{mn} = \dfrac{1}{D\pi^4}\dfrac{P_{mn}}{[(m/a)^2 + (n/b)^2]^2}$ $\quad P_{mn} = \dfrac{4}{ab}\int_0^b\int_0^a p_z(x, y)\,sin(m\pi x/a)\,sin(n\pi y/a)dxdy$ |
| Nadai-Levi solution | $w(x, y) = w_p(x) + w_h(x, y)$; $d^4 w_p/dx^4 = p_z(x)/D$; $P_n = \dfrac{2}{a}\int_0^a p_z(x)\,sin(n\pi x/a)dx$<br>$w_p(x) = \sum_{m=1}^{\infty}\left\{(a/m\pi)^4\dfrac{P_m}{D}sin(m\pi x/a)\right\}$ ; $w_p = \dfrac{p_0 a^4}{24D}[(x/a)^4 - 2(x/a)^3 + (x/a)]$ ;<br>$w_h(x, y) = \sum_{m=1}^{\infty}\{[A_m + D_m(m\pi y/a)]cosh(m\pi y/a) + [C_m + B_m(m\pi y/a)]sinh(m\pi y/a)\}\,sin(m\pi x/a)$ |
| | **Circular plates** |
| Non-dimensional axisymmetric plate | $\widehat{r} = r/R_i$ $\quad \widehat{p_o} = p_z/p_o$ $\quad R = R_0/R_i$ $\quad \widehat{w} = Dw/(p_o R_i^4)$<br>$\dfrac{d\widehat{w}}{d\widehat{r}} = [D(dw/dr)]/(p_o R_i^3)$ $\quad \widehat{\boldsymbol{m}_{rr}} = \boldsymbol{m}_{rr}/(p_o R_i^2)$ $\quad \widehat{\boldsymbol{m}_{\theta\theta}} = \boldsymbol{m}_{\theta\theta}/(p_o R_i^2)$ $\quad \widehat{\boldsymbol{q}_r} = \boldsymbol{q}_r/p_o R_i$<br>$\widehat{I_1}(\widehat{r}) = \widehat{r}^2/2$ $\quad \widehat{I_2}(\widehat{r}) = \widehat{r}^2/4$ $\quad \widehat{I_3}(\widehat{r}) = \widehat{r}^4/16$ $\quad \widehat{I_4}(\widehat{r}) = \widehat{r}^4/64$<br>$\widehat{w}(\widehat{r}) = \widehat{I_4}(\widehat{r}) + (A_1/4)(\widehat{r}^2 ln\widehat{r} - \widehat{r}^2) + (A_2/4)\widehat{r}^2 + A_3 ln\widehat{r} + A_4$<br>$\widehat{r}(d\widehat{w}/d\widehat{r}) = \widehat{I_3}(\widehat{r}) + (A_1/2)(\widehat{r}^2 ln\widehat{r} - \widehat{r}^2/2) + (A_2/2)\widehat{r}^2 + A_3$<br>$\widehat{\boldsymbol{m}}_{rr} = -[\widehat{I_2} - (\widehat{I_3}/\widehat{r}^2)(1-\nu)] - (A_1/2)[(1+\nu)ln\widehat{r} + (1-\nu)/2] - (A_2/2)(1+\nu) + A_3(1-\nu)/\widehat{r}^2$<br>$\widehat{\boldsymbol{m}}_{\theta\theta} = -[(\widehat{I_3}/\widehat{r}^2)(1-\nu) + \nu\widehat{I_2}] - (A_1/2)[(1+\nu)ln\widehat{r} - (1-\nu)/\ ] - (A_2/2)(1+\nu) - A_3(1-\nu)/\widehat{r}^2$<br>$\widehat{\boldsymbol{q}_r} = -(\widehat{I_1} + A_1)/\widehat{r}$ |

# PROBLEMS

## Sections 5.1-5.2

In the problems below $E$ is modulus of elasticity, $\nu$ is the Poisson's ratio, $G$ is the shear modulus of elasticity, $h$ is the plate thickness, $D$ is the plate bending rigidity of the plate, $\rho_m$ is the mass density.

**5.1** Start with the displacement field

$$u = u_o(x,y) - z\left(\frac{\partial w}{\partial x}\right) \qquad v = v_o(x,y) - z\left(\frac{\partial w}{\partial y}\right) \qquad w = w(x,y) \tag{5.60}$$

where, $z$ is measured from the mid plane for a homogeneous plate. Assuming all assumptions in Section 5.2 are valid. Show that the inplane internal forces (stress resultants) are

$$n_{xx} = \frac{Eh}{(1-\nu^2)}\left[\frac{\partial u_o}{\partial x} + \nu\frac{\partial v_o}{\partial y}\right] \qquad n_{yy} = \frac{Eh}{(1-\nu^2)}\left[\frac{\partial v_o}{\partial y} + \nu\frac{\partial u_o}{\partial x}\right] \qquad n_{xy} = Gh\left[\frac{\partial u_o}{\partial y} + \frac{\partial v_o}{\partial x}\right] \tag{5.61}$$

**5.2** Using the results of Problem 5.1, show that the inplane stresses are given by

$$\sigma_{xx} = \frac{n_{xx}}{h} + 12\frac{m_{xx}z}{h^3} \qquad \sigma_{yy} = \frac{n_{yy}}{h} + 12\frac{m_{yy}z}{h^3} \qquad \tau_{xy} = \frac{n_{xy}}{h} + 12\frac{m_{xy}z}{h^3} \tag{5.62}$$

**5.3** A thin, homogeneous, isotropic plate of uniform thickness $h$ has a temperature distribution $T(x,y,z)$ and subjected to a uniform load $p_z(x,y)$. The plate material has $\alpha$ as the coefficient of thermal expansion. The plate is free to expand in the $x$ and $y$ direction and has no inplane load. Show that the differential equation governing the deflection w of the plate is given by

$$D\nabla^4 w = p_z(x,y) - \nabla^2 m_T \qquad \text{where} \qquad m_T(x,y) = \frac{E\alpha}{1-\nu}\int_{-h/2}^{h/2} zT(x,y,z)dz \tag{5.63}$$

**5.4** A cross section of laminated plate made from three materials is shown in Figure 5.4. All three material have the same Poisson's ratio but different modulus of elasticity. The displacement field is given

$$u \approx -(z - z_o)\left(\frac{\partial w}{\partial x}\right) \qquad v \approx -(z - z_o)\left(\frac{\partial w}{\partial y}\right) \qquad w \approx w(x,y)$$

where, $z$ is measured from the mid surface and $z_o$ is the location of the neutral surface. Determine the value of $z_o$ assuming all assumptions in Section 5.2 except for material homogeneity across the thickness are valid.

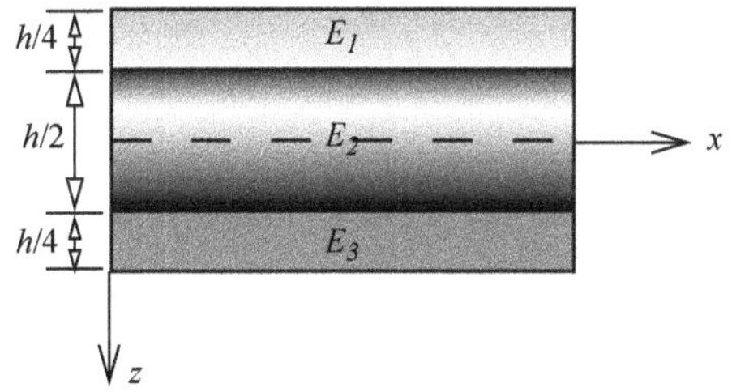


**Fig. P5.4** Laminated Plate.

**5.5** In problem 5.4, assume $E_1 = E_3 = 2E$ and $E_2 = E$. Show that the moment resultants are given by Equation (5.14) with $D$ replaced by $D_L$, where

$$D_L = 5Eh^3/[32(1-\nu^2)]$$

**5.6** Assuming rotational moment of inertia can be neglected show that the differential equation governing vibration of plates is given by

$$D\nabla^4 w + \rho_m h\frac{\partial^2 w}{\partial t^2} = p_z(x,y) \tag{5.64}$$

where, $\rho_m$ is the mass density of the plate material.

**5.7** In Von-Karman region (see limitation 7) the strains we use are

$$\varepsilon_{xx} = \frac{\partial u}{\partial x} + \frac{1}{2}\left(\frac{\partial w}{\partial x}\right)^2 \qquad \varepsilon_{yy} = \frac{\partial v}{\partial y} + \frac{1}{2}\left(\frac{\partial w}{\partial y}\right)^2 \qquad \gamma_{xy} = \frac{\partial u}{\partial y} + \frac{\partial v}{\partial x} + \frac{\partial w}{\partial x}\frac{\partial w}{\partial y} \tag{5.65}$$

Starting with the displacement field of Equation (5.2) and assuming z is measured from the mid-surface and assumptions from 6 through 10 are valid, show that the moment resultants are given by Equation (5.14) and the inplane force resultants are given by the following equations

$$n_{xx} = \frac{Eh}{(1-\nu^2)}\left[\left(\frac{\partial w}{\partial x}\right)^2 + \nu\left(\frac{\partial w}{\partial y}\right)^2\right] \tag{5.66}$$

$$n_{yy} = \frac{Eh}{(1-\nu^2)}\left[\left(\frac{\partial w}{\partial y}\right)^2 + \nu\left(\frac{\partial w}{\partial x}\right)^2\right] \tag{5.67}$$

$$n_{xy} = Gh\left(\frac{\partial w}{\partial x}\frac{\partial w}{\partial y}\right) \tag{5.68}$$

Note moment resultants are unchanged but bending also produces inplane force resultants. This coupling has to be reflected in equilibrium equations. Energy approach is used for deriving the differential equations and the associated boundary conditions.

**5.8** The simply supported rectangular plate shown in Figure 5.10 rests on an elastic foundation with a spring constant per unit area of k. The differential equation governing plate deflection is given by

$$D\nabla^4 w + kw = p_z(x,y) \tag{5.69}$$

Using Navier's approach show

$$W_{mn} = \frac{P_{mn}}{D\pi^4[(m/a)^2 + (n/b)^2]^2 + k} \tag{5.70}$$

where, $W_{mn}$ and $P_{mn}$ are the coefficients in the series defined in Equations (5.28) and (5.28).

**5.9** Write all the boundary conditions for the plate shown in Fig. P5.9 in terms of $E$, $\nu$, $h$, $a$, $b$, $w$, and its derivatives.

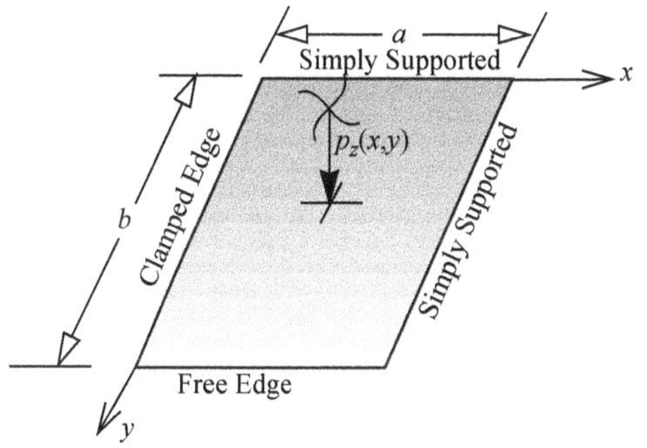


**Fig. P5.9**

**5.10** A tapered plate has a thickness that varies as $h(x) = (2 - x/a)t$ shown in Fig. P5.10. Write all the boundary conditions in terms of $E$, $\nu$, $t$, $a$, $b$, $w$, and its derivatives.

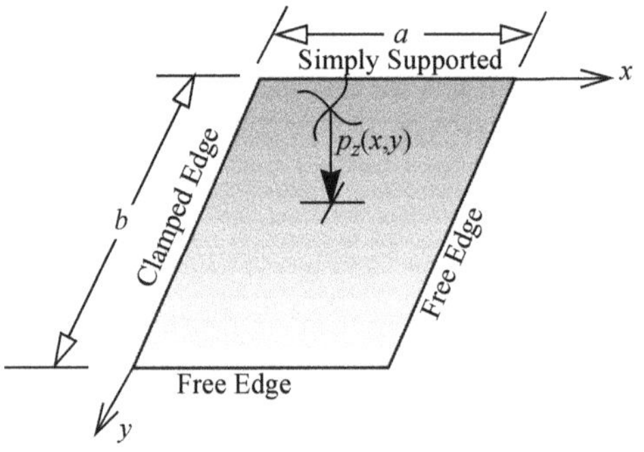


**Fig. P5.10**

**5.11** A tapered plate has a thickness that varies as $h(y) = (2 - y/b)t$ shown in Fig. P5.10. Write all the boundary conditions in terms of $E$, $\nu$, $t$, $a$, $b$, $w$, and its derivatives.

## Sections 5.3-5.4

**5.12** The rectangular plate Figure 5.10 is subjected to a transverse load that varies as $p_z = p_o(x/a)(y/b)(1 - x/a)(1 - y/b)$. Determine (*i*) the deflection $w$ and moment $m_{xx}$ field on the entire plate. (*ii*) the deflection $w$ and moment $m_{xx}$ at the center of the plate using at least 4 terms in the series assuming the plate is square i.e., $b = a$. Use $\nu = 1/3$.

**5.13** The rectangular plate Fig. P5.13 is subjected to a uniform transverse load $p_z = p_o$. Using Nadai-Levy method, determine (*i*) the deflection $w$ field for the entire plate. (*ii*) the deflection $w$ and moment $m_{xx}$ at the center of the plate using at least 4 terms in the series assuming the plate is square i.e., $b = a$. Use $\nu = 1/3$.

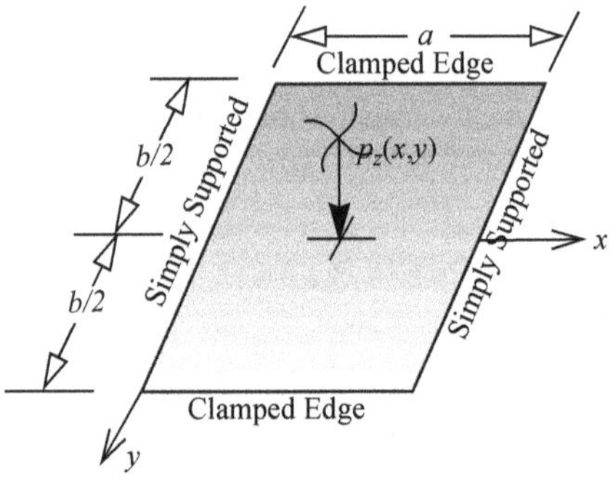


**Fig. P5.13**

## Sections 5.5-5.6

**5.14** The edge view of a solid circular plate that is simply supported and loaded with a uniform distributed load is shown in Fig. P5.14. Determine the deflection and bending moments $m_{rr}$ and $m_{\theta\theta}$ in terms of $D$, $\nu$, $r$, $p_o$, and $R_o$.

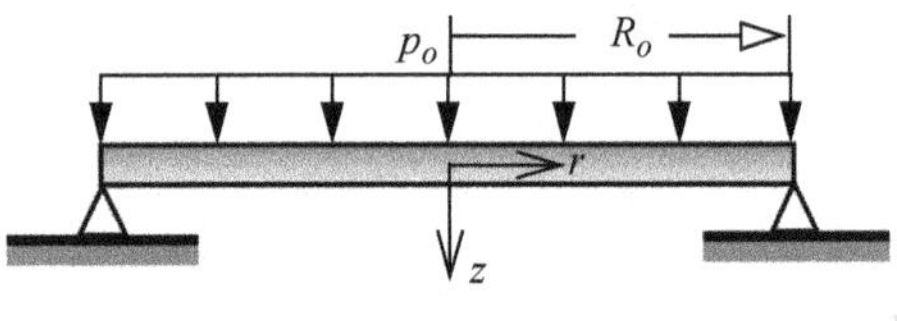


**Fig. P5.14**

**5.15** The edge view of a solid circular plate that is clamped at the outer boundary and loaded with a uniform distributed load is shown in Fig. P5.15. Determine the deflection and bending moments $m_{rr}$ and $m_{\theta\theta}$ in terms of $D$, $\nu$, $r$, $p_o$, and $R_o$.

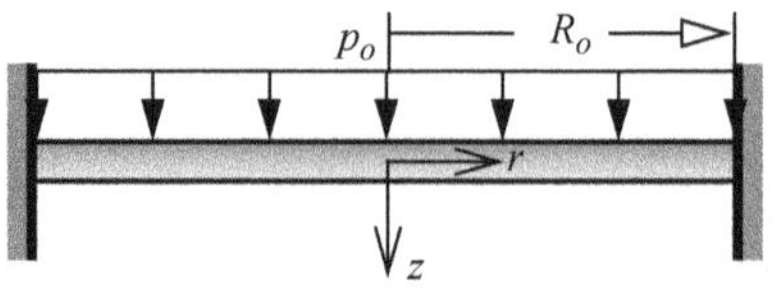


**Fig. P5.15**

**5.16** An edge view of an annular plate that is clamped at the outer edge and loaded with a uniform distributed load is shown in Fig. P5.16. Determine the maximum deflection $w$ and maximum bending moments $m_{rr}$ and $m_{\theta\theta}$ in terms of $D$, $p_o$, and $R_i$. Use $\nu = 1/3$ and $R = R_o/R_i = 2$.

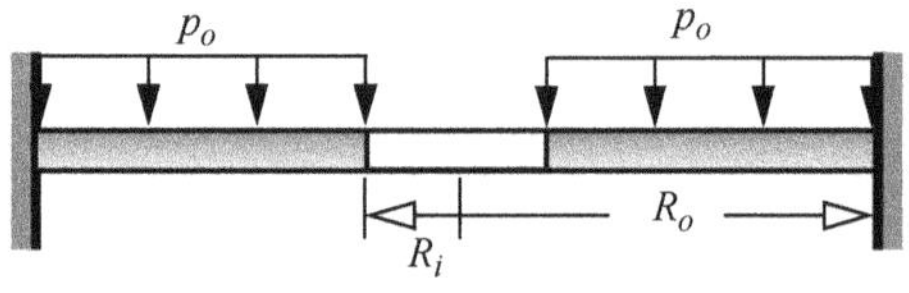


**Fig. P5.16**

**5.17** An edge view of an annular plate that is simply supported on the inner boundary and loaded with a uniform distributed load is shown in Fig. P5.17. Determine the maximum deflection $w$ and maximum bending moments $m_{rr}$ and $m_{\theta\theta}$ in terms of $D$, $p_o$, and $R_i$. Use $\nu = 1/3$ and $R = R_o/R_i = 2$.

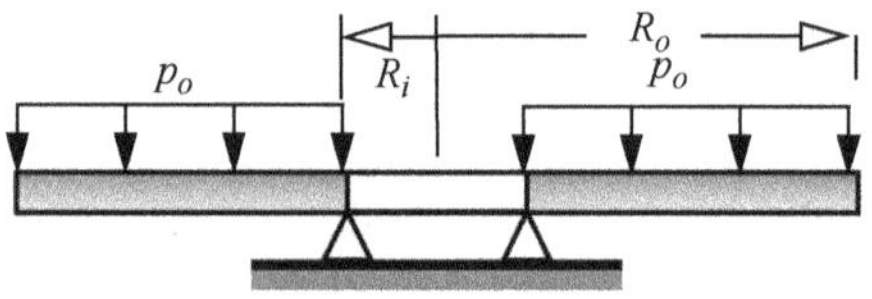


**Fig. P5.17**

**5.18** An edge view of an annular plate that is attached to a rigid shaft, clamped at the outer edge and pulled with a force $P$ as shown Fig. P5.18. Determine the maximum deflection $w$ and maximum bending moments $m_{rr}$ and $m_{\theta\theta}$ in terms of $D$, $P$, and $R_i$. Use $\nu = 1/3$ and $R = R_o/R_i = 2$. [Hint: Let $p_oR_i^2 = P$ in Equation (5.59b), then all the other equations are valid and can be solved with $P = 1$]

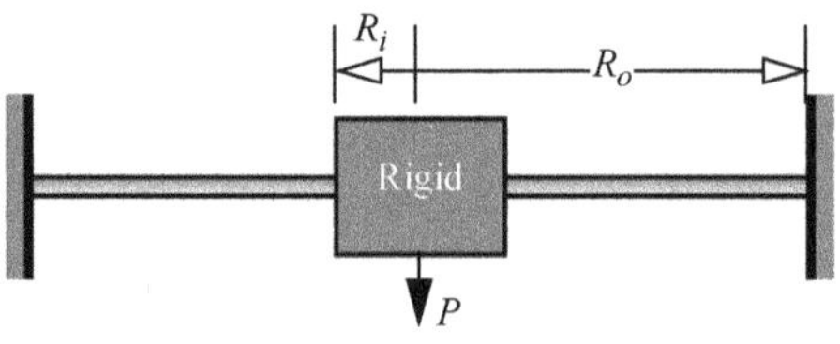


**Fig. P5.18**

# 6 Introduction to Elasticity

## LEARNING OBJECTIVES

1. To become familiar with the basic equations of linear elasticity and the Airy stress function.
2. To understand the application and relationship of the equations of elasticity to mechanics of materials.

Stress and stiffness analysis and design of three-dimensional (and many two-dimensional) machine components cannot be achieved by means of a mechanics of materials approach. Numerical methods are often used, but even these call for an understanding of equations of elasticity. In elasticity, we study how the variables of mechanics (displacements, strains, stresses, internal forces, and moments) vary from point to point on a elastic body. This is in contrast to mechanics of materials approach in which we average the variables of mechanics by approximating their variation across the cross section (rods, shafts, and beams) or thickness (plates and shells) of an elastic body. The field of elasticity is very large and beyond the scope of a single course, let alone a chapter. Here we introduce some basic concepts and equations of elasticity, and use the equations of elasticity to derive formulas for such applications as rotating disks and torsion of non-circular shafts

The basic equations of elasticity are the strain–displacement equations, the generalized version of Hooke's law relating stresses and strains (studied in Chapter 1), the equilibrium equations on stresses, and the compatibility equations for strains, which ensure single-valued solutions for displacements.

An important technique for solving an elasticity problem is the use Airy stress function from which stresses can be derived that satisfy the equilibrium equations. We will see how the Airy stress functions can be used to obtain results for several mechanics of materials problems.

## 6.1 ELASTICITY EQUATIONS

The basic equations of linear elasticity are the strain–displacement equations, the compatibility equations, which are derived from the strain–displacement equations, the generalized version of Hooke's law, and the equilibrium equations (see Barber [1992], Boresi and Chong [1987], and Fung [1965]). These are discussed in this section.

### 6.1.1 Strain-displacement equations

The strain–displacement relationships in cartesian coordinates and polar coordinates derived in Chapter 1, are repeated below for convenience.

$$\varepsilon_{xx} = \frac{\partial u}{\partial x} \qquad \varepsilon_{yy} = \frac{\partial v}{\partial y} \qquad \varepsilon_{zz} = \frac{\partial w}{\partial z} \qquad \gamma_{xy} = \gamma_{yx} = \frac{\partial u}{\partial y} + \frac{\partial v}{\partial x} \qquad \gamma_{yz} = \gamma_{zy} = \frac{\partial v}{\partial z} + \frac{\partial w}{\partial y} \qquad \gamma_{zx} = \gamma_{xz} = \frac{\partial w}{\partial x} + \frac{\partial u}{\partial z} \tag{6.1}$$

$$\varepsilon_{rr} = \frac{\partial u_r}{\partial r} \qquad \varepsilon_{\theta\theta} = \frac{u_r}{r} + \frac{1}{r}\frac{\partial v_\theta}{\partial \theta} \qquad \gamma_{r\theta} = \frac{1}{r}\frac{\partial u_r}{\partial \theta} + \frac{\partial v_\theta}{\partial r} - \frac{v_\theta}{r} \tag{6.2}$$

### 6.1.2 Compatibility equations

Six compatibility equations ensure that the strain fields are such that the displacement fields obtained from integrating Equation (6.1) are single valued. We derive one of the six compatibility equations by taking the derivatives of Equation (6.1) and interchanging their order as shown in Equation (6.3a).

$$\frac{\partial^2 \gamma_{xy}}{\partial x \partial y} = \frac{\partial^2}{\partial x \partial y}\left(\frac{\partial u}{\partial y} + \frac{\partial v}{\partial x}\right) = \frac{\partial^3 u}{\partial x \partial y^2} + \frac{\partial^3 v}{\partial x^2 \partial y} = \frac{\partial^2}{\partial y^2}\left(\frac{\partial u}{\partial x}\right) + \frac{\partial^2}{\partial x^2}\left(\frac{\partial v}{\partial y}\right) \tag{6.3a}$$

Substituting Equation (6.1) in the right-hand side of Equation (6.3a), we obtain Equation (6.3b).

$$\boxed{\frac{\partial^2 \varepsilon_{xx}}{\partial y^2} + \frac{\partial^2 \varepsilon_{yy}}{\partial x^2} = \frac{\partial^2 \gamma_{xy}}{\partial x \partial y}} \qquad \textbf{(6.3b)}$$

The other five compatibility equations can be derived in a similar manner (see Problems 6.1 and 6.2).

### 6.1.3 Plane stress and plane strain

The two-dimensional problems of plane stress and plane strain were discussed in Section 1.8.3. The equations of plane stress and plane strain can be derived from the equations of the generalized Hooke's law [see Equation (1.42)] and are stated in slightly different form for convenience.

For *plane stress* we have

$$\boxed{E\varepsilon_{xx} = \sigma_{xx} - \nu\sigma_{yy} \qquad E\varepsilon_{yy} = \sigma_{yy} - \nu\sigma_{xx} \qquad G\gamma_{xy} = \tau_{xy}} \qquad \textbf{(6.4a)}$$

Alternatively, Equation (6.4a) can be written as

$$\boxed{\sigma_{xx} = E[\varepsilon_{xx} + \nu\varepsilon_{yy}]/(1-\nu^2) \qquad \sigma_{yy} = E[\varepsilon_{yy} + \nu\varepsilon_{xx}]/(1-\nu^2) \qquad \tau_{xy} = G\gamma_{xy}} \qquad \textbf{(6.4b)}$$

For *plane strain* we have

$$\boxed{2G\varepsilon_{xx} = (1-\nu)\sigma_{xx} - \nu\sigma_{yy} \qquad 2G\varepsilon_{yy} = (1-\nu)\sigma_{yy} - \nu\sigma_{xx} \qquad G\gamma_{xy} = \tau_{xy}} \qquad \textbf{(6.5a)}$$

Alternatively, Equation (6.5a) can be written as:

$$\sigma_{xx} = \frac{2G}{(1-2\nu)}[(1-\nu)\varepsilon_{xx} + \nu\varepsilon_{yy}] \qquad \sigma_{yy} = \frac{2G}{(1-2\nu)}[(1-\nu)\varepsilon_{yy} + \nu\varepsilon_{xx}] \qquad \tau_{xy} = G\gamma_{xy} \qquad \textbf{(6.5b)}$$

The above equations are valid for any orthogonal system. By replacing $x$ with $r$ and $y$ with $\theta$, we can obtain the relationship between stresses and strains in polar coordinates.

The equations for plane stress and plane strain are similar in form. It can be verified that we can obtain the equations of plane stress by substituting $\nu/(1+\nu)$ in place of $\nu$ and keeping $G$ unchanged in plane strain equations. This device can be used in obtaining a solution for plane stress from the plane strain solution, as we shall describe in Section 6.2. We record this observation as Equation (6.6) for future use.

$$\boxed{\text{Plane strain} \quad \begin{matrix} \nu \to \nu/(1+\nu) \\ G \to G \end{matrix} \quad \text{Plane stress}} \qquad \textbf{(6.6)}$$

Another alternative to the form of plane stress and strain equations is

$$\varepsilon_{xx} = \frac{(\kappa+1)\sigma_{xx} - (3-\kappa)\sigma_{yy}}{8G} \qquad \varepsilon_{yy} = \frac{(\kappa+1)\sigma_{yy} - (3-\kappa)\sigma_{xx}}{8G} \qquad \text{where} \qquad \kappa = \begin{cases} \dfrac{3-\nu}{1+\nu} & \text{Plane Stress} \\ 3-4\nu & \text{Plane Strain} \end{cases} \qquad \textbf{(6.7a)}$$

Equation (6.7a) can be solved to obtain stresses in terms of strains as shown below.

$$\sigma_{xx} = \frac{G}{(\kappa-1)}[(\kappa+1)\varepsilon_{xx} + (3-\kappa)\varepsilon_{yy}] \qquad \sigma_{yy} = \frac{G}{(\kappa-1)}[(3-\kappa)\varepsilon_{xx} + (\kappa+1)\varepsilon_{yy}] \qquad \textbf{(6.7b)}$$

### 6.1.4 Equilibrium equations

Figure 6.1*a* shows a two-dimensional differential element in plane stress (or plane strain).We assume at each point there are acts body forces $F_x$ and $F_y$, and body moment $C_z$. Body forces are forces per unit volume, such as that due to gravity. In magnetic fields we can also have body moments which have units of moments per unit volume. Figure 6.1*b* shows the free body diagram obtained by multiplying stresses by the areas of the planes and body forces and moments multiplied by the differential volume

By force equilibrium in the $x$ direction in Figure 6.1*b*, we obtain Equation (6.8a).

$$\left(\sigma_{xx} + \frac{\partial\sigma_{xx}}{\partial x}dx\right)(dy\,dz) - (\sigma_{xx})(dy\,dz) + \left(\tau_{yx} + \frac{\partial\tau_{yx}}{\partial y}dy\right)(dx\,dz) - (\tau_{yx})(dx\,dz) + (F_x)(dx\,dy\,dz) = 0$$

$$\boxed{\frac{\partial\sigma_{xx}}{\partial x} + \frac{\partial\tau_{yx}}{\partial y} + F_x = 0} \qquad \textbf{(6.8a)}$$

Similarly by force equilibrium in the $y$ direction in Figure 6.1*b*, we obtain Equation (6.8b).

$$\boxed{\frac{\partial\tau_{xy}}{\partial x} + \frac{\partial\sigma_{yy}}{\partial y} + F_y = 0} \qquad \textbf{(6.8b)}$$

Consider the moment equilibrium about the center of the differential element in Figure 6.1*b*. The moment at the center will be from shear stresses only. We can neglect the terms containing the product of four differentials, since these terms tend to zero faster in the limit than the terms with the products of three differentials. As a result, we obtain Equation (6.8e).

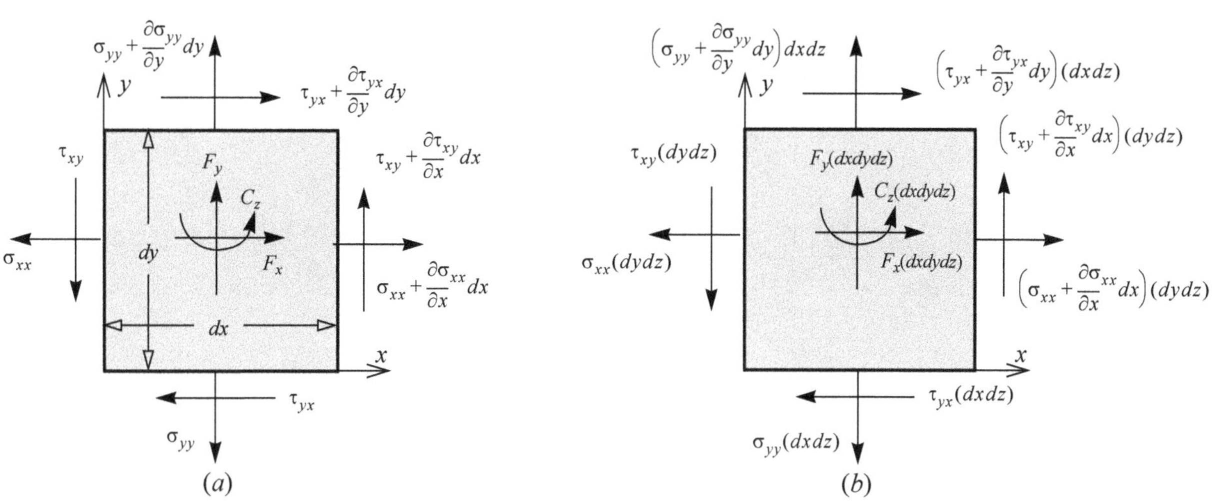


**Figure 6.1** (*a*) A two-dimensional differential element and (*b*) its free body diagram.

$$\left(\tau_{xy}+\frac{\partial\tau_{xy}}{\partial x}dx\right)(dy\,dz)\left(\frac{dx}{2}\right)+(\tau_{xy})(dy\,dz)\left(\frac{dx}{2}\right)-\left(\tau_{yx}+\frac{\partial\tau_{yx}}{\partial y}dy\right)(dx\,dz)\left(\frac{dy}{2}\right)-(\tau_{yx})(dx\,dz)\left(\frac{dy}{2}\right)+C_z dx\,dy\,dz = 0$$

$$(\tau_{xy})(dy\,dz)(dx)-(\tau_{yx})(dx\,dz)(dy)+C_z dx\,dy\,dz = 0$$

$$\boxed{\tau_{xy}+C_z = \tau_{yx}} \tag{6.8c}$$

The above equation shows that the symmetry of shear stresses which we have assumed so far is only possible in absence of body moment. We will continue to assume body moments are zero and not consider its impact in remaining developments.

Equation (6.8a), (6.8b), and (6.8e) are the equilibrium equations at a point of a two-dimensional elastic body in Cartesian coordinates. The equilibrium equations in three dimensions can be obtained by considering the equilibrium of a differential element (cube) in three dimensions and are given below.

$$\boxed{\frac{\partial\sigma_{xx}}{\partial x}+\frac{\partial\tau_{yx}}{\partial y}+\frac{\partial\tau_{zx}}{\partial z}+F_x = 0 \qquad \frac{\partial\tau_{xy}}{\partial x}+\frac{\partial\sigma_{yy}}{\partial y}+\frac{\partial\tau_{zy}}{\partial z}+F_y = 0 \qquad \frac{\partial\tau_{xz}}{\partial x}+\frac{\partial\tau_{yz}}{\partial y}+\frac{\partial\sigma_{zz}}{\partial z}+F_z = 0} \tag{6.8d}$$

$$\boxed{\tau_{xy} = \tau_{yx} \qquad \tau_{yz} = \tau_{zy} \qquad \tau_{zx} = \tau_{xz}} \tag{6.8e}$$

A differential element with stresses in polar coordinates is shown in Figure 6.2*a*. We assume the element extends by $dz$ in the $z$ direction. We multiply the stress by the differential area of the planes on which they are acting to produce a free body diagram of the differential element shown in Figure 6.2*b*. It should be noted that sides *AB* and *CD* are at an angle of $d\theta/2$ to the radial direction. Thus, the forces acting on sides *AB* and *CD* will have to be resolved into components forces in the $r$ and $\theta$ direction.

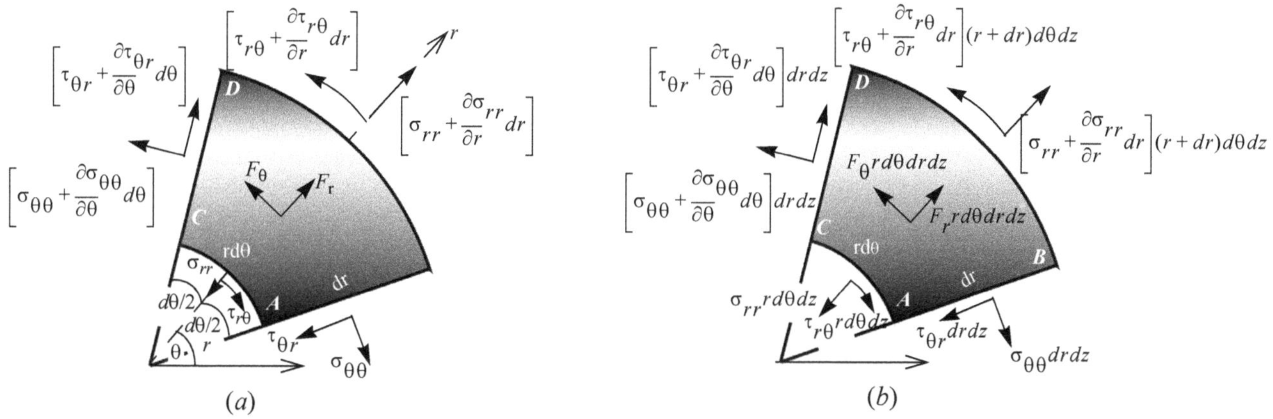


**Figure 6.2** Plane stress infinitesimal (a) stress element. (b) force element in polar coordinates

By equilibrium of forces in the $r$ direction we obtain:

$$\left[\sigma_{rr}+\frac{\partial\sigma_{rr}}{\partial r}dr\right](r+dr)d\theta dz-\sigma_{rr}rd\theta dz-\tau_{\theta r}drdz\cos\left(\frac{d\theta}{2}\right)+\left[\tau_{\theta r}+\frac{\partial\tau_{\theta r}}{\partial\theta}d\theta\right]drdz\cos\left(\frac{d\theta}{2}\right)$$

$$-\left[\sigma_{\theta\theta}+\frac{\partial\sigma_{\theta\theta}}{\partial\theta}d\theta\right]drdz\sin\left(\frac{d\theta}{2}\right)+\sigma_{\theta\theta}drdz\sin\left(\frac{d\theta}{2}\right)+F_r rd\theta drdz = 0$$

We note that $cos(d\theta/2) = 1$ and $sin(d\theta/2) = (d\theta/2)$ for the infinitesimal angle $d\theta$. Substituting and neglecting terms with four differentials as they will tend to zero faster than other terms, we obtain

$$\sigma_{rr}drd\theta dz + \frac{\partial\sigma_{rr}}{\partial r}rdrd\theta dz + \frac{\partial\tau_{\theta r}}{\partial\theta}d\theta drdz - \sigma_{\theta\theta}drdzd\theta + F_r rd\theta drdz = 0$$

$$\boxed{\frac{\partial\sigma_{rr}}{\partial r} + \frac{1}{r}\frac{\partial\tau_{\theta r}}{\partial\theta} + \frac{\sigma_{rr} - \sigma_{\theta\theta}}{r} + F_r = 0} \quad \textbf{(6.9a)}$$

In a similar manner, by equilibrium of forces in the $\theta$ direction we obtain

$$\boxed{\frac{1}{r}\frac{\partial\sigma_{\theta\theta}}{\partial\theta} + \frac{\partial\tau_{r\theta}}{\partial r} + \frac{2\tau_{r\theta}}{r} + F_\theta = 0} \quad \textbf{(6.9b)}$$

By taking moment at point about the center of the differential element we obtain the symmetry of stresses.

$$\boxed{\tau_{r\theta} = \tau_{\theta r}} \quad \textbf{(6.9c)}$$

### 6.1.5 Boundary conditions

To solve the elasticity equations, we need boundary conditions. At a given point on the surface of a body, we must specify either displacement or traction in a given direction. The traction stress vector was given by Equation (1.10). The boundary conditions at a point on a surface are as given below.

$$u = u_0 \quad \text{or} \quad \sigma_{xx}n_x + \tau_{xy}n_y + \tau_{xz}n_z = t_x \quad \textbf{(6.10a)}$$

$$v = v_0 \quad \text{or} \quad \tau_{yx}n_x + \sigma_{yy}n_y + \tau_{yz}n_z = t_y \quad \textbf{(6.10b)}$$

$$w = w_0 \quad \text{or} \quad \tau_{zx}n_x + \tau_{zy}n_y + \sigma_{zz}n_z = t_z \quad \textbf{(6.10c)}$$

where $u_0$, $v_0$, and $w_0$ represent the specified displacement in the $x$, $y$, and $z$ directions, respectively; $t_x$, $t_y$, and $t_z$ represent the specified traction in the $x$, $y$, and $z$ direction, respectively; and $n_x$, $n_y$, and $n_z$ represent the direction cosines of the unit normal to the surface in the $x$, $y$, and $z$ direction, respectively.

## 6.2 AXISYMMETRIC PROBLEMS

In axisymmetric problems the loading, the geometry, and the material properties are all independent of angular location $\theta$. Under the conditions of axisymmetry, the results of displacements, strains, and stresses should also be independent of $\theta$. Thus, all derivatives with respect to $\theta$ will be zero, and all variables are functions of the radial coordinate $r$ only; partial derivatives of $r$ can be written as ordinary derivatives of $r$. Equation (6.2) and Equation (6.9a) can be written as shown below.

$$\varepsilon_{rr} = \frac{\partial u_r}{\partial r} = \frac{du_r}{dr} \qquad \varepsilon_{\theta\theta} = \frac{u_r}{r} \quad \textbf{(6.11a)}$$

$$\frac{d\sigma_{rr}}{dr} + \frac{\sigma_{rr} - \sigma_{\theta\theta}}{r} + F_r = 0 \quad \textbf{(6.11b)}$$

### 6.2.1 Axisymmetric plane strain

For plane strain, Equation (6.5b) can be written as

$$\sigma_{rr} = \frac{2G}{(1-2v)}[(1-v)\varepsilon_{rr} + v\varepsilon_{\theta\theta}] \qquad \sigma_{\theta\theta} = \frac{2G}{(1-2v)}[(1-v)\varepsilon_{\theta\theta} + v\varepsilon_{rr}] \quad \textbf{(6.12a)}$$

Substituting Equation (6.11a) into Equation (6.12a) we obtain

$$\sigma_{rr} = \frac{2G}{(1-2v)}\left[(1-v)\frac{du_r}{dr} + v\frac{u_r}{r}\right] \qquad \sigma_{\theta\theta} = \frac{2G}{(1-2v)}\left[(1-v)\frac{u_r}{r} + v\frac{du_r}{dr}\right] \quad \textbf{(6.12b)}$$

Substituting Equation (6.12b) into Equation (6.11b) and simplifying, we obtain the differential equation below.

$$\frac{d^2u_r}{dr^2} + \frac{1}{r}\frac{du_r}{dr} - \frac{u_r}{r^2} + \frac{(1-2v)}{2(1-v)G}F_r = 0 \quad \text{or} \quad \frac{d}{dr}\left[\frac{1}{r}\frac{d}{dr}(ru_r)\right] + \frac{(1-2v)}{2(1-v)G}F_r = 0 \quad \textbf{(6.13)}$$

The solution to the above equation can be written as shown below.

$$\boxed{u_r = (u_r)_h + (u_r)_p \qquad (u_r)_h = C_1 r + C_2/r} \quad \textbf{(6.14)}$$

where, $(u_r)_p$ is the particular solution that depends upon the value of $F_r$, and $(u_r)_h$ is the homogeneous solution corresponding to $F_r = 0$. The constants $C_1$ and $C_2$ which must be determined from the boundary conditions of a particular application.

### 6.2.2 Axisymmetric plane stress

For plane stress, Equation (6.4b) can be written as

$$\sigma_{rr} = \frac{E}{(1-\nu^2)}[\varepsilon_{rr} + \nu\varepsilon_{\theta\theta}] = \frac{E}{(1-\nu^2)}\left[\frac{du_r}{dr} + \nu\frac{u_r}{r}\right] \tag{6.15a}$$

$$\sigma_{\theta\theta} = \frac{E}{(1-\nu^2)}[\varepsilon_{\theta\theta} + \nu\varepsilon_{rr}] = \frac{E}{(1-\nu^2)}\left[\frac{u_r}{r} + \nu\frac{du_r}{dr}\right] \tag{6.15b}$$

Substituting the Equations (6.15a) and (6.15b) into Equation (6.11b) and simplifying, we obtain a differential equation:

$$\frac{d^2u_r}{dr^2} + \frac{1}{r}\frac{du_r}{dr} - \frac{u_r}{r^2} + \frac{(1-\nu^2)}{E}F_r = 0 \quad \text{or} \quad \frac{\mathrm{d}}{\mathrm{d}r}\left[\frac{1}{r}\frac{\mathrm{d}}{\mathrm{d}r}(ru_r)\right] + \frac{(1-\nu^2)}{E}F_r = 0 \tag{6.16}$$

The solution to Equation (6.16) is given by Equation (6.14), with the particular part of the solution dependent upon the value of $F_r$.

## 6.3 ROTATING DISKS

A grinding wheel or a disk brake can be modeled as a rotating disk. We assume that the disk is thin (plane stress) and is rotating at a constant angular speed $\omega$. This would subject a point of the disk that is at a radial distance $r$ to an acceleration of $\omega^2 r$ in the radial direction. If the mass density of the disk material is $\rho_m$, then the radial body force will be $F_r = \rho_m\omega^2 r$ which we substitute into Equation (6.16) to obtain

$$\frac{\mathrm{d}}{\mathrm{d}r}\left[\frac{1}{r}\frac{\mathrm{d}}{\mathrm{d}r}(ru_r)\right] + \frac{(1-\nu^2)}{E}(\rho_m\omega^2 r) = 0 \tag{6.17}$$

The particular solution to Equation (6.17) is

$$(u_r)_p = -\frac{(1-\nu^2)(\rho_m\omega^2)}{E}\frac{(r^3)}{8} \tag{6.18}$$

The total solution given by Equation (6.14) can now be written as

$$u_r = C_1 r + \frac{C_2}{r} - \frac{(1-\nu^2)(\rho_m\omega^2 r^3)}{8E} \tag{6.19a}$$

Substituting Equation (6.19a) into Equation (6.15a) and (6.15b), we obtain the radial and tangential normal stresses as shown below.

$$\sigma_{rr} = \frac{E}{(1-\nu^2)}\left[C_1(1+\nu) - \frac{C_2(1-\nu)}{r^2}\right] - \frac{(3+\nu)}{8}\rho_m\omega^2 r^2 \tag{6.19b}$$

$$\sigma_{\theta\theta} = \frac{E}{(1-\nu^2)}\left[C_1(1+\nu) + \frac{C_2(1-\nu)}{r^2}\right] - \frac{(1+3\nu)}{8}\rho_m\omega^2 r^2 \tag{6.19c}$$

The constants $C_1$ and $C_2$ in Equations (6.19a), (6.19b) and (6.19c) can now be determined from the boundary conditions for a particular application. Some typical boundary conditions are discussed below.

1. *A solid rotating disk* (Problem 6.12): The outer boundary is stress free, thus $\sigma_{rr}(r = R_o) = 0$. For a solution to be finite at the center ($r = 0$) of a solid disk requires $C_2 = 0$.
2. *A rotating disk with a hole* (Problem 6.13): The inner boundary ($r = R_i$) and outer boundary ($r = R_o$) are stress free. Thus the boundary conditions are $\sigma_{rr}(r = R_i) = 0$ and $\sigma_{rr}(r = R_o) = 0$.
3. *A rotating disk bonded on a rigid shaft:* The outer boundary is stress free, and the point on the inner boundary cannot be displaced. Thus the boundary conditions are $u_r(r = R_i) = 0$ and $\sigma_{rr}(r = R_o) = 0$.

### EXAMPLE 6.1

The maximum rotational speed at which a grinding wheel can operate is called the "bursting speed," since if this speed is exceeded, maximum tensile stress will cause the wheel to burst. Consider a grinding wheel with inner radius $a$, outer radius $2a$, Poisson ratio $\nu = 1/3$, modulus of elasticity $E$, and mass density of $\rho_m$. Obtain a relationship between the burst speed $\omega_{max}$ and the maximum allowable tensile stress $\sigma_{allow}$ in terms $a$, $E$, and $\rho_m$. Assume that the grinding wheel is mounted on a rigid shaft.

**PLAN**

The boundary conditions are the displacement at the inner radius is zero and radial stress on the free surface of outer radius is zero. Using the boundary conditions the constants $C_1$ and $C_2$ can be determined. Now the normal stresses are known and the maximum normal stress can be equated to $\sigma_{allow}$ to obtain the required relationship.

**SOLUTION**

The boundary conditions are:

$$u_r(r = a) = 0 \tag{E1}$$

$$\sigma_{rr}(r = 2a) = 0 \tag{E2}$$

Substituting $\nu = 1/3$ into Equation (6.19a), (6.19b), and (6.19c), we obtain the following equations.

$$u_r = C_1 r + \frac{C_2}{r} - \frac{(1-(1/3)^2)(\rho\omega^2 r^3)}{8E} = C_1 r + \frac{C_2}{r} - \frac{\rho_m \omega^2 r^3}{9E} \tag{E3}$$

$$\sigma_{rr} = \frac{E}{1-(1/3)^2}\left[C_1(1+1/3) - \frac{C_2(1-1/3)}{r^2}\right] - \frac{(3+1/3)}{8}\rho_m\omega^2 r^2 = \frac{3E}{8}\left[4C_1 - \frac{2C_2}{r^2}\right] - \frac{5}{12}\rho_m\omega^2 r^2 \tag{E4}$$

$$\sigma_{\theta\theta} = \frac{E}{1-(1/3)^2}\left[C_1(1+1/3) + \frac{C_2(1-1/3)}{r^2}\right] - \frac{(1+3(1+1/3))}{8}\rho_m\omega^2 r^2 = \frac{3E}{8}\left[4C_1 + \frac{2C_2}{r^2}\right] - \frac{1}{4}\rho_m\omega^2 r^2 \tag{E5}$$

From Equations (E1) and (E2) we obtain the following equations.

$$C_1 a + \frac{C_2}{a} - \frac{\rho_m\omega^2 a^3}{9E} = 0 \quad \text{or} \quad C_1 + \frac{C_2}{a^2} = \frac{\rho_m\omega^2 a^2}{9E} \tag{E6}$$

$$\frac{3E}{8}\left[4C_1 - \frac{2C_2}{(2a)^2}\right] - \frac{5}{12}\rho\omega^2(2a)^2 = 0 \quad \text{or} \quad C_1 - \frac{C_2}{8a^2} = \frac{10}{9E}\rho\omega^2 a^2 \tag{E7}$$

Solving Equation (E6) and (E7), we obtain

$$C_1 = 1.0015\frac{\rho_m\omega^2 a^2}{E} \quad \text{and} \quad C_2 = -0.8765\frac{\rho_m\omega^2 a^4}{E} \tag{E8}$$

Substituting Equation (E8) into Equation (E4) and (E5), we obtain the normal stresses as shown below.

$$\sigma_{rr} = \left[\frac{3}{8}\left\{4(1.0015a^2) - \frac{2(-0.8765a^4)}{r^2}\right\} - \frac{5}{12}r^2\right](\rho_m\omega^2) = \rho_m\omega^2 a^2\left[1.502 + 0.6574\frac{a^2}{r^2} - 0.4167\frac{r^2}{a^2}\right]$$

$$\sigma_{rr} = \left[1.502 + 0.6574\frac{a^2}{r^2} - 0.4167\frac{r^2}{a^2}\right]\rho_m\omega^2 a^2 \tag{E9}$$

$$\sigma_{\theta\theta} = \left[\frac{3}{8}\left\{4(1.0015a^2) + \frac{2(-0.8765a^4)}{r^2}\right\} - \frac{1}{4}r^2\right](\rho_m\omega^2) = \rho_m\omega^2 a^2\left[1.502 - 0.6574\frac{a^2}{r^2} - 0.25\frac{r^2}{a^2}\right]$$

$$\sigma_{\theta\theta} = \left[1.502 - 0.6574\frac{a^2}{r^2} - 0.25\frac{r^2}{a^2}\right]\rho_m\omega^2 a^2 \tag{E10}$$

The maximum normal stress $\sigma_{rr}$ will be at the inner radius. Substituting $r = a$ into Equation (E9) and noting that the stress should be less than the allowable stress, we obtain the burst speed as shown below.

$$\sigma_{rr} = [1.502 + 0.6574 - 0.4167]\rho_m\omega^2 a^2 = 1.7427\rho_m\omega^2 a^2 \le \sigma_{allow} \quad \text{or} \quad \omega \le 0.7575\sqrt{\sigma_{allow}/(\rho_m a^2)} \tag{E11}$$

**ANS.** $\omega_{max} = 0.7575\sqrt{\sigma_{allow}/(\rho_m a^2)}$

**COMMENTS**

1. Materials for grinding wheels are brittle, and thus the use of the maximum tensile stress is appropriate.
2. We determined the maximum normal stress by inspection by recognizing that $\sigma_{\theta\theta}$ decreases with increase of $r$ faster than $\sigma_{rr}$ because of the sign of the second term. Alternatively, we could have plotted Equations (E9) and (E10) and reached the same conclusion.
3. Heat will be generated during a grinding operation, originating at the point of contact. This will generate thermal stresses, which in general will not be axisymmetric.

## 6.4 AIRY STRESS FUNCTION

Many elasticity problems can be solved with significant ease by using the Airy stress function. The function is chosen such that the equilibrium equations in the absence of body forces are implicitly satisfied by the stresses in two dimensions. It may be verified that the stresses defined in Equation (6.20) satisfy Equations (6.8a) and (6.8b) if $F_x = 0$ and $F_y = 0$.

$$\sigma_{xx} = \frac{\partial^2\psi}{\partial y^2} \qquad \sigma_{yy} = \frac{\partial^2\psi}{\partial x^2} \qquad \tau_{xy} = -\left(\frac{\partial^2\psi}{\partial x\partial y}\right) \tag{6.20}$$

The function $\psi$ is called the Airy stress function. To determine $\psi$, we need to satisfy the compatibility condition of Equation (6.3b) and all the boundary conditions.

Substituting Equation (6.20) into Equation (6.7a) we obtain the strains given in Equation (6.21).

$$\varepsilon_{xx} = \frac{1}{8G}\left[(\kappa+1)\frac{\partial^2\psi}{\partial y^2} - (3-\kappa)\frac{\partial^2\psi}{\partial x^2}\right] \qquad \varepsilon_{yy} = \frac{1}{8G}\left[(\kappa+1)\frac{\partial^2\psi}{\partial x^2} - (3-\kappa)\frac{\partial^2\psi}{\partial y^2}\right] \qquad \gamma_{xy} = -\frac{1}{G}\left(\frac{\partial^2\psi}{\partial x\partial y}\right) \tag{6.21}$$

Substituting the above equation into Equation (6.3b) we obtain

$$\frac{\partial^2}{\partial y^2}\left\{\frac{1}{8G}\left[(\kappa+1)\frac{\partial^2\psi}{\partial y^2} - (3-\kappa)\frac{\partial^2\psi}{\partial x^2}\right]\right\} + \frac{\partial^2}{\partial x^2}\left\{\frac{1}{8G}\left[(\kappa+1)\frac{\partial^2\psi}{\partial x^2} - (3-\kappa)\frac{\partial^2\psi}{\partial y^2}\right]\right\} = \frac{\partial^2}{\partial x\,\partial y}\left\{-\frac{1}{G}\left(\frac{\partial^2\psi}{\partial x\partial y}\right)\right\}$$

$$\frac{1}{8}\left\{(\kappa+1)\frac{\partial^4\psi}{\partial y^4} - (3-\kappa)\frac{\partial^4\psi}{\partial x^2\,\partial y^2}\right\} + \frac{1}{8}\left\{(\kappa+1)\frac{\partial^4\psi}{\partial x^4} - (3-\kappa)\frac{\partial^4\psi}{\partial x^2\,\partial y^2}\right\} + \frac{\partial^4\psi}{\partial x^2\,\partial y^2} = 0 \text{ or}$$

$$(\kappa+1)\frac{\partial^4\psi}{\partial y^4} + (\kappa+1)\frac{\partial^4\psi}{\partial x^4} + \frac{\partial^4\psi}{\partial x^2\,\partial y^2}\{8-2(3-\kappa)\} = 0 \qquad \text{or} \qquad (\kappa+1)\left\{\frac{\partial^4\psi}{\partial x^4} + 2\frac{\partial^4\psi}{\partial x^2\,\partial y^2} + \frac{\partial^4\psi}{\partial y^4}\right\} = 0$$

$$\boxed{\nabla^4\psi = \frac{\partial^4\psi}{\partial x^4} + 2\frac{\partial^4\psi}{\partial x^2\,\partial y^2} + \frac{\partial^4\psi}{\partial y^4} = 0} \tag{6.22}$$

The above equation is valid for plane stress and plane strain. We saw the biharmonic operator $\nabla^4$ in Chapter 5. If we can find an Airy stress function that satisfies the biharmonic operator of Equation (6.22), then we have satisfied the equilibrium and compatibility equations. If in the Airy stress function there are some parameters that can be chosen such that the boundary conditions can be met, then we have the complete solution to the boundary value problem. For an arbitrary boundary geometry with arbitrary conditions, the problem is still difficult[1]. We will consider some simple cases for which we can find some solutions.

## 6.5 SOLUTION BY POLYNOMIALS

We can find solutions to Equation (6.22) by substituting polynomials of different order and determining the coefficients of the polynomials. These solutions are independent of the geometry of the body and are used for testing algorithms in numerical methods and for benchmarking in experimental methods. Some polynomials are considered next.

### 6.5.1 Quadratic polynomials

For stresses to be nonzero, the lowest order of polynomial is quadratic. The quadratic polynomial shown below will implicitly satisfy the biharmonic equation.

$$\psi = a_2(x^2/2) + b_2xy + c_2(y^2/2) \tag{6.23a}$$

Substituting Equation (6.23a) into Equation (6.20), we obtain the stresses as

$$\sigma_{xx} = c_2 \qquad \sigma_{yy} = a_2 \qquad \tau_{xy} = -b_2 \tag{6.23b}$$

All three stress components are constant throughout the body, *irrespective of the shape of the body*. For this state of stress to be the solution, the conditions on tractions on the boundary, in accordance with Equations (1.13a) and (1.13b), must be as shown below.

$$S_x = c_2n_x - b_2n_y \qquad S_y = -b_2n_x + a_2n_y \tag{6.23c}$$

where $n_x$ and $n_y$ are the direction cosines of the unit normal to the boundary at a point. For a rectangular body, the state of stress is shown in Figure 6.3.

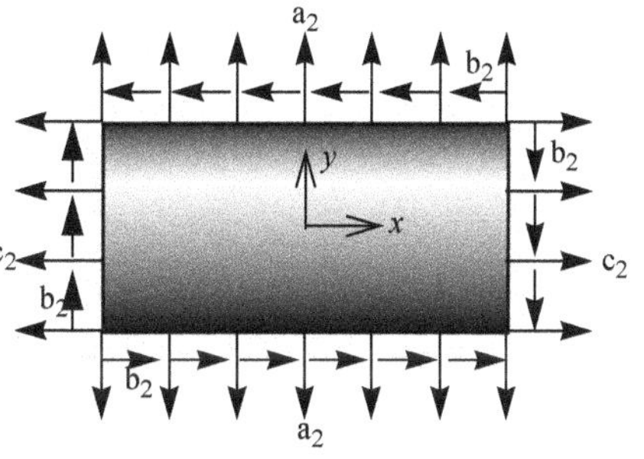


**Figure 6.3** Constant stress state.

Several cases of interest in this constant stress state are briefly described as follows:

- If $b_2 = a_2 = 0$, and $c_2 = \sigma$, we have uniaxial tension.

1. The complex variable method (Muskhelishvili [1963]) and the Fourier transform method (Sneddon [1980]) are two methods that are used in finding more general elasticity solutions.

- If $b_2 = \tau$ and $a_2 = c_2 = 0$, we have a state of pure shear.
- If $b_2 = 0$ and $a_2 = c_2 = \sigma$, we have a hydrostatic state of stress (the normal stress $\sigma$ is the same in all directions).
- If $b_2 = 0$ and $a_2 = -c_2 = \sigma$, we have a state of pure shear in a coordinate system that is 45° to the $xy$ coordinate system.

### 6.5.2 Cubic polynomials

A cubic polynomial will implicitly satisfy the biharmonic equation. We start with the Airy stress function defined as

$$\psi = a_3(x^3/6) + b_3(x^2/2)y + c_3x(y^2/2) + d_3(y^3/6) \tag{6.24a}$$

The stresses can be obtained from Equation (6.20) as

$$\sigma_{xx} = (c_3x + d_3y) \qquad \sigma_{yy} = (a_3x + b_3y) \qquad \tau_{xy} = -(b_3x + c_3y) \tag{6.24b}$$

All three stress components are linear in $x$ and $y$ *irrespective of the shape of the body.* Once more, we consider several subcases for this stress state.

- If $a_3 = b_3 = c_3 = 0$, and $d_3 = \sigma$, we have pure bending of a rectangular cross section as shown in Figure 6.4*a*.
- If $b_3 = c_3 = a_3 = 0$ and $a_3 = \sigma$, we have pure bending of a rectangular cross section as shown in Figure 6.4*b*.

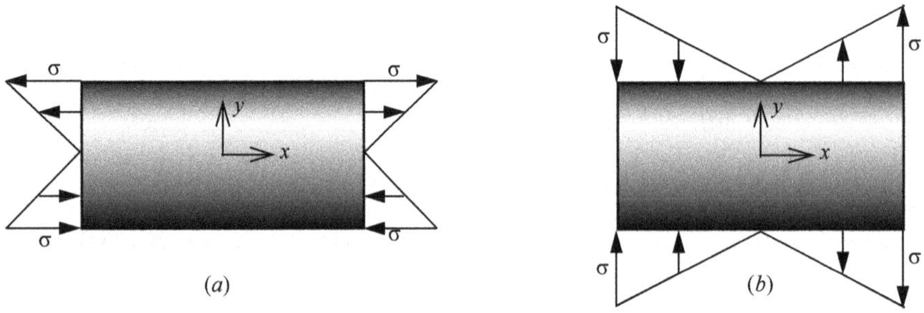


**Figure 6.4** Two states of pure bending with cubic polynomials.

### 6.5.3 Fourth-order polynomials

A fourth-order polynomial will not implicitly satisfy the biharmonic equation, and the relationship between the constants must be determined. We start with the following Airy stress function:

$$\psi = a_4(x^4/24) + b_4(x^3/6)y + c_4(x^2/2)(y^2/2) + d_4x(y^3/6) + e_4(y^4/24) \tag{6.25a}$$

Substituting Equation (6.25a) into Equation (6.22), we obtain the following condition on the constants:

$$a_4 + 2c_4 + e_4 = 0 \tag{6.25b}$$

Substituting Equation (6.25a) and (6.25b) into Equation (6.20) through (6.20), we obtain

$$\sigma_{xx} = c_4(x^2/2) + d_4xy + e_4(y^2/2) \qquad \sigma_{yy} = a_4(x^2/2) + b_4xy + c_4(y^2/2) \qquad \tau_{xy} = -b_4(x^2/2) + c_4xy + d_4(y^2/2) \tag{6.25c}$$

All three stress components are quadratic in $x$ and $y$ *irrespective of the shape of the body.*

## EXAMPLE 6.2

Figure 6.5 shows a cantilever beam with a rectangular cross section. Equation (E1) gives an Airy stress function that could be used.[a] Determine the stress components $\sigma_{xx}$, $\sigma_{yy}$, and $\tau_{xy}$ in terms of $P_1$, $P_2$, $b$, $h$, $x$, $y$, and $L$.

$$\psi = a_1\left[3\frac{y}{h} - \left(\frac{y}{h}\right)^3\right](x - L) + a_2\left(\frac{y}{h}\right)^2 \tag{E1}$$

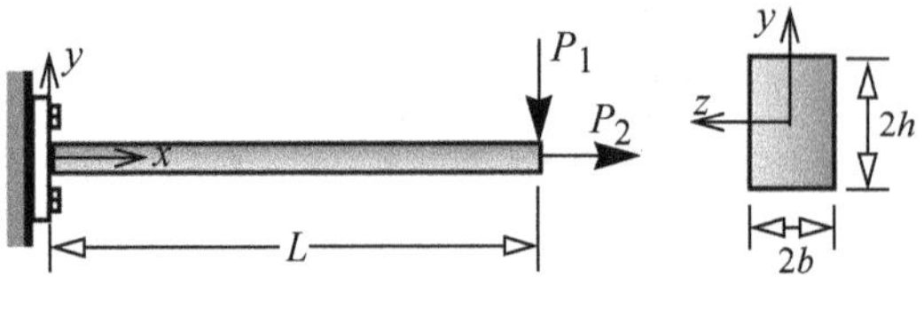


**Figure 6.5** Cantilever beam in Example 6.2.

**PLAN**

From Equation (6.20) we obtain the stress components in terms of the constants $a_1$ and $a_2$. The constants in the stress functions are determined from the boundary conditions on the axial and shear forces at end $B$ and the required stress expression obtained.

**SOLUTION**
Substituting Equation (E1) into Equation (6.20), we obtain

$$\sigma_{xx} = \frac{\partial^2\psi}{\partial y^2} = -6a_1\frac{y(x-L)}{h^3} + 2\frac{a_2}{h^2} \qquad \sigma_{yy} = \frac{\partial^2\psi}{\partial x^2} = 0 \qquad \tau_{xy} = -\left(\frac{\partial^2\psi}{\partial x\,\partial y}\right) = -\left(\frac{3a_1}{h}\right)\left[1-\left(\frac{y}{h}\right)^2\right] \tag{E2}$$

To obtain the boundary conditions at the end of beam, we make an imaginary cut an infinitesimal distance from the end and draw the internal forces and moments in accordance with our sign conventions in Table 2.1 to obtain Figure 6.6. By the equilibrium of forces and moments, we obtain the boundary conditions given below.

$$N(x=L) = P_2 \qquad V_y(x=L) = -P_1 \qquad M_z(x=L) = 0 \tag{E3}$$

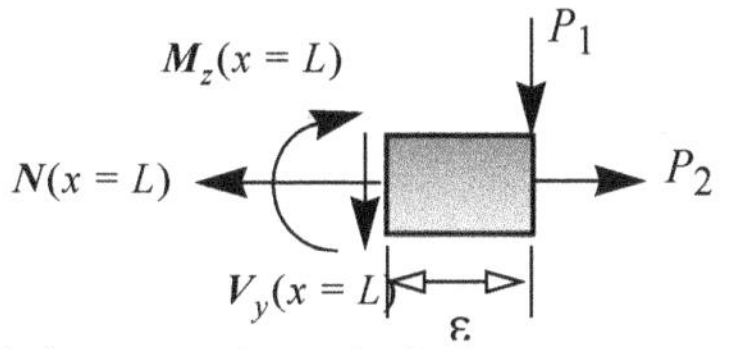


**Figure 6.6** Free body diagram of an infinitesimal element at the end of a beam.
Substituting Equation (E2) into Equations (2.8a-A)and (2.8c-B) and evaluating the result at $x = L$, we obtain the constants as

$$N(x=L) = \int_A \sigma_{xx}dA = \int_{-h}^{h}\left(2\frac{a_2}{h^2}\right)(2b\,dy) = 4\frac{ba_2}{h^2}y\Big|_{-h}^{h} = P_2 \quad \text{or} \quad a_2 = \frac{h}{4b}P_2 \tag{E4}$$

$$V_y = \int_A \tau_{xy}dA = -\left(\frac{3a_1}{h}\right)\int_{-h}^{h}\left[1-\left(\frac{y}{h}\right)^2\right](2b\,dy) = -\left(\frac{6ba_1}{h}\right)\left[y-\frac{y^3}{3h^2}\right]\Big|_{-h}^{h} = -P_1 \quad \text{or} \quad a_1 = \frac{P_1}{4b} \tag{E5}$$

Substituting Equation (E4) and (E5) into Equation (E2), we obtain the following stresses:

$$\textbf{ANS.} \quad \sigma_{xx} = -\left(\frac{3P_1}{4bh^3}\right)y(x-L) + \frac{P_2}{4bh} \qquad \sigma_{yy} = 0 \qquad \tau_{xy} = -\left(\frac{3P_1}{4bh}\right)\left[1-\left(\frac{y}{h}\right)^2\right] \tag{E6}$$

**COMMENTS**

1. Because the stress $\sigma_{xx}$ is uniform across the cross section at $x = L$ the moment condition in Equation (E3) is implicitly met.

2. Because the stress components $\sigma_{yy}$ and $\tau_{xy}$ are zero at $y = \pm h$, the zero traction boundary conditions on top and bottom are implicitly met.

3. The boundary condition of zero traction at all points except where axial force is applied is not met at $x = L$ because the stress components $\sigma_{xx}$ and $\tau_{xy}$ are not zero. We satisfied the boundary condition at $x = L$ in an average sense across the cross section when we used the conditions for equivalent internal forces and moments.

4. Note that $I_{zz} = (2bh^3)/3$, $A = 4bh$, $M_z = P_1(x-L)$, and $N = P_2$. Substituting these values for normal stress under combined loading will result in the expression given by Equation (E6). Similarly, if the bending shear stress is calculated at any point, Equation (E6) will result. This emphasizes that the stress function of Equation (E1) incorporates the approximations of beam theory implicitly. Other stress functions could give other results.

a. The function is an outcome of a method proposed by C.Y. Neou (Boresi and Chong [1987]).

## 6.6 DISPLACEMENTS FROM STRAINS IN 2-D

We have to integrate strain components to obtain displacements. In elasticity there is a general approach of obtaining displacements from strains. We however, will use a simple approach that is similar to the general approach but more revealing of the integration process and is effective for two-dimensional problems.

The strain displacement relationship in two-dimension are given by

$$\varepsilon_{xx} = \frac{\partial u}{\partial x} \qquad \varepsilon_{yy} = \frac{\partial v}{\partial y} \qquad \gamma_{xy} = \gamma_{yx} = \frac{\partial u}{\partial y} + \frac{\partial v}{\partial x} \tag{6.26}$$

The procedure is as follows:

1. Integrate the strain $\varepsilon_{xx}$ with respect to $x$ and add function $f(y)$ to obtain displacement $u(x, y)$.
2. Integrate the strain $\varepsilon_{yy}$ with respect to $y$ and add function $g(x)$ to obtain displacement $v(x, y)$.
3. Substitute $u(x, y)$ and $v(x, y)$ into the strain expression $\gamma_{xy}$. Write all terms that are functions of $x$ on one side of the equal sign and all terms that are functions of $y$ on the other side of the equal sign. This implies that each side must equal to the same constant.
4. The ordinary derivatives of $f(y)$ and $g(x)$ in the above step can be integrated. The constants of integration correspond to rigid body mode discussed in Section 6.6.1 and are determined from the boundary conditions.

**EXAMPLE 6.3**

Obtain the displacement field for the constant stress state given in Equation (6.23b).

**PLAN**

We first obtain the strains using generalized Hooke's law and then follow the procedure described above.

**SOLUTION**

From Equation (6.23b) we have $\sigma_{xx} = c_2$, $\sigma_{yy} = a_2$ , and $\tau_{xy} = -b_2$. We obtain the strains as:

$$\varepsilon_{xx} = \frac{(\kappa+1)}{8G}\sigma_{xx} - \frac{(3-\kappa)}{8G}\sigma_{yy} = \frac{(\kappa+1)}{8G}c_2 - \frac{(3-\kappa)}{8G}a_2 \quad \text{or} \quad \frac{\partial u}{\partial x} = \frac{(\kappa+1)}{8G}c_2 - \frac{(3-\kappa)}{8G}a_2 \tag{E1}$$

$$\varepsilon_{yy} = \frac{(\kappa+1)}{8G}\sigma_{yy} - \frac{(3-\kappa)}{8G}\sigma_{xx} = \frac{(\kappa+1)}{8G}a_2 - \frac{(3-\kappa)}{8G}c_2 \quad \text{or} \quad \frac{\partial v}{\partial y} = \frac{(\kappa+1)}{8G}a_2 - \frac{(3-\kappa)}{8G}c_2 \tag{E2}$$

$$\gamma_{xy} = \tau_{xy}/G = -b_2/G \quad \text{or} \quad \frac{\partial u}{\partial y} + \frac{\partial v}{\partial x} = -\frac{b_2}{G} \tag{E3}$$

1. Integrate Equation (E1) with respect to $x$ to obtain

$$u(x,y) = \left[\frac{(\kappa+1)}{8G}c_2 - \frac{(3-\kappa)}{8G}a_2\right]x + f(y) \tag{E4}$$

2. Integrate Equation (E2) with respect to $y$ to obtain:

$$v(x,y) = \left[\frac{(\kappa+1)}{8G}a_2 - \frac{(3-\kappa)}{8G}c_2\right]y + g(x) \tag{E5}$$

3. Substituting Equations (E4) and (E5) into Equation (E3) we obtain

$$\frac{df}{dy} + \frac{dg}{dx} = -\frac{b_2}{G} \qquad \frac{df}{dy} = -\frac{b_2}{G} - \frac{dg}{dx} \tag{E6}$$

We note that $f(y)$ is only a function of $y$ and $g(x)$ is only a function of $x$. If Equation (E6) is to be valid then each side must equal to the same constant. We represent the constant as $\beta$. We thus obtain two equations

$$\frac{df}{dy} = \beta \quad \text{and} \quad -\frac{b_2}{G} - \frac{dg}{dx} = \beta \quad \text{or} \quad \frac{dg}{dx} = -\frac{b_2}{G} - \beta \tag{E7}$$

Integrating the above equations we obtain:

$$f(y) = \beta y + \alpha_x \qquad g(x) = -\frac{b_2}{G}x - \beta x + \alpha_y \tag{E8}$$

where, $\alpha_x$, $\alpha_y$ are integration constants. Substituting Equations (E8) and (E8) into Equations (E4) and (E5) we obtain the solution of displacements.

$$\textbf{ANS: } u(x,y) = \left[\frac{(\kappa+1)}{8G}c_2 - \frac{(3-\kappa)}{8G}a_2\right]x + \alpha_x + \beta y \qquad v(x,y) = \left[\frac{(\kappa+1)}{8G}a_2 - \frac{(3-\kappa)}{8G}c_2\right]y - \frac{b_2}{G}x + \alpha_y - \beta x \tag{E9}$$

**COMMENTS**

1. The constants $\alpha_x$, $\alpha_y$, and $\beta$ have specific meaning that are discussed in Section 6.6.1.
2. We have obtained the solution to plane stress and plane strain problem as the value of $\kappa$ decides which type of two-dimensional problem we have.

### 6.6.1 Rigid body motion

Integration of strains will result in a rigid body mode in the displacement expressions. The rigid body mode is defined by the constants $\alpha_x$, $\alpha_y$, and $\beta$. Consider $u(x,y) = \alpha_x$—all points on the elastic body move by a constant amount but it produces no strain or stress. In other words, the body translates by an amount $\alpha_x$ in the $x$ direction. Similarly, $\alpha_y$ represents translation in the y direction. The rotation at any point is defined as

$$\omega(x,y) = \frac{1}{2}\left(\frac{\partial u}{\partial y} - \frac{\partial v}{\partial x}\right) \tag{6.27}$$

If we now consider $u(x,y) = \beta y$ and $v(x,y) = -\beta x$, and substitute into the above equation, then we obtain $\omega(x,y) = \beta$. Once more no strain or stress is produced and all points on the elastic body rotate by the same amount of $\beta$. Thus, the constant $\beta$ represents rigid body rotation.

*To eliminate rigid body motion we need to fix the body at least at 3 points, with one point that is not co-linear with the other two.* These support points give us the boundary conditions needed to determine the rigid body constants.

**EXAMPLE 6.4**

Consider the uniaxial tension $\sigma_{xx} = \sigma$ in plane stress and obtain the displacements for the four cases shown in Figure 6.7.

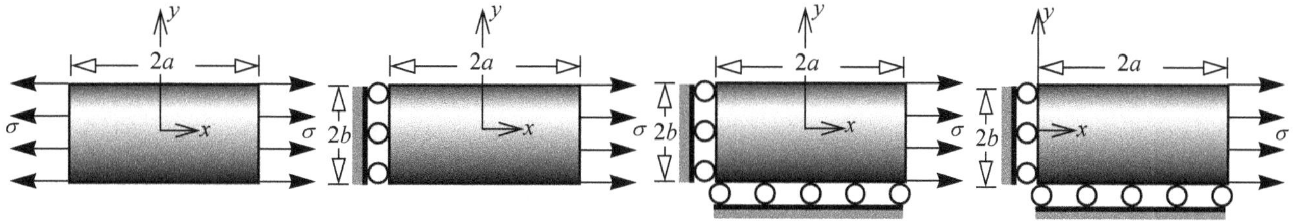


**Figure 6.7**

**PLAN**

We substitute the appropriate constants into Equation (E9) of Example 6.3 to obtain the general displacement field for uniaxial tension. We then evaluate the rigid body constants for each case shown in Figure 6.7.

**SOLUTION**

For plane stress $\kappa = (3-\nu)/(1+\nu)$. We can find the following:

$$\frac{(\kappa+1)}{8G} = \frac{(3-\nu)/(1+\nu)+1}{4E/(1+\nu)} = \frac{1}{E} \qquad \frac{(3-\kappa)}{8G} = \frac{3-(3-\nu)/(1+\nu)}{4E/(1+\nu)} = \frac{-\nu}{E} \tag{E1}$$

From Equation (6.23b) we have $c_2 = \sigma$, and $a_2 = b_2 = 0$. Substituting these values and Equation (E1) into Equation (E9) of Example 6.3, we obtain

$$u(x,y) = (\sigma/E)x + \alpha_x + \beta y \qquad v(x,y) = -\nu(\sigma/E)y + \alpha_y - \beta x \tag{E2}$$

**Case I:** There are no boundary conditions on any displacements and the body is free to translate in any direction and rotate about any point. Thus, the rigid body constants cannot be determined. The displacement field is given by Equation (E2).

**ANS.** $u(x,y) = (\sigma/E)x + \alpha_x + \beta y \qquad v(x,y) = -\nu(\sigma/E)y + \alpha_y - \beta x$

**Case II:** The boundary condition is $u(-a,y) = 0$. From Equation (E2) we obtain

$$(\sigma/E)(-a) + \alpha_x + \beta y = 0 \qquad \text{or} \qquad \alpha_x = (\sigma/E)a \qquad \beta = 0 \tag{E3}$$

Substituting Equation (E3) into Equation (E2) we obtain the displacement field as:

**ANS.** $u(x,y) = (\sigma/E)(x+a) \qquad v(x,y) = -\nu(\sigma/E)y + \alpha_y$

**Case III**: The boundary conditions are $u(-a,y) = 0$ and $v(x,-b) = 0$. The condition on $u$ will yield the constants given in Equation (E3). From Equation (E2) we obtain

$$-\nu(\sigma/E)(-b) + \alpha_y - \beta x = 0 \qquad \text{or} \qquad \alpha_y = -\nu(\sigma/E)b \qquad \beta = 0 \tag{E4}$$

Substituting Equations (E3) and (E4) into Equation (E2) we obtain the displacement field as:

**ANS.** $u(x,y) = (\sigma/E)(x+a) \qquad v(x,y) = -\nu(\sigma/E)(y+b)$

**Case IV**: The boundary conditions are $u(0,y) = 0$ and $v(x,-b) = 0$. The boundary condition on v will yield the constants in Equation (E4). From boundary condition on $u$ and Equation (E2) we obtain:

$$\alpha_x + \beta y = 0 \qquad \text{or} \qquad \alpha_x = 0 \qquad \beta = 0 \tag{E5}$$

Substituting Equations (E5) and (E4) into Equation (E2) we obtain the displacement field as:

**ANS.** $u(x,y) = (\sigma/E)x \qquad v(x,y) = -\nu(\sigma/E)(y+b)$

**COMMENTS**

1. The example demonstrates that the rigid body constants depend upon the location of support as well as on the coordinate system in use. Case III and IV have no rigid body motion, but the difference in the origin of coordinate system gives different expressions for the displacement field $u(x,y)$.
2. In Case I, the body was unsupported and we could not determine the constants. In Case II the body was free to move in $y$-direction and we could not determine $\alpha_y$.
3. These simple cases can be used for testing numerical algorithms as the analytical solutions are simple.

## 6.7 TORSION OF NON-CIRCULAR SHAFTS

In Chapter 1 we saw the theory for the torsion of circular shafts. In this section we develop a more general theory in which there is no limitation on cross-sectional shape. Saint-Venant was the first to develop a theory for the torsion for non-circular shafts. Prandtl later developed an alternative based on Airy's stress function approach. We will use both theories. We will use Prandtl's approach for obtaining the stresses and Saint-Venant's to obtain the deformation.

### 6.7.1 Saint-Venant's method

Saint-Venant observed that a displacement field for a cross section of a non-circular shaft under torsion shown in Figure 6.8 should account for the following:

(i) The cross section would warp under torsion.

(ii) The only nonzero stress components would be $\tau_{xy}$ and $\tau_{xz}$, which for isotropic materials implies that the only nonzero strain components would be $\gamma_{xy}$ and $\gamma_{xz}$.

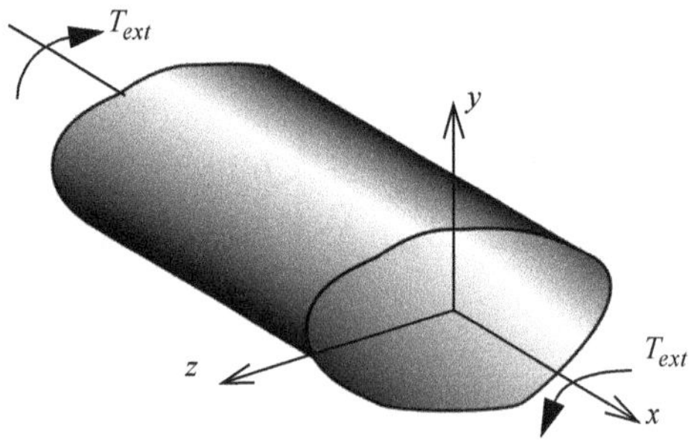


**Figure 6.8** Torsion of a non-circular shaft.

Saint-Venant proposed a displacement field that included a warping function $\chi$ and a form that would result in shear strains $\gamma_{xy}$ and $\gamma_{xz}$ only, as given below

$$u = \chi(y,z)\frac{d\phi}{dx} \qquad v = -xz\frac{d\phi}{dx} \qquad w = xy\frac{d\phi}{dx} \tag{6.28}$$

where $u$, v, and $w$ are the displacements in the $x$, $y$, and $z$ direction, respectively; $d\phi/dx$ is a *constant* representing the rate of twist per unit length; and the warping function $\chi(y, z)$ describes the movement of points out of the cross-sectional plane

With the displacement field known, the logic schematized in Figure 2.7 can be used to obtain the required formulas, as shown in the discussion that follows.

Substituting Equation (6.28) into Equation (6.1) we obtain the nonzero strain components given below.

$$\gamma_{xy} = \frac{\partial}{\partial y}\left[\chi(y,z)\frac{d\phi}{dx}\right] + \frac{\partial}{\partial x}\left[-xz\frac{d\phi}{dx}\right] = \frac{d\phi}{dx}\left[\frac{\partial\chi}{\partial y} - z\right] \qquad \gamma_{xz} = \frac{\partial}{\partial x}\left[xy\frac{d\phi}{dx}\right] + \frac{\partial}{\partial z}\left[\chi(y,z)\frac{d\phi}{dx}\right] = \frac{d\phi}{dx}\left[y + \frac{\partial\chi}{\partial z}\right] \tag{6.29}$$

From Hooke's law for isotropic materials, the shear stresses can be written as shown below.

$$\tau_{xy} = G\left(\frac{\partial\chi}{\partial y} - z\right)\frac{d\phi}{dx} \qquad \tau_{xz} = G\left(\frac{\partial\chi}{\partial z} + y\right)\frac{d\phi}{dx} \tag{6.30}$$

The shear stresses can be replaced by an equivalent internal torque as shown in Figure 6.9. By equating the moment about the origin, we obtain Equation (6.31).

$$T = \int_A (y\tau_{xz} - z\tau_{xy})dA \tag{6.31}$$

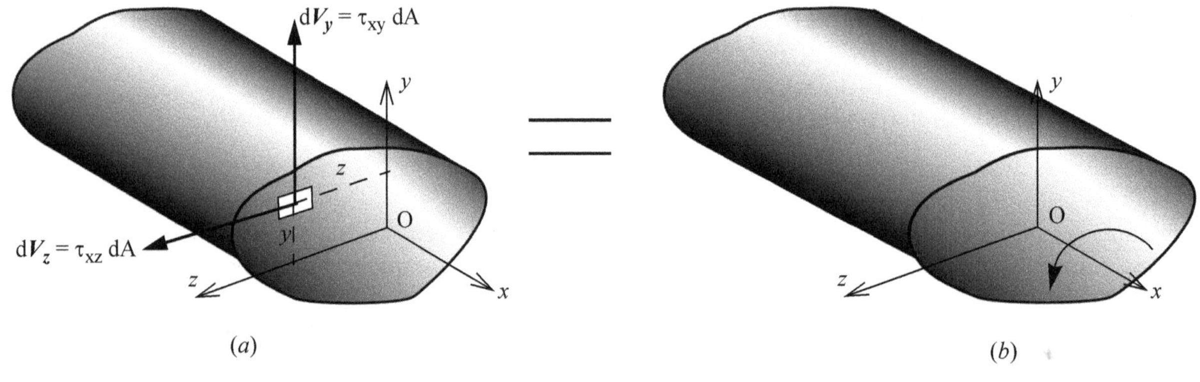


**Figure 6.9** Static equivalency for non-circular shafts.

We could substitute Equation (6.30) into Equation (6.31) and obtain other formulas, but at this stage we will switch to Prandtl's method, which results in a similar form of equations.

## 6.7.2 Prandtl's method

The nonzero stress components $\tau_{xy}$ and $\tau_{xz}$ that will be generated to resist the external torque must satisfy the equilibrium equations. In the absence of body forces, Equation (6.8d) can be written as

$$\frac{\partial\tau_{yx}}{\partial y} + \frac{\partial\tau_{zx}}{\partial z} = 0 \qquad \frac{\partial\tau_{xy}}{\partial x} = 0 \qquad \frac{\partial\tau_{xz}}{\partial x} = 0 \tag{6.32}$$

Equation (6.32) implies that shear stresses $\tau_{xy}$ and $\tau_{xz}$ cannot be functions of $x$, which is possible as long as *there is no distributed torque on the shaft*. To satisfy Equation (6.32), we define a stress function as

$$\boxed{\tau_{yx} = \tau_{xy} = \frac{\partial\psi}{\partial z} \qquad \tau_{zx} = \tau_{xz} = -\left(\frac{\partial\psi}{\partial y}\right)} \tag{6.33}$$

The boundary condition on the surface of the non-circular shaft is zero traction. Substituting = 0, $\sigma_{xx} = 0$, and Equation (6.33) into Equation (6.10a), we obtain

$$\frac{\partial\psi}{\partial z}n_y - \frac{\partial\psi}{\partial y}n_z = 0 \tag{6.34}$$

The outward normal to the surface of the non-circular shaft will lie in the $y$-$z$ plane. Figure 6.10 shows the geometry by which we can relate the direction cosines of the unit normal to the surface geometry. Noting that the tangent to the surface is 90° to the unit normal (i.e., $\lambda_z = 90° + \theta_z$), we obtain Equation (6.35a) and (6.35b).

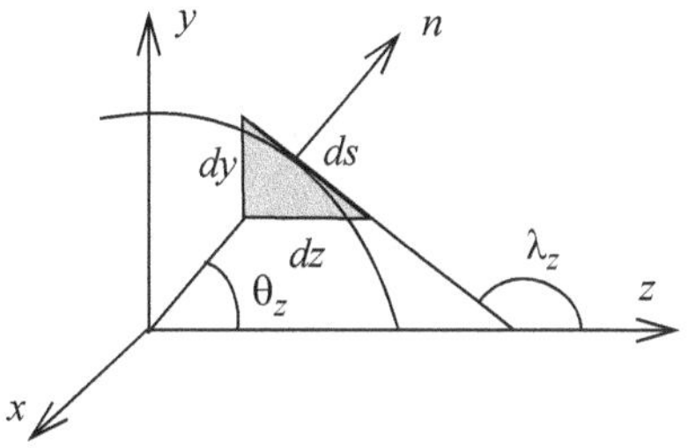


**Figure 6.10** Direction cosines of a unit normal.

$$n_y = sin\ \theta_z = sin(\lambda_z - 90°) = -cos\ \lambda_z = -dz/ds \quad \text{or} \quad n_y = -(dz/ds) \tag{6.35a}$$

$$n_z = cos\ \theta_z = cos\ (\lambda_z - 90°) = sin\ \lambda_z = dy/ds \quad \text{or} \quad n_z = dy/ds \tag{6.35b}$$

Substituting Equation (6.35a) and (6.35b) into Equation (6.34), we obtain Equation (6.36).

$$-\left(\frac{\partial\psi}{\partial z}\right)\left(\frac{dz}{ds}\right) - \frac{\partial\psi}{\partial y}\left(\frac{dy}{ds}\right) = 0 \quad \text{or} \quad -\left(\frac{d\psi}{ds}\right) = 0 \tag{6.36}$$

Equation (6.36) implies that $\psi$ is a constant on the shaft surface, that is, on the boundary of the cross section. This conclusion plays a critical role in selection of the stress function, and we record it for future use.

$$\boxed{\text{The stress function } \psi \text{ must be constant on the boundary of the cross section of the non-circular shaft.}} \tag{6.37}$$

Substituting Equation (6.33) and (6.33) into Equation (6.31) and integrating, we obtain Equation (6.38),

$$T = \iint_A \left(y\left[-\left(\frac{\partial\psi}{\partial y}\right)\right] - z\left[\frac{\partial\psi}{\partial z}\right]\right)dy\ dz = -\oint[y\psi]dz - \oint[z\psi]dy + \iint_A[\psi + \psi]dy\ dz \tag{6.38}$$

where $\oint$ represents integration on the closed boundary defining the cross section. From Equation (6.37), $\psi$ is constant on the boundary of the cross section. The contour integrals can be written as

$$-\oint[y\psi]dz - \oint[z\psi]dy = -\psi\oint[y\ dz + z\ dy] = -\psi\oint d(yz) = 0 \tag{6.39}$$

The integral in Equation (6.39) is zero because if we start at any point on the boundary, go around it, and return to the starting point, the $y$ and $z$ coordinates do not change. Substituting Equation (6.39) into Equation (6.38), we obtain the final result for the internal torque as shown in Equation (6.40).

$$\boxed{T = 2\iint_A \psi dy\ dz} \tag{6.40}$$

To obtain the displacements from the stresses in Equation (6.33), we could first obtain strains and then integrate the partial derivatives. We choose an easier alternative, as described in the discussion that follows.

We substitute Equation (6.30) into Equation (6.33) to obtain

$$\frac{\partial\psi}{\partial z} = G\left(\frac{\partial\chi}{\partial y} - z\right)\frac{d\phi}{dx} \qquad \left(\frac{\partial\psi}{\partial y}\right) = -G\left(\frac{\partial\chi}{\partial z} + y\right)\frac{d\phi}{dx} \tag{6.41a}$$

Taking the partial derivative of first equation with respect to $z$ and the partial derivative of second equation with respect to $y$ and adding, we obtain

$$\boxed{\frac{\partial^2\psi}{\partial y^2} + \frac{\partial^2\psi}{\partial z^2} = -2G\frac{d\phi}{dx}} \tag{6.42}$$

In obtaining Equation (6.42), we made use of the fact that $d\phi/dx$ and $G$ are constants. We now have all the equations we need to solve problems of torsion of non-circular shafts. The next section describes the solution procedure.

### 6.7.3 Procedure for solving problems of torsion of non-circular shafts

**Step 1** For a given cross-sectional shape, obtain a stress function $\psi$ that is constant on the boundary of the cross section in accordance with Equation (6.37).

**Step 2** Determine the constant in the stress function in terms of the internal torque $T$ by integrating over the cross section from Equation (6.40).

**Step 3** Determine the shear stresses from Equation (6.33).

**Step 4** Determine the rate of twist $d\phi/dx$ from Equation (6.42).

## EXAMPLE 6.5

A shaft has an elliptical cross section as shown in Figure 6.11. Determine the equations for maximum shear stress $\tau_{max}$ at a cross section and the relative rotation $(\phi_2 - \phi_1)$ of the cross sections at points $x_1$ and $x_2$ along the length of the shaft in terms of internal torque $T$, shear modulus $G$, $a$, $b$, $x_1$, and $x_2$.

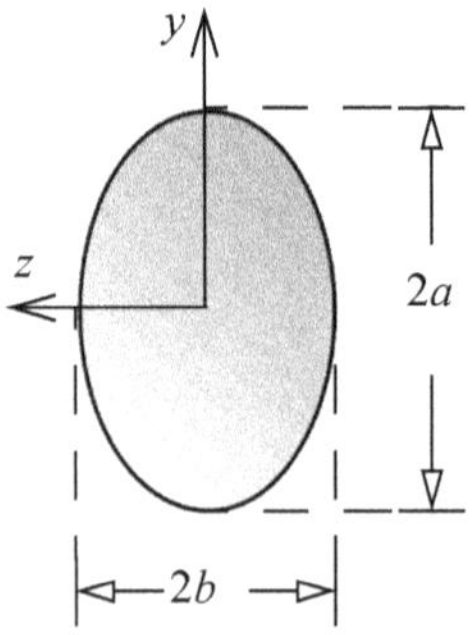


**Figure 6.11** Elliptical cross section for Example 6.5.

**PLAN**

The equation of an ellipse can be written so that all terms are on the left and there is a zero on the right. If the stress function $\psi$ is chosen equal to the terms on the left-hand side of the ellipse equation and multiplied by any arbitrary constant, the stress function $\psi$ will give a zero value at any point on the boundary. Using this stress function the required quantities can be determined by means of the steps outlined in Section 6.7.3.

**SOLUTION**

**Step 1** The stress function $\psi$, which will have a constant value on the boundary, can be written as

$$\psi = K(y^2/a^2 + z^2/b^2 - 1) \tag{E1}$$

where $K$ is a constant to be determined.

**Step 2** Substituting Equation (E1) into Equation (6.40) and integrating, we obtain

$$T = 2\iint_A K(y^2/a^2 + z^2/b^2 - 1)dy\,dz = 2K[I_{zz}/a^2 + I_{yy}/b^2 - A] \tag{E2}$$

where $I_{yy}$ and $I_{zz}$ are the area moments of inertia of an ellipse and $A$ is the cross-sectional area of an ellipse. Substituting $I_{yy} = (\pi ab^3)/4$, $I_{zz} = (\pi ba^3)/4$, and $A = \pi ab$ into Equation (E2), we obtain

$$T = 2K[\pi a^3 b/4a^2 + \pi ab^3/4b^2 - \pi ab] = -K\pi ab \quad \text{or} \quad K = -T/\pi ab \tag{E3}$$

Substituting Equation (E3) into Equation (E1), we obtain

$$\psi = -(T/\pi ab)(y^2/a^2 + z^2/b^2 - 1) \tag{E4}$$

**Step 3** The shear stress components can be obtained by substituting Equation (E4) into Equation (6.33) to obtain

$$\tau_{yx} = \tau_{xy} = -(2Tz)/\pi ab^3 \qquad \tau_{zx} = \tau_{xz} = (2Ty)/\pi a^3 b \tag{E5}$$

Since $a > b$, the magnitude of the maximum shear stress will be given by Equation (E5) when $z = b$ as shown by Equation (E6).

$$\textbf{ANS.} \quad \tau_{max} = (2T)/\pi ab^2 \tag{E6}$$

**Step 4** The rate of rotation can be found by substituting Equation (E4) into Equation (6.42) as shown below

$$-2G\frac{d\phi}{dx} = \left(-\frac{T}{\pi ab}\right)\left(\frac{2}{a^2} + \frac{2}{b^2}\right) \quad \text{or} \quad \frac{d\phi}{dx} = \frac{T(a^2+b^2)}{G\pi a^3 b^3} \tag{E7}$$

We note that because the rate of rotation is a constant, it can be written as $d\phi/dx = (\phi_2 - \phi_1)/(x_2 - x_1)$ to obtain

$$\textbf{ANS.} \quad \phi_2 - \phi_1 = T(a^2+b^2)(x_2 - x_1)/(G\pi a^3 b^3) \tag{E8}$$

**COMMENTS**

1. The torque in Equation (E6) and (E8) is an internal torque. It can be related to the external torque as demonstrated next, in Example 6.6.
2. For a circular shaft $a = b$. Substituting $a = b$ into Equation (E6) and (E8), we obtain Equation (E9).

$$\tau_{max} = 2T/\pi a^3 \qquad \phi_2 - \phi_1 = 2T(x_2 - x_1)/(G\pi a^4) \tag{E9}$$

The values of $\tau_{max}$ and $\phi_2 - \phi_1$ are the same as those that would be obtained from the theory of the torsion of circular shafts.

3. If we compare Equation (E7) with Equation (3.8-T), we can say that the torsional rigidity of the elliptical shaft is $G\pi a^3 b^3/(a^2+b^2)$.

## EXAMPLE 6.6

An aluminum ($G = 4000$ ksi) elliptical shaft is loaded as shown in Figure 6.12. Determine the maximum shear stress in the shaft and the rotation of section $D$ with respect to the rotation of section $A$.

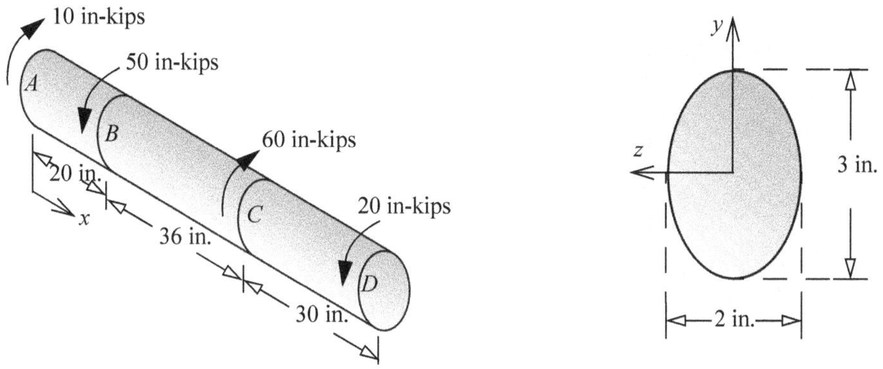


**Figure 6.12** Elliptical shaft for Example 6.6.

**PLAN**

The internal torque in each segment of the shaft can be found by making an imaginary cut and drawing free body diagrams. The equations of the maximum shear stress and relative rotation of two sections were developed in Example 6.5 in terms of the internal torque. By substituting the internal torque for each segment, the maximum shear stress and relative rotation of segment ends can be found, and the desired results can be calculated from these values.

**SOLUTION**

The maximum shear stress and relative rotation for an elliptical shaft were determined in Example 6.5 and are written below.

$$\tau_{max} = \frac{2T}{\pi a b^2} \qquad \phi_2 - \phi_1 = \frac{T(a^2+b^2)(x_2-x_1)}{G\pi a^3 b^3} \tag{E1}$$

The internal torque can be written using discontinuity function as

$$T = 10 - 50\langle x-20\rangle^0 + 60\langle x-56\rangle^0 \tag{E2}$$

Using the above equation, the internal torque in each section can be written as

$$T_{AB} = 10 \text{ in}\cdot\text{kips} \qquad T_{BC} = -40 \text{ in}\cdot\text{kips} \qquad T_{CD} = 20 \text{ in}\cdot\text{kips} \tag{E3}$$

With $a$ and $b$ constant for the entire shaft, the maximum shear stress will be in segment $BC$, where the internal torque has the largest magnitude. Substituting $a = 1.5$ in, $b = 1$ in, and $T_{BC}$ from Equation (E3) into Equation (E1), we obtain the magnitude of maximum shear stress as shown in Equation (E4).

$$\tau_{max} = \left|\frac{2T_{BC}}{\pi a b^2}\right| = \frac{2(40)}{\pi(1.5)(1)^2} = 16.98 \text{ ksi} \tag{E4}$$

Substituting $G = 4000$ ksi, $a = 1.5$ in, $b = 1$ in, and the internal torques in Equation (E3) into Equation (E1), we obtain the relative rotations of each segment, as shown below.

$$\phi_B - \phi_A = \frac{T_{AB}(a^2+b^2)(x_B-x_A)}{G\pi a^3 b^3} = \frac{(10)(1.5^2+1^2)(20)}{(4000)\pi(1.5)^3 1^3} = 15.326(10^{-3}) \text{ rad} \tag{E5}$$

$$\phi_C - \phi_B = \frac{T_{BC}(a^2+b^2)(x_C-x_B)}{G\pi a^3 b^3} = \frac{(-40)(1.5^2+1^2)(36)}{(4000)\pi(1.5)^3 1^3} = -110.347(10^{-3}) \text{ rad} \tag{E6}$$

$$\phi_D - \phi_C = \frac{T_{CD}(a^2+b^2)(x_D-x_C)}{G\pi a^3 b^3} = \frac{(20)(1.5^2+1^2)(30)}{(4000)\pi(1.5)^3 1^3} = 45.978(10^{-3}) \text{ rad} \tag{E7}$$

Adding Equation (E5), (E6), and (E7), we obtain the relative rotation of section $D$ with respect to section $A$ as shown below.

$$\phi_D - \phi_A = [15.326 - 110.347 + 45.978](10^{-3}) = -49.04(10^{-3}) \text{ rad} \tag{E8}$$

**ANS.** $\tau_{max} = 17$ ksi ; $\phi_D - \phi_A = 0.049$ rad CW

**COMMENTS**

The problem once more emphasizes that the equilibrium equations relating internal forces and moments to external forces and moments are independent of kinematics. The change in kinematics impacted the relationship of rotation and maximum stress to internal torque.

## 6.8 MEMBRANE ANALOGY

Prandtl realized that the deflection of membrane under pressure has the same differential equation as torsion of non-circular bars. Thus, an analogy exists that can be used for determining the variables in torsion of non-circular bars by looking at membrane deflection. This analogy has proved to be important in experimentally determining the torsional rigidity of complex cross sectional shapes, such as cross section of a wing of an aircraft.

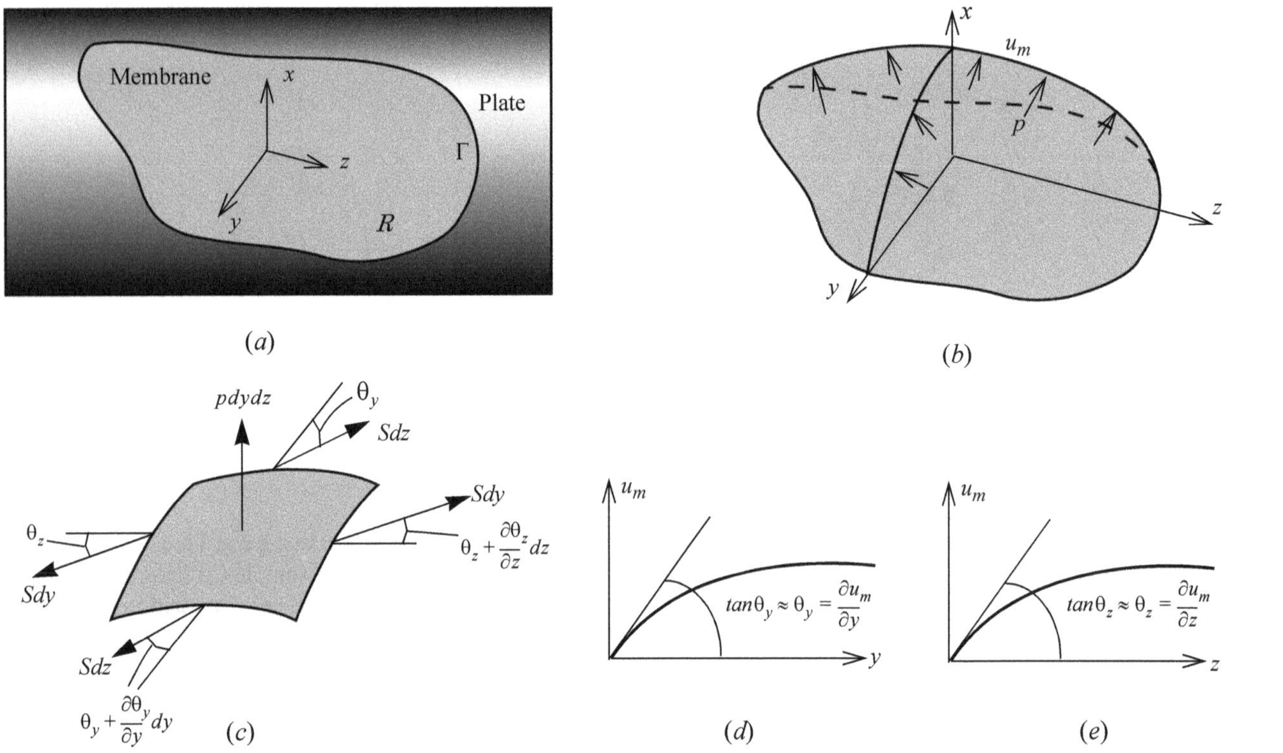


**Figure 6.13** Membrane deflection

Consider a plate with a hole of the same shape as the cross section of a non-circular bar. Hole does not need to be of the same dimension as the cross section of the shaft, provided an appropriate calibration is conducted. Stretch a membrane across it ensuring by appropriate means that the membrane does not deflect at the hole boundary.We represent this membrane region as $R$ with a boundary $\Gamma$ as shown in Figure 6.13*a*. Apply a pressure $p$ underneath the membrane, causing it to deflects by $u_m$ as shown in Figure 6.13*b*. The tension per unit length in the membrane ($S$) is uniform in all directions. Furthermore the tension acts in the tangent direction of deflected membrane at all points as shown in Figure 6.13*c*. The tangent directions can be related to membrane deflections as shown in Figures 6.13*d* and *e*. In Figures 6.13*d* and *e*, it is assumed that the tangent angles are small and hence the Tangent function can be replaced by its arguments. By equilibrium of forces in the $x$ direction in Figure 6.13*c*, we obtain:

$$(Sdy)sin\left(\theta_z + \frac{\partial\theta_z}{\partial z}dz\right) - (Sdy)sin(\theta_z) + (Sdz)sin\left(\theta_y + \frac{\partial\theta_y}{\partial y}dy\right) - (Sdz)sin(\theta_y) + pdzdy = 0$$

Noting that for small angles the we can replace the Sine function by its argument we obtain:

$$Sdy\left(\theta_z + \frac{\partial\theta_z}{\partial z}dz\right) - Sdy(\theta_z) + Sdz\left(\theta_y + \frac{\partial\theta_y}{\partial y}\right) - Sdz(\theta_y) + pdzdy = 0 \qquad \text{or} \qquad \frac{\partial\theta_z}{\partial z} + \frac{\partial\theta_y}{\partial y} = -\frac{p}{S}$$

$$\boxed{\nabla^2 u_m = \frac{\partial^2 u_m}{\partial y^2} + \frac{\partial^2 u_m}{\partial z^2} = -\left(\frac{p}{S}\right) \qquad \text{in } R} \tag{6.43}$$

$$\boxed{u_m = 0 \qquad \text{on } \Gamma} \tag{6.44}$$

We see that Poisson's equation represents the Prandtl's formulation of the torsion problem and the deflection of a membrane. The differential equation can be solved using zero deflection condition on the boundary which is analogous to zero value of $\psi$ on the boundary. Thus, the boundary value problems for torsion of non-circular shafts and deflection of membrane are similar and the analogy between the two problems can now be drawn as shown in Table 6.1. It should be mentioned that Poisson's equation is obtained in many fields: thermal problems, diffusion problems, electrostatic, etc. Membrane analogy can be used in all of these applications of Poisson's equation.

Membrane deflection can be used to determine the torque carrying capacity of a cross section by determining the volume under the membrane and doubling it. The experiment can also tell us where the torsional shear stresses will be maximum by looking at the maximum slopes in the membrane deflected shape. The membrane analogy can also be used to obtain approximate analytical solutions, particularly for thin walled open sections as will be shown in Section 6.9.

**Table 6.1** Prandtl membrane analogy

| Membrane deflection problem. | Prandtl torsion problem. |
|---|---|
| $\frac{\partial^2 u_m}{\partial y^2}+\frac{\partial^2 u_m}{\partial z^2} = -\left(\frac{p}{S}\right)$ | $\frac{\partial^2 \psi}{\partial y^2}+\frac{\partial^2 \psi}{\partial z^2} = -2G\frac{d\phi}{dx}$ |
| $u_m$ | $\psi$ |
| $p$ | $2(d\phi/dx)$ |
| $1/S$ | $G$ |
| $\partial u_m/\partial z$ | $\tau_{yx} = \tau_{xy} = \partial\psi/\partial z$ |
| $\partial u_m/\partial y$ | $\tau_{zx} = \tau_{xz} = -\partial\psi/\partial y$ |
| Volume beneath the membrane = $\iint_A u_m dydz$ | $T = 2\iint_A \psi dydz$ |

## 6.8.1 Membrane analogy for cross section with holes

The governing differential equation and the analogy does not change when the cross section has a hole. However, there is no deformation in the hole and the stresses on the boundary of the hole are zero. According to the analogy in Table 6.1, zero stresses implies zero slope of the membrane at the hole boundary. Thus, to model the hole we cover the membrane with a flat weightless rigid plate of the same shape as the hole. The rigid plate rises (translates) in the $x$ direction as shown in Figure 6.14. At points $B$ and $C$ the slope of the membrane deflection are zero, modeling the zero stress condition of the hole boundary. Once more using the analogy of Table 6.1, the torsional rigidity and maximum torsional shear stresses can be found. Thus, experimentally we can determine torsional properties of very complex cross sections.

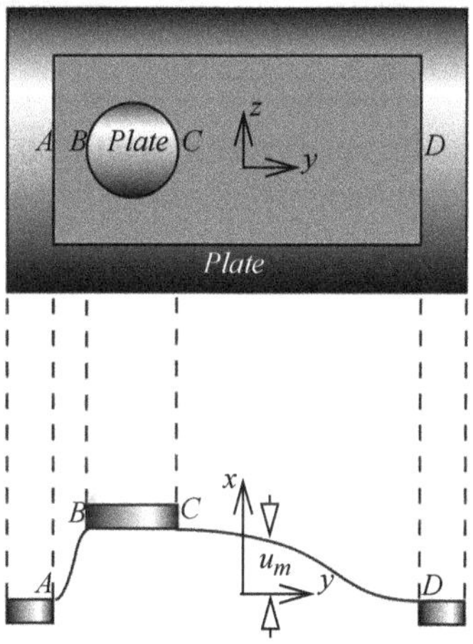


**Figure 6.14** Membrane with a hole.

## EXAMPLE 6.7

Using membrane analogy for a circular shaft in torsion prove $T = G(\pi R^4/2)(d\phi/dx)$.

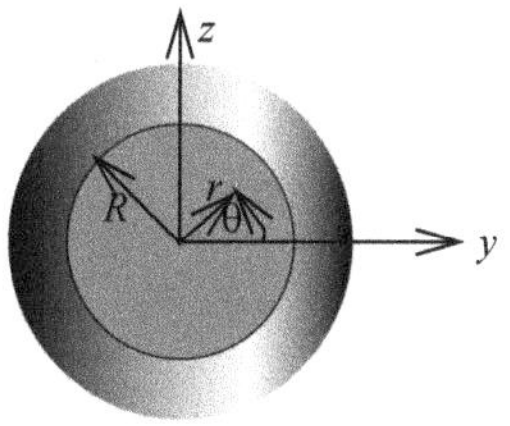


**Figure 6.15** Circular geometry in Example 6.7.

**PLAN**

We write Equation (6.43) in polar coordinates using Equation (5.44f). We note the deflection of the membrane is an axisymmetric problem. We solve the differential equation using the analogy shown in Table 6.1 to obtain the appropriate formulas.

**SOLUTION**

Using Equation (5.44f) we can write Equation (6.43) as given below.

$$\nabla^2 u_m = \frac{1}{r}\frac{\partial}{\partial r}\left[r\frac{\partial u_m}{\partial r}\right]+\frac{1}{r^2}\frac{\partial^2 u_m}{\partial \theta^2} = -\left(\frac{p}{S}\right) \qquad \text{(E1)}$$

We note that a circular membrane under uniform pressure is axisymmetric, that is, $u_m$ does not depend upon angular position and is only a function of $r$. Setting the derivative with respect to $\theta$ to zero we obtain

$$\frac{1}{r}\frac{d}{dr}\left[r\frac{du_m}{dr}\right] = -\left(\frac{p}{S}\right) \tag{E2}$$

Integrating twice we obtain:

$$r\frac{du_m}{dr} = \left(-\frac{p}{S}\right)\frac{r^2}{2} + C_1 \qquad u_m = -\left(\frac{p}{S}\right)\left(\frac{r^2}{4}\right) + C_1 \ln r + C_2 \tag{E3}$$

At the center ($r = 0$) the displacement has to be finite, hence $C_1 = 0$. At the boundary ($r = R$) the deflection should be zero, hence

$$C_2 = (p/S)(R^2/4) \tag{E4}$$

The deflection solution is:

$$u_m = (p/4S)(R^2 - r^2) \tag{E5}$$

The volume ($V$) under the deflected membrane can be found as:

$$V = \int_0^R u_m(2\pi r)\, dr = \left(\frac{\pi p}{2S}\right)\int_0^R (R^2 r - r^3)\, dr = \left(\frac{\pi p}{2S}\right)\left(R^2\frac{r^2}{2} - \frac{r^4}{4}\right)\Bigg|_0^R = \left(\frac{\pi p R^4}{8S}\right) \tag{E6}$$

From analogy in Table 6.1, we replace $V$ by $T/2$, $p$ by $2(d\phi/dx)$, and $1/S$ by $G$ in (E6) to obtain:

$$\frac{T}{2} = \left(\frac{\pi R^4}{8}\right)G\left(2\frac{d\phi}{dx}\right) \quad \text{or} \quad T = G\left(\frac{\pi R^4}{2}\right)\frac{d\phi}{dx} = GK\frac{d\phi}{dx} \tag{E7}$$

In the above equation we note that for a solid circular shaft the polar moment of inertia is $K = (\pi R^4)/2$. Equation (E7) is our torsional formula for circular shafts.

**COMMENT**

The example demonstrate that many mechanics of materials equations can be obtained from equations of elasticity.

## 6.9 TORSION OF THIN-WALLED OPEN SECTION

Obtaining analytical formulas for torsion of thin-walled open sections is difficult. However, by use of membrane analogy we can get approximate results and can visualize the torsional response of the cross section.

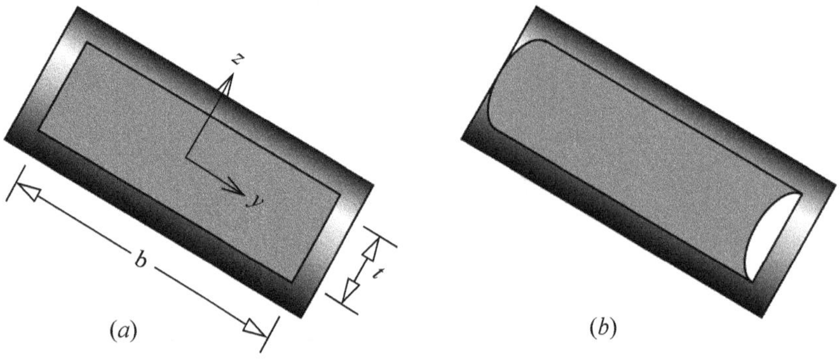


**Figure 6.16** Membrane analogy for thin-walled open section.

To develop the approximate formulas we will first consider a *thin* rectangular cross section. By *thin* we imply that the long side of rectangle is significantly greater than the thickness, i.e., $b \gg t$ in Figure 6.16*a*. Usually an order of magnitude quantifies the word 'significantly greater' in engineering, thus $b > 10t$ at least. With this requirement we expect that the membrane deflection will be dominated by $z$ and does not change significantly with $y$, that is, $u_m(z)$. With this assumption we can write Equation (6.43) as

$$\frac{d^2 u_m}{dz^2} = -\left(\frac{p}{S}\right) \tag{6.45}$$

Integrating Equation (6.45) we obtain:

$$u_m = -(pz^2)/(2S) + C_1 z + C_2 \tag{6.46a}$$

The integration constants can be obtained by noting that the membrane deflection is zero on the boundary in the thickness direction, that is, $u_m(z = \pm t/2) = 0$. Using these conditions we find that $C_1 = 0$ and $C_2 = (pt^2)/(8S)$. The membrane deflection equation thus is:

$$u_m = \frac{p}{2S}\left(\frac{t^2}{4} - z^2\right) \quad \textbf{(6.46b)}$$

Equation (6.46b) is an equation of a parabola. The approximate deflected shape of the membrane is shown in Figure 6.16*b*. Clearly at $y = \pm b/2$ the approximate membrane deflected shape is incorrect. We will discuss this end effect later but we continue under the assumption that the error introduced at the end is small.

The volume under the membrane is the area under the parabola times the length and can be determined as:

$$V = b\int_{-t/2}^{t/2} u_m\, dz = \frac{bp}{2S}\int_{-t/2}^{t/2}\left(\frac{t^2}{4} - z^2\right) dz = \frac{bpt^3}{12S} \quad \textbf{(6.47)}$$

Noting the analogy in Table 6.1 we replace $V$ by $\boldsymbol{T}/2$, $p$ by $2\frac{d\phi}{dx}$, and $\frac{1}{S}$ by $G$ in Equation (6.47) to obtain:

$$\boldsymbol{T} = G\frac{bt^3}{3}\frac{d\phi}{dx} \quad \text{or}$$

$$\boxed{\boldsymbol{T} = G\frac{bt^3}{3}\frac{d\phi}{dx} = GK\frac{d\phi}{dx} \qquad \text{where} \qquad K = bt^3/3} \quad \textbf{(6.48)}$$

For other shapes of cross sections there will be different values of K, but will be obtained from use of the above formula. GK is the *torsional rigidity.* It must however be emphasized that K is not the polar moment of inertia as was the case for circular cross sections.

The torsional shear stress is related to the slope of the deflected membrane curve. In our approximation we are only permitting slope in the z direction. Taking the partial derivative with respect to z of Equation (6.46b), we obtain:

$$\frac{\partial u_m}{\partial z} = -\left(\frac{pz}{S}\right) \quad \textbf{(6.49a)}$$

Using the analogy in Table 6.1, we replace $V$ by $\boldsymbol{T}/2$, $p$ by $2(d\phi/dx)$, and $1/S$ by $G$ in Equation (6.49a) to obtain:

$$\tau_{xy} = -2Gz(d\phi/dx) \quad \textbf{(6.49b)}$$

In thin-walled closed sections we assumed that the torsional shear stress in the thickness direction was uniform. But for thin-walled open section, Equation (6.49b) shows that the torsional shear stress varies linearly in the thickness direction. We also note that $y$ is in the direction of the center line. So we will drop the subscripts and hence forth represent the maximum torsional shear stress as $\tau$ in the direction of the tangent to the center line. This will permit us to generalize our results to arbitrary shaped thin open sections.The maximum torsional shear stress will be at edge of the thickness, i.e., at $z = \pm t/2$. We substitute $d\phi/dx$ from Equation (6.48) to obtain the magnitude of maximum torsional shear stress as:

$$\boxed{\tau_{max} = \boldsymbol{T}t/K} \quad \textbf{(6.50)}$$

Equation (6.46b) was derived under the assumption $b \gg t$. Thus all our formulas in this section are subject to this limitation. When a rectangular cross-section does not meet this limitation, then different formulas can still be obtained using membrane analogy. For thin cross sections of different shapes we will use the above approach assuming that the membrane does form a parabola in the thickness directions as demonstrated in Example 6.8.

## 6.9.1 End effects on rectangular cross sections

Consider a corner of a rectangular cross section as shown in Figure 6.17. As the outside surfaces of the bar are free surfaces, the stresses $\tau_{yx}$ and $\tau_{zx}$ are zero, hence by symmetry of torsional shear stresses $\tau_{xy}$ and $\tau_{xz}$ are zero. Both these components reach

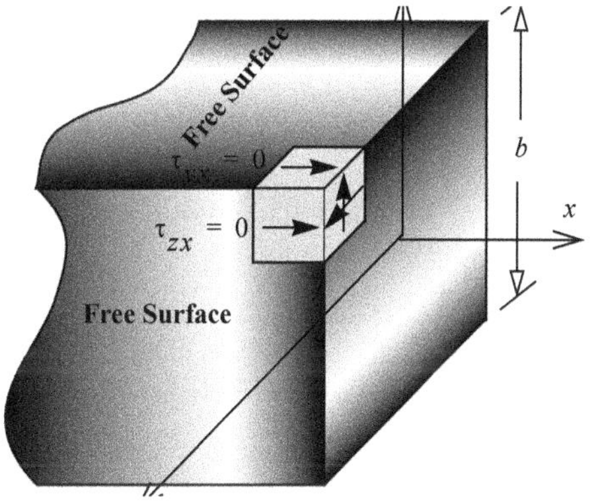


**Figure 6.17** End effect on torsional shear stress.

a maximum value at the mid point of each side which we conclude by visualizing the slope of deflected shape of a membrane. Thus, there is a transition region near towards the edge of $y = \pm b/2$ in Figure 6.16 where the torsional shear stress $\tau_{xz}$, which we neglected is non-zero. We assume that this region is small and its impact can be neglected.

## EXAMPLE 6.8

Obtain the torsional rigidity and maximum torsional shear stress for the thin open cross section of uniform thickness $t$ shown in Figure 6.18. Assume $t \ll a$ and gap at $D$ is of negligible thickness. Report the answer in terms of internal torque $T$, shear modulus $G$, thickness $t$, and parameter $a$.

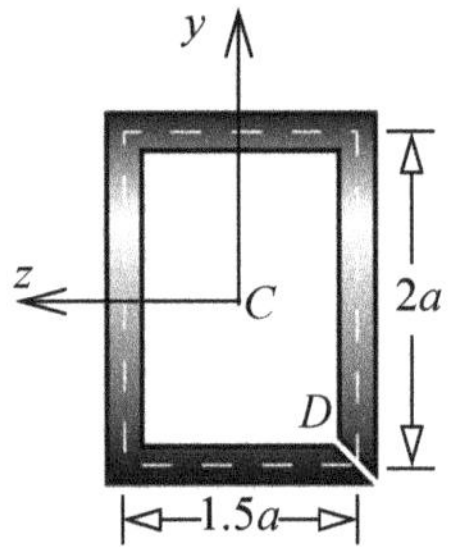


**Figure 6.18** Geometry in Example 6.8.

**PLAN**

We assume each straight segment of the membrane will deflect like a parabola and calculate the total volume.

**SOLUTION**

The total volume under the deflected membrane is the sum of the volume of each segment. Using Equation (6.47), we obtain:

$$V = 2\left[\frac{(1.5a)pt^3}{12S}\right] + 2\left[\frac{(2a)pt^3}{12S}\right] = \frac{7apt^3}{12S} \quad \text{(E1)}$$

Noting the analogy in Table 6.1, we replace $V$ by $T/2$, $p$ by $2(d\phi/dx)$, and $1/S$ by $G$, we obtain:

$$\frac{T}{2} = \left(\frac{7at^3}{12}\right)G\left(2\frac{d\phi}{dx}\right) \quad \text{or} \quad T = G\left[\frac{7at^3}{3}\right]\left(\frac{d\phi}{dx}\right) \quad \text{(E2)}$$

The torsional rigidity is:

$$\textbf{ANS: } GK = (7/3)Gat^3 \quad \text{(E3)}$$

The parameter $K = (7at^3)/3$. The maximum torsional shear stress from Equation (6.50) is thus:

$$\tau = \frac{Tt}{K} = \frac{Tt}{(7at^3)/3} \qquad \textbf{ANS: } \tau = 3T/(7at^2) \quad \text{(E4)}$$

**COMMENTS**

1. The torsional stress and rate of rotation of thin-walled closed sections is governed (Vable[2013]) by the following equations

$$\tau = T/(2tA_E) \qquad d\phi/dx = [T/(4A_E^2G)]\oint ds/t \quad \textbf{(6.51)}$$

where, $A_E$ is the enclosed area by the center line and the line integral is along the perimeter of the center line. We note that the enclosed area is $A_E = 3a^2$. The thickness is uniform, thus $\oint ds/t = (1/t)\oint ds = 7a/t$. Substituting these values in Equation (E5) we obtain:

$$\tau_{closed} = T/(6a^2t) \qquad d\phi/dx = (T/36a^4G)[7a/t] = T/(GK)_{closed} \qquad \text{where} \qquad (GK)_{closed} = G[36a^3t/7] \quad \text{(E5)}$$

2. If we take the ratio of Equations (E5) to (E3) we obtain:

$$(GK)_{closed}/(GK) = 2.2(a/t)^2 \quad \text{(E6)}$$

In the above equation $a \gg t$, and thus the torsional rigidity of a closed section is at least two orders of magnitude greater than that of an open section with the same geometry and dimensions.

3. If we take the ratio of Equations (E5) to (E4) we obtain:

$$\tau_{closed}/\tau = 0.3889(t/a) \quad \text{(E7)}$$

The above equation implies that the maximum torsional shear stress for a closed section is at least one order of magnitude smaller than that of an open section with the same geometry and dimensions.

4. The above two comments highlight that if a cross section will have significant torsional loads then design a closed section. However, if torsional loads are small and are unavoidable by-product of the loading and geometry then the formulas for torsion of open section provide a good estimate of the stiffness and strength.

## 6.10 CLOSURE

The kinematic equations relating strains and displacements, the compatibility equations on strains, the generalized version of Hooke's law, the equilibrium equations, and the boundary conditions on tractions or displacements are the equations of elasticity that were introduced in this chapter. From arguments of axisymmetry, the equations of elasticity can be used to obtain results for plane strain problems of thick long cylinders and for plane stress problems of thin disks or thick short cylinders with no axial load. Equilibrium equations on stresses are intrinsically satisfied by the Airy stress function, a powerful concept whose one application is for the torsion of non-circular shafts.

## 6.11 SYNOPSIS OF EQUATIONS

| | |
|---|---|
| | **Equations of Elasticity** |
| Strains | $\varepsilon_{xx} = \frac{\partial u}{\partial x}$ $\varepsilon_{yy} = \frac{\partial v}{\partial y}$ $\varepsilon_{zz} = \frac{\partial w}{\partial z}$; $\gamma_{xy} = \gamma_{yx} = \frac{\partial u}{\partial y} + \frac{\partial v}{\partial x}$ $\gamma_{yz} = \gamma_{zy} = \frac{\partial v}{\partial z} + \frac{\partial w}{\partial y}$ $\gamma_{zx} = \gamma_{xz} = \frac{\partial w}{\partial x} + \frac{\partial u}{\partial z}$<br>$\varepsilon_{rr} = \frac{\partial u_r}{\partial r}$ $\varepsilon_{\theta\theta} = \frac{u_r}{r} + \frac{1}{r}\frac{\partial v_\theta}{\partial \theta}$ $\gamma_{r\theta} = \frac{1}{r}\frac{\partial u_r}{\partial \theta} + \frac{\partial v_\theta}{\partial r} - \frac{v_\theta}{r}$ |
| Compatibility | $\frac{\partial^2 \varepsilon_{xx}}{\partial y^2} + \frac{\partial^2 \varepsilon_{yy}}{\partial x^2} = \frac{\partial^2 \gamma_{xy}}{\partial x \partial y}$ |
| Plane Stress | $E\varepsilon_{xx} = \sigma_{xx} - \nu\sigma_{yy}$ $E\varepsilon_{yy} = \sigma_{yy} - \nu\sigma_{xx}$ $G\gamma_{xy} = \tau_{xy}$<br>$\sigma_{xx} = E[\varepsilon_{xx} + \nu\varepsilon_{yy}]/(1-\nu^2)$ $\sigma_{yy} = E[\varepsilon_{yy} + \nu\varepsilon_{xx}]/(1-\nu^2)$ $\tau_{xy} = G\gamma_{xy}$ |
| Plane Strain | $2G\varepsilon_{xx} = (1-\nu)\sigma_{xx} - \nu\sigma_{yy}$ $2G\varepsilon_{yy} = (1-\nu)\sigma_{yy} - \nu\sigma_{xx}$ $G\gamma_{xy} = \tau_{xy}$<br>$\sigma_{xx} = \frac{2G}{(1-2\nu)}[(1-\nu)\varepsilon_{xx} + \nu\varepsilon_{yy}]$ $\sigma_{yy} = \frac{2G}{(1-2\nu)}[(1-\nu)\varepsilon_{yy} + \nu\varepsilon_{xx}]$ $\tau_{xy} = G\gamma_{xy}$ |
| Equilibrium | $\frac{\partial \sigma_{xx}}{\partial x} + \frac{\partial \tau_{yx}}{\partial y} + \frac{\partial \tau_{zx}}{\partial z} + F_x = 0$ $\frac{\partial \tau_{xy}}{\partial x} + \frac{\partial \sigma_{yy}}{\partial y} + \frac{\partial \tau_{zy}}{\partial z} + F_y = 0$ $\frac{\partial \tau_{xz}}{\partial x} + \frac{\partial \tau_{yz}}{\partial y} + \frac{\partial \sigma_{zz}}{\partial z} + F_z = 0$<br>$\tau_{xy} = \tau_{yx}$ $\tau_{yz} = \tau_{zy}$ $\tau_{zx} = \tau_{xz}$<br>$\frac{\partial \sigma_{rr}}{\partial r} + \frac{1}{r}\frac{\partial \tau_{\theta r}}{\partial \theta} + \frac{\sigma_{rr} - \sigma_{\theta\theta}}{r} + F_r = 0$ $\frac{1}{r}\frac{\partial \sigma_{\theta\theta}}{\partial \theta} + \frac{\partial \tau_{r\theta}}{\partial r} + \frac{2\tau_{r\theta}}{r} + F_\theta = 0$ $\tau_{r\theta} = \tau_{\theta r}$ |
| Boundary Conditions | $u = u_0$ or $\sigma_{xx}n_x + \tau_{xy}n_y + \tau_{xz}n_z = t_x$<br>$v = v_0$ or $\tau_{yx}n_x + \sigma_{yy}n_y + \tau_{yz}n_z = t_y$<br>$w = w_0$ or $\tau_{zx}n_x + \tau_{zy}n_y + \sigma_{zz}n_z = t_z$ |
| | **Axisymmetric** |
| | $\varepsilon_{rr} = \frac{du_r}{dr}$ $\varepsilon_{\theta\theta} = \frac{u_r}{r}$ $\frac{d\sigma_{rr}}{dr} + \frac{\sigma_{rr} - \sigma_{\theta\theta}}{r} + F_r = 0$ $u_r = (u_r)_h + (u_r)_p$ $(u_r)_h = C_1 r + C_2/r$ |
| Plane Strain | $\sigma_{rr} = \frac{2G}{(1-2\nu)}\left[(1-\nu)\frac{du_r}{dr} + \nu\frac{u_r}{r}\right]$ $\sigma_{\theta\theta} = \frac{2G}{(1-2\nu)}\left[(1-\nu)\frac{u_r}{r} + \nu\frac{du_r}{dr}\right]$ $\frac{d}{dr}\left[\frac{1}{r}\frac{d}{dr}(ru_r)\right] + \frac{(1-2\nu)}{2(1-\nu)G}F_r = 0$ |
| Plane Stress | $\sigma_{rr} = \frac{E}{(1-\nu^2)}\left[\frac{du_r}{dr} + \nu\frac{u_r}{r}\right]$ $\sigma_{\theta\theta} = \frac{E}{(1-\nu^2)}\left[\frac{u_r}{r} + \nu\frac{du_r}{dr}\right]$ $\frac{d}{dr}\left[\frac{1}{r}\frac{d}{dr}(ru_r)\right] + \frac{(1-\nu^2)}{E}F_r = 0$ |
| Rotating Disk | $u_r = C_1 r + \frac{C_2}{r} - \frac{(1-\nu^2)(\rho_m\omega^2 r^3)}{8E}$<br>$\sigma_{rr} = \frac{E}{(1-\nu^2)}\left[C_1(1+\nu) - \frac{C_2(1-\nu)}{r^2}\right] - \frac{(3+\nu)}{8}\rho_m\omega^2 r^2$ $\sigma_{\theta\theta} = \frac{E}{(1-\nu^2)}\left[C_1(1+\nu) + \frac{C_2(1-\nu)}{r^2}\right] - \frac{(1+3\nu)}{8}\rho_m\omega^2 r^2$ |
| | **Airy Stress Function** |
| Stresses | $\sigma_{xx} = \partial^2\psi/\partial y^2$ $\sigma_{yy} = \partial^2\psi/\partial x^2$ $\tau_{xy} = -(\partial^2\psi/\partial x\partial y)$ |
| Differential equation | $\nabla^4\psi = 0$ |
| | **Torsion of non-circular shafts** |
| Torsion | $\boldsymbol{T} = \int_A (y\tau_{xz} - z\tau_{xy})dA$ |
| Saint Venant's Method | $u = \chi(y,z)(d\phi/dx)$ $v = -xz(d\phi/dx)$ $w = xy(d\phi/dx)$;<br>$\tau_{xy} = G(\partial\chi/\partial y - z)(d\phi/dx)$ $\tau_{xz} = G(\partial\chi/\partial z + y)(d\phi/dx)$ |
| Prandtl's Method | $\tau_{xy} = (\partial\psi/\partial z)$ $\tau_{xz} = -(\partial\psi/\partial y)$; $\nabla^2\psi = -2G(d\phi/dx)$; $\boldsymbol{T} = 2\iint_A \psi\, dy\, dz$ |
| Membrane Analogy | $u_m \rightarrow \psi$ $p \rightarrow 2(d\phi/dx)$ $1/S \rightarrow G$ $\partial u_m/\partial z \rightarrow \tau_{yx}$ $\partial u_m/\partial y \rightarrow -\tau_{xz}$<br>Volume beneath the membrane $\rightarrow \boldsymbol{T}/2$ |
| Thin-walled open sections | $\boldsymbol{T} = GK(d\phi/dx)$ $K = bt^3/3$ $\tau_{max} = \boldsymbol{T}t/K$ |

## PROBLEMS

### Section 6.1

**6.1** Starting with $\partial^2\gamma_{yz}/\partial y\,\partial z$, derive the compatibility expression of Equation (6.52a).

$$\frac{\partial^2\varepsilon_{yy}}{\partial z^2}+\frac{\partial^2\varepsilon_{zz}}{\partial y^2}=\frac{\partial^2\gamma_{yz}}{\partial y\,\partial z} \qquad \textbf{(6.52a)}$$

**6.2** Starting with $\partial^2\varepsilon_{zz}/\partial x\,\partial y$, derive the expression of Equation (6.52b).

$$\frac{\partial^2\varepsilon_{zz}}{\partial x\,\partial y}=\frac{1}{2}\frac{\partial}{\partial z}\left[\frac{\partial\gamma_{yz}}{\partial x}+\frac{\partial\gamma_{zx}}{\partial y}-\frac{\partial\gamma_{xy}}{\partial z}\right] \qquad \textbf{(6.52b)}$$

**6.3** Show that the Equation (6.4a) of plane stress and Equation (6.5a) of plane strain can be written in the compact form given by Equation (6.7a).

**6.4** Show that Equation (6.7a) can be written as Equation (6.7b).

### Section 6.2

**6.5** A thick cylinder (plane strain) is subjected to internal and external pressure as shown in Figure 6.5. Obtain the normal stresses $\sigma_{rr}$, $\sigma_{\theta\theta}$ and radial displacements $u_r$ at any point $r$ in terms of $R_o$, $R_i$, $p_o$, $p_i$, $G$, $\nu$, and $r$.

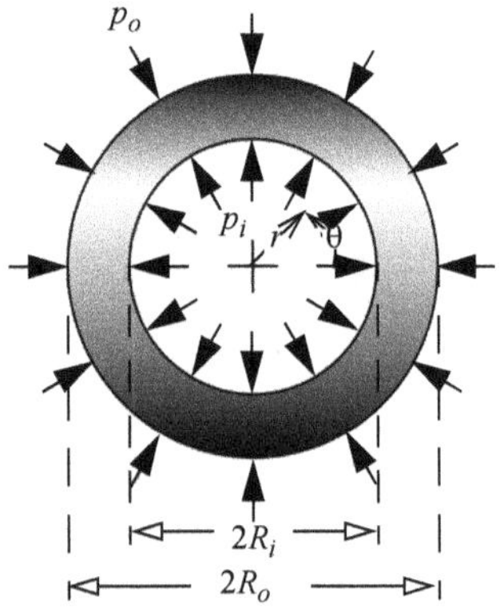


**Fig. P6.5**

**6.6** A thin disk (plane stress) is subjected to internal and external pressure as shown in Figure 6.5. Obtain the normal stresses $\sigma_{rr}$, $\sigma_{\theta\theta}$ and radial displacements $u_r$ at any point $r$ in terms of $R_o$, $R_i$, $p_o$, $p_i$, $G$, $\nu$, and $r$.

**6.7** A steel cylinder with an inside diameter of 8 inches and an outside diameter of 12 inches is subjected to an internal pressure of 12 ksi. Determine (a) the maximum tensile stress in the cylinder and (b) the radial and tangential stresses in the middle (i.e., at $r = 5$ in).

**6.8** A thick-walled cylinder having an inner radius of 6 inches is to be subjected to an internal pressure of 15 ksi. The maximum allowable tensile stress is not to exceed 25 ksi. Determine the thickness of the cylinder.

**6.9** A circular hole of radius $R$ in an infinite plate is subjected to a uniform radial pressure $p_i$. The infinite boundary is stress free. Obtain the radial and tangential stress in terms of the radial coordinate $r$, $R$, and $p_i$.

**6.10** A stress free circular hole of radius $R$ in an infinite plate is subjected to a uniform radial tensile stress of $\sigma_\infty$ at infinity. Obtain the radial and tangential stress in terms of the radial coordinate $r$, $R$, and $\sigma_\infty$.

**6.11** A steel cylinder with an inside diameter of 8 inches and an outside diameter of 12 inches has a yield stress of 30 ksi. Determine the maximum internal pressure that the cylinder can hold if von Mises stress is not to exceed yield stress.

### Section 6.3

**6.12** A thin solid disk of radius $a$, modulus of elasticity $E$, Poisson ratio $\nu$, and mass density $\rho_m$ is rotating at an angular speed of $\omega$ as shown in Fig. P6.12. Determine the radial and tangential normal stresses in terms of $a$, $E$, $\nu$, $\rho_m$, $\omega$, and $r$.

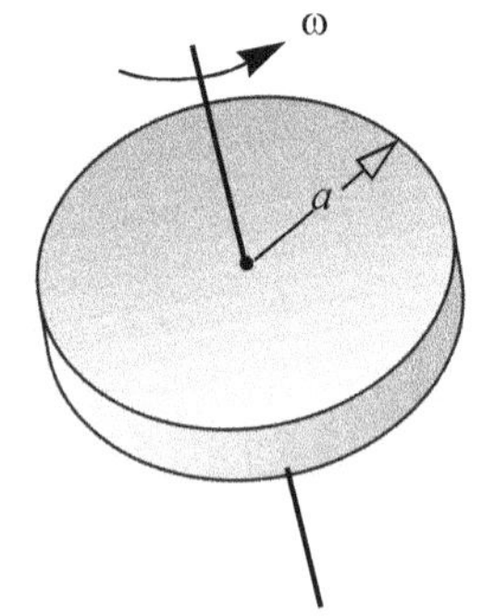


**Fig. P6.12**

**6.13** A thin disk with of radius $2a$ has a hole of radius $a$, modulus of elasticity $E$, Poisson ratio $\nu$, and mass density $\rho_m$; it is rotating at an angular speed of $\omega$ as shown in Fig. P6.13. Assume that the inner and outer surfaces are stress free. Determine the radial and tangential normal stresses in terms of $a$, $E$, $\nu$, $\rho_m$, $\omega$, and $r$.

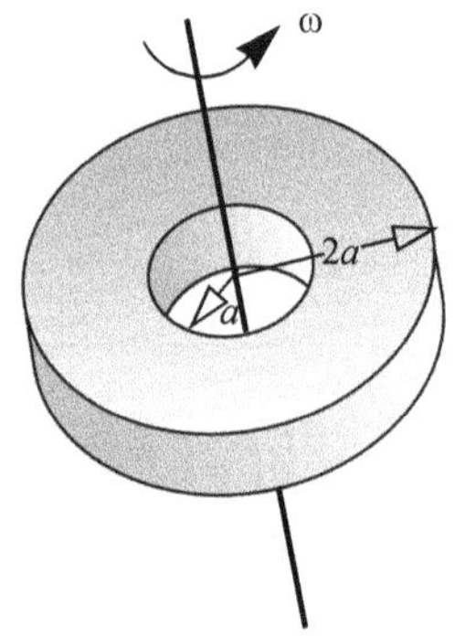


**Fig. P6.13**

**6.14** A grinding wheel has an allowable stress of 12 ksi, a modulus of elasticity of 1500 ksi, a Poisson ratio of 1/3, and a specific weight of 0.1 lb/in$^3$. The inner radius of the wheel is 3 in, and the outer radius is 6 in. Determine the maximum rotational speed (burst speed) of the grinding wheel.

### Sections 6.4-6.6

**6.15** Starting with the fifth-order polynomial given, determine the relationship between the constants for the biharmonic function to be satisfied.

$$\psi=\frac{a_5}{20}x^5+\frac{b_5}{12}x^4y+\frac{c_5}{6}x^3y^2+\frac{d_5}{6}x^2y^3+\frac{e_5}{12}xy^4+\frac{f_5}{20}y^5$$

**6.16** Determine the displacement field for the geometry shown in Fig. P6.16 and state of stress $\sigma_{xx} = \sigma y$, $\sigma_{yy} = \sigma x$, $\tau_{xy} = 0$.

Assume plane stress, a modulus of elasticity E and Poisson's ratio ν. The rectangle is $2a$ units long in the $x$ direction and $2b$ units long in the $y$-direction. Report your answer in terms of σ,$E$, ν, $x$, $y$, and the rigid body constants $\alpha_x$, $\alpha_y$, and β.

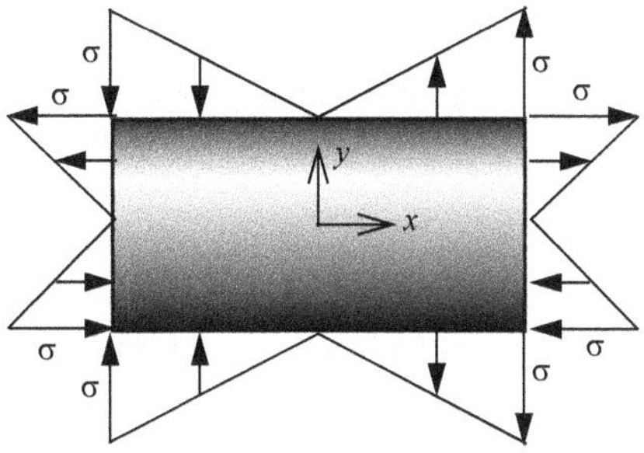


**Fig. P6.16**

**6.17** Determine the displacement field for the geometry shown in Fig. P6.16 and state of stress $\sigma_{xx} = 0$, $\sigma_{yy} = 0$, and $\tau_{xy} = \tau$ given below. Assume plane stress, a modulus of elasticity E and Poisson's ratio ν. The rectangle is $2a$ units long in the $x$ direction and $2b$ units long in the $y$-direction. Report your answer in terms of σ,$E$, ν, $x$, $y$, $a$, $b$ and the rigid body constants $\alpha_x$, $\alpha_y$, and β.

**6.18** An elastic body with a modulus of elasticity E and Poisson's ratio ν is supported as shown in Fig. P6.18. The Airy stress function for the elastic body was determined to be

$\psi = E(y^3 - 3xy^2)/3$. Assuming plane stress, determine the stresses $\sigma_{xx}$; $\sigma_{yy}$, $\tau_{xy}$, and displacements u and v in terms of $E$, ν, $x$, $y$, $a$, and $b$.

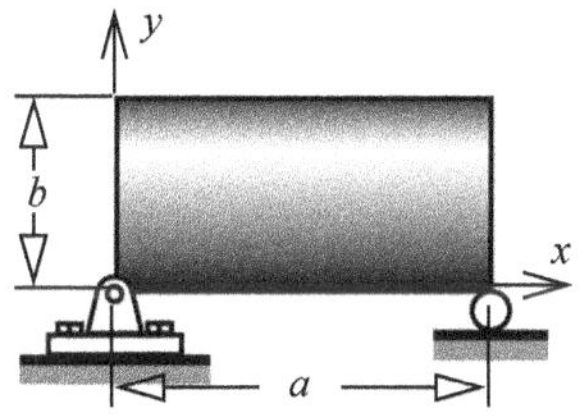


**Fig. P6.18**

**6.19** An elastic body with a modulus of elasticity E and Poisson's ratio ν is supported as shown in Fig. P6.18. The Airy stress function for the elastic body was determined to be

$\psi = E(x^3 - 3x^2y)/3$. Assuming plane stress, determine the stresses $\sigma_{xx}$; $\sigma_{yy}$, $\tau_{xy}$, and displacements u and v in terms of $E$, ν, $x$, $y$, $a$, and $b$.

## Sections 6.7-6.8

**6.20** Two solid elliptical steel ($G_{st}$ = 80 GPa) shafts and a solid elliptical bronze ($G_{Cu/Sn}$ = 40 GPa) shaft are securely connected by a coupling at $C$. The major diameter of the elliptical cross section is 100 mm, and the minor diameter is 80 mm. A torque of $T$ = 10 kN · m is applied to the rigid wheel $B$ as shown in Fig. P6.20. The coupling plates cannot rotate relative to each other. Determine the angle of rotation of the wheel $B$ due to the applied torque and the maximum shear stress in the shaft. Use the results of Example 6.5 to solve the problem.

**6.21** The stress function for the cross section shown in Fig. P6.21, shaped as an equilateral triangle, is given by the equation

$$\psi = K[y + z\sqrt{3} - 2h/3] \times [y - z\sqrt{3} - 2h/3][y + h/3]$$

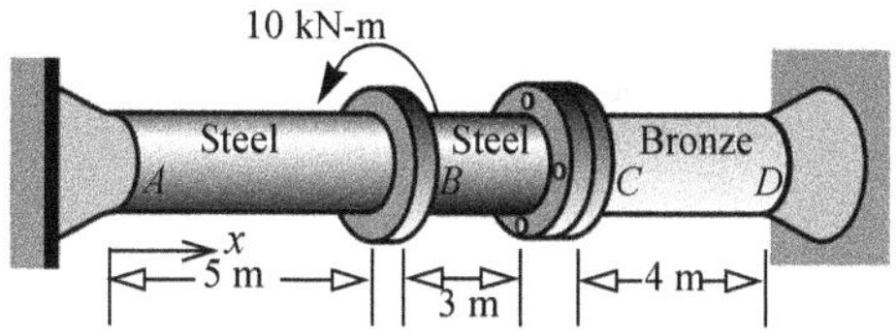


**Fig. P6.20**

where $K$ is a constant to be determined. Determine the maximum shear stress and the rate of twist in terms of internal torque $T$, shear modulus $G$, and $h$.[Hint: Substitute $y + h/3 = \eta$ and $h - z\sqrt{3} = \xi$. in the integral]

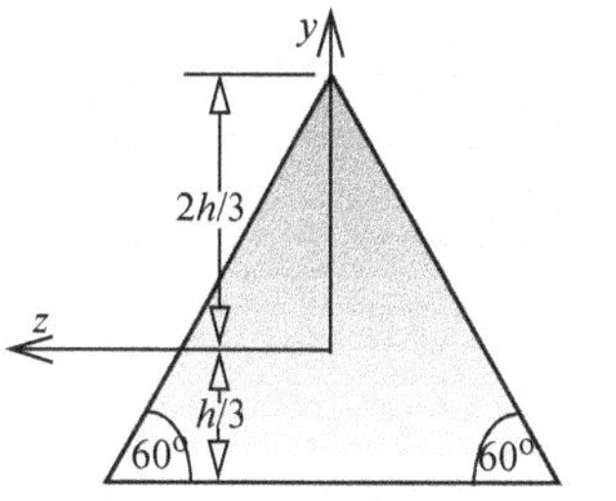


**Fig. P6.21**

**6.22** The shaft shown in Fig. P6.22 has the cross section of an equilateral triangle. Each side of the triangle is 200 mm, and the shear modulus of elasticity is 70 GPa. Use the results of Problem 6.21 to determine the maximum shear stress in the shaft and the rotation of section $D$ with respect to section $A$.

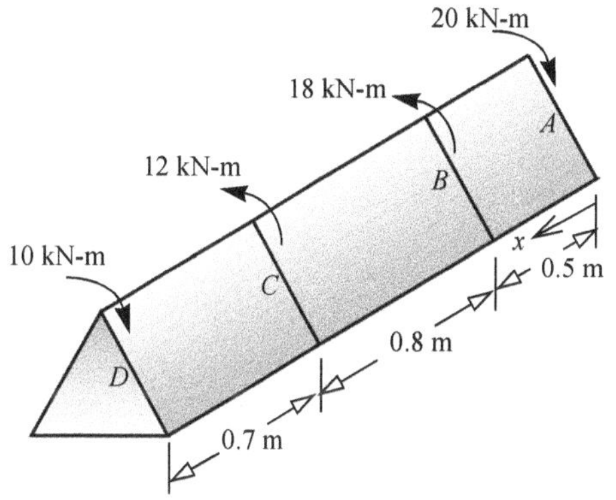


**Fig. P6.22**

## Sections 6.9

**6.23** The thin-walled open section shown in Fig. P6.23 is subjected to a torque of 5 in-kips. Determine the torsional rigidity and maximum torsional shear stress. Use shear modulus of elasticity of G = 12,000 ksi.

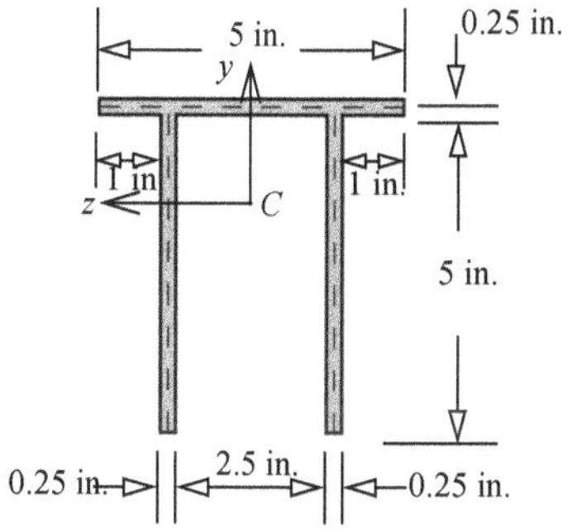


**Fig. P6.23**

**6.24** Determine the torsional rigidity and maximum torsional shear stress due to torsion for the thin open section shown in Figure

6.25. Assume $t \ll a$. Report the answer in terms of internal torque ***T***, shear modulus *G*, thickness *t*, and parameter *a*.

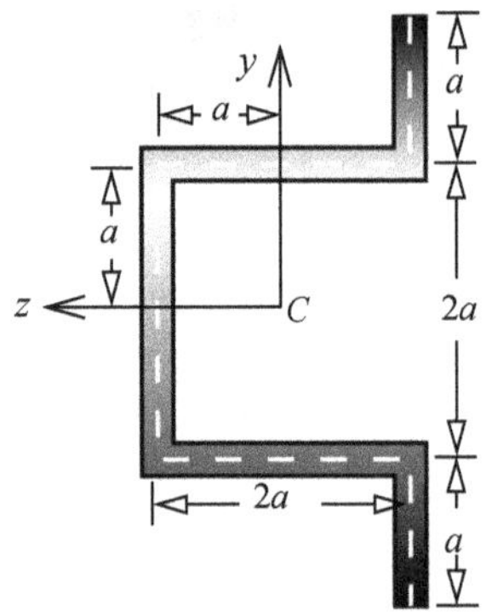


**Fig. P6.24**

**6.25** Determine the torsional rigidity and maximum torsional shear stress due to torsion for the thin open section shown in Figure 6.25. Assume $t \ll R$ and gap at O is of negligible thickness. Report the answer in terms of internal torque ***T***, shear modulus *G*, thickness *t*, and radius *R*.

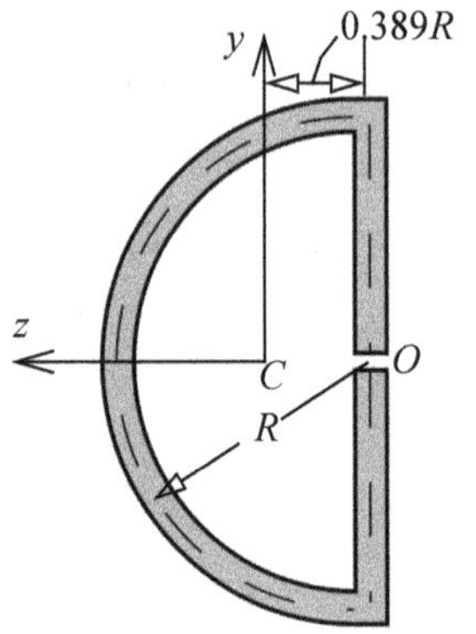


**Fig. P6.25**

# 7 Variational and Energy Methods

## LEARNING OBJECTIVE

1. Understand the concepts in variational calculus.
2. Understand the use of variational calculus to obtain boundary value problems in mechanics of materials.
3. Understand the use of variational calculus in approximate methods of Rayleigh-Ritz and Finite Element Method.

Strain energy, the potential of work, and potential energy are all functions of the displacements. The displacements are functions of the position coordinates. Thus, strain energy, the potential of work, and potential energy are functions of functions. Such functions of functions are called **functionals**. **Variational calculus** (Lanczos [1986]) is the branch of mathematics in which the maximum and minimum values of functionals are determined. Designing structures for minimum weight is a problem in optimization in which the weight is the functional that is minimized. Industrial engineering contains many applications of variational calculus involving minimizing cost, time, and so on.

The principles used in energy methods are a sub-class of the principles defined in variational calculus. We will use the theorem of minimum potential energy to obtain the boundary value problems first for linear problems and then for geometric and material nonlinear problems. We will also use theorem of minimum potential energy to develop the approximate methods of Rayleigh-Ritz and Finite Element Method.

## 7.1 BASIC CONCEPTS IN VARIATIONAL CALCULUS

Our vocabulary has lot of superlative terms like largest, smallest, quickest, least, most, that is, words that reflect extreme values of a given attribute. But how do we determine an extreme value?

Suppose we wish to decide who is the tallest person in a room? What do we do? We compare heights of two people and decide who is taller. We next compare the taller of the two people with the next person and repeat this till we conclude Jill is the tallest person in the room. Suppose we now ask who is the tallest person in the building? We once more compare the heights of people in the building and conclude that Jack is the tallest person. We next ask who is the tallest person in the university is and go through the exercise of comparing heights of all people in the university. This example highlights two important points.

1. To find an extreme value we need to compare.
2. The comparison is in the immediate neighborhood and will give us only a local extremum value and not a global extremum.

The second point has tremendous impact on solution procedures of non-linear problems. Because global extremum (say a minimum value) cannot be guaranteed unless some additional conditions are specified.

Now suppose we wish to find the highest point in our city. We can conduct the height comparison in two ways.

1. We can walk to different points and measure the elevation— we will call it the d-process; so d$x$, d$s$, d$u$ represent the **actual movement** along a path (curve).
2. We can conduct a thought experiment. For example, without moving, we ask the question, if we go to that point will the elevation increase or decrease? This imaginary movement is called the **virtual movement** or the δ-process.

Figure 7.1 shows the actual displacement of beam under uniform load and a thought experiment (virtual displacement) we may consider in an effort to find the actual displacement. The functions and its derivatives are related in the actual displacements as we move along the curve. But in the virtual displacement we can change the function and it derivatives (slope) independently. In many aspects the d-process and the δ-process behave mathematically in a similar manner, but the difference between the two is critical and will be elaborated further a little later. Note the virtual displacement in Figure 7.1*b* satisfies the zero displacement condition at the support and is a continuous function. We will soon make use of this observation.

**Figure 7.1** Difference between actual and virtual displacement.

When a variable in a set cannot be represented as a function of the other variables then it is said to be **independent**. If all $a$'s in the equation below are non-zero, then we could solve one of the $u$'s in terms of others and hence by definition the set is not independent.

$$a_1u_1 + a_2u_2 + a_3u_3 \bullet \quad \bullet \quad \bullet \quad \bullet \quad + a_nu_n = 0 \tag{7.1}$$

Alternatively stated,

$$\boxed{\text{If } u_1, u_2, u_3, \ldots u_n \text{ are independent functions then Equation (7.1) implies } a_i = 0 \qquad i = 1 \text{ to } n} \tag{7.2}$$

Any set of independent variables (parameters) that describes the system geometry are called the **generalized coordinates.** Displacements of points, rotation of lines, coefficients of a polynomial representing displacements are some examples of generalized coordinates. The space spanned by the generalized coordinates is called the **configuration space**. Any condition that limits the change in geometry in the configuration space is called the **kinematic condition.** A point cannot move, a line cannot rotate, are some examples of kinematic conditions.

Functions that are *continuous* and satisfy *all* the *kinematic boundary conditions* are called **kinematically admissible functions**. The actual displacement solution is always a kinematically admissible function. But unlike actual displacements, the kinematically admissible functions can result in forces and moments that do not satisfy the equilibrium or boundary conditions on forces and moments. *Virtual displacement must always be kinematically admissible.*

### 7.1.1 Extremum and stationary Values

Suppose we want to find the minimum of a function of independent variables $u_1, u_2, u_3 \ldots u_n$ as shown below.

$$F = F(u_1, u_2, \quad \bullet \quad \bullet \quad \bullet \quad + u_n) \tag{7.3a}$$

We consider a virtual change in the configuration space, that is, space spanned by the independent variables $u_1, u_2, u_3 \ldots u_n$. The total virtual change $\delta F$ is the sum of slope multiplied by virtual change in each direction.

$$\delta F = \frac{\partial F}{\partial u_1}\delta u_1 + \frac{\partial F}{\partial u_2}\delta u_2 + \frac{\partial F}{\partial u_3}\delta u_3 \bullet \quad \bullet \quad \bullet \quad \bullet \quad + \quad + \frac{\partial F}{\partial u_n}\delta u_n \tag{7.3b}$$

$\delta F$ is called the **first variation** of $F$. If $F$ is to be a minimum at a point in the configuration space, then this change of $\delta F$ must be zero.

$$\delta F = \frac{\partial F}{\partial u_1}\delta u_1 + \frac{\partial F}{\partial u_2}\delta u_2 + \frac{\partial F}{\partial u_3}\delta u_3 \bullet \quad \bullet \quad \bullet \quad \bullet \quad + \quad + \frac{\partial F}{\partial u_n}\delta u_n = 0 \tag{7.3c}$$

If $u_1, u_2, u_3 \ldots u_n$ are independent variables then we are free to move in any direction. So if we only walk in $u_1$ (all other virtual displacement are zero) then we have $\partial F / \partial u_1 = 0$. In a similar manner we can walk in each of the directions and conclude:

$$\frac{\partial F}{\partial u_i} = 0 \qquad i = 1, 2, 3 \bullet \quad \bullet \quad \bullet n \tag{7.4}$$

$\delta F$ is also zero at maximum and at saddle point. To decide whether we have minimum, maximum, or saddle point, that is an extreme value (extremum), we need to consider the sign of **second variation** of $F$, that is sign of $\delta^2 F$. A function is said to have a **stationary value** if it does not change in the immediate neighborhood of $\delta F = 0$. Thus, to decide an extreme value we must consider both $\delta F = 0$ and sign of $\delta^2 F$, but for stationary value we need to consider only $\delta F = 0$.

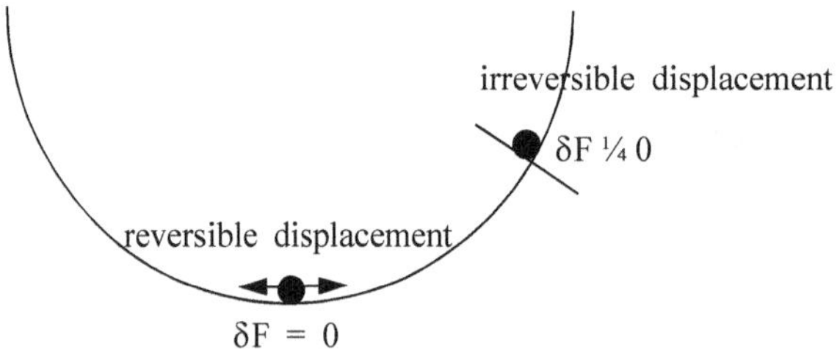


**Figure 7.2** Reversible and irreversible displacements.

Figure 7.2 brings out another difference between the extremum and stationary value. The marble at the bottom of the trough satisfies conditions for extremum and stationary value. But a marble prevented from rolling down to the bottom shows an extreme value (lowest point) but $\delta F$ will not be zero. In comparing heights, marble at the lowest point can be moved in any direction, that is, the displacement is **reversible.** But when the marble is prevented from rolling down to the lowest point, comparison of height is only possible in one direction, that is, displacement is irreversible. *For stationary values we need the virtual displacement to be reversible.*

### 7.1.2 Functionals

Regular calculus considers the variation of a function. Variational calculus considers the variation of functionals. We will consider some properties of functionals in this section. These properties will be used to simplify algebra in subsequent sections.

A **function** $u(x)$ is a rule of correspondence such that for all $x$ in $D$ there is assigned a unique element $u(x)$ in $R$. A **functional** $F[u(x)]$ is a rule of correspondence such that for all $u(x)$ in $R$ there is assigned a unique element $F[u(x)]$ in $\Omega$ In other words, a functional is a function of a function.

A **linear functional** $l(u)$ is one that satisfies the relationship in Equation (7.5a).

$$l(\alpha_1 u + \alpha_2 v) = \alpha_1 l(u) + \alpha_2 l(v) \tag{7.5a}$$

where $\alpha_1$ and $\alpha_2$ are any scalars. Work potential discussed in Section 7.2 is a linear functional of displacements.

A **bilinear functional** $B(u, v)$ is one that is a linear functional in each of its arguments of $u$ and v, as shown in Equation (7.5b).

$$B(\alpha_1 u_1 + \alpha_2 u_2, v) = \alpha_1 B(u_1, v) + \alpha_2 B(u_2, v) \qquad B(u, \alpha_1 v_1 + \alpha_2 v_2) = \alpha_1 B(u, v_1) + \alpha_2 B(u, v_2) \tag{7.5b}$$

A **symmetric bilinear functional** is a bilinear functional that is symmetric with respect to its arguments, as shown in Equation (7.5c). Strain energy discussed in Section 7.3 is a symmetric bilinear functional.

$$B(u, v) = B(v, u) \tag{7.5c}$$

In the above definitions of linear and bilinear functionals, $u$ and v can be vectors, that is, each of the argument can have many components and the property of the functionals are applicable to each of the component.

## 7.2 WORK

Work is done by a force if the point at which the force is applied moves. If the point at which force $\vec{F}$ is applied moves through an infinitesimal distance $d\vec{u}$, then the work is defined as

$$dW = \vec{F} \cdot d\vec{u} \tag{7.6}$$

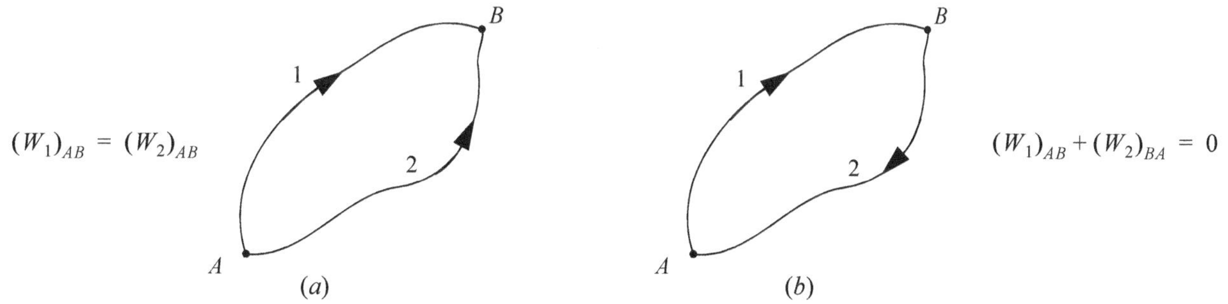


**Figure 7.3** Conservative work.

Integrating the work expression in Equation (7.6) along the path of movement gives the total work done by the force. Figure 7.3 shows two ways in which work done is conserved. If the start and end points are the same and the work is path independent, then the total work done must be zero as the work expended in moving a point forward is recovered (conserved) when the point moves back to the starting point. Work done by a force is **conserved** if it is path independent. Clearly, if friction is present, the work done to overcome friction must vary with the length of path; hence the frictional force is nonconservative. In a similar manner, if a body deforms plastically, the work done to create permanent deformation will not be recovered, and once more we have a nonconservative system. Thus, elastic deformation is a conservative deformation, while plastic deformation is nonconservative. Elastic deformation can, however, be linear or nonlinear. Rubber has nonlinear stress–strain curve. Work done in stretching rubber is recovered when the forces are released and the rubber returns to the undeformed position. Thus, there is a distinction between nonconservative and nonlinear systems.

- Nonlinear systems and nonconservative systems are two independent descriptions of a system.

Equation (7.6) shows that work is a scalar quantity. Hence work done by different forces and moments can be added to obtain the total work done by the forces acting on a structure. We now consider work done by forces and moments on axial rod, shafts, beams, plates, and in an elastic plane which we can later add to obtain the total work done on the structure.

**Axial Member**: Consider an axial member subjected to a distributed force per unit length of $p_x(x)$. Thus, the force acting on a differential element will be $p_x(x)dx$. In addition, we assume there are $m$ concentrated forces $F_q$ acting at points $x_q$. Multiplying these forces by the displacement of the points and adding we obtain the total work done on the axial member $W_A$ as shown by the Equation (7.7a). The integration is over the length $L$ of the axial member. In Equation (7.7a), we assumed that $p_x(x)$, $u(x)$, $F_q$, and $u(x_q)$ are all positive in the positive $x$ direction. We note that the $W_A$ is a linear functional of $u$ as shown in the equation.

**Torsion of shaft:** In a manner similar to an axial member, a shaft subjected to a distributed torque per unit length $t(x)$, and $m$ concentrated torques $T_q$ applied at points $x_q$ will yield the total work done on the shaft $W_T$ as shown by Equation (7.7b). The quantities $t(x)$, $\phi(x)$, $T_q$, and $\phi(x_q)$, are all positive counter-clockwise with respect to the $x$ axis.

**Bending of symmetric beams**: The work $W_B$ done on the symmetric beam in a similar manner is given by the Equation (7.7c). Where, v and $p_y(x)$ are the displacement and the distributed force per unit length in the $y$ direction, $F_q$ are the transverse forces acting at points $x_q$, and $M_q$ are bending moments acting at points $x_q$. Displacements and forces are positive in the positive $y$ directions, and moments and slopes are positive counter-clockwise about the $z$ direction (bending axis).

**Bending of thin plates**: For thin plates, we will consider only the transverse distributed force per unit area $p_z(x, y)$ acting in the positive $z$ direction. Line loads, and concentrated forces and moments can be incorporated but results in long expressions which do not add to the concepts discussed. Defining $w(x, y)$ as the positive displacement in the z direction, we obtain the work done on the plate $W_P$ as an integral over the area $A$ of the plate as shown in Equation (7.7d).

**Elastic plane:** We consider only body forces $F_x(x, y)$ and $F_y(x, y)$ for plane stress elasticity problems. We assume the uniform thickness of the plane is $h$, and the displacements in $x$ and $y$ directions are $u(x, y)$ and $v(x, y)$. The work done in plane stress elasticity $W_E$ is given by the Equation (7.7e).

**Table 7.1** Work Expressions

| | Work | |
|---|---|---|
| Axial | $W_A = \int_L p_x(x)u(x)dx + \sum_{q=1}^{m} F_q u(x_q) = l(u)$ | **(7.7a)** |
| Torsion of circular shafts | $W_T = \int_L t(x)\phi(x)dx + \sum_{q=1}^{m} T_q \phi(x_q) = l(\phi)$ | **(7.7b)** |
| Symmetric bending of beams | $W_B = \int_L p_y(x)v(x)dx + \sum_{q=1}^{m_1} F_q v(x_q) + \sum_{q=1}^{m_2} M_q \frac{dv}{dx}(x_q) = l(v)$ | **(7.7c)** |
| Bending of thin plates | $W_P = \iint_A p_z(x, y)w(x, y)dxdy = l(w)$ | **(7.7d)** |
| Plane stress elasticity | $W_E = h\iint_A [F_x(x, y)u(x, y) + F_y(x, y)v(x, y)]dxdy = l(u, v)$ | **(7.7e)** |

## 7.3 STRAIN ENERGY

The change in internal energy in a body during deformation is called the **strain energy**. The energy per unit volume is called the **strain energy density** and is the area under the stress–strain curve up to the point of deformation.
Equation (7.8) shows the relationship between the strain energy and the strain energy density,

$$U = \int_V U_0 \, dV \tag{7.8}$$

where $U$ is the strain energy, $U_0$ is the strain energy density, and $V$ is the volume of the body. Noting that the strain energy density is the area under the curve shown in Figure 7.4, we obtain

$$U_0 = \int_0^{\varepsilon} \sigma \, d\varepsilon \tag{7.9}$$

Equation (7.9) shows that strain energy density has the same dimensions as that of stress because strain is dimensionless. But the units of strain energy density are *units of energy per unit volume,* which are different from those of stress. The units for strain energy density are Newton-meters per cubic meter (N · m/m$^3$), joules per cubic meter (J/m$^3$), inch-pounds per cubic inch (in · lb/in$^3$), and foot-pounds per cubic foot (ft · lb/ft$^3$).

Another related concept is the **complementary strain energy density** ($\overline{U}_0$), shown in Figure 7.4 and defined as follows:

$$\overline{U}_0 = \int_0^{\sigma} \varepsilon \, d\sigma \tag{7.10}$$

Most engineering structures are designed for linear elastic materials. In the linear region, the area under the stress–strain curve is a triangle, and we obtain Equation (7.11) for the strain energy density function.

$$U_0 = \sigma\varepsilon/2 \tag{7.11}$$

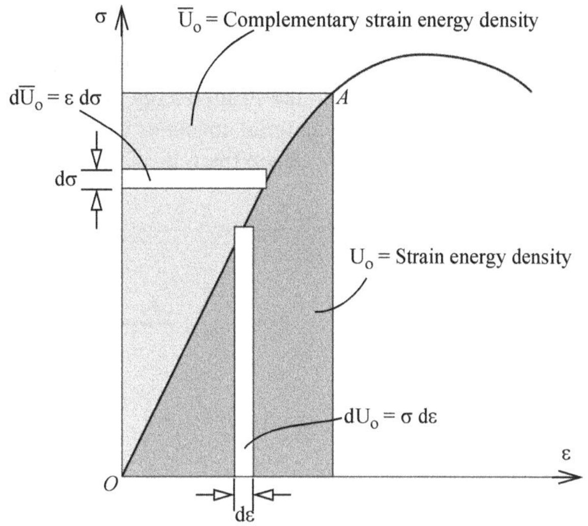


**Figure 7.4** Energy densities.

If instead of a curve representing normal stress vs. normal strain, we have a curve of shear stress vs. shear strain, then we will have a similar expression for strain energy density in terms of shear stress and shear strain, as shown by Equation (7.12).

$$U_0 = \int_0^{\gamma} \tau \, d\gamma = \tau\gamma/2 \tag{7.12}$$

Strain energy, hence strain energy density, is a scalar quantity. We can add the strain energy density due to individual stress and strain components to obtain Equation (7.13) for the total *linear strain energy density* during deformation.

$$\boxed{U_0 = \frac{1}{2}[\sigma_{xx}\varepsilon_{xx} + \sigma_{yy}\varepsilon_{yy} + \sigma_{zz}\varepsilon_{zz} + \tau_{xy}\gamma_{xy} + \tau_{yz}\gamma_{yz} + \tau_{zx}\gamma_{zx}]} \tag{7.13}$$

In the above equations the strain is the mechanical strain. If thermal strains are present then they need to be subtracted from the total strain as shown below.

$$U_0 = \frac{1}{2}[\sigma_{xx}(\varepsilon_{xx} - \alpha\Delta T) + \sigma_{yy}(\varepsilon_{yy} - \alpha\Delta T) + \sigma_{zz}(\varepsilon_{zz} - \alpha\Delta T) + \tau_{xy}\gamma_{xy} + \tau_{yz}\gamma_{yz} + \tau_{zx}\gamma_{zx}] \tag{7.14}$$

## 7.3.1 Strain energy in symmetric bending of beams

We now consider the strain energy expressions for symmetric bending about the *z* axis. Expressions for axial and torsion can be obtained in a similar manner and is left as an exercise for the reader (see problems 7.1 and 7.2).

When symmetric sections are bent about the *z* axis, there are two nonzero stress components, $\sigma_{xx}$ and $\tau_{xy}$. We shall consider strain energy due to each separately. Substituting $\sigma_{xx} = E\varepsilon_{xx}$ and $\varepsilon_{xx} = -y(d^2v/dx^2)$ in Equation (7.13) and noting that $(d^2v/dx^2)$ is just a function of *x* and does not change across the cross-section, we obtain the bending strain energy $U_B$ as

$$U_B = \int_V \frac{1}{2}E\varepsilon_{xx}^2 \, dV = \int_L \left[\int_A \frac{1}{2}E\left(y\frac{d^2v}{dx^2}\right)^2 dA\right]dx = \int_L \left[\frac{1}{2}\left(\frac{d^2v}{dx^2}\right)^2 \int_A Ey^2 dA\right]dx \tag{7.15a}$$

where, *L* is the length of the beam and *A* is the cross sectional area. Assuming material is homogeneous across the cross section, that is, *E* does not change in the integral, we obtain

$$\int_A Ey^2 dA = E\int_A y^2 dA = EI_{zz} \tag{7.15b}$$

Substituting Equation (7.15b) into Equation (7.15a) we obtain Equation (7.17c). For composite symmetric beams, the integral over the area *A* would be replaced by an integral over each material area, and the value of the integral in Equation (7.15b) would be the sum of the bending rigidities of all the materials.

The strain energy due to shear in bending is $U_S = (1/2)\int \tau_{xy}\gamma_{xy} dV$. We note that the maximum shear stress $\tau_{xy}$ and shear strain $\gamma_{xy}$ are an order of magnitude smaller than the maximum normal stress $\sigma_{xx}$ and the maximum normal strain $\varepsilon_{xx}$. Thus, $U_S$ will be two orders of magnitude smaller than $U_B$ and can be neglected in our calculations.

## 7.3.2 Strain energy in bending of thin plates

From Chapter 5, on bending of thin plates with no inplane forces, we have the following equations for strains and stresses.

$$\varepsilon_{xx} = \frac{\partial u}{\partial x} = -z\frac{\partial^2 w}{\partial x^2} \qquad \varepsilon_{yy} = \frac{\partial v}{\partial y} = -z\frac{\partial^2 w}{\partial y^2} \qquad \gamma_{xy} = \frac{\partial u}{\partial y} + \frac{\partial v}{\partial x} = -2z\frac{\partial^2 w}{\partial x \partial y} \tag{7.16a}$$

$$\sigma_{xx} = \frac{E}{(1-\nu^2)}(\varepsilon_{xx} + \nu\varepsilon_{yy}) \qquad \sigma_{yy} = \frac{E}{(1-\nu^2)}(\varepsilon_{yy} + \nu\varepsilon_{xx}) \qquad \tau_{xy} = G\gamma_{xy} \tag{7.16b}$$

The transverse shear stresses $\tau_{xz}$ and $\tau_{yz}$ are an order of magnitude less than the inplane stresses, hence the strain energy due to these shear stresses will be negligible compared to the strain energy due to the inplane stresses and we neglect it. For simplicity we assume that the plate is homogeneous, hence material constants can be taken outside the integrals. Substituting Equations (7.16a) and (7.16b) into Equation (7.13) and (7.8) we obtain the following equations for the plate strain energy $U_P$.

$$U_P = \frac{1}{2}\int_V\left[\frac{E}{(1-\nu^2)}(\varepsilon_{xx} + \nu\varepsilon_{yy})\varepsilon_{xx} + \frac{E}{(1-\nu^2)}(\varepsilon_{yy} + \nu\varepsilon_{xx})\varepsilon_{yy} + \frac{E\gamma_{xy}^2}{2(1+\nu)}\right]dV$$

$$U_P = \frac{1}{2}\frac{E}{(1-\nu^2)}\int_V\left[\varepsilon_{xx}^2 + \varepsilon_{yy}^2 + 2\nu\varepsilon_{xx}\varepsilon_{yy} + \frac{(1-\nu)}{2}\gamma_{xy}^2\right]dV$$

$$U_P = \frac{1}{2}\frac{E}{(1-\nu^2)}\int_V\left[z^2\left(\frac{\partial^2 w}{\partial x^2}\right)^2 + z^2\left(\frac{\partial^2 w}{\partial y^2}\right)^2 + 2\nu z^2\left(\frac{\partial^2 w}{\partial x^2}\right)\left(\frac{\partial^2 w}{\partial y^2}\right) + 2z^2(1-\nu)\left(\frac{\partial^2 w}{\partial x\partial y}\right)^2\right]dV$$

We can write the volume integral as integral over the area ($A$) and integral over the thickness from ($-h/2$) to ($h/2$), where $h$ is the plate thickness. We further note that the curvatures are not functions of $z$, hence can be taken outside the integral over thickness. We obtain

$$U_P = \frac{1}{2}\frac{E}{(1-\nu^2)}\left[\iint_A\left\{\left(\frac{\partial^2 w}{\partial x^2}\right)^2\int_{-h/2}^{h/2} z^2 dz + \left(\frac{\partial^2 w}{\partial y^2}\right)^2\int_{-h/2}^{h/2} z^2 dz + 2\nu\left(\frac{\partial^2 w}{\partial x^2}\right)\left(\frac{\partial^2 w}{\partial y^2}\right)\int_{-h/2}^{h/2} z^2 dz + 2(1-\nu)\left(\frac{\partial^2 w}{\partial x\partial y}\right)^2\int_{-h/2}^{h/2} z^2 dz\right\}dxdy\right]$$

Noting that the plate's bending rigidity is given by $D = Eh^3/[12(1-\nu^2)]$, we obtain Equation (7.17d).

**Table 7.2** Strain Energy

| | Strain Energy | |
|---|---|---|
| Axial | $U_A = \frac{1}{2}\int_L EA\left(\frac{du}{dx}\right)^2 dx$ | **(7.17a)** |
| Torsion of circular shafts | $U_T = \frac{1}{2}\int_L GJ\left(\frac{d\phi}{dx}\right)^2 dx$ | **(7.17b)** |
| Symmetric bending of beams | $U_B = \frac{1}{2}\int_L EI_{zz}\left(\frac{d^2 v}{dx^2}\right)^2 dx$ | **(7.17c)** |
| Thin Plates | $U_P = \frac{D}{2}\iint_A\left\{\left(\frac{\partial^2 w}{\partial x^2}\right)^2 + \left(\frac{\partial^2 w}{\partial y^2}\right)^2 + 2\nu\left(\frac{\partial^2 w}{\partial x^2}\right)\left(\frac{\partial^2 w}{\partial y^2}\right) + 2(1-\nu)\left(\frac{\partial^2 w}{\partial x\partial y}\right)^2\right\}dxdy$ | **(7.17d)** |
| Plane Stress Elasticity | $U_E = \frac{Eh}{2(1-\nu^2)}\iint_A\left[\left\{\left(\frac{\partial u}{\partial x}\right)^2 + 2\nu\left(\frac{\partial u}{\partial x}\right)\left(\frac{\partial v}{\partial y}\right) + \left(\frac{\partial v}{\partial y}\right)^2\right\} + \frac{(1-\nu)}{2}\left(\frac{\partial u}{\partial y} + \frac{\partial v}{\partial x}\right)^2\right]dxdy$ | **(7.17e)** |

### 7.3.3 Strain energy in plane stress elasticity

Equation (6.4a) and Equation (6.4b) for plane stress are re-written below for convenience.

$$E\varepsilon_{xx} = \sigma_{xx} - \nu\sigma_{yy} \qquad E\varepsilon_{yy} = \sigma_{yy} - \nu\sigma_{xx} \qquad G\gamma_{xy} = \tau_{xy} \tag{7.18a}$$

$$\sigma_{xx} = E[\varepsilon_{xx} + \nu\varepsilon_{yy}]/(1-\nu^2) \qquad \sigma_{yy} = E[\varepsilon_{yy} + \nu\varepsilon_{xx}]/(1-\nu^2) \qquad \tau_{xy} = G\gamma_{xy} \tag{7.18b}$$

For plane stress elasticity problems all stresses with subscripts $z$ are zero. From Equation (7.13) we obtain

$$U_0 = \frac{1}{2}[\sigma_{xx}\varepsilon_{xx} + \sigma_{yy}\varepsilon_{yy} + \tau_{xy}\gamma_{xy}] \tag{7.19a}$$

Substituting Equation (7.18b) into Equation (7.19a), and then substituting the strains in terms of displacements, we obtain

$$U_0 = \frac{1}{2}\left[\frac{E}{(1-\nu^2)}\{\varepsilon_{xx}^2 + 2\nu\varepsilon_{xx}\varepsilon_{yy} + \varepsilon_{yy}^2\} + \frac{E}{2(1+\nu)}\gamma_{xy}^2\right]$$

$$U_0 = \frac{E}{2(1-\nu^2)}\left[\left\{\left(\frac{\partial u}{\partial x}\right)^2 + 2\nu\left(\frac{\partial u}{\partial x}\right)\left(\frac{\partial v}{\partial y}\right) + \left(\frac{\partial v}{\partial y}\right)^2\right\} + \frac{(1-\nu)}{2}\left(\frac{\partial u}{\partial y} + \frac{\partial v}{\partial x}\right)^2\right] \tag{7.20}$$

We assume the uniform thickness of the plane is $h$, thus the differential volume is $dV = hdxdy$. Substituting Equation (7.20) into Equation (7.8), we obtain the plane stress elasticity strain energy $U_E$ as shown in Equation (7.17e).

### 7.3.4 Strain energy in form of bilinear functional

Table 7.3 shows the strain energy written as a bilinear functional. By substituting $u_1 = u_2 = u$, $\phi_1 = \phi_2 = \phi$, $v_1 = v_2 = v$, and $w_1 = w_2 = w$ into the equations of Table 7.3 we obtain the equations of Table 7.2. Note the bilinear functional is symmetric in all cases shown in Table 7.3.

**Table 7.3** Bilinear functional form of strain energy.

| | Strain Energy | |
|---|---|---|
| Axial | $U_A = \frac{1}{2}\int_L \left[EA\frac{du_1}{dx}\frac{du_2}{dx}\right]dx = \frac{1}{2}B(u_1, u_2)$ | **(7.21a)** |
| Torsion of circular shafts | $U_T = \frac{1}{2}\int_L \left[GJ\frac{d\phi_1}{dx}\frac{d\phi_2}{dx}\right]dx = \frac{1}{2}B(\phi_1, \phi_2)$ | **(7.21b)** |
| Symmetric bending of beams | $U_B = \frac{1}{2}\int_L \left[EI_{zz}\frac{d^2 v_1}{dx^2}\frac{d^2 v_2}{dx^2}\right]dx = \frac{1}{2}B(v_1, v_2)$ | **(7.21c)** |
| Thin plates | $U_P = \frac{D}{2}\iint_A \left[\frac{\partial^2 w_1}{\partial x^2}\frac{\partial^2 w_2}{\partial x^2} + \frac{\partial^2 w_1}{\partial y^2}\frac{\partial^2 w_2}{\partial y^2} + v\left(\frac{\partial^2 w_1}{\partial x^2}\frac{\partial^2 w_2}{\partial y^2} + \frac{\partial^2 w_2}{\partial x^2}\frac{\partial^2 w_1}{\partial y^2}\right) + 2(1-v)\frac{\partial^2 w_1}{\partial x\partial y}\frac{\partial^2 w_2}{\partial x\partial y}\right]dx\,dy = \frac{1}{2}B(w_1, w_2)$ | **(7.21d)** |
| Plane stress elasticity | $U_E = \frac{Eh}{2(1-v^2)}\iint_A \left[\left\{\frac{\partial u_1}{\partial x}\frac{\partial u_2}{\partial x} + v\left(\frac{\partial u_2}{\partial x}\frac{\partial v_1}{\partial y} + \frac{\partial u_1}{\partial x}\frac{\partial v_2}{\partial y}\right) + \frac{\partial v_1}{\partial y}\frac{\partial v_2}{\partial y}\right\} + \frac{(1-v)}{2}\left(\frac{\partial u_1}{\partial y} + \frac{\partial v_1}{\partial x}\right)\left(\frac{\partial u_2}{\partial y} + \frac{\partial v_2}{\partial x}\right)\right]dx\,dy$ $= \frac{1}{2}B(u_1, v_1, u_2, v_2)$ | **(7.21e)** |

## 7.4 VIRTUAL WORK

Virtual work methods are applicable to linear and nonlinear systems, to conservative as well as nonconservative systems. The principle of virtual work is deceptively simple. It states:

$$\boxed{\text{The total virtual work done on a body at equilibrium is zero.}} \tag{7.22}$$

Virtual work implies it is not actual work but work done by actual forces in moving points through virtual displacements, or, virtual forces moving through actual displacement. We will focus only on virtual displacements and the work associated with it. Symbolically virtual work can be written as

$$\delta W = 0 \tag{7.23}$$

The total virtual work can be divided into work done by external and internal forces. Since the internal forces are always opposed to the external forces, the internal virtual work will always be opposite in sign to the external virtual work. We rewrite Equation (7.23) as

$$\delta W_{ext} = \delta W_{int} \tag{7.24}$$

## 7.5 MINIMUM POTENTIAL ENERGY

We define the potential energy function $\Omega$ as

$$\Omega = U - W \tag{7.25}$$

where $U$ is the strain energy and $W$ is the work potential of a force. Unlike work, which is a concept associated with any force (conservative or nonconservative) that moves, the "work potential of a force" is associated with conservative forces only and implies that there is a potential function from which such a force can be obtained. Gravitational forces, electromagnetic forces, spring forces are some examples of conservative forces that can be obtained from a potential function. If we assume that the external forces are acting on an *elastic* (linear or nonlinear) structure and are conservative, then the "work potential of a force" is calculated as the work term in Table 7.1. Thus, from our perspective, the work potential and the work done by the force are calculated similarly, provided the external forces are conservative and are applied to elastic systems.

To obtain the statement of the theorem of minimum potential energy, we consider the statement of virtual work as given by Equation (7.24). Restricting ourselves to conservative elastic systems, we can state that the internal virtual work is the variation in elastic strain energy during deformation ($\delta W_{int} = \delta U$) and the external virtual work is the variation in the work potential of the force ($\delta W_{ext} = \delta W$). Thus, from Equation (7.24) we obtain $\delta W_{int} - \delta W_{ext} = \delta U - \delta W$ or

$$\boxed{\delta\Omega = 0} \tag{7.26}$$

Equation (7.26) implies that at equilibrium, the virtual variation in the potential energy function is zero—which occurs where the slopes of the potential energy function with respect to the parameters defining the potential function are zero.

The theorem of minimum potential energy can be stated as follows.

> Of all the kinematically admissible displacement functions, the actual displacement function is the one that minimizes the potential energy function at stable equilibrium.

It needs to be emphasized that there are many kinematically admissible displacement functions, and there is no requirement that these functions satisfy the equilibrium equations or the boundary conditions on forces and moments. The actual displacement is kinematically admissible and satisfies all the equilibrium conditions and the static boundary conditions. Thus, if we choose an arbitrary kinematically admissible function and calculate the potential energy function, the value so obtained will always be greater than the value of the potential energy function at equilibrium. In other words, we approach the potential energy function value at equilibrium from above. Thus, if we have two approximations for the displacement functions, the one that gives the lower potential energy is the better approximation. A corollary to the preceding statement is that if we add another term (increase the degrees of freedom) in an approximation, the potential energy value can only decrease, that is, improve the accuracy of the approximation. We record the following observations.

- The better approximation of displacement function is the one that yields the lower potential energy.
- The greater the degrees of freedom, the lower will be the potential energy for a given set of kinematically admissible functions.

From Table 7.1, we saw that the work is represented by a linear functional: $W = l(u)$. From Table 7.3, we saw that the strain energy is represented by a symmetric bilinear functional: $U = B(u, u)/2$. Substituting these expressions into Equation (7.25) we obtain the potential energy expression as:

$$\Omega = \frac{1}{2}B(u, u) - l(u) \tag{7.27}$$

We will make use of the above form in Rayleigh-Ritz and finite element method discussed latter.

## 7.6 MATHEMATICAL PRELIMINARIES

The first variations of line, area, and volume integrals will generate terms inside the integral of functions and its derivatives. After the process of variation the functions and their derivatives will no longer be independent, and hence we will need to transfer derivatives from one term onto another inside the integrals. The transfer of derivatives is accomplished: in one-dimension by integration by parts; in two-dimension by Green's formula; and in three-dimension by Gauss divergence formula (Kreyszig [1979]). We briefly review the mathematics that we will need for finding the stationary values of line, area, and volume integrals.

*Integration by parts* is given by the equation below.

$$\int_a^b f\frac{dg}{dx}dx = fg\Big|_a^b - \int_a^b \frac{df}{dx}g\,dx \tag{7.28}$$

where $f(x)$ and $g(x)$ are two continuous functions with continuous first derivative on the line bounded by $x = a$ and $x = b$.

*Green's formulas* are given by the equations below.

$$\iint_A f\frac{\partial g}{\partial x}dx\,dy = \oint_\Gamma fg\,dy - \iint g\frac{\partial f}{\partial x}dx\,dy \tag{7.29a}$$

$$\iint_A f\frac{\partial g}{\partial y}dx\,dy = -\oint_\Gamma fg\,dx - \iint_A g\frac{\partial f}{\partial y}dx\,dy \tag{7.29b}$$

where $f(x, y)$ and $g(x, y)$ are two continuous functions with continuous first derivative in the area $A$ that is bounded by the curve $\Gamma$ as shown in Figure 7.5*a*.

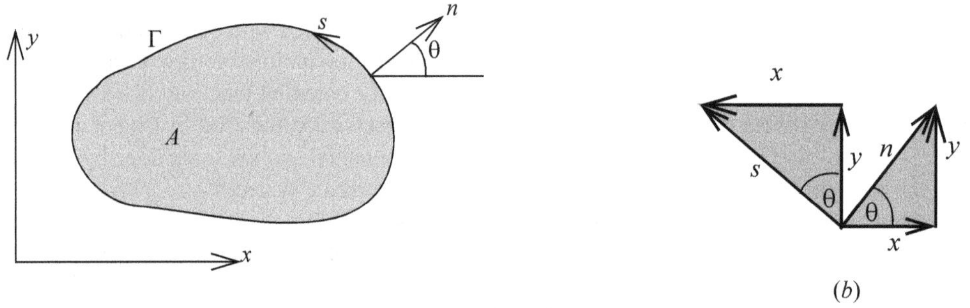


**Figure 7.5** (a) Generic geometry (b) Coordinate transformation.

Equations (7.29a) and (7.29b) are useful for rectangular geometries. For smooth arbitrary geometries there is an alternative form that we will use. This is described next.

Figure 7.5*b* shows the geometry by which we can relate the Cartesian coordinates $(x, y)$ and to normal and tangential coordinates $(n, s)$ as shown in the equations that follow.

$$x = n\cos\theta - s\sin\theta \qquad y = n\sin\theta + s\cos\theta \tag{7.29c}$$

Equation (7.29c) can be solved for $n$ and $s$ to obtain

$$n = x\cos\theta + y\sin\theta \qquad s = -x\sin\theta + y\cos\theta \tag{7.29d}$$

We define the direction cosines of the unit normal as

$$n_x = \frac{\partial n}{\partial x} = \cos\theta = \frac{\partial s}{\partial y} \qquad n_y = \frac{\partial n}{\partial y} = \sin\theta = -\frac{\partial s}{\partial x} \tag{7.29e}$$

If a point is restricted to the boundary, then $dn = 0$; from Equation (7.29c) we obtain

$$dx = -\sin\theta ds = -n_y ds \qquad dy = \cos\theta ds = n_x ds \tag{7.29f}$$

Substituting Equation (7.29f) into Equations (7.29a) and (7.29b) we obtain another form of Green's formula as

$$\boxed{\iint_A f\frac{\partial g}{\partial x}dx\,dy = \oint_\Gamma (fg)n_x ds - \iint_A g\frac{\partial f}{\partial x}dx\,dy} \tag{7.30a}$$

$$\boxed{\iint_A f\frac{\partial g}{\partial y}dx\,dy = \oint_\Gamma (fg)n_y dy - \iint_A g\frac{\partial f}{\partial y}dx\,dy} \tag{7.30b}$$

Using chain rule, we can write the following identities for later use.

$$\frac{\partial f}{\partial x} = \frac{\partial f}{\partial n}\frac{\partial n}{\partial x} + \frac{\partial f}{\partial s}\frac{\partial s}{\partial x} = \frac{\partial f}{\partial n}n_x - \frac{\partial f}{\partial s}n_y \qquad \text{and} \qquad \frac{\partial f}{\partial y} = \frac{\partial f}{\partial n}\frac{\partial n}{\partial y} + \frac{\partial f}{\partial s}\frac{\partial s}{\partial y} = \frac{\partial f}{\partial n}n_y + \frac{\partial f}{\partial s}n_x \tag{7.31a}$$

In the course of development, we will be taking derivative of the direction cosines with respect to $s$. We consider these derivatives next.

$$\frac{\partial n_x}{\partial s} = -\sin\theta\left(\frac{\partial\theta}{\partial s}\right) = -n_y\left(\frac{\partial\theta}{\partial s}\right) = -\left(\frac{n_y}{R_{cur}}\right) \tag{7.31b}$$

$$\frac{\partial n_y}{\partial s} = \cos\theta\left(\frac{\partial\theta}{\partial s}\right) = n_x\left(\frac{\partial\theta}{\partial s}\right) = \frac{n_x}{R_{cur}} \tag{7.31c}$$

where, $R_{cur}$ is the radius of curvature of the boundary at the point under consideration and can be function of $s$.

*Gauss divergence formula* is given by the equation below.

$$\iiint_T \left[\frac{\partial u_x}{\partial x} + \frac{\partial u_y}{\partial y} + \frac{\partial u_z}{\partial z}\right]dV = \iint_S [u_x n_x + u_y n_y + u_z n_z]dA \tag{7.32}$$

where, $u_x$, $u_y$, and $u_z$ are the components of a vector function which is continuous with continuous first derivatives in a region $T$ bounded by a smooth surface $S$; and $n_x$, $n_y$, and $n_z$ are the direction cosines of a unit normal on surface $S$. To develop formulas in which we transfer derivatives $x$, $y$, or $z$ from one function to another we let each of the components of $u$ be equal to product of two continuous functions $fg$ while the other two components are zero. This produces the formulas below.

$$\boxed{\begin{aligned} &\iiint_T \frac{\partial f}{\partial x}g\,dV = \iint_S fgn_x\,dA - \iiint_T f\frac{\partial g}{\partial x}dV \qquad \iiint_T \frac{\partial f}{\partial y}g\,dV = \iint_S fgn_y\,dA - \iiint_T f\frac{\partial g}{\partial y}dV \\ &\iiint_T \frac{\partial f}{\partial z}g\,dV = \iint_S fgn_z\,dA - \iiint_T f\frac{\partial g}{\partial z}dV \end{aligned}} \tag{7.33}$$

## 7.7 STATIONARY VALUE OF A DEFINITE LINE INTEGRAL

Definite line integrals arise in potential energy equations for one dimensional structural members. By theorem of minimum potential energy, that is, first variation of potential energy, we can obtain the equilibrium equations. As we shall see, we also obtain all possible boundary conditions. In other words, we can obtain the complete boundary value problem from the stationary value of definite line integrals representing potential energy.

In Section 7.7.1 we consider a functional with $u$ and its first derivative. Next we consider first and second derivative in the functional in Section 7.7.3. Finally, we generalize to the $r^{\text{th}}$ order derivative in the functional in Section 7.7.4.

We will be taking derivatives of the functionals with respect to the derivatives of $u$. To simplify algebraic manipulation we introduce the following notation

$$u^{(0)} = u \qquad u^{(i)} = du/dx \qquad u^{(ii)} = d^2u/dx^2 \qquad \bullet\ \bullet \qquad u^{(r)} = d^ru/dx^r \tag{7.34}$$

### 7.7.1 Stationary value of a functional with first order derivatives

We will develop the ideas by first considering a functional that depends on the function $u$ and its first derivative.

$$I(u) = \int_a^b H(u^{(i)}, u, x)\ dx \tag{7.35}$$

We consider the first variation of $I(u)$ and observe that the limits are fixed; hence the variation can be taken inside the integral to obtain the following.

$$\delta I(u) = \delta\int_a^b H(u^{(i)}, u, x)\ dx = \int_a^b \delta H(u^{(i)}, u, x)\ dx \tag{7.36a}$$

Note the process of variation is a process of considering virtual displacement. Figure 7.1 emphasized that function and derivative are independent in virtual displacement. Furthermore, we can consider virtual displacement without changing $x$ in Figure 7.1, which implies that variation in $x$ is not considered during virtual displacement. Hence we obtain

$$\delta I(u) = \int_a^b \left[\frac{\partial H}{\partial u^{(i)}}\delta u^{(i)} + \frac{\partial H}{\partial u}\delta u\right]\ dx \tag{7.36b}$$

Once we have considered virtual displacement, we are now on a specific curve and function and its derivative are related, that is, no longer independent. If we are to draw any conclusion by setting $\delta I(u) = 0$, then we need to obtain an expression in only $u$—we perform integration by parts.

We further note the following

$$\delta u^{(i)} = \delta\left(\frac{du}{dx}\right) = \frac{d(\delta u)}{dx} \tag{7.36c}$$

Substituting Equation (7.36c) into Equation (7.36b), performing integration by parts, and setting $\delta I(u) = 0$, we obtain

$$\delta I(u) = \int_a^b \left[\frac{\partial H}{\partial u^{(i)}}\frac{d}{dx}(\delta u) + \frac{\partial H}{\partial u}\delta u\right]\ dx = 0$$

$$\delta I(u) = \int_a^b \left[\frac{\partial H}{\partial u} - \frac{d}{dx}\left(\frac{\partial H}{\partial u^{(i)}}\right)\right]\delta u\ dx + \left.\frac{\partial H}{\partial u^{(i)}}\delta u\right|_a^b = 0 \tag{7.36d}$$

There are two possible ways by which we can meet the condition in Equation (7.36d).

*Possibility 1*: We meet the condition $\delta I(u) = 0$ in the average or overall sense. This lead to approximate methods as the condition is not satisfied at each and every point between $a$ and $b$. We will consider this approach in Sections 7.12 and 7.13.

*Possibility 2:* We require that $\delta I(u) = 0$ at each and every point between $a$ and $b$. This results in boundary value problem discussed in Section 7.7.2.

Before moving to the next section we record the following observations from the above derivation.

- During the process of variation, the function and its derivative are independent.
- After the process of variation, the function and its derivative are no longer independent.
- Integration by parts will generate an expression only in terms of variation of the function.

**(7.37)**

### 7.7.2 Boundary value problem

We first consider the integral in Equation (7.36d) to be zero. For it to be zero, the integrand must be zero at all $x$. $\delta u(x)$ cannot be zero as it is the virtual displacement. Thus the term in bracket is zero at each and every $x$ as shown below.

$$\text{Differential Equation:}\quad \boxed{\frac{\partial H}{\partial u} - \frac{d}{dx}\left(\frac{\partial H}{\partial u^{(i)}}\right) = 0 \qquad a < x < b} \tag{7.38a}$$

The above equation is called the *Euler-Lagrange* equation. We now consider the boundary term in Equation (7.36d). At each boundary end, the term must be zero. So either $\partial H/\partial u^{(i)} = 0$ or $\delta u$ must be zero at each end as shown below.

$$\text{Boundary Conditions:}\quad \boxed{[\partial H/\partial u^{(i)} = 0 \quad \text{or} \quad \delta u = 0] \quad \text{at } x = a \text{ and at } x = b} \tag{7.38b}$$

Equations (7.38a) and (7.38b) represent the complete boundary value problem associated with the functional in Equation (7.35). We derived the boundary value problem not from differential calculus but from variational calculus. The elegance of obtaining boundary value problem statement from variational principle will be demonstrated in several examples.

We record the following observations that we will generalize in Section 7.7.4.

- Equation (7.38a) will result in a second order differential equation because the highest derivative $u^{(i)}$ will be differentiated in the second term. When we use the potential energy for axial or torsion as the functional, we will obtain the second order differential equations (2.16-A) and (2.16-T).

- If the functional $H$ is a quadratic in $u$ and its derivatives, then Equation (7.38a) will result in a linear differential equation. However, if the functional $H$ is cubic or higher in $u$ and its derivatives, then we will obtain a nonlinear differential equation.
- The boundary condition $\partial H / \partial u^{(i)} = 0$ will result in a condition on first order derivative. We will obtain condition on the internal axial force or torsion when we use the potential energy for axial or torsion as the functional.
- The boundary condition $\delta u = 0$ implies that $u$ is specified, but it could be a non-zero value.
- If the functional contains more than one variable ($u$), say $u_i$, then we could replace $u$ with $u_i$ in the above equations. In other words, the above equations are applicable for each $u_i$

### 7.7.3 Stationary value of a functional with second order derivatives

In this section we obtain the boundary value problem from a functional with first and second order derivatives shown below.

$$I(u) = \int_a^b H(u^{(ii)}, u^{(i)}, u, x)\, dx \tag{7.39a}$$

We take the first variation $\delta I(u)$ and then by successive integration by parts we transfer derivatives as shown below.

$$\delta I(u) = \int_a^b \left[ \frac{\partial H}{\partial u^{(ii)}} \delta u^{(ii)} + \frac{\partial H}{\partial u^{(i)}} \delta u^{(i)} + \frac{\partial H}{\partial u} \delta u \right] dx = \int_a^b \left[ \frac{\partial H}{\partial u^{(ii)}} \frac{\mathrm{d}}{\mathrm{d}x}(\delta u^{(i)}) + \frac{\partial H}{\partial u^{(i)}} \delta u^{(i)} + \frac{\partial H}{\partial u} \delta u \right] dx$$

$$\delta I(u) = \frac{\partial H}{\partial u^{(ii)}} \delta u^{(i)} \bigg|_a^b + \int_a^b \left[ \left( \frac{\partial H}{\partial u^{(i)}} - \frac{\mathrm{d}}{\mathrm{d}x}\left( \frac{\partial H}{\partial u^{(ii)}} \right) \right) \delta u^{(i)} + \frac{\partial H}{\partial u} \delta u \right] dx$$

$$\delta I(u) = \frac{\partial H}{\partial u^{(ii)}} \delta u^{(i)} \bigg|_a^b + \int_a^b \left[ \left\{ \frac{\partial H}{\partial u^{(i)}} - \frac{\mathrm{d}}{\mathrm{d}x}\left( \frac{\partial H}{\partial u^{(ii)}} \right) \right\} \frac{\mathrm{d}}{\mathrm{d}x}(\delta u) + \frac{\partial H}{\partial u} \delta u \right] dx$$

$$\delta I(u) = \frac{\partial H}{\partial u^{(ii)}} \delta u^{(i)} \bigg|_a^b + \left\{ \frac{\partial H}{\partial u^{(i)}} - \frac{\mathrm{d}}{\mathrm{d}x}\left( \frac{\partial H}{\partial u^{(ii)}} \right) \right\} \delta u \bigg|_a^b + \int_a^b \left[ \frac{\partial H}{\partial u} - \frac{\mathrm{d}}{\mathrm{d}x} \left\{ \frac{\partial H}{\partial u^{(i)}} - \frac{\mathrm{d}}{\mathrm{d}x}\left( \frac{\partial H}{\partial u^{(ii)}} \right) \right\} \right] \delta u dx$$

$$\delta I(u) = \frac{\partial H}{\partial u^{(ii)}} \delta u^{(i)} \bigg|_a^b + \left\{ \frac{\partial H}{\partial u^{(i)}} - \frac{\mathrm{d}}{\mathrm{d}x}\left( \frac{\partial H}{\partial u^{(ii)}} \right) \right\} \delta u \bigg|_a^b + \int_a^b \left[ \frac{\partial H}{\partial u} - \frac{\mathrm{d}}{\mathrm{d}x}\left( \frac{\partial H}{\partial u^{(i)}} \right) + \frac{\mathrm{d}^2}{\mathrm{d}x^2}\left( \frac{\partial H}{\partial u^{(ii)}} \right) \right] \delta u dx = 0 \tag{7.39b}$$

Each term in the above equation must equal zero and we obtain the boundary value problem shown below.

Differential Equation: $$\boxed{\frac{\partial H}{\partial u} - \frac{\mathrm{d}}{\mathrm{d}x}\left( \frac{\partial H}{\partial u^{(i)}} \right) + \frac{\mathrm{d}^2}{\mathrm{d}x^2}\left( \frac{\partial H}{\partial u^{(ii)}} \right) = 0 \qquad a < x < b} \tag{7.40a}$$

Boundary Conditions:

$$\frac{\partial H}{\partial u^{(ii)}} = 0 \quad \text{or} \quad \delta u^{(i)} = 0$$

and

$$\frac{\partial H}{\partial u^{(i)}} - \frac{\mathrm{d}}{\mathrm{d}x}\left( \frac{\partial H}{\partial u^{(ii)}} \right) = 0 \quad \text{or} \quad \delta u = 0$$

at $x = a$ and at $x = b$ **(7.40b)**

We record the following observations that we will generalize in Section 7.7.4.

- Equation (7.40a) will result in a fourth order differential equation because the highest derivative $u^{(ii)}$ will be differentiated twice in the third term. When we use the potential energy for symmetric bending of beams as the functional then we will obtain the fourth order differential equation (2.16-B).
- If the functional $H$ is a quadratic in $u$ and its derivatives, then Equation (7.38a) will result in a linear differential equation. However, if the functional $H$ is cubic or higher in $u$ and its derivatives, then we will obtain a non-linear differential equation.
- The boundary condition $\partial H / \partial u^{(ii)} = 0$ will result in a condition on second order derivative. When we use the potential energy for symmetric bending of beams as the functional we will obtain a condition on the internal bending moment.

- The boundary condition on $\partial H/\partial u^{(i)} - d(\partial H/\partial u^{(ii)})/dx$ will result in a condition on third order derivative. When we use the potential energy for symmetric bending of beams as the functional we will obtain a condition on the internal shear force.
- The boundary conditions $\delta u^{(i)} = 0$ and $\delta u = 0$ implies that $u^{(i)}$ and $u$ are specified, but they could be a non-zero values.
- If the functional contains more than one variable ($u$), say $u_i$, then we could replace $u$ with $u_i$ in the above equations. In other words, the above equations are applicable for each $u_i$.

### 7.7.4 Generalization

In this section we extend the results to a functional containing $r^{\text{th}}$ order derivative and define the functional as

$$I(u) = \int_a^b H(u^{(r)}, u^{(r-1)}, u^{(r-1)}, \quad \cdot \cdot \cdot \quad, u^{(i)}, u^{(0)}, x)\ dx \tag{7.41a}$$

Using the three observations made in Equation (7.37) we take the first variation $\delta I(u)$ and then by integration by parts transfer one of the derivatives as shown below.

$$\delta I(u) = \sum_{q=0}^{r+1} \int_a^b \left(\frac{\partial H}{\partial u^{(q)}}\right) \delta u^{(q)} dx = \sum_{q=1}^{r+1} \left(\frac{\partial H}{\partial u^{(q)}}\right) \delta u^{(q-1)} \Bigg|_a^b - \sum_{q=1}^{r+1} \int_a^b \frac{\partial}{\partial x}\left(\frac{\partial H}{\partial u^{(q)}}\right) \delta u^{(q-1)} dx + \int_a^b H \delta u^{(0)} dx \tag{7.41b}$$

We note that the integration by parts changes the sign as we transfer derivative from one term to the other. We transfer all the derivatives from $\delta u^{(r)}$ till we obtain just $\delta u^{(0)} = \delta u$ in the integral. Setting $\delta I(u) = 0$, we obtain the differential equation and all possible boundary conditions as shown in Figure 7.6. The observations we made in earlier sections can now be generalized as given below.

Differential equation: $$\frac{\partial H}{\partial u^{(0)}} - \frac{\mathrm{d}}{\mathrm{d}x}\left(\frac{\partial H}{\partial u^{(i)}}\right) + \frac{\mathrm{d}^2}{\mathrm{d}x^2}\left(\frac{\partial H}{\partial u^{(ii)}}\right) - \frac{\mathrm{d}^3}{\mathrm{d}x^3}\left(\frac{\partial H}{\partial u^{(iii)}}\right) + \quad \cdot \cdot \cdot \quad + (-1)^r \frac{\mathrm{d}^r}{\mathrm{d}x^r}\left(\frac{\partial H}{\partial u^{(r)}}\right) = 0 \tag{7.42a}$$

Boundary conditions at $x = a$ and at $x = b$

$$\left.\begin{array}{lcl}
-\left(\frac{\partial H}{\partial u^{(i)}}\right) + \frac{d}{dx}\left(\frac{\partial H}{\partial u^{(ii)}}\right) - \frac{d^2}{dx^2}\left(\frac{\partial H}{\partial u^{(iii)}}\right) + \bullet \bullet + (-1)^r \frac{d^{r-1}}{dx^{r-1}}\left(\frac{\partial H}{\partial u^{(r)}}\right) = 0 & \text{or} & \delta u = 0 \\
\left(\frac{\partial H}{\partial u^{(ii)}}\right) - \frac{\mathrm{d}}{\mathrm{d}x}\left(\frac{\partial H}{\partial u^{(iii)}}\right) + \bullet \bullet + (-1)^r \frac{d^{r-2}}{dx^{r-2}}\left(\frac{\partial H}{\partial u^{(r)}}\right) = 0 & \text{or} & \delta u^{(i)} = 0 \\
-\left(\frac{\partial H}{\partial u^{(iii)}}\right) + \bullet \bullet + (-1)^r \frac{d^{r-3}}{dx^{r-3}}\left(\frac{\partial H}{\partial u^{(r)}}\right) = 0 & \text{or} & \delta u^{(ii)} = 0 \\
\bullet = 0 & \text{or} & = 0 \\
\bullet = 0 & \text{or} & = 0 \\
\bullet = 0 & \text{or} & = 0 \\
\frac{\partial H}{\partial u^{(r)}} = 0 & \text{or} & \delta u^{(r-1)} = 0
\end{array}\right\} \tag{7.42b}$$

**Figure 7.6** Boundary Value Problem

1. If the highest derivative in the functional is $r$, then the differential equation will be of order $2r$. Thus, from the strain energy expressions for axial and torsion in Table 7.2 we see that $r = 1$, hence the differential equation would be second order as we saw in Chapter 2. The strain energy for beam bending shows $r = 2$ and we will get a fourth order differential equation.
2. The boundary conditions with the variation symbol of $\delta$ have derivatives from 0 to $r$-1. These quantities must be continuous as we are taking their variation. The derivatives are called the **principal derivatives.** The boundary conditions are called kinematic boundary conditions in mechanics of materials. **Essential boundary conditions** and **primary variables** (in place of principal derivatives) are more general names that are acquiring increasing popularity. In beam bending, deflection and slope are the primary variables, while deflection in axial and rotation in torsion are the primary variables. The primary variables are our generalized displacements. Thus, principal derivatives, primary variables, and generalized displacement refer to the same variables.

3. The boundary conditions that have derivatives of the functionals that vary from $r$ to $2r$-1 are our internal forces and moments and are called **statical** variables in mechanics of material. A more general name of these variables is **secondary variables.** The boundary conditions on these variables are called **statical boundary conditions** in mechanics of material. A more general name is **natural boundary conditions**. In beam bending these will result in boundary conditions on internal bending moment (derivative order 2) and internal shear force (derivative order 3).
4. If the functional contains more than one variable ($u$), say $u_i$, then we could replace $u$ with $u_i$ in the above equations. In other words, the above equations are applicable for each $u_i$.
5. If the functional is quadratic in $u$ and its derivative, then the boundary value problem will be linear. However, the above equations are applicable to any functional.

## EXAMPLE 7.1

Obtain the boundary value problem for symmetric bending of beams using variational principles. Assume there are only distributed force $p_y$ and no concentrated force or moment acting on the beam.

**PLAN Method I**

We can use potential energy as the functional and set the first variation of it to zero to obtain the boundary value problem.

**SOLUTION**

Using Equations (7.7d) and (7.25), we write the potential energy of the beam as shown below.

$$\Omega = U - W = \frac{1}{2}\int_L EI_{zz}\left(\frac{d^2\mathrm{v}}{dx^2}\right)^2 dx - \int_L p_y(x)\mathrm{v}(x)dx = \frac{1}{2}\int_0^L EI_{zz}[\mathrm{v}^{(2)}]^2 dx - \int_0^L p_y\mathrm{v}dx \tag{E1}$$

where we used $\mathrm{v}^{(r)} = d^r\mathrm{v}/dx^r$ and assumed the length of the beam is $L$.
Taking the first variation of the potential energy we obtain

$$\delta\Omega = \int_0^L EI_{zz}\mathrm{v}^{(2)}\delta\mathrm{v}^{(2)}dx - \int_0^L p_y\delta\mathrm{v}dx = \int_0^L EI_{zz}\mathrm{v}^{(2)}\frac{d}{dx}(\delta\mathrm{v}^{(1)})dx - \int_0^L p_y\delta\mathrm{v}dx \tag{7.42c}$$

Integrating by parts two times we obtain

$$\delta\Omega = [EI_{zz}\mathrm{v}^{(2)}\delta\mathrm{v}^{(1)}]\Big|_0^L - \int_0^L EI_{zz}\mathrm{v}^{(2)}\delta\mathrm{v}^{(1)}dx + \int_0^L p_y\delta\mathrm{v}dx = [EI_{zz}\mathrm{v}^{(2)}\delta\mathrm{v}^{(1)}]\Big|_0^L - \int_0^L \frac{d}{dx}(EI_{zz}\mathrm{v}^{(2)})\frac{d}{dx}(\delta\mathrm{v}^{(0)})dx - \int_0^L p_y\delta\mathrm{v}dx$$

$$\delta\Omega = [EI_{zz}\mathrm{v}^{(2)}\delta\mathrm{v}^{(1)}]\Big|_0^L - \left[\frac{d}{dx}(EI_{zz}\mathrm{v}^{(2)})(\delta\mathrm{v}^{(0)})\right]\Bigg|_0^L + \int_0^L \frac{d^2}{dx^2}(EI_{zz}\mathrm{v}^{(2)})(\delta\mathrm{v}^{(0)})dx - \int_0^L p_y\delta\mathrm{v}dx \tag{E2}$$

We note that $M_z = EI_{zz}\mathrm{v}^{(2)}$ and $V_y = -d(EI_{zz}\mathrm{v}^{(2)})/dx$. Substituting and setting $\delta\Omega = 0$, we obtain

$$\delta\Omega = \left[M_z\delta\left(\frac{d\mathrm{v}}{dx}\right)\right]\Bigg|_0^L + [V_y(\delta\mathrm{v})]\Big|_0^L + \int_0^L\left[\frac{d^2}{dx^2}\left(EI_{zz}\frac{d^2\mathrm{v}}{dx^2}\right) - p_y\right](\delta\mathrm{v})dx = 0 \tag{E3}$$

Each term has to be zero in the above equation. For the term inside the integral to be zero for all $x$, the following has to hold

$$\text{Differential Equation: } \frac{d^2}{dx^2}\left(EI_{zz}\frac{d^2\mathrm{v}}{dx^2}\right) - p_y = 0 \qquad 0 < x < L \tag{E4}$$

$$\text{Boundary Condition 1: } M_z = 0 \quad \text{or} \quad \delta\left(\frac{d\mathrm{v}}{dx}\right) \quad \text{at } x = 0 \text{ and } x = L \tag{E5}$$

$$\text{Boundary Condition 2: } V_y = 0 \quad \text{or} \quad \delta\mathrm{v} \quad \text{at } x = 0 \text{ and } x = L \tag{E6}$$

Equations (E4) through (E6) represent the boundary value problem for symmetric beam bending in absence of concentrated force and moment.

### COMMENTS

1. The variation symbol on $\delta(d\mathrm{v}/dx)$ and $\delta\mathrm{v}$, implies $d\mathrm{v}/dx$ and v must be specified, that is, they do not vary. The specified value however, could be zero or non-zero.
2. In Equation (E4), the bending rigidly $EI_{zz}$ can be a function of $x$.
3. Suppose a concentrated moment $M_{ext}$ and a concentrated force $P_{ext}$ are applied at $x = L$. Then the work term would include

$$M_{ext}\left(\frac{d\mathrm{v}}{dx}(L)\right) + P_{ext}\mathrm{v}(L) \tag{E7}$$

In such a case Equation (E3) would be

$$\delta\Omega = \left[M_z\delta\frac{d\mathrm{v}}{dx}\right]\Bigg|_0^L + [V_y(\delta\mathrm{v})]\Big|_0^L + \int_0^L\left[\frac{d^2}{dx^2}\left(EI_{zz}\frac{d^2\mathrm{v}}{dx^2}\right) - p_y\right](\delta\mathrm{v})dx - M_{ext}\delta\left(\frac{d\mathrm{v}}{dx}(L)\right) - P_{ext}\delta\mathrm{v}(L) = 0 \tag{E8}$$

Further suppose $dv/dx$ and v are specified at $x = 0$, such as in a cantilever beam, then the above equation will yield

$$\delta\Omega = [M_z(L) - M_{ext}]\delta\left(\frac{dv}{dx}(L)\right) + [V_y(L) - P_{ext}]\delta v(L) + V_y(0)\delta v(0) + \int_0^L \left[\frac{d^2}{dx^2}\left(EI_{zz}\frac{d^2v}{dx^2}\right) - p_y\right](\delta v)dx = 0 \qquad \text{(E9)}$$

The above equation implies that $M_z(L) - M_{ext} = 0$ and $V_y(L) - P_{ext} = 0$, which will be the boundary conditions at $x = L$.

4. Suppose there is a concentrated moment $M_a$ and a concentrated force $P_a$ are applied at $x = a$. One approach would be to include the work as we did in Equation (E7). In such a case the integral from zero to $L$ will have to be written as sum of integral from 0 to $a$-ε and $a$+ε to $L$, and then $\varepsilon \to 0$ to account for the discontinuity in shear force and bending moment. An easier alternative would be to write the distributed load using discontinuity function as $p_y + P_a\langle x-a\rangle^{-1} + M_a\langle x-a\rangle^{-2}$. The work is now included in the distributed force and it will change the differential equation in Equation (E4).
5. Comments 3 and 4 emphasize that as long as the potential energy functional is correctly defined, we will get the correct boundary value problem statement.

**PLAN Method II**

Using potential energy as the functional and Equations (7.42a) and (7.42b) we can obtain the boundary value problem.

**SOLUTION**

Comparing Equation (E1) to Equation (7.41a), we note $r = 2$, $a = 0$, and $b = L$. We can write the following functional

$$H(v^{(2)}, v^{(1)}, v^{(0)}, x) = \frac{1}{2}EI_{zz}[v^{(2)}]^2 - p_y v^{(0)} \qquad \text{(E10)}$$

The derivatives of the functional are

$$\frac{\partial H}{\partial v^{(0)}} = -p_y \qquad \frac{\partial H}{\partial v^{(1)}} = 0 \qquad \frac{\partial H}{\partial v^{(2)}} = EI_{zz}[v^{(2)}] = EI_{zz}\frac{d^2v}{dx^2} \qquad \text{(E11)}$$

For $r = 2$, we can write Equation (7.42a) and substitute Equation (E11) to obtain the differential equation below.

$$\frac{\partial H}{\partial v^{(0)}} - \frac{d}{dx}\left(\frac{\partial H}{\partial v^{(1)}}\right) + \frac{d^2}{dx^2}\left(\frac{\partial H}{\partial v^{(2)}}\right) = 0$$

$$\text{Differential Equation: } -p_y + \frac{d^2}{dx^2}EI_{zz}[v^{(2)}] = 0 \quad \text{or} \quad \frac{d^2}{dx^2}\left(EI_{zz}\frac{d^2v}{dx^2}\right) - p_y = 0 \qquad 0 < x < L \qquad \text{(E12)}$$

The first two equations of Equation (7.42b) for $r = 2$ can be written, and Equation (E11) substituted to obtain the boundary conditions as shown below.

$$-\left(\frac{\partial H}{\partial v^{(1)}}\right) + \frac{d}{dx}\left(\frac{\partial H}{\partial v^{(2)}}\right) = 0 \quad \text{or} \quad \delta v = 0 \quad \text{at } x = 0 \text{ and } x = L$$

$$\text{Boundary Condition 1: } \frac{d}{dx}\left(EI_{zz}\frac{d^2v}{dx^2}\right) = -V_y = 0 \quad \text{or} \quad \delta v = 0 \quad \text{at } x = 0 \text{ and } x = L \qquad \text{(E13)}$$

$$\left(\frac{\partial H}{\partial v^{(2)}}\right) = 0 \quad \text{or} \quad \delta v^{(1)} = 0 \quad \text{at } x = 0 \text{ and } x = L$$

$$\text{Boundary Condition 2: } EI_{zz}\frac{d^2v}{dx^2} = M_z = 0 \quad \text{or} \quad \delta\frac{dv}{dx} \quad \text{at } x = 0 \text{ and } x = L \qquad \text{(E14)}$$

Equations (E12) through (E14) represent the boundary value problem for symmetric beam bending in absence of and concentrated force and moment

**COMMENT**

1. Method I show the details of the process of obtaining boundary value problem from a functional. Method II is procedural and has less algebra. It is recommended that both approaches be used in the initial attempts to understand the concepts of variational principles and its application.

## 7.8 STATIONARY VALUE OF A DEFINITE AREA INTEGRAL

In finding the stationary value of line integral we followed the three steps shown in Equation (7.37), that included the transfer of derivatives by integration by parts. In stationary value of an area integral the transfer of derivative is accomplished by using Green's theorem as presented in Section 7.6.

### 7.8.1 Stationary value of a functional with first order derivatives

In this section we obtain the boundary value problem from a functional with first derivatives shown below. We introduce the notation that a comma implies partial derivative with regard to the coordinate that follows. Thus, $u_{,x} = \partial u/\partial x$ and

$u_{,y} = \partial u/\partial y$. *A subscript without a comma implies a component and with a comma implies derivative.* Using this notation, we define the functional and consider its first variation below.

$$I(u) = \iint_A H(u_{,x}, u_{,y}, u, x, y)dxdy \qquad \textbf{(7.43a)}$$

$$\delta I(u) = \iint_A \left[\left(\frac{\partial H}{\partial u_{,x}}\right)\delta u_{,x} + \left(\frac{\partial H}{\partial u_{,y}}\right)\delta u_{,y} + \left(\frac{\partial H}{\partial u}\right)\delta u\right]dxdy = \iint_A \left[\left(\frac{\partial H}{\partial u_{,x}}\right)\frac{\partial}{\partial x}(\delta u) + \left(\frac{\partial H}{\partial u_{,y}}\right)\frac{\partial}{\partial y}(\delta u) + \left(\frac{\partial H}{\partial u}\right)\delta u\right]dxdy \qquad \textbf{(7.43b)}$$

Using Equation (7.30a), we transfer the derivatives in the first two terms and set the first variation to zero as shown below.

$$\delta I(u) = \oint_\Gamma \left[\frac{\partial H}{\partial u_{,x}}n_x + \frac{\partial H}{\partial u_{,y}}n_y\right]\delta u ds + \iint_A \left[\frac{\partial H}{\partial u} - \frac{\partial}{\partial x}\left(\frac{\partial H}{\partial u_{,x}}\right) - \frac{\partial}{\partial y}\left(\frac{\partial H}{\partial u_{,y}}\right)\right]\delta u dxdy = 0 \qquad \textbf{(7.43c)}$$

Each term in the first variation must be zero at all points.Thus, we can obtain the boundary value problem written below.

$$\text{Differential Equation: } \boxed{\frac{\partial H}{\partial u} - \frac{\partial}{\partial x}\left(\frac{\partial H}{\partial u_{,x}}\right) - \frac{\partial}{\partial y}\left(\frac{\partial H}{\partial u_{,y}}\right) = 0 \qquad x, y \text{ in } A} \qquad \textbf{(7.44a)}$$

$$\text{Boundary Conditions: } \boxed{\left(\frac{\partial H}{\partial u_{,x}}\right)n_x + \left(\frac{\partial H}{\partial u_{,y}}\right)n_y = 0 \quad \text{or} \quad \delta u = 0 \qquad x, y \text{ on } \Gamma} \qquad \textbf{(7.44b)}$$

If the functional in Equation (7.43a) contains more than one variable then Equations (7.44a) and (7.44b) must be written for each of the variables as demonstrated in Example 7.2.

In Equation (7.44b) it is possible that in part of the boundary the primary variable is specified and in another part of the boundary the secondary variable is specified. If the functional $I$ is associated with the *Laplace* operator such as in heat conduction, then specifying the primary variable will be specifying temperature and specifying the secondary variable will be specifying heat.

## EXAMPLE 7.2

Obtain the boundary value problem for an elastic plane in plane stress with only body forces.

**PLAN**

We write the potential energy functional and use Equations (7.44a) and (7.44b) to obtain the boundary value problem.

**SOLUTION**

Substituting Equations (7.7e) and (7.17e) into Equation (7.25) we obtain the potential energy as

$$\Omega(u, \mathrm{v}) = \frac{Eh}{2(1-\nu^2)}\iint_A \left[\left\{\left(\frac{\partial u}{\partial x}\right)^2 + 2\nu\left(\frac{\partial u}{\partial x}\right)\left(\frac{\partial \mathrm{v}}{\partial y}\right) + \left(\frac{\partial \mathrm{v}}{\partial y}\right)^2\right\} + \frac{(1-\nu)}{2}\left(\frac{\partial u}{\partial y} + \frac{\partial \mathrm{v}}{\partial x}\right)^2\right]dxdy$$
$$-h\iint_A [F_x(x, y)u(x, y) + F_y(x, y)\mathrm{v}(x, y)]dxdy \qquad \text{(E1)}$$

The functional in Equation (7.43a) can be written as

$$H(u_{,x}, \mathrm{v}_{,x}, u_{,y}, \mathrm{v}_{,y}, u, \mathrm{v}, x, y) = \frac{Eh}{2(1-\nu^2)}\left[\{u_{,x}^2 + 2\nu u_{,x}\mathrm{v}_{,y} + \mathrm{v}_{,y}^2\} + \frac{(1-\nu)}{2}(u_{,y} + \mathrm{v}_{,x})^2\right] - h[F_x u + F_y \mathrm{v}] \qquad \text{(E2)}$$

The derivatives of the functional $F$ can be written as shown below.

$$\partial H/\partial u = -hF_x \qquad \partial H/\partial \mathrm{v} = -hF_y \qquad \text{(E3)}$$

$$\frac{\partial H}{\partial u_{,x}} = \frac{Eh}{(1-\nu^2)}[u_{,x} + \nu \mathrm{v}_{,y}] \qquad \frac{\partial H}{\partial u_{,y}} = \frac{Eh(1-\nu)}{2(1-\nu^2)}[u_{,y} + \mathrm{v}_{,x}] \qquad \text{(E4)}$$

$$\frac{\partial H}{\partial \mathrm{v}_{,x}} = \frac{Eh(1-\nu)}{2(1-\nu^2)}[u_{,y} + \mathrm{v}_{,x}] \qquad \frac{\partial H}{\partial \mathrm{v}_{,y}} = \frac{Eh}{(1-\nu^2)}[\nu u_{,x} + \mathrm{v}_{,y}] \qquad \text{(E5)}$$

Substituting Equations (E3), (E4), and (E5) into Equation (7.44a) we obtain two differential equations shown below.

$$-hF_x - \frac{\partial}{\partial x}\left(\frac{Eh}{(1-\nu^2)}[u_{,x} + \nu \mathrm{v}_{,y}]\right) - \frac{\partial}{\partial y}\left(\frac{Eh(1-\nu)}{2(1-\nu^2)}[u_{,y} + \mathrm{v}_{,x}]\right) = 0$$

$$\text{Differential Equation 1: } \frac{E}{(1-\nu^2)}\left\{\frac{\partial^2 u}{\partial x^2} + \frac{(1-\nu)}{2}\frac{\partial^2 u}{\partial y^2} + \frac{(1+\nu)}{2}\frac{\partial^2 \mathrm{v}}{\partial x \partial y}\right\} + F_x = 0 \qquad x, y \text{ in } A \qquad \text{(E6)}$$

$$-hF_y - \frac{\partial}{\partial x}\left(\frac{Eh(1-\nu)}{2(1-\nu^2)}\left[\frac{\partial u}{\partial y} + \mathrm{v}_{,x}\right]\right) - \frac{\partial}{\partial y}\left(\frac{Eh}{(1-\nu^2)}\left[\nu\frac{\partial u}{\partial x} + \mathrm{v}_{,y}\right]\right) = 0 \qquad \text{(E7)}$$

Differential Equation 2: $$\frac{E}{(1-\nu^2)}\left\{\frac{(1+\nu)}{2}\frac{\partial^2 u}{\partial x \partial y}+\frac{(1-\nu)}{2}\frac{\partial^2 \mathrm{v}}{\partial x^2}+\frac{\partial^2 \mathrm{v}}{\partial y^2}\right\}+F_y = 0 \qquad x, y \text{ in } A \tag{E8}$$

Substituting Equations (E3), (E4), and (E4) into Equation (7.44b) we obtain the boundary conditions shown below.

$$\frac{Eh}{(1-\nu^2)}[u_{,x}+\nu \mathrm{v}_{,y}]n_x+\frac{Eh(1-\nu)}{2(1-\nu^2)}[u_{,y}+\mathrm{v}_{,x}]n_y = 0 \qquad \text{or} \qquad \delta u = 0$$

Boundary Condition 1: $$\left[\frac{\partial u}{\partial x}+\nu\frac{\partial \mathrm{v}}{\partial y}\right]n_x+\frac{(1-\nu)}{2}\left[\frac{\partial u}{\partial y}+\frac{\partial \mathrm{v}}{\partial x}\right]n_y = 0 \qquad \text{or} \qquad \delta u = 0 \qquad x, y \text{ on } \Gamma \tag{E9}$$

$$\frac{Eh(1-\nu)}{2(1-\nu^2)}[u_{,y}+\mathrm{v}_{,x}]n_x+\frac{Eh}{(1-\nu^2)}[\nu u_{,x}+\mathrm{v}_{,y}]n_y = 0 \qquad \text{or} \qquad \delta \mathrm{v} = 0$$

Boundary Condition 2: $$\frac{(1-\nu)}{2}\left[\frac{\partial u}{\partial y}+\frac{\partial \mathrm{v}}{\partial x}\right]n_x+\left[\nu\frac{\partial u}{\partial x}+\frac{\partial \mathrm{v}}{\partial y}\right]n_y = 0 \qquad \text{or} \qquad \delta \mathrm{v} = 0 \qquad x, y \text{ on } \Gamma \tag{E10}$$

**COMMENTS**

1. This example demonstrates that equations obtained from variational principles for one variable can be used for multiple variables.
2. We obtained the boundary value problem in terms of displacements. An alternative form in stresses can be obtained by recognizing the terms Equations (E4) and (E5) are as follows

$$\frac{\partial H}{\partial u_{,x}} = \sigma_{xx} \qquad \frac{\partial H}{\partial u_{,y}} = \tau_{xy} = \frac{\partial H}{\partial \mathrm{v}_{,x}} \qquad \frac{\partial H}{\partial \mathrm{v}_{,y}} = \sigma_{yy} \tag{E11}$$

If we substitute the above in Equations (7.44a) and (7.44b) we obtain the boundary value problem in terms of stresses as

Differential Equations: $$\frac{\partial \sigma_{xx}}{\partial x}+\frac{\partial \tau_{yx}}{\partial y}+hF_x = 0 \qquad \frac{\partial \tau_{xy}}{\partial x}+\frac{\partial \sigma_{yy}}{\partial y}+hF_y = 0 \tag{E12}$$

Boundary Conditions $[\sigma_{xx}n_x+\tau_{xy}n_y = 0 \text{ or } \delta u = 0]$ and $[\tau_{yx}n_x+\sigma_{yy}n_y = 0 \text{ or } \delta \mathrm{v} = 0]$ $x, y$ on $\Gamma$ (E13)

The equilibrium equation we saw as Equations (6.8a) and (6.8b) are given in Equation (E12). The boundary conditions we saw as Equations (6.10a) and (6.10b) are given in Equation (E13).

## 7.9 FUNCTIONALS WITH SECOND ORDER DERIVATIVES

In this section we obtain the boundary value problem from a functional with first and second order derivatives. In Section 7.9.1 we first consider the derivation in Cartesian coordinates that is applicable to areas of rectangular geometries and then in Section 7.9.2 we consider the more general case of geometries of arbitrary shapes in which we use normal and tangential coordinate systems for describing the boundary conditions.

### 7.9.1 Boundary value problem for rectangular geometries

We will use the rectangular geometry shown in Figure 7.7 to obtain the boundary value problem from the functional given below.

$$I(u) = \iint_A H(u_{,xx}, u_{,xy}, u_{,yy}, u_{,x}, u_{,y}, u, x, y)\,dx\,dy \tag{7.45}$$

where, $u_{,xx} = \partial^2 u/\partial x^2$, $u_{,yy} = \partial^2 u/\partial y^2$, and $u_{,xy} = \partial^2 u/\partial x \partial y$.

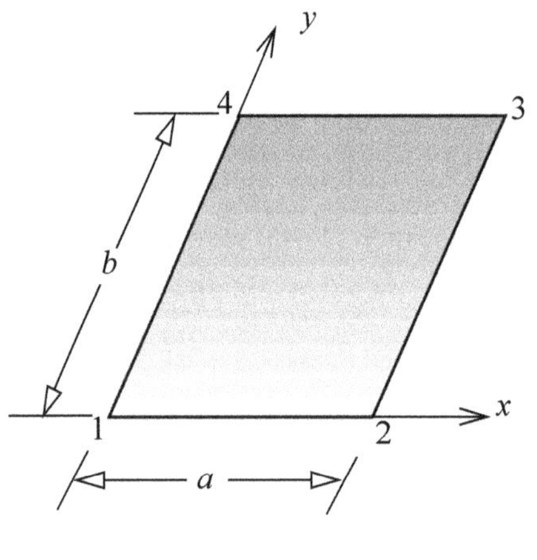


**Figure 7.7** Rectangular geometry.

We take the first variation of the functional shown in Equation (7.45) to obtain

$$\delta I(u) = \iint_A \left[\left(\frac{\partial H}{\partial u}\right)\delta u + \left(\frac{\partial H}{\partial u_{,x}}\right)\delta u_{,x} + \left(\frac{\partial H}{\partial u_{,y}}\right)\delta u_{,y} + \left(\frac{\partial H}{\partial u_{,xx}}\right)\delta u_{,xx} + \left(\frac{\partial H}{\partial u_{,yy}}\right)\delta u_{,yy} + \left(\frac{\partial H}{\partial u_{,xy}}\right)\delta u_{,xy}\right] dx\,dy \qquad \textbf{(7.46)}$$

The integrals with first derivatives are straight forward, but those with second order derivatives, particularly the cross derivatives with $x$ and $y$, have several steps that need to be elaborated. Hence, we evaluate the integral in Equation (7.46) as sum of six integrals.

The first integral that has the term $(\partial H/\partial u)\delta u$ is labeled $\delta I_1(u)$ and it needs no additional steps. In the next two integrals we transfer derivatives using Equations (7.29a) and (7.29b) to obtain the following.

$$\delta I_2(u) = \iint_A \left[\left(\frac{\partial H}{\partial u_{,x}}\right)\delta u_{,x}\right] dx\,dy = \iint_A \left[\left(\frac{\partial H}{\partial u_{,x}}\right)\frac{\partial}{\partial x}(\delta u)\right] dx\,dy = \oint\left(\frac{\partial H}{\partial u_{,x}}\right)\delta u\,dy - \iint_A \left[\frac{\partial}{\partial x}\left(\frac{\partial H}{\partial u_{,x}}\right)\delta u\right] dx\,dy \qquad \textbf{(7.46a)}$$

$$\delta I_3(u) = \iint_A \left(\frac{\partial H}{\partial u_{,y}}\right)\delta u_{,y}\,dx\,dy = \iint_A \left(\frac{\partial H}{\partial u_{,y}}\right)\frac{\partial}{\partial y}(\delta u)\,dx\,dy = -\oint\left(\frac{\partial H}{\partial u_{,y}}\right)\delta u\,dx - \iint_A \left[\frac{\partial}{\partial y}\left(\frac{\partial H}{\partial u_{,y}}\right)\delta u\right] dx\,dy \qquad \textbf{(7.46b)}$$

In the next integral, using Equation (7.29a) we twice transfer derivatives with respect to $x$ as shown below.

$$\delta I_4(u) = \iint_A \left[\left(\frac{\partial H}{\partial u_{,xx}}\right)\delta u_{,xx}\right] dx\,dy = \iint_A \left[\left(\frac{\partial H}{\partial u_{,xx}}\right)\frac{\partial}{\partial x}(\delta u_{,x})\right] dx\,dy = \oint\left(\frac{\partial H}{\partial u_{,xx}}\right)\delta u_{,x}\,dy - \iint_A \left[\frac{\partial}{\partial x}\left(\frac{\partial H}{\partial u_{,xx}}\right)\frac{\partial}{\partial x}(\delta u)\right] dx\,dy$$

$$\delta I_4(u) = \oint\left\{\left(\frac{\partial H}{\partial u_{,xx}}\right)\delta u_{,x} - \frac{\partial}{\partial x}\left(\frac{\partial H}{\partial u_{,xx}}\right)\delta u\right\} dy + \iint_A \left[\frac{\partial^2}{\partial x^2}\left(\frac{\partial H}{\partial u_{,xx}}\right)\delta u\right] dx\,dy \qquad \textbf{(7.46c)}$$

In a similar manner we can transfer derivatives with respect to $y$ using Equation (7.29b) to obtain

$$\delta I_5(u) = \iint_A \left[\left(\frac{\partial H}{\partial u_{,yy}}\right)\delta u_{,yy}\right] dx\,dy = -\oint\left\{\left(\frac{\partial H}{\partial u_{,yy}}\right)\delta u_{,y} - \frac{\partial}{\partial y}\left(\frac{\partial H}{\partial u_{,yy}}\right)\delta u\right\} dx + \iint_A \left[\frac{\partial^2}{\partial y^2}\left(\frac{\partial H}{\partial u_{,yy}}\right)\delta u\right] dx\,dy \qquad \textbf{(7.46d)}$$

In the last integral we first transfer the derivative with respect to $x$ using Equation (7.29a) then we transfer derivative with respect to $y$ using Equation (7.29b) as shown below.

$$\delta I_6(u) = \iint_A \left[\left(\frac{\partial H}{\partial u_{,xy}}\right)\frac{\partial}{\partial x}(\delta u_{,y})\right] dx\,dy = \oint\left(\frac{\partial H}{\partial u_{,xy}}\right)\delta u_{,y}\,dy - \iint_A \left[\frac{\partial}{\partial x}\left(\frac{\partial H}{\partial u_{,xy}}\right)\delta u_{,y}\right] dx\,dy$$

$$\delta I_6(u) = \oint\left(\frac{\partial H}{\partial u_{,xy}}\right)\frac{\partial}{\partial y}(\delta u)\,dy - \iint_A \left[\frac{\partial}{\partial x}\left(\frac{\partial H}{\partial u_{,xy}}\right)\frac{\partial}{\partial y}(\delta u)\right] dx\,dy$$

$$\delta I_6(u) = \oint\left(\frac{\partial H}{\partial u_{,xy}}\right)\frac{\partial}{\partial y}(\delta u)\,dy + \oint\frac{\partial}{\partial x}\left(\frac{\partial H}{\partial u_{,xy}}\right)\delta u\,dx + \iint_A \left[\frac{\partial^2}{\partial x\partial y}\left(\frac{\partial H}{\partial u_{,xy}}\right)\delta u\right] dx\,dy \qquad \textbf{(7.46e)}$$

The first integral in equation above can be further evaluated by integration by parts. We note that the boundary integral in the above equation with just $dy$ refer to the boundaries $x = 0$ and $x = a$ in Figure 7.7.

$$\oint\left(\frac{\partial H}{\partial u_{,xy}}\right)\frac{\partial}{\partial y}(\delta u)\,dy = \int_{y_2}^{y_3}\left(\frac{\partial H}{\partial u_{,xy}}\right)\frac{\partial}{\partial y}(\delta u)\,dy + \int_{y_4}^{y_1}\left(\frac{\partial H}{\partial u_{,xy}}\right)\frac{\partial}{\partial y}(\delta u)\,dy$$

$$\oint\left(\frac{\partial H}{\partial u_{,xy}}\right)\frac{\partial}{\partial y}(\delta u)\,dy = \left(\frac{\partial H}{\partial u_{,xy}}\right)\delta u\Big|_{y_2}^{y_3} - \int_{y_2}^{y_3}\frac{\partial}{\partial y}\left(\frac{\partial H}{\partial u_{,xy}}\right)\delta u\,dy + \left(\frac{\partial H}{\partial u_{,xy}}\right)\delta u\Big|_{y_4}^{y_1} - \int_{y_4}^{y_1}\frac{\partial}{\partial y}\left(\frac{\partial H}{\partial u_{,xy}}\right)\delta u\,dy$$

$$\oint\left(\frac{\partial H}{\partial u_{,xy}}\right)\frac{\partial}{\partial y}(\delta u)\,dy = \left(\frac{\partial H}{\partial u_{,xy}}\right)_3\delta u_3 - \left(\frac{\partial H}{\partial u_{,xy}}\right)_2\delta u_2 + \left(\frac{\partial H}{\partial u_{,xy}}\right)_1\delta u_1 - \left(\frac{\partial H}{\partial u_{,xy}}\right)_4\delta u_4 - \oint\frac{\partial}{\partial y}\left(\frac{\partial H}{\partial u_{,xy}}\right)\delta u\,dy$$

$$\oint\left(\frac{\partial H}{\partial u_{,xy}}\right)\frac{\partial}{\partial y}(\delta u)\,dy = -\sum_{k=1}^{4}(-1)^k\left(\frac{\partial H}{\partial u_{,xy}}\right)_k\delta u_k - \oint\frac{\partial}{\partial y}\left(\frac{\partial H}{\partial u_{,xy}}\right)\delta u\,dy \qquad \textbf{(7.46f)}$$

where, the subscripts 1 through 4 refer to the four corners in Figure 7.7. Substituting Equation (7.46f) into Equation (7.46e) we obtain

$$\delta I_6(u) = -\sum_{k=1}^{4}(-1)^k\left(\frac{\partial H}{\partial u_{,xy}}\right)_k\delta u_k + \oint\frac{\partial}{\partial x}\left(\frac{\partial H}{\partial u_{,xy}}\right)\delta u\,dx - \oint\left[\frac{\partial}{\partial y}\left(\frac{\partial H}{\partial u_{,xy}}\right)\right]\delta u\,dy + \iint_A \left[\frac{\partial^2}{\partial x\partial y}\left(\frac{\partial H}{\partial u_{,xy}}\right)\delta u\right] dx\,dy \qquad \textbf{(7.46g)}$$

Substituting the evaluated integrals given by Equations (7.46a), (7.46b), (7.46c), (7.46d), and (7.46g) into (7.46) and collecting terms we obtain the final form of the first variation of the functional as shown below.

$$\delta I(u) = \iint_A \left[\frac{\partial^2}{\partial x^2}\left(\frac{\partial H}{\partial u_{,xx}}\right) + \frac{\partial^2}{\partial y^2}\left(\frac{\partial H}{\partial u_{,yy}}\right) + \frac{\partial^2}{\partial x \partial y}\left(\frac{\partial H}{\partial u_{,xy}}\right) - \frac{\partial}{\partial x}\left(\frac{\partial H}{\partial u_{,x}}\right) - \frac{\partial}{\partial y}\left(\frac{\partial H}{\partial u_{,y}}\right) + \left(\frac{\partial H}{\partial u}\right)\right]\delta u dx dy$$

$$+ \oint \left(\frac{\partial H}{\partial u_{,xx}}\right)\delta u_{,x} dy - \oint \left[\frac{\partial}{\partial x}\left(\frac{\partial H}{\partial u_{,xx}}\right) + \frac{\partial}{\partial y}\left(\frac{\partial H}{\partial u_{,xy}}\right) - \left(\frac{\partial H}{\partial u_{,x}}\right)\right]\delta u dy \qquad \textbf{(7.47)}$$

$$- \oint \left(\frac{\partial H}{\partial u_{,yy}}\right)\delta u_{,y} dx + \oint \left[\frac{\partial}{\partial y}\left(\frac{\partial H}{\partial u_{,yy}}\right) + \frac{\partial}{\partial x}\left(\frac{\partial H}{\partial u_{,xy}}\right) - \left(\frac{\partial H}{\partial u_{,y}}\right)\right]\delta u dx - \sum_{k=1}^{4} (-1)^k \left(\frac{\partial H}{\partial u_{,xy}}\right)_k \delta u_k$$

We note that the boundary integrals in the above equation with just $dy$ refer to the boundaries $x = 0$ and $x = a$ in Figure 7.7, and boundary integrals with just $dx$ refer to the boundaries $y = 0$ and $y = b$. We set $\delta I(u) = 0$ and note that each term must be zero and obtain the boundary value problem as shown below.

Differential Equation:
$$\frac{\partial^2}{\partial x^2}\left(\frac{\partial H}{\partial u_{,xx}}\right) + \frac{\partial^2}{\partial y^2}\left(\frac{\partial H}{\partial u_{,yy}}\right) + \frac{\partial^2}{\partial x \partial y}\left(\frac{\partial H}{\partial u_{,xy}}\right) - \frac{\partial}{\partial x}\left(\frac{\partial H}{\partial u_{,x}}\right) - \frac{\partial}{\partial y}\left(\frac{\partial H}{\partial u_{,y}}\right) + \left(\frac{\partial H}{\partial u}\right) = 0 \qquad \textbf{(7.48a)}$$

Boundary Conditions at $x = 0$ and $x = a$: **(7.48b)**
$$\frac{\partial H}{\partial u_{,xx}} = 0 \quad \text{or} \quad \delta u_{,x} = 0$$
and
$$\frac{\partial}{\partial x}\left(\frac{\partial H}{\partial u_{,xx}}\right) + \frac{\partial}{\partial y}\left(\frac{\partial H}{\partial u_{,xy}}\right) - \left(\frac{\partial H}{\partial u_{,x}}\right) = 0 \quad \text{or} \quad \delta u = 0$$

Boundary Conditions at $y = 0$ and $y = b$: **(7.48c)**
$$\frac{\partial H}{\partial u_{,yy}} = 0 \quad \text{or} \quad \delta u_{,y} = 0$$
and
$$\frac{\partial}{\partial y}\left(\frac{\partial H}{\partial u_{,yy}}\right) + \frac{\partial}{\partial x}\left(\frac{\partial H}{\partial u_{,xy}}\right) - \left(\frac{\partial H}{\partial u_{,y}}\right) = 0 \quad \text{or} \quad \delta u = 0$$

Conditions at each corner:
$$\left(\frac{\partial H}{\partial u_{,xy}}\right)_k = 0 \quad \text{or} \quad \delta u_k = 0 \qquad k = 1 \text{ to } 4 \qquad \textbf{(7.48d)}$$

## EXAMPLE 7.3

Obtain the boundary value problem for a thin homogeneous plate bending with only transverse distributed forces and correlate the variables to those introduced in plate theory.

**PLAN**

We write the potential energy functional for thin plate bending and use Equations (7.48a) through (7.48d) to obtain the boundary value problem.

**SOLUTION**

Substituting Equations (7.7d) and (7.17d) into Equation (7.25), we obtain the potential energy as

$$\Omega(w) = \frac{D}{2}\iint_A \left\{\left(\frac{\partial^2 w}{\partial x^2}\right)^2 + \left(\frac{\partial^2 w}{\partial y^2}\right)^2 + 2\nu\left(\frac{\partial^2 w}{\partial x^2}\right)\left(\frac{\partial^2 w}{\partial y^2}\right) + 2(1-\nu)\left(\frac{\partial^2 w}{\partial x \partial y}\right)^2\right\}dx dy - \iint_A p_z w dx dy \qquad \text{(E1)}$$

The functional in Equation (7.45) can be written as

$$H(w_{,xx}, w_{,xy}, w_{,yy}, w_{,x}, w_{,y}, w, x, y) = \frac{D}{2}[w_{,xx}^2 + w_{,yy}^2 + 2\nu w_{,xx} w_{,yy} + 2(1-\nu) w_{,xy}^2] - p_z w \qquad \text{(E2)}$$

The derivatives of the functional $H$ can be written as shown below.

$$\frac{\partial H}{\partial w} = -p_z \qquad \frac{\partial H}{\partial w_{,x}} = 0 \qquad \frac{\partial H}{\partial w_{,y}} = 0 \qquad \text{(E3)}$$

$$\frac{\partial H}{\partial w_{,xx}} = D[w_{,xx} + \nu w_{,yy}] \qquad \frac{\partial H}{\partial w_{,yy}} = D[w_{,yy} + \nu w_{,xx}] \qquad \frac{\partial H}{\partial w_{,xy}} = 2D(1-\nu) w_{,xy} \qquad \text{(E4)}$$

Comparing Equation (E4) to Equation (5.14) we obtain the following

$$\frac{\partial H}{\partial w_{,xx}} = -\boldsymbol{m}_{xx} \qquad \frac{\partial H}{\partial w_{,yy}} = -\boldsymbol{m}_{yy} \qquad \frac{\partial H}{\partial w_{,xy}} = -2\boldsymbol{m}_{xy} \qquad \text{(E5)}$$

We evaluate the terms in Equations (7.48b) and (7.48c) and use the fact that the plate is homogeneous in $x$ and $y$ direction, that is, $D$ and $\nu$ are not function of $x$ and $y$.

$$\frac{\partial}{\partial x}\left(\frac{\partial H}{\partial w_{,xx}}\right)+\frac{\partial}{\partial y}\left(\frac{\partial H}{\partial w_{,xy}}\right)-\frac{\partial}{\partial x}\left(\frac{\partial H}{\partial w_{,x}}\right) = \frac{\partial}{\partial x}\left[D\left(\frac{\partial^2 w}{\partial x^2}+\nu\frac{\partial^2 w}{\partial y^2}\right)\right]+\frac{\partial}{\partial y}\left[2D(1-\nu)\frac{\partial^2 w}{\partial x\partial y}\right] = D\left[\frac{\partial^3 w}{\partial x^3}+(2-\nu)\frac{\partial^3 w}{\partial x\partial y^2}\right] \quad \text{(E6)}$$

$$\frac{\partial}{\partial y}\left(\frac{\partial H}{\partial w_{,yy}}\right)+\frac{\partial}{\partial x}\left(\frac{\partial H}{\partial w_{,xy}}\right)-\frac{\partial}{\partial y}\left(\frac{\partial H}{\partial w_{,y}}\right) = \frac{\partial}{\partial x}\left[D(1-\nu)\frac{\partial^2 w}{\partial x\partial y}\right]\frac{\partial}{\partial y}\left[D\left(\frac{\partial^2 w}{\partial y^2}+\nu\frac{\partial^2 w}{\partial x^2}\right)\right] = D\left[(2-\nu)\frac{\partial^3 w}{\partial x^2\partial y}+\frac{\partial^3 w}{\partial y^3}\right] \quad \text{(E7)}$$

Comparing Equations (E6) and Equations (E7) with Equations (5.21a) and (5.21b) we obtain

$$\frac{\partial}{\partial x}\left(\frac{\partial H}{\partial w_{,xx}}\right)+\frac{\partial}{\partial y}\left(\frac{\partial H}{\partial w_{,xy}}\right)-\frac{\partial}{\partial x}\left(\frac{\partial H}{\partial w_{,x}}\right) = -V_x \qquad \frac{\partial}{\partial y}\left(\frac{\partial H}{\partial w_{,yy}}\right)+\frac{\partial}{\partial x}\left(\frac{\partial H}{\partial w_{,xy}}\right)-\frac{\partial}{\partial y}\left(\frac{\partial H}{\partial w_{,y}}\right) = -V_y \quad \text{(E8)}$$

Substituting Equation (E4) into Equations (7.48a) we obtain.

$$\frac{\partial^2}{\partial x^2}[D(w_{,xx}+\nu w_{,yy})]+\frac{\partial^2}{\partial y^2}[D(w_{,yy}+\nu w_{,xx})]+\frac{\partial^2}{\partial x\partial y}[D(1-\nu)w_{,xy}]-p_z = 0 \quad \text{(E9)}$$

Given the plate is homogeneous in $x$ and $y$ direction, that is, $D$ and $\nu$ are not function of $x$ and $y$, we obtain the differential equation shown below.

$$D\frac{\partial^2}{\partial x^2}\left[\frac{\partial^2 w}{\partial x^2}+\nu\frac{\partial^2 w}{\partial y^2}\right]+D\frac{\partial^2}{\partial x^2}\left[\frac{\partial^2 w}{\partial y^2}+\nu\frac{\partial^2 w}{\partial x^2}\right]+D(1-\nu)\frac{\partial^2}{\partial x\partial y}\left[\frac{\partial^2 w}{\partial x\partial y}\right]-p_z = 0$$

$$\text{Differential equation: } \boxed{D\left[\frac{\partial^4 w}{\partial x^4}+2\frac{\partial^4 w}{\partial x^2\partial y^2}+\frac{\partial^4 w}{\partial y^4}\right]-p_z = 0 \qquad x, y \text{ in } A} \quad \text{(E10)}$$

The above differential equation is same as in Equations (5.20).
Substituting Equations (E5) and (E8) into Equations (7.48b), (7.48c), and (7.48d) we obtain the boundary conditions as shown below.

Boundary Conditions at $x = 0$ and $x = a$: $\boxed{\begin{array}{lll} \boldsymbol{m}_{xx} = 0 & \text{or} & \delta(\partial u/\partial x) = 0 \\ \text{and} & & \\ V_x = 0 & \text{or} & \delta u = 0 \end{array}}$ (E11)

Boundary Conditions at $y = 0$ and $y = b$: $\boxed{\begin{array}{lll} \boldsymbol{m}_{yy} = 0 & \text{or} & \delta(\partial u/\partial y) = 0 \\ \text{and} & & \\ V_y = 0 & \text{or} & \delta u = 0 \end{array}}$ (E12)

Conditions at each corner: $\boxed{(\boldsymbol{m}_{xy})_k = 0 \quad \text{or} \quad \delta u_k = 0 \quad k = 1 \text{ to } 4}$ (E13)

The boundary conditions in Equations (E11) through (E13) are the same as the boundary conditions in Equations (5.23).

**COMMENTS**

1. This example demonstrates the elegance of obtaining boundary value problem using variational calculus. The internal moments, the equivalent shear force on the boundary, and the corner conditions gets defined naturally without elaborate arguments. Obtaining similar results for curved boundary though possible, will be algebraically lengthy and tedious. In the next section we will develop formulas for curved boundaries using variational principles.
2. These derivations can be further shortened by use of indicial notation discussed in Chapter 8.

## 7.9.2 Boundary value problem for geometries with curvilinear boundaries

In this section we obtain the boundary value problem from the functional below for curvilinear boundaries. The functional below is slightly different than the one in Equation (7.45) as it contains $u_{,xy}$ and $u_{,yx}$, which though equal need to be differentiated in order to maintain the symmetry with $x$ and $y$ in the derivation. This difference is also necessary for writing the results in indicial notation discussed in Chapter 8.

$$I(u) = \iint_A H(u_{,xx}, u_{,xy}, u_{,yx}, u_{,yy}, u_{,x}, u_{,y}, u, x, y)dxdy \quad \textbf{(7.49)}$$

Taking the first variation of the above functional we obtain

$$\delta I(u) = \iint_A\left[\left(\frac{\partial H}{\partial u_{,xx}}\right)\delta u_{,xx}+\left(\frac{\partial H}{\partial u_{,xy}}\right)\delta u_{,xy}+\left(\frac{\partial H}{\partial u_{,yx}}\right)\delta u_{,yx}+\left(\frac{\partial H}{\partial u_{,yy}}\right)\delta u_{,yy}+\left(\frac{\partial H}{\partial u_{,x}}\right)\delta u_{,x}+\left(\frac{\partial H}{\partial u_{,y}}\right)\delta u_{,y}+\left(\frac{\partial H}{\partial u}\right)\delta u\right]dxdy \quad \textbf{(7.50)}$$

We have seen in Examples 7.2 and 7.3, the derivatives of the functional $H$ result in quantities that have physical meaning such as stress, bending moments, and shear force. In the derivation that follows we will define variables to reduce writing lengthy expressions but the variables will have physical meaning in context of plate bending. The formulas developed will however be applicable to any functional containing second order derivatives.

Equation (7.50) is re-written and the derivatives transfered using Equations (7.30a) and (7.30b) as shown below.

$$\delta I(u) = \iint_A \left\{ \left(\frac{\partial H}{\partial u_{,xx}}\right)\frac{\partial(\delta u_{,x})}{\partial x} + \left(\frac{\partial H}{\partial u_{,xy}}\right)\frac{\partial(\delta u_{,x})}{\partial y} + \left(\frac{\partial H}{\partial u_{,yx}}\right)\frac{\partial(\delta u_{,y})}{\partial x} + \left(\frac{\partial H}{\partial u_{,yy}}\right)\frac{\partial(\delta u_{,y})}{\partial y} \right\} dx\,dy$$

$$+ \iint_A \left\{ \left(\frac{\partial H}{\partial u_{,x}}\right)\frac{\partial(\delta u)}{\partial x} + \left(\frac{\partial H}{\partial u_{,y}}\right)\frac{\partial(\delta u)}{\partial y} + \left(\frac{\partial H}{\partial u}\right)\delta u \right\} dx\,dy$$

$$\delta I(u) = \oint_\Gamma \left\{ \left[\left(\frac{\partial H}{\partial u_{,xx}}\right)n_x + \left(\frac{\partial H}{\partial u_{,xy}}\right)n_y\right]\delta u_{,x} + \left[\left(\frac{\partial H}{\partial u_{,yx}}\right)n_x + \left(\frac{\partial H}{\partial u_{,yy}}\right)n_y\right]\delta u_{,y} \right\} ds$$

$$+ \iint_A \left\{ \left[\left(\frac{\partial H}{\partial u_{,x}}\right) - \frac{\partial}{\partial x}\left(\frac{\partial H}{\partial u_{,xx}}\right) - \frac{\partial}{\partial y}\left(\frac{\partial H}{\partial u_{,xy}}\right)\right]\frac{\partial(\delta u)}{\partial x} + \left[\left(\frac{\partial H}{\partial u_{,y}}\right) - \frac{\partial}{\partial x}\left(\frac{\partial H}{\partial u_{,yx}}\right) - \frac{\partial}{\partial y}\left(\frac{\partial H}{\partial u_{,yy}}\right)\right]\frac{\partial(\delta u)}{\partial y} + \left(\frac{\partial H}{\partial u}\right)\delta u \right\} dx\,dy \qquad \textbf{(7.51a)}$$

We first consider the boundary term only. On the boundary, from Equations (7.31a) and (7.31b) we obtain

$$\delta u_{,x} = \frac{\partial(\delta u)}{\partial x} = \frac{\partial(\delta u)}{\partial n}n_x - \frac{\partial(\delta u)}{\partial s}n_y = (\delta u_{,n})n_x - \frac{\partial(\delta u)}{\partial s}n_y$$
$$\delta u_{,y} = \frac{\partial(\delta u)}{\partial y} = \frac{\partial(\delta u)}{\partial n}n_y + \frac{\partial(\delta u)}{\partial s}n_x = (\delta u_{,n})n_y + \frac{\partial(\delta u)}{\partial s}n_x \qquad \textbf{(7.51b)}$$

Substituting the above two equations in the boundary term, we obtain

$$\delta I_{boundary} = \oint_\Gamma \left\{ \left[\left(\frac{\partial H}{\partial u_{,xx}}\right)n_x + \left(\frac{\partial H}{\partial u_{,xy}}\right)n_y\right]n_x + \left[\left(\frac{\partial H}{\partial u_{,yx}}\right)n_x + \left(\frac{\partial H}{\partial u_{,yy}}\right)n_y\right]n_y \right\}(\delta u_{,n})ds$$

$$+ \oint_\Gamma \left\{ \left[\left(\frac{\partial H}{\partial u_{,xx}}\right)n_x + \left(\frac{\partial H}{\partial u_{,xy}}\right)n_y\right](-n_y) + \left[\left(\frac{\partial H}{\partial u_{,yx}}\right)n_x + \left(\frac{\partial H}{\partial u_{,yy}}\right)n_y\right]n_x \right\}\frac{\partial(\delta u)}{\partial s}ds \qquad \textbf{(7.51c)}$$

We introduce the following notation.

$$\boxed{\begin{aligned} \boldsymbol{m}_{nn} &= \left(\frac{\partial H}{\partial u_{,xx}}\right)n_x^2 + \left(\frac{\partial H}{\partial u_{,xy}} + \frac{\partial H}{\partial u_{,yx}}\right)n_x n_y + \left(\frac{\partial H}{\partial u_{,yy}}\right)n_y^2 \\ \boldsymbol{m}_{nt} &= -\left(\frac{\partial H}{\partial u_{,xx}} - \frac{\partial H}{\partial u_{,yy}}\right)n_x n_y + n_x^2\left(\frac{\partial H}{\partial u_{,yx}}\right) - n_y^2\left(\frac{\partial H}{\partial u_{,xy}}\right) \end{aligned}} \qquad \textbf{(7.52)}$$

Substituting the above equations into Equation (7.51c) and integrating with respect to $s$, we obtain

$$\delta I_{boundary} = \oint_\Gamma \boldsymbol{m}_{nn}(\delta u_{,n})ds + \oint_\Gamma \boldsymbol{m}_{nt}\frac{\partial(\delta u)}{\partial s}ds = \oint_\Gamma \boldsymbol{m}_{nn}(\delta u_{,n})ds + (\boldsymbol{m}_{nt}\delta u)\big|_\Gamma - \oint_\Gamma \frac{\partial \boldsymbol{m}_{nt}}{\partial s}\delta u\,ds \qquad \textbf{(7.53a)}$$

The term $(\boldsymbol{m}_{nt}\delta u)\big|_\Gamma$ represents the integration on the whole boundary, that is, we start from a point and return back to it. Thus, the term will be zero, unless $\boldsymbol{m}_{nt}$ is discontinuous at some point on the boundary. This discontinuity will occur at each corner point on the boundary due to the discontinuity in the direction cosines of the normal as shown in Figure 7.8. We represent the discontinuity in $\boldsymbol{m}_{nt}$ at each corner as $(\Delta \boldsymbol{m}_{nt})_C$. Equation (7.53a) can now be written as

$$\delta I_{boundary} = \oint_\Gamma \left[\boldsymbol{m}_{nn}(\delta u_{,n}) - \frac{\partial \boldsymbol{m}_{nt}}{\partial s}\delta u\right]ds + (\Delta \boldsymbol{m}_{nt})_C\,\delta u_C \qquad \textbf{(7.53b)}$$

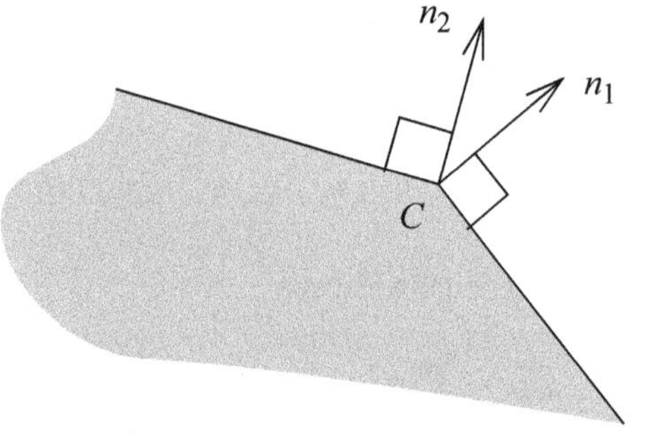


**Figure 7.8** Discontinuity in the unit normal at a corner.

We introduce the following notation

$$-q_x = \left(\frac{\partial H}{\partial u_{,x}}\right) - \frac{\partial}{\partial x}\left(\frac{\partial H}{\partial u_{,xx}}\right) - \frac{\partial}{\partial y}\left(\frac{\partial H}{\partial u_{,xy}}\right) \qquad -q_y = \left(\frac{\partial H}{\partial u_{,y}}\right) - \frac{\partial}{\partial x}\left(\frac{\partial H}{\partial u_{,yx}}\right) - \frac{\partial}{\partial y}\left(\frac{\partial H}{\partial u_{,yy}}\right) \tag{7.54}$$

Substituting Equation (7.53b) for the boundary term and Equation (7.54) into Equation (7.51a) we obtain

$$\delta I(u) = \oint_\Gamma \left[ m_{nn}(\delta u_{,n}) - \frac{\partial m_{nt}}{\partial s}\delta u \right] ds + (\Delta m_{nt})_C \delta u_C + \iint_A \left\{ (-q_x)\frac{\partial}{\partial x}(\delta u) + (-q_y)\frac{\partial}{\partial y}(\delta u) + \left(\frac{\partial F}{\partial u}\right)\delta u \right\} dx\,dy \tag{7.55a}$$

Using Equations (7.30a) and Equations (7.30b), we transfer the derivatives to obtain

$$\delta I(u) = \oint_\Gamma \left[ m_{nn}(\delta u_{,n}) - \frac{\partial m_{nt}}{\partial s}\delta u \right] ds + (\Delta m_{nt})_C \delta u_C - \oint_\Gamma (q_x n_x + q_y n_y)\delta u\, ds + \iint_A \left\{ \frac{\partial q_x}{\partial x} + \frac{\partial q_y}{\partial y} + \left(\frac{\partial F}{\partial u}\right) \right\}(\delta u)\,dx\,dy$$

$$\delta I(u) = \iint_A \left\{ \frac{\partial q_x}{\partial x} + \frac{\partial q_y}{\partial y} + \left(\frac{\partial H}{\partial u}\right) \right\}(\delta u)\,dx\,dy + \oint_\Gamma [m_{nn}(\delta u_{,n})]\,ds - \oint_\Gamma \left[ \left( \frac{\partial m_{nt}}{\partial s} + q_x n_x + q_y n_y \right)\delta u \right] ds + (\Delta m_{nt})_C \delta u_C = 0 \tag{7.55b}$$

We define the following

$$q_n = q_x n_x + q_y n_y \qquad V_n = q_n + \frac{\partial m_{nt}}{\partial s} \tag{7.55c}$$

For the above equation to be valid each term in the equation must be zero and we obtain the boundary value problem written below.

Differential equation: $$\frac{\partial q_x}{\partial x} + \frac{\partial q_y}{\partial y} + \left(\frac{\partial H}{\partial u}\right) = 0 \qquad x, y \text{ in } A \tag{7.56a}$$

Boundary Conditions:
$$\left.\begin{array}{lll} m_{nn} = 0 & \text{or} & \delta\left(\frac{\partial u}{\partial n}\right) = 0 \\ \text{and} & & \\ V_n = 0 & \text{or} & \delta u = 0 \end{array}\right\} \quad x, y \text{ on } \Gamma \tag{7.56b}$$

Corner Condition: $$(\Delta m_{nt})_C = 0 \qquad \text{or} \qquad \delta u_C \qquad x, y \text{ on corner } \Gamma_C \tag{7.56c}$$

If the functional in Equation (7.45) contains more than one variable then Equations (7.56a), (7.56b), and (7.56c) must be written for each of the variables.We substitute $m_{nt}$ from Equation (7.52) into $V_n$ in Equation (7.55c) to obtain

$$V_n = q_n + \frac{\partial}{\partial s}\left[ -\left(\frac{\partial H}{\partial u_{,xx}} - \frac{\partial H}{\partial u_{,yy}}\right) n_x n_y + n_x^2\left(\frac{\partial H}{\partial u_{,yx}}\right) - n_y^2\left(\frac{\partial H}{\partial u_{,xy}}\right) \right]$$

$$V_n = q_n - \frac{\partial}{\partial s}\left(\frac{\partial H}{\partial u_{,xx}} - \frac{\partial H}{\partial u_{,yy}}\right) n_x n_y - \left(\frac{\partial H}{\partial u_{,xx}} - \frac{\partial H}{\partial u_{,yy}}\right)\left\{ \frac{\partial n_x}{\partial s} n_y + n_x \frac{\partial n_y}{\partial s} \right\}$$
$$+ \frac{\partial}{\partial s}\left(\frac{\partial H}{\partial u_{,yx}}\right) n_x^2 + 2\left(\frac{\partial H}{\partial u_{,yx}}\right) n_x \frac{\partial n_x}{\partial s} - \frac{\partial}{\partial s}\left(\frac{\partial H}{\partial u_{,xy}}\right) n_y^2 - 2\left(\frac{\partial}{\partial u_{,xy}}\frac{\partial H}{\partial u_{,xy}}\right)\left(n_y \frac{\partial n_y}{\partial s}\right) \tag{7.57a}$$

Substituting Equation (7.31b) into the above equation we obtain

$$V_n = q_n - \frac{\partial}{\partial s}\left(\frac{\partial H}{\partial u_{,xx}} - \frac{\partial H}{\partial u_{,yy}}\right) n_x n_y + \frac{\partial}{\partial s}\left(\frac{\partial H}{\partial u_{,yx}}\right) n_x^2 - \frac{\partial}{\partial s}\left(\frac{\partial H}{\partial u_{,xy}}\right) n_y^2 - \left(\frac{\partial H}{\partial u_{,xx}} - \frac{\partial H}{\partial u_{,yy}}\right)\frac{n_x^2 - n_y^2}{R_{cur}} - \left(\frac{\partial H}{\partial u_{,xy}} + \frac{\partial H}{\partial u_{,yx}}\right)\frac{2 n_x n_y}{R_{cur}} \tag{7.57b}$$

In the next example we resolve Example 7.3 on plate bending but a plate with curvilinear boundaries to demonstrate the application of the formulas in this section.

## EXAMPLE 7.4

Obtain the boundary value problem for thin plate bending with only transverse distributed forces and correlate the variables to those introduced in plate theory.

**PLAN**

We write the potential energy functional for thin plate bending and use Equations (7.56a) through (7.56c) to obtain the boundary value problem.

**SOLUTION**

Substituting Equations (7.7d) and (7.17d) into Equation (7.25), we obtain the potential energy as

$$\Omega(w) = \frac{D}{2}\iint_A \left\{\left(\frac{\partial^2 w}{\partial x^2}\right)^2 + \left(\frac{\partial^2 w}{\partial y^2}\right)^2 + 2\nu\left(\frac{\partial^2 w}{\partial x^2}\right)\left(\frac{\partial^2 w}{\partial y^2}\right) + (1-\nu)\left[\left(\frac{\partial^2 w}{\partial x \partial y}\right)^2 + \left(\frac{\partial^2 w}{\partial y \partial x}\right)^2\right]\right\} dx\,dy - \iint_A p_z w\,dx\,dy \tag{E1}$$

The functional in Equation (7.45) can be written as

$$H(w_{,xx}, w_{,xy}, w_{,yx}, w_{,yy}, w_{,x}, w_{,y}, w, x, y) = \frac{D}{2}[w_{,xx}{}^2 + w_{,yy}{}^2 + 2\nu w_{,xx}w_{,yy} + (1-\nu)(w_{,xy}{}^2 + w_{,yx}{}^2)] - p_z w \tag{E2}$$

The derivatives of the functional $H$ can be written as shown below.

$$\frac{\partial H}{\partial w} = -p_z \qquad \frac{\partial H}{\partial w_{,x}} = 0 \qquad \frac{\partial H}{\partial w_{,y}} = 0 \tag{E3}$$

$$\frac{\partial H}{\partial w_{,xx}} = D[w_{,xx} + \nu w_{,yy}] \qquad \frac{\partial H}{\partial w_{,yy}} = D[w_{,yy} + \nu w_{,xx}] \qquad \frac{\partial H}{\partial w_{,xy}} = D(1-\nu)w_{,xy} \qquad \frac{\partial H}{\partial w_{,yx}} = D(1-\nu)w_{,yx} \tag{E4}$$

Comparing Equation (E4) to Equation (5.14) we obtain the bending moments of the plate theory as shown below.

$$\frac{\partial H}{\partial w_{,xx}} = -\boldsymbol{m}_{xx} \qquad \frac{\partial H}{\partial w_{,yy}} = -\boldsymbol{m}_{yy} \qquad \frac{\partial H}{\partial w_{,xy}} = -\boldsymbol{m}_{xy} \qquad \frac{\partial H}{\partial w_{,yx}} = -\boldsymbol{m}_{yx} \qquad \boldsymbol{m}_{xy} = \boldsymbol{m}_{yx} \tag{E5}$$

Substituting the derivatives of the functionals in Equations (7.52) and (7.54)we obtain the following.

$$\boldsymbol{m}_{nn} = -[\boldsymbol{m}_{xx}n_x^2 + \boldsymbol{m}_{yy}n_y^2 + 2\boldsymbol{m}_{xy}n_x n_y] \tag{E6}$$

$$\boldsymbol{m}_{nt} = -[-(\boldsymbol{m}_{xx} - \boldsymbol{m}_{yy})n_x n_y + (n_x^2 - n_y^2)\boldsymbol{m}_{xy}] \tag{E7}$$

$$\boldsymbol{q}_x = -\frac{\partial}{\partial x}\left[D\left(\frac{\partial^2 w}{\partial x^2} + \nu\frac{\partial^2 w}{\partial y^2}\right)\right] - \frac{\partial}{\partial y}\left[D(1-\nu)\frac{\partial^2 w}{\partial x \partial y}\right] = -D\left[\frac{\partial^3 w}{\partial x^3} + \frac{\partial^3 w}{\partial x \partial y^2}\right] = -D\frac{\partial}{\partial x}(\nabla^2 w) \tag{E8}$$

$$\boldsymbol{q}_y = -\frac{\partial}{\partial x}\left[D(1-\nu)\frac{\partial^2 w}{\partial x \partial y}\right] - \frac{\partial}{\partial y}\left[D\left(\frac{\partial^2 w}{\partial y^2} + \nu\frac{\partial^2 w}{\partial x^2}\right)\right] = -D\left[\frac{\partial^3 w}{\partial x^2 \partial y} + \frac{\partial^3 w}{\partial y^3}\right] = -D\frac{\partial}{\partial y}(\nabla^2 w) \tag{E9}$$

Comparing Equations (E8) and (E9) with Equation (5.19) we see that the Equation (7.54) from variational calculus gives us the shear force in plate bending.

From Equation (7.55c) and Equation (7.57a) we obtain

$$\boldsymbol{q}_n = -D\left[\frac{\partial}{\partial x}(\nabla^2 w)n_x + \frac{\partial}{\partial y}(\nabla^2 w)n_y\right] = -D\frac{\partial}{\partial n}(\nabla^2 w) \tag{E10}$$

$$V_n = -D\frac{\partial}{\partial n}(\nabla^2 w) + n_x n_y \frac{\partial}{\partial s}(\boldsymbol{m}_{xx} - \boldsymbol{m}_{yy}) - (n_x^2 - n_y^2)\frac{\partial \boldsymbol{m}_{xy}}{\partial s} + (\boldsymbol{m}_{xx} - \boldsymbol{m}_{yy})\frac{n_x^2 - n_y^2}{R_{cur}} + \boldsymbol{m}_{xy}\frac{2n_x n_y}{R_{cur}} \tag{E11}$$

From Equations (7.56a) through Equations (7.56c) we obtain the boundary value problem as shown below.

$$\text{Differential equation: } D\left[\frac{\partial^4 w}{\partial x^4} + 2\frac{\partial^4 w}{\partial x^2 \partial y^2} + \frac{\partial^4 w}{\partial y^4}\right] - p_z = 0 \qquad x, y \text{ in } R \tag{E12}$$

$$\text{Boundary Condition 1: } \boldsymbol{m}_{nn} = 0 \quad \text{or} \quad \delta\left(\frac{\partial u}{\partial n}\right) = 0 \qquad x, y \text{ on } S \text{ and} \tag{E13}$$

$$\text{Boundary Condition 2: } V_n = 0 \quad \text{or} \quad \delta u = 0 \qquad x, y \text{ on } S$$

$$\text{Corner Condition: } (\Delta \boldsymbol{m}_{nt})_C = 0 \quad \text{or} \quad \delta u_C \qquad x, y \text{ on corner } S_C \tag{E14}$$

**COMMENTS**

1. Consider a boundary $x$ = constant on a rectangular plate. On this boundary, the normal is in the $x$ direction, thus $|n_x| = 1$ and $n_y = 0$ .The $s$ direction is same as the $y$ direction. Furthermore, $R_{cur} = \infty$ as the boundary is straight. Substituting these values into Equation (E11) we obtain

$$V_n = V_x = -D\frac{\partial}{\partial x}(\nabla^2 w) - \frac{\partial \boldsymbol{m}_{xy}}{\partial y} = -D\left[\frac{\partial^3 w}{\partial x^3} + (2-\nu)\frac{\partial^3 w}{\partial x \partial y^2}\right] \tag{E15}$$

The above equation is the same as Equation (5.21a).

2. In a square plate, the direction cosines will be zero, plus or minus one. Hence, $\boldsymbol{m}_{nn}$ will be either $-\boldsymbol{m}_{xx}$ or $-\boldsymbol{m}_{yy}$ and $\boldsymbol{m}_{nt}$ will be plus or minus $\boldsymbol{m}_{xy}$. The minus sign is a consequence of our definition of static equivalency in Equation (5.8).
3. These derivations can be further shortened by use of indicial notation discussed in Chapter 8. See Example 8.5.

## 7.10 STATIONARY VALUE OF A DEFINITE VOLUME INTEGRAL

Gauss divergence formula presented in Section 7.6 is used to transfer derivatives in finding the stationary value of volume integral. The rest of the process is as before for finding the stationary values of line and area integral. We will restrict ourselves to functional with first order derivatives as higher order derivatives lead to excessive algebra.

We consider a volume $T$ bounded by surface $S$. Using the notation $u_{,x} = \partial u/\partial x$, $u_{,y} = \partial u/\partial y$, and $u_{,z} = \partial u/\partial z$, we define the functional and consider its first variation below.

$$I(u) = \iiint_T H(u_{,x}, u_{,y}, u_{,z}, u, x, y, z)dxdydz \tag{7.58a}$$

$$\delta I(u) = \iiint_T \left[\left(\frac{\partial H}{\partial u_{,x}}\right)\delta u_{,x} + \left(\frac{\partial H}{\partial u_{,y}}\right)\delta u_{,y} + \left(\frac{\partial H}{\partial u_{,z}}\right)\delta u_{,z} + \left(\frac{\partial H}{\partial u}\right)\delta u\right]dxdydz$$

$$\delta I(u) = \iiint_T \left[\left(\frac{\partial H}{\partial u_{,x}}\right)\frac{\partial}{\partial x}(\delta u) + \left(\frac{\partial H}{\partial u_{,y}}\right)\frac{\partial}{\partial y}(\delta u) + \left(\frac{\partial H}{\partial u_{,z}}\right)\frac{\partial}{\partial z}(\delta u) + \left(\frac{\partial H}{\partial u}\right)\delta u\right]dxdydz$$

Using Equation (7.33), we transfer the derivatives in the first three terms and set the first variation to zero as shown below.

$$\delta I(u) = \iint_S \left[\frac{\partial H}{\partial u_{,x}}n_x + \frac{\partial H}{\partial u_{,y}}n_y + \frac{\partial H}{\partial u_{,z}}n_z\right]\delta u dS + \iiint_T \left[\frac{\partial H}{\partial u} - \frac{\partial}{\partial x}\left(\frac{\partial H}{\partial u_{,x}}\right) - \left(\frac{\partial}{\partial y}\left(\frac{\partial H}{\partial u_{,y}}\right) - \frac{\partial}{\partial z}\left(\frac{\partial H}{\partial u_{,z}}\right)\right)\right]\delta u dxdydz = 0 \tag{7.58b}$$

Each term in the first variation must be zero at all points.Thus, we can obtain the boundary value problem written below.

Differential Equation: $$\frac{\partial H}{\partial u} - \frac{\partial}{\partial x}\left(\frac{\partial H}{\partial u_{,x}}\right) - \frac{\partial}{\partial y}\left(\frac{\partial H}{\partial u_{,y}}\right) - \frac{\partial}{\partial z}\left(\frac{\partial H}{\partial u_{,z}}\right) = 0 \qquad x, y, z \text{ in } T \tag{7.59a}$$

Boundary Conditions: $$\left(\frac{\partial H}{\partial u_{,x}}\right)n_x + \left(\frac{\partial H}{\partial u_{,y}}\right)n_y + \left(\frac{\partial H}{\partial u_{,z}}\right)n_z = 0 \quad \text{or} \quad \delta u = 0 \qquad x, y, z \text{ on S} \tag{7.59b}$$

If the functional in Equation (7.58a) contains more than one variable then Equations (7.59a) and (7.59b) must be written for each of the variables.

## 7.11 NONLINEARITIES

The three most common nonlinear problems in structural analysis are those that arise from material, geometry, and contact. Geometric nonlinearities arise from large deformation—we will consider these in Section 7.11.1. Material nonlinearity arises from the nonlinear relationships between stresses and strains—we incorporated these into analysis in Chapter 2 and will look at it from energy perspective in Section 7.11.2.Contact nonlinearities arise when the contact region between two surfaces change due to applied loads—it requires modifying variational equations and will not be considered any further.

### 7.11.1 Geometric nonlinearity

In Section 1.5 we saw that there are two definitions of large strains. Lagrangian strain is computed from deformation by using the original *undeformed geometry* as the reference geometry. Eulerian strain is computed from deformation by using the *final deformed* geometry as the reference geometry. We will derive expressions for these two strains in Section 8.5, but in this chapter we will take a given large strain expression and incorporate it into analysis by dropping the small strain approximation for only one-dimensional structural members. Analysis of plates will be done in the next chapter as indicial notation reduce the tedious algebra that arises from incorporating large strain into two dimensional analysis.

We will develop expression for potential energy using the given large strain. Minimum potential energy is applicable to linear and nonlinear system provided the system is conservative. So even though we are considering large deformation, we shall assume we are still in the elastic region. The potential energy incorporating large strain will give us a new functional. We can then use variational calculus equations that minimize a functional to obtain the boundary value problem. This is elaborated in the next example for beams and in Example 8.6. for plates.

**EXAMPLE 7.5**

Lagrangian strain with moderately large rotation for symmetric beam bending is given as

$$\varepsilon_{xx} = \frac{du}{dx} + \frac{1}{2}\left(\frac{dv}{dx}\right)^2 \tag{7.60}$$

Assume the displacement field is $u(x) = u_o - y dv/dx$, where $u_o$ and v are the axial and bending displacements that are function of $x$ only. All assumptions of classical beam theory except small strain approximation are valid. $p_x(x)$ and $p_y(x)$ are the distributed loads in $x$ and $y$ directions. Obtain the stress formula and the statement of boundary value problems for displacements $u_o$ and v.

**PLAN**

We will follow the logic described in Chapter 2 to develop the stress formula and use potential energy and variational equations of Section 7.7.3 to develop the boundary value problem.

**SOLUTION**

1. *Strains*: Substituting the displacement field into Equation (7.60) we obtain

$$\varepsilon_{xx} = \frac{du_o}{dx} - y\frac{d^2v}{dx^2} + \frac{1}{2}\left(\frac{dv}{dx}\right)^2 \tag{E1}$$

2. *Stresses*: By Hooke's law we obtain

$$\sigma_{xx} = E\varepsilon_{xx} = E\left[\frac{du_o}{dx} - y\frac{d^2v}{dx^2} + \frac{1}{2}\left(\frac{dv}{dx}\right)^2\right] \tag{E2}$$

3. *Internal forces and Moments*: By static equivalency, assuming material is homogeneous across the cross section, and $y$ is measured from the centroid of the cross section we obtain

$$N = \int_A \sigma_{xx}dA = \int_A E\left(\frac{du_o}{dx} - y\frac{d^2v}{dx^2} + \frac{1}{2}\left(\frac{dv}{dx}\right)^2\right)dA = \frac{du_o}{dx}\int_A EdA - \frac{d^2v}{dx^2}\int_A EydA + \frac{1}{2}\left(\frac{dv}{dx}\right)^2\int_A EdA = EA\left[\frac{du_o}{dx} + \frac{1}{2}\left(\frac{dv}{dx}\right)^2\right] \tag{E3}$$

$$M_z = -\int_A y\sigma_{xx}dA = -\int_A Ey\left[\frac{du_o}{dx} - y\frac{d^2v}{dx^2} + \frac{1}{2}\left(\frac{dv}{dx}\right)^2\right]dA = -\frac{du_o}{dx}\int_A EydA + \frac{d^2v}{dx^2}\int_A Ey^2dA + \frac{1}{2}\left(\frac{dv}{dx}\right)^2\int_A EydA = EI_{zz}\frac{d^2v}{dx^2} \tag{E4}$$

4. *Stress Formula*: Substituting Equations (E3) and (E4) into Equation (E2) we obtain our result for stress.

**ANS.** $\sigma_{xx} = N/A - (M_z y)/I_{zz}$

5. *Potential Energy*: The work and strain energy can be found and potential energy written as shown below.

$$W = \int_L (p_x u_o + p_y v)dx \tag{E5}$$

$$U = \frac{1}{2}\int_L\left[\int_A \sigma_{xx}\varepsilon_{xx}dA\right]dx = \frac{1}{2}\int_L\left[\int_A E\left\{\frac{du_o}{dx} - y\frac{d^2v}{dx^2} + \frac{1}{2}\left(\frac{dv}{dx}\right)^2\right\}^2 dA\right]dx$$

$$U = \frac{1}{2}\int_L\left[\int_A E\left\{\left(\frac{du_o}{dx}\right)^2 + y^2\left(\frac{d^2v}{dx^2}\right)^2 + \frac{1}{4}\left(\frac{dv}{dx}\right)^4 - 2y\frac{du_o}{dx}\frac{d^2v}{dx^2} - y\frac{d^2v}{dx^2}\left(\frac{dv}{dx}\right)^2 + \frac{du_o}{dx}\left(\frac{dv}{dx}\right)^2\right\}dA\right]dx$$

$$U = \frac{E}{2}\int_L\left[A\left\{\left(\frac{du_o}{dx}\right)^2 + \frac{1}{4}\left(\frac{dv}{dx}\right)^4 + \frac{du_o}{dx}\left(\frac{dv}{dx}\right)^2\right\} + I_{zz}\left(\frac{d^2v}{dx^2}\right)^2\right]dx \tag{E6}$$

$$\Omega = \frac{E}{2}\int_L\left[A\left\{\left(\frac{du_o}{dx}\right)^2 + \frac{1}{4}\left(\frac{dv}{dx}\right)^4 + \frac{du_o}{dx}\left(\frac{dv}{dx}\right)^2\right\} + I_{zz}\left(\frac{d^2v}{dx^2}\right)^2\right]dx - \int_L (p_x u_o + p_y v)dx \tag{E7}$$

6. *Functional and its derivatives*: The highest order of derivative is two in potential energy. We write the potential energy as a functional described in Section 7.7.3 and take its derivatives.

$$\Omega = \int_0^L H(u_o^{(i)}, u_o, v^{(ii)}, v^{(i)}, v, x)\ dx \ \text{where} \tag{E8}$$

$$H(u_o^{(i)}, u_o, v^{(ii)}, v^{(i)}, v, x) = \frac{EA}{2}\left\{(u_o^{(i)})^2 + \frac{1}{4}(v^{(i)})^4 + u_o^{(i)}(v^{(i)})^2\right\} + \frac{EI_{zz}}{2}(v^{(ii)})^2 - (p_x u_o + p_y v) \tag{E9}$$

$$\frac{\partial H}{\partial u_o^{(i)}} = EAu_o^{(i)} + \frac{EA}{2}(v^{(i)})^2 = EA\left[\frac{du_o}{dx} + \frac{1}{2}\left(\frac{dv}{dx}\right)^2\right] \qquad \frac{\partial H}{\partial u_o} = -p_x \tag{E10}$$

$$\frac{\partial H}{\partial v^{(ii)}} = EI_{zz}v^{(ii)} = EI_{zz}\frac{d^2v}{dx^2} \qquad \frac{\partial H}{\partial v^{(i)}} = \frac{EA}{2}(v^{(i)})^3 + EAu_o^{(i)}(v^{(i)}) = EA\left[\frac{du_o}{dx} + \frac{1}{2}\left(\frac{dv}{dx}\right)^2\right]\frac{dv}{dx} \qquad \frac{\partial H}{\partial v} = -p_y \tag{E11}$$

7. *Boundary value problem*: The two differential equations can be written as shown below

$$\frac{d}{dx}\left(\frac{\partial H}{\partial u_o^{(i)}}\right) - \frac{\partial H}{\partial u_o} = \frac{d}{dx}\left\{EA\left[\frac{du_o}{dx} + \frac{1}{2}\left(\frac{dv}{dx}\right)^2\right]\right\} - p_x = 0$$

$$\text{Differential equation 1: } \frac{d}{dx}\left\{EA\left[\frac{du_o}{dx} + \frac{1}{2}\left(\frac{dv}{dx}\right)^2\right]\right\} = p_x \tag{E12}$$

$$\frac{d^2}{dx^2}\left(\frac{\partial H}{\partial v^{(ii)}}\right) - \frac{d}{dx}\left(\frac{\partial H}{\partial v^{(i)}}\right) + \frac{\partial H}{\partial v} = \frac{d^2}{dx^2}\left(EI_{zz}\frac{d^2 v}{dx^2}\right) - \frac{d}{dx}\left(EA\left[\frac{du_o}{dx} + \frac{1}{2}\left(\frac{dv}{dx}\right)^2\right]\frac{dv}{dx}\right) - p_y = 0$$

$$\frac{d^2}{dx^2}\left(EI_{zz}\frac{d^2 v}{dx^2}\right) - \frac{d}{dx}\left\{EA\left[\frac{du_o}{dx} + \frac{1}{2}\left(\frac{dv}{dx}\right)^2\right]\right\}\frac{dv}{dx} - A\left[\frac{du_o}{dx} + \frac{1}{2}\left(\frac{dv}{dx}\right)^2\right]\frac{d^2 v}{dx^2} = p_y \qquad \text{(E13)}$$

Substituting Equation (E12) into Equation (E13) we obtain

$$\text{Differential equation 2: } \frac{d^2}{dx^2}\left(EI_{zz}\frac{d^2 v}{dx^2}\right) - EA\left[\frac{du_o}{dx} + \frac{1}{2}\left(\frac{dv}{dx}\right)^2\right]\frac{d^2 v}{dx^2} = p_y + p_x\frac{dv}{dx} \qquad \text{(E14)}$$

Using Equations (E3) and (E4) we note the following before writing the boundary conditions.

$$\frac{\partial H}{\partial u_o^{(i)}} = \boldsymbol{N} \qquad \frac{\partial H}{\partial v^{(ii)}} = \boldsymbol{M}_z \qquad \frac{d}{dx}\left(\frac{\partial H}{\partial v^{(ii)}}\right) - \frac{\partial H}{\partial v^{(i)}} = \frac{d\boldsymbol{M}_z}{dx} - \boldsymbol{N}\frac{dv}{dx} \qquad \text{(E15)}$$

$$\text{Boundary Condition 1: } \boldsymbol{N} = 0 \quad \text{or} \quad \delta u_o \quad \text{at } x = 0 \text{ and } x = L \qquad \text{(E16)}$$

$$\text{Boundary Condition 2: } \boldsymbol{M}_z = 0 \quad \text{or} \quad \delta\left(\frac{dv}{dx}\right) = 0 \quad \text{at } x = 0 \text{ and } x = L \qquad \text{(E17)}$$

$$\text{Boundary Condition 3: } \frac{d\boldsymbol{M}_z}{dx} - \boldsymbol{N}\frac{dv}{dx} = 0 \quad \text{or} \quad \delta v = 0 \quad \text{at } x = 0 \text{ and } x = L \qquad \text{(E18)}$$

The two differential equations and the three boundary conditions constitute the boundary value problem.

**COMMENTS**

1. The example highlights that the variational calculus equations that define the differential equations and boundary conditions are applicable to linear and non-linear problems.
2. The axial and bending problem cannot be decoupled for large deflection of beams.
3. Clearly the differential equations are nonlinear. However, consider the case when $p_x = 0$. From differential equation 1 and Equation (E3) we see that $N$ is a constant. Substituting this into differential equation 2, we obtain a linear differential equation given below

$$\frac{d^2}{dx^2}\left(EI_{zz}\frac{d^2 v}{dx^2}\right) = p_y + N\frac{d^2 v}{dx^2} \qquad \text{(E19)}$$

If the applied axial load $P$ is compressive then $N = -P$ and we obtain the buckling equation given in Problem 4.15.

4. For the case $p_x = 0$, if the distance between the two ends of the beam is not fixed, that is, one of the end is free to move in axial direction then $N = 0$. In such a case beam bending problem can be solved like a linear problem and nonlinearity has no impact unless $u_o$ needs to be determined.
5. Figure 7.9 shows how axial force generates force in the transverse direction for large deflection of beams. In a similar manner the distributed force in the $x$ direction will generate a component $p_x dv/dx$. The equilibrium equations can be written that incorporates the vertical components from axial forces and Equations (E3) and (E4) substituted to obtain the differential equations. But it is significantly simpler to obtain the differential equations from potential energy as demonstrated in this example.

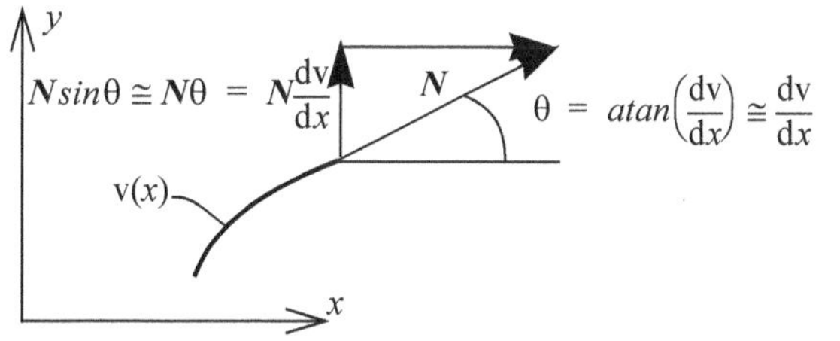


**Figure 7.9** Component of axial force in the vertical direction during large deflection of beams.

### 7.11.2 Material nonlinearity

Given the fact the minimum potential energy is applicable to only conservative system, that is elastic systems, the only material nonlinearity we will consider is the one in which stress-strain relationship is given by the power law model of Section 1.8.7. The equations describing the stress–strain relationship are given by Equation (1.49) which is written below for convenience.

$$\sigma = \begin{cases} E\varepsilon^n & \varepsilon \geq 0 \\ -E(-\varepsilon)^n & \varepsilon < 0 \end{cases} \quad \text{and} \quad \tau = \begin{cases} G\gamma^n & \gamma \geq 0 \\ -G(-\gamma)^n & \gamma < 0 \end{cases} \qquad \textbf{(7.61)}$$

We can substitute the above equation into Equation (7.9) and noting that strain energy is always positive; we obtain the strain energy density given below

$$U_0 = \begin{cases} E\varepsilon^{n+1}/(n+1) & \varepsilon \ge 0 \\ E(-\varepsilon)^{n+1}/(n+1) & \varepsilon < 0 \end{cases} \qquad U_0 = \begin{cases} G\gamma^{n+1}/(n+1) & \gamma \ge 0 \\ G(-\gamma)^{n+1}/(n+1) & \gamma < 0 \end{cases} \tag{7.62}$$

We can substitute the strain displacement equations for one-dimensional structural members in the strain energy density, perform the integration over the volume and obtain strain energy expressions over the length of the member. This process is similar to the one demonstrated for linear strain energy in Section 7.3.1. We can then write the work potential and the potential energy, which we can then minimize to obtain the boundary value problems. The process is demonstrated for symmetric beams in the next example.

## EXAMPLE 7.6

The hollow square cross section of a beam is shown in Figure 7.10. The stress-strain equation of beam material is given by $\sigma = E\varepsilon^{0.5}$. The beam is subjected to a transverse distributed force $p_y(x)$. Obtain the boundary value problem for the beam deflection v if all assumptions of classical beam are valid except for the Hooke's law.

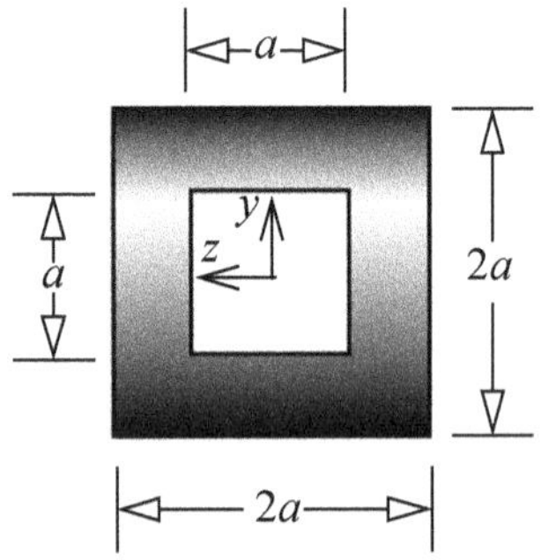


**Figure 7.10** Symmetric beam cross section of Example 7.6.

**PLAN**

We substitute the bending normal strain into Equation (7.62), and integrate it over the volume to obtain the strain energy over length of the beam. We write the potential energy and minimize it using equations in Section 7.8.1 to obtain the boundary value problem.

**SOLUTION**

1. *Bending strain*: The bending normal strain is given by

$$\varepsilon_{xx} = -y\frac{d^2v}{dx^2} \tag{E1}$$

2. *Strain energy density:* We substitute the above equation into Equation (7.62) with $n = 0.5$ and obtain

$$U_0 = \begin{cases} \dfrac{E}{1.5}\left(-y\dfrac{d^2v}{dx^2}\right)^{1.5} & y < 0 \\ \dfrac{E}{1.5}\left(y\dfrac{d^2v}{dx^2}\right)^{1.5} & y > 0 \end{cases} \tag{E2}$$

3. *Potential energy*: As the cross section is symmetric about the z axis, we can perform integration on the top half and doubling it to obtain the strain energy as shown below.

$$U = \int_L \left[\int_A U_0 dA\right] dx = 2\int_L \left[\int_0^{a/2} \frac{E}{1.5}\left(y\frac{d^2v}{dx^2}\right)^{1.5}(ady) + \int_{a/2}^{a} \frac{E}{1.5}\left(y\frac{d^2v}{dx^2}\right)^{1.5}(2ady)\right] dx$$

$$U = 2\int_L \frac{aE}{1.5}\left(\frac{d^2v}{dx^2}\right)^{1.5}\left[\int_0^{a/2} y^{1.5}dy + 2\int_{(a/2)}^{a} y^{1.5}dy\right] dx = \frac{2aE}{1.5}\int_L \left(\frac{d^2v}{dx^2}\right)^{1.5}\left[\frac{y^{2.5}}{2.5}\Big|_0^{a/2} + 2\frac{y^{2.5}}{2.5}\Big|_{a/2}^{a}\right] dx$$

$$U = \frac{2aE}{3.75}\int_L \left(\frac{d^2v}{dx^2}\right)^{1.5}\left[\left(\frac{a}{2}\right)^{2.5} + 2\left(a^{2.5} - \left(\frac{a}{2}\right)^{2.5}\right)\right] dx = \frac{2a^{3.75}E}{3.75}\int_L \left(\frac{d^2v}{dx^2}\right)^{1.5}\left[2 - \left(\frac{1}{2}\right)^{2.5}\right] dx = 0.9724a^{3.75}E\int_L \left(\frac{d^2v}{dx^2}\right)^{1.5} dx$$

The work potential is

$$W = \int_L p_y v dx \tag{E3}$$

The potential energy is

$$V = 0.9724a^{3.75}E\int_L \left(\frac{d^2v}{dx^2}\right)^{1.5} dx - \int_L p_y v\, dx \tag{E4}$$

4. *Functional and its derivatives*: We write the potential energy as a functional described in Section 7.7.3 and take its derivatives.

$$V = \int_0^L H(v^{(ii)}, v^{(i)}, v, x)\, dx \text{ where} \tag{E5}$$

$$H(v^{(ii)}, v^{(i)}, v, x) = 0.9724a^{3.75}E\left(\frac{d^2v}{dx^2}\right)^{1.5} - p_y v \tag{E6}$$

The derivatives of the functional are.

$$\frac{\partial H}{\partial v^{(ii)}} = 0.9724a^{3.75}E(1.5)\left(\frac{d^2v}{dx^2}\right)^{0.5} = 1.4586a^{3.75}E\left(\frac{d^2v}{dx^2}\right)^{0.5} \qquad \frac{\partial H}{\partial v^{(i)}} = 0 \qquad \frac{\partial H}{\partial v} = -p_y \tag{E7}$$

5. *Boundary value problem*: The differential equation is

$$\frac{d^2}{dx^2}\left(\frac{\partial H}{\partial v^{(ii)}}\right) - \frac{d}{dx}\left(\frac{\partial H}{\partial v^{(i)}}\right) + \frac{\partial H}{\partial v} = 1.4586a^{3.75}E\frac{d^2}{dx^2}\left[\left(\frac{d^2v}{dx^2}\right)^{0.5}\right] - p_y = 0 \tag{E8}$$

$$\text{Differential equation: } 1.4586a^{3.75}E\frac{d^2}{dx^2}\left[\left(\frac{d^2v}{dx^2}\right)^{0.5}\right] = p_y \tag{E9}$$

The boundary conditions are

$$\frac{\partial H}{\partial v^{(ii)}} = 1.4586a^{3.75}E\left(\frac{d^2v}{dx^2}\right)^{0.5} = 0 \qquad \text{or} \qquad \delta(v^{(i)}) = 0$$

$$\text{Boundary Condition 1: } \left(\frac{d^2v}{dx^2}\right)^{0.5} = 0 \qquad \text{or} \qquad \delta\left(\frac{dv}{dx}\right) = 0 \qquad \text{at } x = 0 \text{ and } x = L \tag{E10}$$

$$\frac{d}{dx}\left(\frac{\partial H}{\partial v^{(ii)}}\right) - \frac{\partial H}{\partial v^{(i)}} = 1.4586a^{3.75}E\frac{d}{dx}\left[\left(\frac{d^2v}{dx^2}\right)^{0.5}\right] = 0 \qquad \text{or} \qquad \delta v = 0 \qquad \text{at } x = 0 \text{ and } x = L$$

$$\text{Boundary Condition 2: } \frac{d}{dx}\left[\left(\frac{d^2v}{dx^2}\right)^{0.5}\right] = 0 \qquad \text{or} \qquad \delta v = 0 \qquad \text{at } x = 0 \text{ and } x = L \tag{E11}$$

**COMMENTS**

1. This example demonstrates the use of variational calculus equations in deriving boundary value problem for structural member having a nonlinear stress-strain curve. The key step is to obtain the strain energy for the given member. This approach can also be applied to axial and torsion problems.
2. If we calculated the equivalent internal moment, we would find that $M_z = 1.4586a^{3.75}E(d^2v/dx^2)^{0.5}$. With this observation the boundary conditions are our usual boundary conditions that are on moment or slope, and on shear force or deflection.

## 7.12 RAYLEIGH-RITZ METHOD

Rayleigh-Ritz method is applicable to conservative systems that may be linear or nonlinear. Rayleigh-Ritz method is a formal process of minimizing the potential energy using a series of kinematically admissible displacement functions to produce a set of algebraic equations in the unknown constants of the series. We shall develop the set of algebraic equations using the property of bilinear and linear functionals.

Suppose $f_j$ are a set of independent kinematically admissible functions. These $f_j$ could be functions of just one variable $x$ in one dimension, of functions of two variables (Cartesian or polar coordinates) in two dimensions, or of three variables in three dimensions (Cartesian. cylindrical, or spherical coordinates). We represent the displacement by the series below.

$$u = \sum_{j=1}^{n} C_j f_j \tag{7.63a}$$

where, $C_j$ are constants to be determined. Note $C_j$ are the generalized coordinates as the variation of them represents the variation of the displacement curve, that is, virtual displacement.

The potential energy of Equation (7.27) can be written in the following form

$$\Omega = \frac{1}{2}B(u_1, u_2) - l(u_1) \tag{7.63b}$$

where, $u_1$ and $u_2$ are represented as shown below.

$$u_1(x) = \sum_{j=1}^{n} C_j f_j(x) \quad \text{and} \quad u_2(x) = \sum_{k=1}^{n} C_k f_k(x) \tag{7.63c}$$

Substituting $u_1$ in the linear functional and bilinear functional and using Equations (7.5a) and (7.5b) we obtain

$$l(u_1) = l\left(\sum_{j=1}^{n} C_j f_j\right) = \sum_{j=1}^{n} C_j l(f_j) \tag{7.63d}$$

$$B(u_1, u_2) = B\left(\sum_{j=1}^{n} C_j f_j, u_2\right) = \sum_{j=1}^{n} C_j B(f_j, u_2) \tag{7.63e}$$

Substituting $u_2$ in the bilinear functional and using Equation (7.5b) we obtain

$$B(u_1, u_2) = \sum_{j=1}^{n} C_j B\left(f_j, \sum_{k=1}^{n} C_k f_k\right) = \sum_{j=1}^{n} C_j \sum_{k=1}^{n} C_k B(f_j, f_k) \tag{7.63f}$$

$$B(u_1, u_2) = \sum_{j=1}^{n} \sum_{k=1}^{n} C_j C_k B(f_j, f_k) \tag{7.63g}$$

The potential energy can now be written as

$$\Omega = \frac{1}{2}\sum_{j=1}^{n} \sum_{k=1}^{n} C_j C_k B(f_j, f_k) - \sum_{j=1}^{n} C_j l(f_j) \tag{7.63h}$$

The potential energy now is a functional of the generalized coordinates $C_j$. We take the first variation of it and set it equal to zero to minimize the potential energy.

$$\delta\Omega = \frac{1}{2}\sum_{j=1}^{n} \sum_{k=1}^{n} [\delta C_j C_k B(f_j, f_k) + C_j \delta C_k B(f_j, f_k)] - \sum_{j=1}^{n} \delta C_j l(f_j) \tag{7.63i}$$

For the second term, we can interchange the summation over *j* and *k* and write the above equation as

$$\delta\Omega = \frac{1}{2}\left\{\sum_{j=1}^{n} \sum_{k=1}^{n} [\delta C_j C_k B(f_j, f_k)] + \sum_{k=1}^{n} \sum_{j=1}^{n} [\delta C_j C_k B(f_k, f_j)]\right\} - \sum_{j=1}^{n} \delta C_j l(f_j)$$

As all our bilinear functional are symmetric in their arguments, that is, $B(f_j, f_k) = B(f_k, f_j)$, we obtain the following

$$\delta\Omega = \frac{1}{2}\sum_{j=1}^{n} \sum_{k=1}^{n} [\delta C_j C_k \{B(f_j, f_k) + B(f_j, f_k)\}] - \sum_{j=1}^{n} \delta C_j l(f_j) = \sum_{j=1}^{n} \sum_{k=1}^{n} \delta C_j C_k (B(f_j, f_k)) - \sum_{j=1}^{n} \delta C_j l(f_j)$$

$$\delta\Omega = \sum_{j=1}^{n} \delta C_j \left\{\sum_{k=1}^{n} [B(f_j, f_k) C_k - l(f_j)]\right\} = 0 \tag{7.63j}$$

The above equations must be satisfied for all variations of $\delta C_j$, we obtain

$$\boxed{\sum_{k=1}^{n} B(f_j, f_k) C_k - l(f_j) = 0 \qquad j = 1 \text{ to } n} \tag{7.64}$$

The above equation represents *n* equilibrium equations that must be satisfied by the constants $C_j$. Let $C_j^*$ represent the solution of the algebraic equations (the values at equilibrium), that is

$$\sum_{k=1}^{n} B(f_j, f_k) C_k^* = l(f_j) \tag{7.65a}$$

Equation (7.63h) can be written in terms of $C_j^*$ and Equation (7.65a) substituted to obtain

$$\Omega^* = \frac{1}{2}\sum_{j=1}^{n} C_j^* \left\{\sum_{k=1}^{n} B(f_j, f_k) C_k^*\right\} - \sum_{j=1}^{n} C_j^* l(f_j) = \frac{1}{2}\sum_{j=1}^{n} C_j^* \{l(f_j)\} - \sum_{j=1}^{n} C_j^* l(f_j) = -\frac{1}{2}\sum_{j=1}^{n} C_j^* l(f_j) \tag{7.65b}$$

The term $\sum_{j=1}^{n} C_j^* l(f_j)$ represents the *work potential* $W^*$ at equilibrium. Thus, $\Omega^* = -W^*/2$. We write this observation for future use.

*At equilibrium, the potential energy of the system is negative of half the work potential.*

Symbolically the set of algebraic equations in Equation (7.64) and potential energy at equilibrium can be written in matrix form as

$$[K]\{C\} = \{R\} \qquad \text{where} \qquad K_{jk} = B(f_j, f_k) \qquad R_j = l(f_j) \tag{7.66a}$$

$$\Omega^* = -\frac{1}{2}\sum_{j=1}^{n} C_j^* R_j = -\frac{1}{2}\{C^*\}^T\{R\} \quad \textbf{(7.66b)}$$

$[K]$ is called the stiffness matrix and because the bilinear functional is symmetric, the stiffness matrix is symmetric. We will make use of the above equations in the examples that follow.

## EXAMPLE 7.7

A beam and its loading are shown in Figure 7.11. Use the Rayleigh-Ritz method with one and two parameters to determine the deflection at $x = 0.25L$, $x = 0.5L$, $x = 0.75L$, and $x = L$, and the potential energy function. Compare your results with the analytical solution. Assume that $EI$ is constant for the beam.

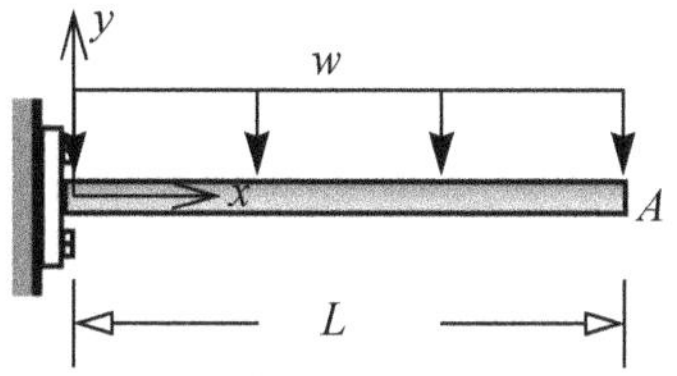


**Figure 7.11** Beam for Example 7.7.

**PLAN**

The linear and bilinear functionals for beam bending can be identified from Tables 7.1 and 7.3.The kinematic boundary conditions for the deflection and slope are zero at $x = 0$. We can determine the set of kinematically admissible functions that satisfy the kinematic boundary conditions. The matrix elements and the elements of the right-hand-side vector can be found from Equation (7.66a). The algebraic equations can be solved for the constants for each case (one and two parameters) and the required quantities calculated.

**SOLUTION**

From Tables 7.1 and 7.3, the linear and bilinear functionals for beam bending can be written as

$$l(v_1) = \int_0^L p_y(x)v_1(x)dx = -w\int_0^L v_1(x)dx \quad \text{(E1)}$$

$$B(v_1, v_2) = \int_L \left[EI\frac{d^2v_1}{dx^2}\frac{d^2v_2}{dx^2}\right]dx = \int_0^L \left[EI\frac{d^2v_1}{dx^2}\frac{d^2v_2}{dx^2}\right]dx \quad \text{(E2)}$$

The kinematic boundary conditions for this bending problem are

$$v(0) = 0 \qquad \frac{dv}{dx}(0) = 0 \quad \text{(E3)}$$

We are looking for functions that go to zero at $x = 0$; $x$ to any exponent would satisfy this requirement. However, a constant or linear function of $x$ will not correspond to any beam deformation. Hence $x^2$ and greater exponent powers of $x$ will satisfy all requirements and hence we have the following kinematically admissible displacement field:

$$v(x) = \sum_{j=1}^{n} C_j x^{j+1} = \sum_{j=1}^{n} C_j f_j \qquad \text{where} \qquad f_j = x^{j+1} \quad \text{(E4)}$$

From Equation (7.66a) we can write

$$K_{jk} = B(f_j, f_k) = \int_0^L \left[EI_{zz}\frac{d^2f_j}{dx^2}\frac{d^2f_k}{dx^2}\right]dx \qquad R_j = -w\int_0^L f_j dx \quad \text{(E5)}$$

The first and second derivatives of the functions $f_j$ can be written as shown below.

$$f_j = x^{j+1} \qquad \frac{df_j}{dx} = (j+1)x^j \qquad \frac{d^2f_j}{dx^2} = (j+1)(j)x^{j-1} \quad \text{(E6)}$$

*One parameter* ($n = 1$): From Equation (E5) we have

$$K_{11} = (EI)\int_0^L\left(\frac{d^2f_1}{dx^2}\right)\left(\frac{d^2f_1}{dx^2}\right)dx = (EI)\int_0^L (2)(2)dx = 4EIL \quad \text{(E7)}$$

$$R_1 = -w\int_0^L x^2 dx = -(wL^3/3) \quad \text{(E8)}$$

From Equation (7.66a) we obtain

$$4(EI)LC_1^* = -\left(\frac{wL^3}{3}\right) \qquad \text{or} \qquad C_1^* = -\left(\frac{wL^2}{12EI}\right) \quad \text{(E9)}$$

The potential energy can be found from Equation (7.66b) as shown below.

$$\Omega^* = -\left(\frac{1}{2}\right)C_1^*R_1 = -\frac{1}{2}\left(\frac{wL^2}{12EI}\right)\left(\frac{wL^3}{3}\right) = -\left(\frac{w^2L^5}{72EI}\right) = -0.0139\left(\frac{w^2L^5}{EI}\right) \quad \text{(E10)}$$

The constant $C_1^*$ can be substituted into Equation (E4) and the deflection evaluated at $x = 0.25L$, $x = 0.5L$, $x = 0.75L$, and $x = L$. The results are shown in Table 7.4.

*Two parameters* ($n = 2$): We calculate $K_{11}$ and $R_1$ as for the one-parameter solution. The rest of the quantities can be found as shown below.

$$K_{12} = (EI)\int_0^L \left(\frac{d^2 f_1}{dx^2}\right)\left(\frac{d^2 f_2}{dx^2}\right) dx = (EI)\int_0^L (2)(6x)dx = 6(EI)L^2 \quad \text{(E11)}$$

$$K_{22} = (EI)\int_0^L \left(\frac{d^2 f_2}{dx^2}\right)\left(\frac{d^2 f_2}{dx^2}\right) dx = (EI)\int_0^L (6x)(6x)dx = 12(EI)L^3 \quad \text{(E12)}$$

$$R_2 = -w\int_0^L f_2(x)dx = -w\int_0^L x^3 dx = -(wL^4/4) \quad \text{(E13)}$$

Noting that $K_{21} = K_{12}$, we obtain

$$\begin{bmatrix} 4(EI)L & 6(EI)L^2 \\ 6(EI)L^2 & 12(EI)L^3 \end{bmatrix} \begin{Bmatrix} C_1^* \\ C_2^* \end{Bmatrix} = \begin{Bmatrix} -wL^3/3 \\ -wL^4/4 \end{Bmatrix} \quad \text{(E14)}$$

Solving Equation (E14), we obtain the values of $C_1$ and $C_2$ as shown in Equation (E15).

$$C_1^* = -5wL^2/(24EI) \quad \text{and} \quad C_2^* = wL/(12EI) \quad \text{(E15)}$$

The potential energy can be found from Equation (7.66b) as shown below.

$$\Omega = -\left(\frac{1}{2}\right)(C_1^* R_1 + C_2^* R_2) = -\frac{1}{2}\left[\left(\frac{5wL^2}{24EI}\right)\left(\frac{wL^3}{3}\right) - \left(\frac{wL}{12EI}\right)\left(\frac{wL^4}{4}\right)\right] = -\left(\frac{7wL^5}{288EI}\right) = -0.0243\frac{wL^5}{EI} \quad \text{(E16)}$$

The constants $C_1^*$ and $C_2^*$ can be substituted in Equation (E4) and the deflection evaluated at $x = 0.25L$, $x = 0.5L$, $x = 0.75L$, and $x = L$. The results are given in Table 7.4.

*Analytical solution*: The analytical solution for the deflection can be obtained by integration and is given below.

$$\mathrm{v}(x) = -(w/24EI)(x^4 - 4Lx^3 + 6L^2x^2) \quad \text{(E17)}$$

The deflection at $x = 0.25L$, $x = 0.5L$, $x = 0.75L$, and $x = L$ can be found from above equation as shown in Table 7.4.
The strain energy in bending can be found as

$$U_B = \frac{1}{2}EI\int_0^L \left(\frac{d^2\mathrm{v}}{dx^2}\right)^2 dx = \frac{w^2}{8EI}\int_0^L (x-L)^4 dx = \left(\frac{w^2L^5}{40EI}\right) \quad \text{(E18)}$$

The work potential can be found as

$$W = \int_0^L (-w)\left[-\left(\frac{w}{24EI}\right)(x^4 - 4Lx^3 + 6L^2x^2)\right]dx = \frac{w^2}{24EI}\left(\frac{x^5}{5} - 4L\frac{x^4}{4} + 6L^2\frac{2x^3}{3}\right)\Bigg|_0^L = \frac{w^2L^5}{20EI} \quad \text{(E19)}$$

The potential energy can be found as

$$\Omega = U_B - W_B = \frac{w^2L^5}{40EI} - \frac{w^2L^5}{20EI} = -\frac{w^2L^5}{40EI} = -0.025\left(\frac{w^2L^5}{EI}\right) \quad \text{(E20)}$$

The results of the deflection and potential energy are given in Table 7.4. The percentage difference between Rayleigh-Ritz's solution and analytical values were calculated by using equation below.

$$\%\text{diff} = \left|\frac{\text{analytical value} - \text{calculated value}}{\text{analytical value}}\right| \times 100 \quad \text{(E21)}$$

**Table 7.4** Results of Example 7.7

| | Deflection: $\mathrm{v}/\left(\frac{wL^4}{EI}\right)$ | | | | | | | | Potential Energy $\Omega/\left(\frac{w^2L^5}{EI}\right)$ | |
|---|---|---|---|---|---|---|---|---|---|---|
| | $\mathrm{v}\left(\frac{L}{4}\right)$ | % diff | $\mathrm{v}\left(\frac{L}{2}\right)$ | % diff | $\mathrm{v}\left(\frac{3L}{4}\right)$ | % diff | $\mathrm{v}(L)$ | % diff | Value | % diff |
| n = 1 | -0.0052 | 60.5 | -0.0208 | 52.94 | -0.0469 | 43.86 | -0.0833 | 33.3 | -0.0139 | 44.44 |
| n = 2 | -0.0117 | 11.1 | -0.0417 | 5.88 | -0.0820 | 1.75 | -0.125 | 0 | -0.0243 | 2.78 |
| Analytical | -0.0132 | | -0.0443 | | -0.0835 | | -0.125 | | -0.025 | |

## COMMENTS

1. The improvement of results with two parameters over the solution by one parameter may be less dramatic for more complex problems than those shown in Table 7.4.

2. Notice that there is no difference between the analytical results and two-parameter results at $x = L$, but at $x = L/4$ there is an 11.1% difference. Great care must be taken in drawing conclusions of improvement from point values of displacements and stresses.
3. Table 7.4 shows that the potential energy decreases with an increase in parameters. With the addition of parameters, the potential energy either decreases or remains the same, but it will never increase. Thus, the decrease in potential energy is a surer measure of improvement in accuracy than the use of values at a point.
4. For three parameters, we would get the analytical results of this problem because the three kinematic functions $x^2$, $x^3$, and $x^4$ in the approximation are three terms in the analytical solution of Equation (E17).
5. Analytical results are not same as exact results because an analytical solution also starts with a model that is an approximation of reality. Beam theory is an approximate theory.

## EXAMPLE 7.8

A square plate is uniformly loaded and is simply supported on all sides as shown in Figure 7.12. Using Rayleigh-Ritz method determine deflection $w$ at the center of the plate and the potential energy for each case.

*Case* I: $w(x,y) = C_1 \sin\pi(x/a)\sin\pi(y/a)$

*Case* II: $w(x,y) = C_1 \sin\pi(x/a)\sin\pi(y/a) + C_2\sin\pi(2x/a)\sin\pi(y/a) + C_3\sin\pi(x/a)\sin\pi(2y/a)$

*Case* III: $w(x,y) = C_1 \sin\pi(x/a)\sin\pi(y/a) + C_2\sin\pi(3x/a)\sin\pi(y/a) + C_3\sin\pi(x/a)\sin\pi(3y/a)$

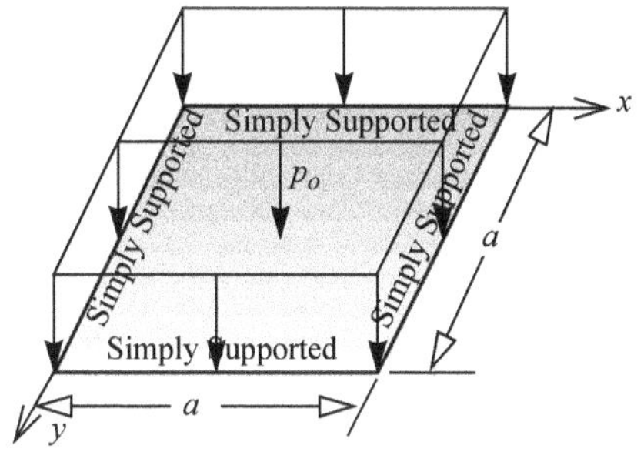


**Figure 7.12** Uniformly loaded, simply supported, square plate

**PLAN**

The linear and bilinear functionals for thin plate bending can be identified from Tables 7.1 and 7.3. For the given kinematic displacement function, the matrix elements and the elements of the right-hand-side vector can be found from Equation (7.66a). The algebraic equations can be solved for the constants for each case and the required quantities calculated

**SOLUTION**

From Tables 7.1 and 7.3, the linear and bilinear functionals for thin plate can be written as

$$l(w_1) = \iint_A p_z(x,y)w_1(x,y)dxdy = p_o\int_0^a\int_0^a w_1 dxdy \tag{E1}$$

$$B(w_1,w_2) = D\int_0^a\int_0^a\left[\frac{\partial^2 w_1}{\partial x^2}\frac{\partial^2 w_2}{\partial x^2} + \frac{\partial^2 w_1}{\partial y^2}\frac{\partial^2 w_2}{\partial y^2} + \nu\left(\frac{\partial^2 w_1}{\partial x^2}\frac{\partial^2 w_2}{\partial y^2} + \frac{\partial^2 w_2}{\partial x^2}\frac{\partial^2 w_1}{\partial y^2}\right) + 2(1-\nu)\frac{\partial^2 w_1}{\partial x\partial y}\frac{\partial^2 w_2}{\partial x\partial y}\right]dxdy \tag{E2}$$

From Equation (7.66a) we can write

$$K_{jk} = D\int_0^a\int_0^a\left[\frac{\partial^2 f_j}{\partial x^2}\frac{\partial^2 f_k}{\partial x^2} + \frac{\partial^2 f_j}{\partial y^2}\frac{\partial^2 f_k}{\partial y^2} + \nu\left(\frac{\partial^2 f_j}{\partial x^2}\frac{\partial^2 f_k}{\partial y^2} + \frac{\partial^2 f_k}{\partial x^2}\frac{\partial^2 f_j}{\partial y^2}\right) + 2(1-\nu)\frac{\partial^2 f_j}{\partial x\partial y}\frac{\partial^2 f_k}{\partial x\partial y}\right]dxdy \qquad R_j = p_o\int_0^a\int_0^a f_j dxdy \tag{E3}$$

We substitute $\pi x/a = \theta$ and write the orthogonality condition. The integral with respect to $y$ will also yield the same result.

$$\int_0^a \sin m\pi(x/a)\sin n\pi(x/a)dx = \frac{a}{\pi}\int_0^\pi \sin m\theta\sin n\theta d\theta = \begin{cases} 0 & m \neq n \\ a/2 & m = n\end{cases} \tag{E4}$$

$$\int_0^a \cos m\pi(x/a)\cos n\pi(x/a)dx = \frac{a}{\pi}\int_0^\pi \cos m\theta\cos n\theta d\theta = \begin{cases} 0 & m \neq n \\ a/2 & m = n\end{cases} \tag{E5}$$

*Case* 1: The kinematic function is

$$f_1 = \sin\pi(x/a)\sin\pi(y/a) \tag{E6}$$

The derivatives of $f_1$ can be evaluated as follows

$$\frac{\partial^2 f_1}{\partial x^2} = -\left(\frac{\pi^2}{a^2}\right)\sin\pi\left(\frac{x}{a}\right)\sin\pi\left(\frac{y}{a}\right) \qquad \frac{\partial^2 f_1}{\partial y^2} = -\left(\frac{\pi^2}{a^2}\right)\sin\pi\left(\frac{x}{a}\right)\sin\pi\left(\frac{y}{a}\right) \qquad \frac{\partial^2 f_j}{\partial x\partial y} = -\left(\frac{\pi^2}{a^2}\right)\cos\pi\left(\frac{x}{a}\right)\cos\pi\left(\frac{y}{a}\right) \tag{E7}$$

We obtain from Equation (E3) the following.

$$R_1 = p_o\int_0^a\int_0^a sin\pi(x/a)sin\pi(y/a)dxdy = p_o[-(a/\pi)cos\pi(x/a)]|_0^a[-(a/\pi)cos\pi(y/a)]|_0^a = 4p_oa^2/\pi^2 \quad (E8)$$

$$K_{11} = D\int_0^a\int_0^a\{-(\pi^2/a^2)sin\pi(x/a)sin\pi(y/a)\}^2dxdy + D\int_0^a\int_0^a\{-(\pi^2/a^2)sin\pi(x/a)sin\pi(y/a)\}^2dxdy$$

$$+ D\nu\int_0^a\int_0^a\{-(\pi^2/a^2)sin\pi(x/a)sin\pi(y/a)\}^2dxdy + D\nu\int_0^a\int_0^a\{-(\pi^2/a^2)sin\pi(x/a)sin\pi(y/a)\}^2dxdy$$

$$+ 2(1-\nu)D\int_0^a\int_0^a\{-(\pi^2/a^2)cos\pi(x/a)cos\pi(y/a)\}^2dxdy$$

$$K_{11} = D\frac{\pi^4}{a^4}\left(\frac{a}{2}\right)\left(\frac{a}{2}\right) + D\frac{\pi^4}{a^4}\left(\frac{a}{2}\right)\left(\frac{a}{2}\right) + D\nu\frac{\pi^4}{a^4}\left(\frac{a}{2}\right)\left(\frac{a}{2}\right) + D\nu\frac{\pi^4}{a^4}\left(\frac{a}{2}\right)\left(\frac{a}{2}\right) + 2D(1-\nu)\frac{\pi^4}{a^4}\left(\frac{a}{2}\right)\left(\frac{a}{2}\right) = \frac{D\pi^4}{a^2} \quad (E9)$$

For one parameter, the constant $C_1^*$ can be found as:

$$C_1^* = R_1/K_{11} = 4p_oa^4/(D\pi^6) \quad (E10)$$

The deflection at the center of the plate are solution are

$$w\left(\frac{a}{2},\frac{a}{2}\right) = \frac{4p_oa^4}{D\pi^6}sin\left(\frac{\pi}{2}\right)sin\left(\frac{\pi}{2}\right) = \frac{4p_oa^4}{D\pi^6} = 4.1606(10^{-3})\frac{p_oa^4}{D} \quad (E11)$$

The potential energy can be found from Equation (7.66b) as shown below.

$$\Omega^* = -\left(\frac{1}{2}\right)C_1^*R_1 = -\frac{1}{2}\left(\frac{4p_oa^4}{D\pi^6}\right)\left(\frac{4p_oa^2}{\pi^2}\right) = -\left(\frac{8p_o^2a^6}{D\pi^8}\right) = -0.8431(10^{-3})\left(\frac{p_o^2a^6}{D}\right) \quad (E12)$$

**ANS.** $w(a/2, a/2) = 4.1606(10^{-3})p_oa^4/D;\ \Omega^* = -0.8431(10^{-3})p_o^2a^6/D$

*Case* II: The kinematic functions are

$$f_1 = sin\pi(x/a)sin\pi(y/a) \qquad f_2 = sin2\pi(x/a)sin\pi(y/a) \qquad f_3 = sin\pi(x/a)sin2\pi(y/a) \quad (E13)$$

The derivatives of $f_1$ are given in Equation (E7). The derivatives of the other functions are can be evaluated as follows

$$\frac{\partial^2 f_2}{\partial x^2} = -\left(\frac{4\pi^2}{a^2}\right)sin2\pi\left(\frac{x}{a}\right)sin\pi\left(\frac{y}{a}\right) \qquad \frac{\partial^2 f_2}{\partial y^2} = -\left(\frac{\pi^2}{a^2}\right)sin2\pi\left(\frac{x}{a}\right)sin\pi\left(\frac{y}{a}\right) \qquad \frac{\partial^2 f_2}{\partial x\partial y} = -\left(\frac{2\pi^2}{a^2}\right)cos2\pi\left(\frac{x}{a}\right)cos\pi\left(\frac{y}{a}\right) \quad (E14)$$

$$\frac{\partial^2 f_3}{\partial x^2} = -\left(\frac{\pi^2}{a^2}\right)sin\pi\left(\frac{x}{a}\right)sin2\pi\left(\frac{y}{a}\right) \qquad \frac{\partial^2 f_2}{\partial y^2} = -\left(\frac{4\pi^2}{a^2}\right)sin\pi\left(\frac{x}{a}\right)sin2\pi\left(\frac{y}{a}\right) \qquad \frac{\partial^2 f_2}{\partial x\partial y} = -\left(\frac{2\pi^2}{a^2}\right)cos\pi\left(\frac{x}{a}\right)cos2\pi\left(\frac{y}{a}\right) \quad (E15)$$

$R_1$ is same as in case I. The remaining terms can be found as shown below.

$$R_2 = p_o\int_0^a\int_0^a sin2\pi\left(\frac{x}{a}\right)sin\pi\left(\frac{y}{a}\right)dxdy = p_o\left[-\frac{a}{2\pi}cos2\pi\left(\frac{x}{a}\right)\right]\Bigg|_0^a\left[-\frac{a}{\pi}cos\pi\left(\frac{y}{a}\right)\right]\Bigg|_0^a = 0 \quad (E16)$$

$$R_3 = p_o\int_0^a\int_0^a sin\pi\left(\frac{x}{a}\right)sin2\pi\left(\frac{y}{a}\right)dxdy = p_o\left[-\frac{a}{\pi}cos\pi\left(\frac{x}{a}\right)\right]\Bigg|_0^a\left[-\frac{a}{2\pi}cos2\pi\left(\frac{y}{a}\right)\right]\Bigg|_0^a = 0 \quad (E17)$$

$K_{11}$ is same as in case I. The off-diagonal terms will contain integrals such as shown below

$$\int_0^a sin\pi(x/a)sin2\pi(x/a)dx \quad \text{or} \quad \int_0^a sin\pi(y/a)sin2\pi(y/a)dy$$

$$\int_0^a cos\pi(x/a)cos2\pi(x/a)dx \quad \text{or} \quad \int_0^a cos\pi(y/a)cos2\pi(y/a)dy \quad (E18)$$

All the above integrals will result in zero values because of the orthogonality condition given by Equation (E4). Hence

$$K_{12} = K_{21} = 0 \qquad K_{13} = K_{31} = 0 \qquad K_{23} = K_{32} = 0 \quad (E19)$$

$K_{22}$ and $K_{33}$ will be non-zero and can be determined. Thus, the stiffness matrix is a diagonal matrix and the solution is

$$C_1^* = R_1/K_{11} = p_oa^4/(D\pi^4) \qquad C_2^* = R_2/K_{22} = 0 \qquad C_3^* = R_3/K_{33} = 0 \quad (E20)$$

The above values of the constants results in the same solution as in case I.

**ANS.** $w(a/2, a/2) = 4.1606(10^{-3})p_oa^4/D;\ \Omega^* = -0.8431(10^{-3})p_o^2a^6/D$

*Case* III: The kinematic function is

$$f_1 = sin\pi(x/a)sin\pi(y/a) \qquad f_2 = sin3\pi(x/a)sin\pi(y/a) \qquad f_3 = sin\pi(x/a)sin3\pi(y/a) \quad (E21)$$

The derivatives of $f_1$ are given in Equation (E7). The derivatives of the other functions are can be evaluated as follows

$$\frac{\partial^2 f_2}{\partial x^2} = -\left(\frac{9\pi^2}{a^2}\right) sin 3\pi\left(\frac{x}{a}\right) sin\pi\left(\frac{y}{a}\right) \quad \frac{\partial^2 f_2}{\partial y^2} = -\left(\frac{\pi^2}{a^2}\right) sin 3\pi\left(\frac{x}{a}\right) sin\pi\left(\frac{y}{a}\right) \quad \frac{\partial^2 f_2}{\partial x \partial y} = -\left(\frac{3\pi^2}{a^2}\right) cos 3\pi\left(\frac{x}{a}\right) cos\pi\left(\frac{y}{a}\right) \tag{E22}$$

$$\frac{\partial^2 f_3}{\partial x^2} = -\left(\frac{\pi^2}{a^2}\right) sin\pi\left(\frac{x}{a}\right) sin 3\pi\left(\frac{y}{a}\right) \quad \frac{\partial^2 f_3}{\partial y^2} = -\left(\frac{9\pi^2}{a^2}\right) sin\pi\left(\frac{x}{a}\right) sin 3\pi\left(\frac{y}{a}\right) \quad \frac{\partial^2 f_3}{\partial x \partial y} = -\left(\frac{3\pi^2}{a^2}\right) sin\pi\left(\frac{x}{a}\right) sin 3\pi\left(\frac{y}{a}\right) \tag{E23}$$

$R_1$ is same as in case I. The remaining terms can be found as shown below.

$$R_2 = p_o\int_0^a\int_0^a sin 3\pi\left(\frac{x}{a}\right) sin\pi\left(\frac{y}{a}\right) dx dy = p_o\left[-\frac{a}{3\pi} cos 3\pi\left(\frac{x}{a}\right)\right]_0^a\left[-\frac{a}{\pi} cos\pi\left(\frac{y}{a}\right)\right]_0^a = \frac{4p_o a^2}{3\pi^2} \tag{E24}$$

$$R_3 = p_o\int_0^a\int_0^a sin\pi\left(\frac{x}{a}\right) sin 3\pi\left(\frac{y}{a}\right) dx dy = p_o\left[-\frac{a}{\pi} cos\pi\left(\frac{x}{a}\right)\right]_0^a\left[-\frac{a}{3\pi} cos 3\pi\left(\frac{y}{a}\right)\right]_0^a = \frac{4p_o a^2}{3\pi^2} \tag{E25}$$

$K_{11}$ is same as in case I. The off-diagonal terms will once more contain integrals that will be zero and once more the off diagonal terms will be zero as in Equation (E19). By the symmetry in $x$ and $y$ terms $K_{22} = K_{33}$ and can be evaluated as shown below.

$$K_{22} = K_{33} = D\int_0^a\int_0^a\left\{-\left(\frac{9\pi^2}{a^2}\right) sin\pi\left(\frac{3x}{a}\right) sin\pi\left(\frac{y}{a}\right)\right\}^2 dx dy + D\int_0^a\int_0^a\left\{-\left(\frac{\pi^2}{a^2}\right) sin 3\pi\left(\frac{x}{a}\right) sin\pi\left(\frac{y}{a}\right)\right\}^2 dx dy$$

$$+ D\nu\int_0^a\int_0^a\left(\frac{9\pi^2}{a^2}\right)\left\{sin 3\pi\left(\frac{x}{a}\right) sin\pi\left(\frac{y}{a}\right)\right\}^2 dx dy + D\nu\int_0^a\int_0^a\left(\frac{9\pi^2}{a^2}\right)\left\{sin 3\pi\left(\frac{x}{a}\right) sin\pi\left(\frac{y}{a}\right)\right\}^2 dx dy$$

$$+ 2(1-\nu)D\int_0^a\int_0^a\left\{-\left(\frac{3\pi^2}{a^2}\right) cos 3\pi\left(\frac{x}{a}\right) cos\pi\left(\frac{y}{a}\right)\right\}^2 dx dy$$

$$K_{22} = K_{33} = D\frac{81\pi^4}{a^4}\left(\frac{a}{2}\right)\left(\frac{a}{2}\right) + D\frac{\pi^4}{a^4}\left(\frac{a}{2}\right)\left(\frac{a}{2}\right) + D\nu\frac{9\pi^4}{a^4}\left(\frac{a}{2}\right)\left(\frac{a}{2}\right) + D\nu\frac{9\pi^4}{a^4}\left(\frac{a}{2}\right)\left(\frac{a}{2}\right) + 2D(1-\nu)\frac{9\pi^4}{a^4}\left(\frac{a}{2}\right)\left(\frac{a}{2}\right)$$

$$K_{22} = K_{33} = D\frac{\pi^4}{4a^2}[81 + 1 + 18\nu + 18 - 18\nu] = D\frac{25\pi^4}{a^2}$$

For the diagonal stiffness matrix the solution is

$$C_1^* = \frac{R_1}{K_{11}} = \frac{4p_o a^4}{D\pi^6} \quad C_2^* = \frac{R_2}{K_{22}} = \frac{4p_o a^4}{75D\pi^6} \quad C_3^* = \frac{R_3}{K_{33}} = \frac{4p_o a^4}{75D\pi^6} \tag{E26}$$

The deflection at the center of the plate is

$$w\left(\frac{a}{2},\frac{a}{2}\right) = \frac{4p_o a^4}{D\pi^6}\left[sin\left(\frac{\pi}{2}\right) sin\left(\frac{\pi}{2}\right) + \frac{1}{75} sin\left(\frac{3\pi}{2}\right) sin\pi\left(\frac{\pi}{2}\right) + \frac{1}{75} sin\left(\frac{\pi}{2}\right) sin\pi\left(\frac{3\pi}{2}\right)\right] = \frac{292p_o a^4}{75D\pi^6} = 4.0497(10^{-3})\frac{p_o a^4}{D} \tag{E27}$$

The potential energy can be found from Equation (7.66b) as shown below.

$$\Omega^* = -\frac{1}{2}\sum_{j=1}^{3} C_j^* R_j = -\frac{1}{2}\left(\frac{4p_o a^2}{\pi^2}\right)\left(\frac{4p_o a^4}{D\pi^6}\right)\left[1 + \frac{1}{(3)(75)} + \frac{1}{(3)(75)}\right] = -\left(\frac{1816p_o^2 a^6}{225D\pi^8}\right) = -0.8506(10^{-3})\frac{p_o^2 a^6}{D} \tag{E28}$$

**ANS.** $w(a/2, a/2) = 4.0497(10^{-3})p_o a^4/D$; $\Omega^* = -0.8506(10^{-3})p_o^2 a^6/D$

## COMMENTS

1. In case II, the addition of two more terms did not improve the solution. This emphasizes that the Rayleigh-Ritz's method chooses the best possible solution, ignoring terms that will not improve the results.
2. The potential energy for case III is more negative than for the other two cases, thus indicating the results are improving.
3. The first three terms for center deflection by Navier's solution that are shown in Table 5.3 are reproduced below for convenience.

| No. of terms | m | n | $\frac{w_{max}(10^{-3})}{(p_o a^4)/D}$ |
|---|---|---|---|
| 1 | 1 | 1 | 4.1606 |
| 2 | 3 | 1 | 4.1052 |
| 3 | 1 | 3 | 4.0497 |

We note case I corresponds to $m = 1$ and $n = 1$ and Rayleigh Ritz's method produce identical results. In Navier's solution, even values of $m$ and $n$ resulted in zero value for the constants, the same as in case II. Case III gave us the same values as Navier's solution when the three terms are included.

4. Comment 3 shows that when the set of approximating kinematic functions are the same, then Rayleigh Ritz's method produces the same result as Navier's solution. Thus, method based on variational principle produce the same results as the solution of boundary value problems, but the perspective of variational calculus (Eulerian view) is very different than that of differential calculus (Newtonian view).

## 7.13 FINITE ELEMENT METHOD

The finite element method is a versatile numerical method that is ubiquitous in stress analysis and in the design of machines and structures. A discussion of finite element method that does justice to it can cover few courses. Our objective is much more limited. It is to show how Rayleigh-Ritz method can be used in formulation of one of the versions of the finite element method.

The finite element method began as a matrix method of analysis in structures[1]. There are two versions of it: the **stiffness method** and the **flexibility method.** To elaborate the difference between the two methods visualize a truss in which the displacements of pins are unknown. If we now use the "method of joints" and write the equilibrium of forces at each joint, we will obtain a set of algebraic equations in which the unknowns are the pin displacements and the right-hand-side quantities are the external forces acting on the pin—this is the stiffness method as the matrix in the algebraic equation is the stiffness matrix. Now in the truss, suppose we use the internal forces as the unknowns and write the compatibility equations that relate the displacements of the individual truss members to the displacement of the pin. We will then obtain a set of algebraic equations in the unknown internal forces and the right-hand-side quantities will be the displacements of the pins—this is the flexibility method as the matrix in the algebraic equation is the flexibility matrix. The flexibility method can be derived by considering minimum complimentary potential energy, which is similar to the potential energy with strain energy replaced by complimentary strain energy. The stiffness method is derived from minimum potential energy, which is formalized as the Rayleigh-Ritz method. Commercial computer programs are usually based on the stiffness method. We will only consider Rayleigh-Ritz method application in the formulation of the stiffness method.

The finite element method formulation presented here is very similar to the Rayleigh-Ritz method with one important difference: the kinematically admissible displacement functions in the finite element method are defined piecewise continuously over small (finite) domains; these are the **elements**. The boundary points of the elements are called **nodes**, although nodes can also be points inside the element. The constants multiplying the piecewise kinematically admissible functions are the displacements of the nodes, and the kinematically admissible functions are called interpolation functions because they can be used to interpolate the values of displacements between the nodes. The representation of a structure by elements and nodes is called a **mesh**. A mesh with boundary conditions, applied loads, and material property is called a **model**. A model is a finite element representation of a real-life problem, and the accuracy of the model's predictions is determined by the assumptions and limitations that are made in constructing the finite element model and the errors introduced in solving the model by numerical methods.

The use of piecewise kinematically admissible functions changes the perspective with which we view and solve a problem by means of the finite element method. To elaborate this perspective, consider a structure made up of axial members, circular shafts, symmetric beams, and other members such as curved beams, plates, and shells. Potential energy is a scalar quantity and can be written as the sum of the potential energy of the $n$ individual structural members $\Omega^{[i]}$ as shown below.

$$\Omega = \sum_{i=1}^{n} \Omega^{[i]} \tag{7.67}$$

Equation (7.67) is valid for structural members of all types, irrespective of orientation. We could thus develop the potential energy in matrix form for each member separately. That is, we could develop matrices at the element level in a local coordinate system without regard to how a member is used in the structure. The individual local matrices, called **element stiffness matrices,** could be assembled by using Equation (7.67) to form the **global stiffness matrix** of the entire structure. This perspective of reducing the *complexity of analyzing large structures to the analysis of simple individual members* (*elements*) is what makes the finite element method such a versatile and popular tool in structural (and engineering application) analysis.

Equation (7.67) represents an assembly process in which continuity of the primary variables has to be ensured. The matrix generated has a defined structure, with lots of zero's in the matrix. There are multitudes of solution procedures that exploit the structure to produce very efficient solution procedures of the algebraic equations. We will however, restrict ourselves to only generating the element stiffness matrix and leave it to the reader to read books on finite element method to learn the assembly and solution techniques in finite element method.

In Section 7.7.4, we saw that for kinematically admissible functions, the principal derivatives, that is, the primary variables must be continuous. The primary variables are the generalized displacements. The order of derivatives that must be continuous depends upon the highest order of derivative in the functional. When the highest derivative in the functional is $r$, then highest order of derivative in primary variable that must be continuous is $r$-1. Thus, in axial, torsion, plane elasticity, where the order of derivative in the functional was $r = 1$, then we only need to ensure the continuity of the displacement. In beam bending and plate bending, the functional had second order derivatives, and the primary variables (generalized displacements) that must

1. Today the finite element method is used extensively in solving engineering problems. It is viewed as a numerical method for solving partial differential equations, and algebraic equations are obtained from an approach called the "weak form." which is equivalent to the variational approach we have developed in this chapter. See Reddy [1993] for additional details.

be continuous are displacement and slopes. **Lagrange polynomials** ensure continuity of the function (displacements) at the nodes and **Hermite polynomials** ensure continuity of the function and its derivatives at the nodes. The requirement to ensure continuity of the generalized displacement on the boundary is that only the nodes on the boundary appear in the equation representing the generalized displacement. This is because the nodal values of generalized displacements are independent variables and thus a node that is not on the boundary can change the equation of generalized displacements independently thus violating the continuity requirements. When the kinematically admissible functions satisfy the continuity at the nodes and the boundary then the element is called a **conforming element**. When continuity is only ensured at the nodes but not ensured across the boundary then the element is called **non-conforming element**. Non-conforming elements are often used when continuity of higher order derivatives is needed which may result in large number of unknowns that may or may not improve accuracy of the solution.

We have seen Timoshenko beams and Mindlin-Rissner plates, which permit shear stress in transverse direction, are higher order theories. These theories can simulate solutions of classical beam and plate theories. The major difficulty with these theories is in getting analytical solution. But if the objective is to get a numerical solution then we can use these theories and reap the benefit of the fact that continuity is needed only of the functions and not its derivatives. Thus, we can use Lagrange polynomials with higher order theories to obtain results for the classical beam and plate theory. We will restrict our discussion to Lagrange polynomials but will use Hermite polynomials for kinematically admissible functions that will be given without explanation of how they were constructed.

## 7.13.1 Lagrange polynomials in one dimension

Lagrange polynomials ensure continuity of generalized displacements at the nodes. The polynomials are introduced by using the axial members to provide the motivation and practical relevance of these functions in the FEM.

Figure 7.13*a* shows an axial member (element) with two nodes. The axial displacements $u^{[1]}$ and $u^{[2]}$ are the degrees of freedom (generalized displacements) in terms of which we plan to derive the element stiffness matrix. We will use the *superscript with square bracket* to designate generalized displacements or generalized forces at the *node of an element.* The element in Figure 7.13*a* has 2 degrees of freedom, and so we choose a linear function $u(x) = C_1 + C_2 x$ with two unknown parameters. We note that at $x = x_1$ the displacement $u(x_1) = u^{[1]}$, and at $x = x_2$ the displacement $u(x_2) = u^{[2]}$. The constants $C_1$ and $C_2$ can be solved in terms of $u^{[1]}$ and $u^{[2]}$ and substituted into the linear function representation to obtain Equation (7.68a),

$$u(x) = u^{[1]}\left(\frac{x - x_2}{x_1 - x_2}\right) + u^{[2]}\left(\frac{x - x_1}{x_2 - x_1}\right) = u^{[1]}\mathcal{L}_1(x) + u^{[2]}\mathcal{L}_2(x) = \sum_{i=1}^{2} u^{[i]}\mathcal{L}_i(x) \qquad \textbf{(7.68a)}$$

$$\mathcal{L}_1(x) = \left(\frac{x - x_2}{x_1 - x_2}\right) \quad \text{and} \quad \mathcal{L}_2(x) = \left(\frac{x - x_1}{x_2 - x_1}\right) \qquad \textbf{(7.68b)}$$

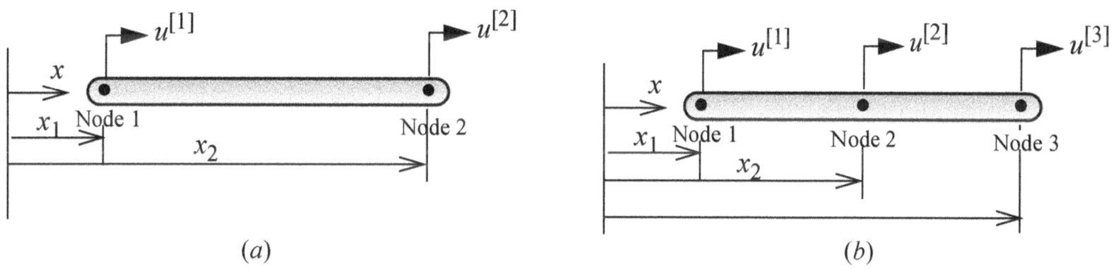


**Figure 7.13** Axial (*a*) linear and (*b*) quadratic elements.

A linear representation of displacement is sufficient if the forces are applied only at the element end and only if the cross sectional area does not change across the element. If the axial member has a distributed load, or if the member is tapered, then the axial displacement is no longer linear inside the element. A quadratic or higher-order polynomial may converge to the actual solution faster than a linear element. Figure 7.13*b* shows an element with three nodes. With 3 degrees of freedom, we can start with a quadratic displacement function $u(x) = C_1 + C_2 x + C_3 x^2$, solve the constant in terms of the nodal displacement, and obtain an equation analogous to Equation (7.68a). This process would be tedious for higher-order polynomials. So we use an alternative approach. We represent the displacement in the element by Equation (7.69a),

$$u(x) = \sum_{j=1}^{n} u^{[j]}\mathcal{L}_j(x) \qquad \textbf{(7.69a)}$$

where $n$ is the degrees of freedom (number of nodes, in this case) of the element that can be used for representing the $(n-1)$ order of polynomials. Now at the $j$th node, the displacement $u(x_j) = u_j$, and we obtain Equation (7.69b).

$$u(x_k) = \sum_{j=1}^{n} u^{[j]}\mathcal{L}_j(x_k) = u^{(k)} \qquad \textbf{(7.69b)}$$

For Equation (7.69b) to be true, the property given in Equation (7.69c) must hold.

$$\mathcal{L}_j(x_k) = \begin{cases} 1 & j = k \\ 0 & j \neq k \end{cases} \qquad \textbf{(7.69c)}$$

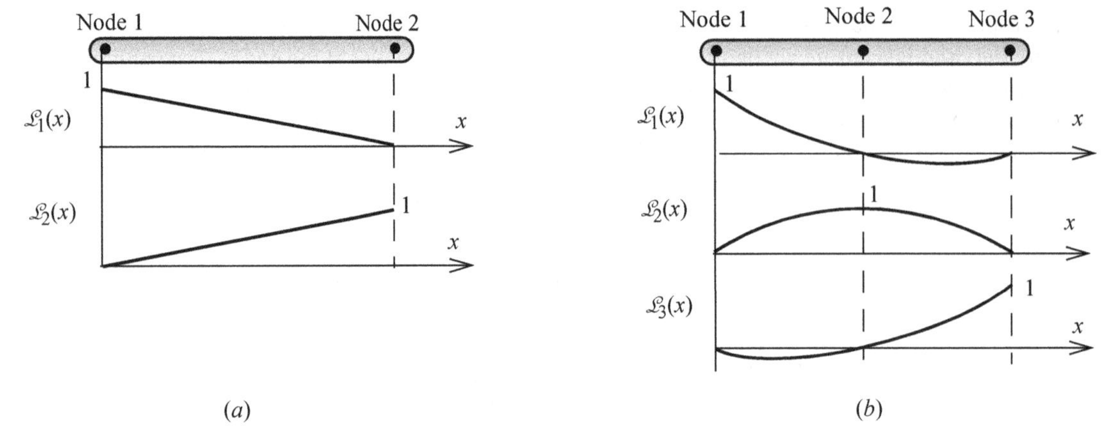


**Figure 7.14** (a) Linear and (b) quadratic Lagrange polynomials.

Equation (7.69c) implies that the polynomials $\mathcal{L}_j(x)$ are such that the value is 1 on its own *j*th node and zero at other nodes. Figure 7.14 shows the approximate plots for linear and quadratic $\mathcal{L}_j(x)$ that meet this requirement.

Now consider $\mathcal{L}_1$ in a quadratic polynomial. If we represent $\mathcal{L}_1(x) = a_1(x - x_2)(x - x_3)$, its value is zero at nodes 2 and 3. We can now determine the constant $a_1$ such that $\mathcal{L}_1$ at node 1 is equal to one, and we obtain Equation (7.70).

$$\mathcal{L}_1(x) = \left(\frac{x - x_2}{x_1 - x_2}\right)\left(\frac{x - x_3}{x_1 - x_3}\right) \tag{7.70}$$

In a similar manner, we can start with $\mathcal{L}_2(x) = a_2(x - x_3)(x - x_1)$ and $\mathcal{L}_3(x) = a_3(x - x_1)\,(x - x_2)$ and determine the value of $a_2$ and $a_3$ such that $\mathcal{L}_2$ and $\mathcal{L}_3$ at nodes 2 and 3, respectively, have a value of one to obtain Equation (7.70a).

$$\mathcal{L}_2(x) = \left(\frac{x - x_1}{x_2 - x_1}\right)\left(\frac{x - x_3}{x_2 - x_3}\right) \quad \text{and} \quad \mathcal{L}_3(x) = \left(\frac{x - x_1}{x_3 - x_1}\right)\left(\frac{x - x_2}{x_3 - x_2}\right) \tag{7.70a}$$

The process we used to obtain the polynomials for the quadratic can now be generalized to obtain Equation (7.71),

$$\mathcal{L}_j(x) = \Pi_{\substack{k=1 \\ j \neq k}}^{n}\left[\frac{(x - x_k)}{(x_j - x_k)}\right] \tag{7.71}$$

where $\Pi_{j=1}^{n}[\cdots]$ represents the product of the terms in square brackets. The functions defined by Equation (7.71) are called *Lagrange polynomials.*

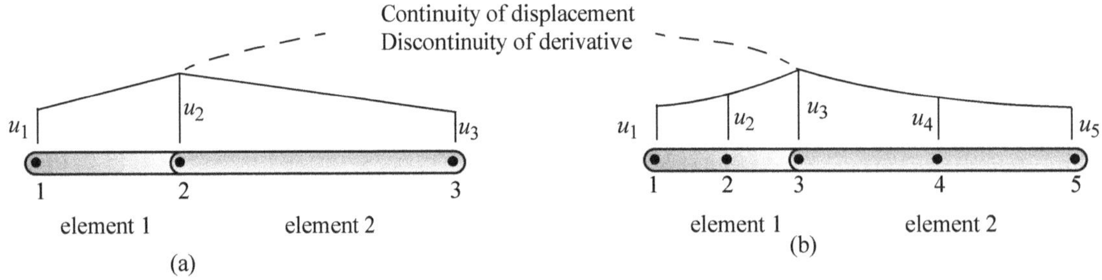


**Figure 7.15** Possible variation of displacement for (a) linear and (b) quadratic elements.

Functions represented by Lagrange polynomials will be continuous at the element ends. Inside the elements, all orders of derivatives are defined as polynomials are continuous. However, the continuity of the derivative of the function cannot be ensured at the element end irrespective of the order of polynomials when Lagrange polynomials are used for representing the function. Figure 7.15 shows a possible variation of a displacement field represented by Lagrange polynomials over two adjoining elements. Displacements at nodes are independent parameters that can have any value; hence the variation shown in Figure 7.15 is a possibility. As can be seen from Figure 7.15, the continuity of the displacement is maintained, but its first derivative is not continuous for either the linear or the quadratic element at the element end node.

## 7.13.2 Natural coordinates

Non-dimensional coordinates are called **natural coordinates** in finite element method. We will use $\xi$ to represent the natural coordinates in one dimension and $L$ to represent the characteristic length of the element. When $0 \le \xi \le 1$ the transformation is $\xi = x/L$ and the origin is the left most end of the element. When $-1 \le \xi \le 1$ the transformation is $\xi = 2x/L$ and the origin is in the middle of the element. We will use $\xi = x/L$. One of the important uses of natural coordinates is that the Lagrange polynomials can be used for approximating the geometry as well as the generalized displacements. Think of a curved beam that may or may not be circular. We could approximate the curved beam with several small beam segments in which the $x$ and $y$ coordinates could be represented as

$$x = \sum_{j=1}^{n} x_j \mathcal{L}_j(\xi) \qquad y = \sum_{j=1}^{n} y_j \mathcal{L}_j(\xi) \tag{7.72}$$

where, $x_j$ and $y_j$ are the coordinates of the nodes and $n$ is the polynomial order. **Isoparametric elements** are elements in which geometric transformation and the generalized displacements are approximated by the same interpolation functions. Another benefit of using natural coordinates and isoparametric elements is the use of numerical integration when the integrand becomes complicated due to geometric transformation. The use of isoparametric elements, natural coordinates, and numerical integration is sometimes the only choice in two and three dimensions for modeling complex geometries.

We can write the linear and quadratic Lagrange polynomials in natural coordinates as shown below.

$$\text{Linear} \qquad \mathcal{L}_1(\xi) = 1-\xi \qquad \mathcal{L}_2(\xi) = \xi \tag{7.73a}$$

$$\text{Quadratic} \qquad \mathcal{L}_1(\xi) = (1-\xi)(1-2\xi) \qquad \mathcal{L}_2(\xi) = 4(1-\xi)\xi \qquad \mathcal{L}_3(\xi) = -\xi(1-2\xi) \tag{7.73b}$$

### 7.13.3 Vector arithmetic

The Lagrange polynomials in two and three dimensions using non-dimensional coordinates will require areas of triangles and volumes of tetrahedrons in terms of coordinates of points. Vector arithmetic provides a simple way of obtaining formulas for areas and volumes in terms of coordinates of points— reviewed in this section.

Let $\vec{A}$, $\vec{B}$, and $\vec{C}$ be three vectors. From vector arithmetic we know that the cross product $\vec{A} \times \vec{B}$ results in a vector whose magnitude is the area of the parallelogram and direction is perpendicular to the plane formed by the two vectors as shown in Figure 7.16*a*. The area of the triangle is half of the area of parallelogram formed by the two vectors. We further know that the scalar triple product $(\vec{A} \times \vec{B}) \bullet \vec{C}$ results in a quantity whose magnitude is the volume of the parallelepiped and the volume of the tetrahedron is one-sixth of the parallelepiped as shown in Figure 7.16*b*. The order of multiplication in cross product is in counter-clockwise direction with respect to the positive $z$-direction.

**Figure 7.16** (*a*) Parallelogram formed by cross product. (*b*) Parallelepiped formed by scalar triple product.

For finding the area of triangle we assume the vectors $\vec{A}$ and $\vec{B}$ are in the $x$-$y$ plane. We represent the unit vectors in $x$, $y$, and $z$ direction as $\vec{i}$, $\vec{j}$, and $\vec{k}$, respectively. We can write the vectors in terms of the coordinates of the three points shown in Figure 7.16*a*, take the cross product of the two vectors and divide by two to get the area of the triangle as shown below.

$$\vec{A} = (x_2-x_1)\vec{i} + (y_2-y_1)\vec{j} \qquad \vec{B} = (x_3-x_1)\vec{i} + (y_3-y_1)\vec{j}$$

$$\text{Area of triangle 1-2-3} = \vec{A} \times \vec{B}/2 = [(x_2-x_1)(y_3-y_1)-(x_3-x_1)(y_2-y_1)]/2 \tag{7.74}$$

For finding the volume of the tetrahedron we write the three vectors in terms of the coordinates of the four points shown in Figure 7.16*b*, take the scalar triple product and divide by six to obtain the volume of the tetrahedron as shown below.

$$\vec{A} = (x_2-x_1)\vec{i} + (y_2-y_1)\vec{j} + (z_2-z_1)\vec{k} \qquad \vec{B} = (x_3-x_1)\vec{i} + (y_3-y_1)\vec{j} + (z_3-z_1)\vec{k}$$

$$\vec{C} = (x_4-x_1)\vec{i} + (y_4-y_1)\vec{j} + (z_4-z_1)\vec{k}$$

$$\text{Volume of tetrahedron 1-2-3-4} = \frac{1}{6}(\vec{A} \times \vec{B}) \bullet \vec{C} = \frac{1}{6}\begin{vmatrix} (x_2-x_1) & (y_2-y_1) & (z_2-z_1) \\ (x_3-x_1) & (y_3-y_1) & (z_3-z_1) \\ (x_4-x_1) & (y_4-y_1) & (z_3-z_1) \end{vmatrix} \tag{7.75}$$

where | | represent the determinant, which can be expanded to give the volume of tetrahedron in terms of vertices coordinates.

### 7.13.4 Lagrange polynomials in two dimensions

The triangular element is the simplest element that can be used for modeling regions with curved boundary. A linear representation of generalized coordinate is shown below.

$$u(x) = C_0 + C_1 x + C_2 y \tag{7.76}$$

The constants $C$'s can be found in terms of the nodal values of $u$. The functions multiplying the nodal values of $u$ would be the Lagrange polynomials in two dimensions. We will develop the Lagrange polynomials in two dimensional natural coordinates called **area coordinates.** Figure 7.17 shows a triangle with point $P$ whose coordinates are $(x, y)$. The areas of the three triangles formed by joining point $P$ to the vertices can be found. Line $AB$ is parallel to side $JK$. The area $A_I$ will be the same irrespective of the location of point $P$ on line $AB$. This unique value of area $A_I$ makes the area coordinates possible. $L_I$, $L_J$, $L_K$ are the area coordinates defined below. Since a point in two dimensions can have only two coordinates, we obtain the identity shown that emphasize that only two of the three area coordinates are independent.

$$L_I = \frac{A_I}{A} \qquad L_J = \frac{A_J}{A} \qquad L_K = \frac{A_K}{A} \qquad L_I + L_J + L_K = 1 \tag{7.77a}$$

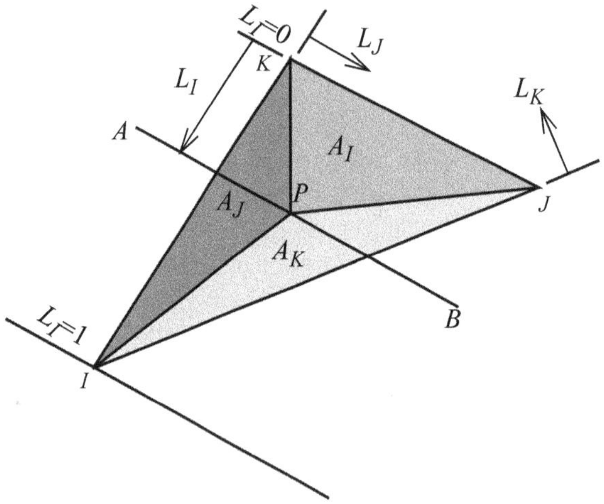


**Figure 7.17** Area coordinates

The areas $A_I$, $A_J$, and $A_K$ can be found using Equation (7.74) and are given below.

$$A_I = [(x_J - x)(y_K - y) - (x_K - x)(y_J - y)]/2 \qquad A_J = [(x_K - x)(y_I - y) - (x_I - x)(y_K - y)]/2 \tag{7.77b}$$

$$A_K = [(x_I - x)(y_J - y) - (x_J - x)(y_I - y)]/2 \qquad A = [(x_J - x_I)(y_K - y_I) - (x_K - x_I)(y_J - y_I)]/2 \tag{7.77c}$$

The derivatives of the area coordinates are

$$\frac{\partial L_I}{\partial x} = \frac{y_{JK}}{2A} \qquad \frac{\partial L_I}{\partial y} = \frac{-x_{JK}}{2A} \qquad \text{where} \qquad x_{JK} = x_J - x_K \text{ and } y_{JK} = y_J - y_K \tag{7.78a}$$

$$\frac{\partial L_J}{\partial x} = \frac{y_{KI}}{2A} \qquad \frac{\partial L_J}{\partial y} = \frac{-x_{KI}}{2A} \qquad \text{where} \qquad x_{KI} = x_K - x_I \text{ and } y_{KI} = y_K - y_I \tag{7.78b}$$

$$\frac{\partial L_K}{\partial x} = \frac{y_{IJ}}{2A} \qquad \frac{\partial L_K}{\partial y} = \frac{-x_{IJ}}{2A} \qquad \text{where} \qquad x_{IJ} = x_I - x_J \text{ and } y_{IJ} = y_I - y_J \tag{7.78c}$$

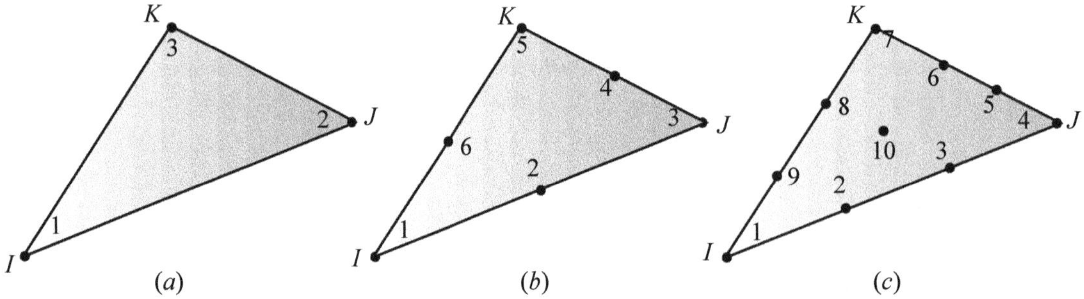


**Figure 7.18** (a) Linear, (b) Quadratic and (C) Cubic triangular elements.

Figure 7.18 shows linear, quadratic and cubic triangular elements. The interpolation functions can be written using the property that the value of Lagrange polynomial at its own node is 1 and at other nodes it is 0. The Lagrange polynomials for linear and quadratic are given below and the cubic polynomial is left to the reader to obtain (see Problem 7.27). In Figure 7.18, we assume all nodes are equally spaced on the triangle.

$$\text{Linear} \qquad \mathcal{L}_1 = L_I \qquad \mathcal{L}_2 = L_J \qquad \mathcal{L}_3 = L_K \tag{7.79a}$$

$$\text{Quadratic} \quad \begin{matrix} \mathcal{L}_1 = L_I(2L_I - 1) & \mathcal{L}_3 = L_J(2L_J - 1) & \mathcal{L}_5 = L_K(2L_K - 1) \\ \mathcal{L}_2 = 4L_IL_J & \mathcal{L}_4 = 4L_JL_K & \mathcal{L}_6 = 4L_KL_I \end{matrix} \tag{7.79b}$$

The following identities (Reddy[1993]) will be useful in calculation of element stiffness matrix and the right hand side vector.

$$\iint_A L_I^m L_J^n L_K^p dx dy = (2A)\frac{m!n!p!}{(m+n+p+2)!} \qquad \int_a^b L_I^m L_J^n L_K^p \, ds = (b-a)\frac{m!n!p!}{(m+n+p+1)!} \tag{7.80}$$

## 7.13.5 Element stiffness matrix and right hand side vector

We can use Equation (7.66a) to find the terms in the stiffness matrix and right hand side vector as we did in Rayleigh-Ritz with $f_j = \mathcal{L}_j$ to obtain

$$K_{jk} = B(\mathcal{L}_j, \mathcal{L}_k) \qquad R_j = l(\mathcal{L}_j) \tag{7.81}$$

The use of the above equation will be demonstrated in Example 7.9. However, when a node has more than one degree of freedom, then the approach causes some difficulties. One way of overcoming this difficulty is to cast the bilinear functional as a product of several matrices—a common practice in finite element method. We will however use an alternative approach based on the idea of variation.

We represent the element displacement vector as $\{u^{[e]}\}^T = \{u^{[1]}, u^{[2]}, \cdots u^{[n]}\}^T$, then the element potential energy can be written as

$$\Omega^{[e]} = U^{[e]} - W^{[e]} \qquad U^{[e]} = \frac{1}{2}\{u^{[e]}\}^T[K^{[e]}]\{u^{[e]}\} \qquad W^{[e]} = \{R^{[e]}\}^T\{u^{[e]}\} \tag{7.82a}$$

By taking the first and second variation of strain energy and noting that the stiffness matrix is symmetric we obtain

$$\delta U^{[e]} = \frac{1}{2}[\{\delta u^{[e]}\}^T[K^{[e]}]\{u^{[e]}\} + \{u^{[e]}\}^T[K^{[e]}]\{\delta u^{[e]}\}] = \frac{1}{2}\{u^{[e]}\}^T([K^{[e]}] + [K^{[e]}]^T)\{\delta u^{[e]}\} = \{u^{[e]}\}^T[K^{[e]}]\{\delta u^{[e]}\} \tag{7.82b}$$

$$\delta^2 U^{[e]} = \{\delta u^{[e]}\}^T[K^{[e]}]\{\delta u^{[e]}\} = \sum_{i=1}^{n}\sum_{j=1}^{n} \delta u^{[e]} K_{ij} \delta u^{[e]} \tag{7.82c}$$

We can also write

$$\delta^2 U^{[e]} = \sum_{i=1}^{n}\sum_{j=1}^{n} \frac{\partial^2 U^{[e]}}{\partial u^{[e]} \delta u^{(j)}} \delta u^{[i]} \delta u^{[j]} \tag{7.82d}$$

Equating Equations (7.82c) and (7.82d) and noting $u_i^{(e)}$ are independent variables we obtain

$$\boxed{K_{ij} = \frac{\partial^2 U^{[e]}}{\partial u^{[i]} \partial u^{[j]}}} \tag{7.83a}$$

In a similar manner by taking the first variation of work we can show

$$\boxed{R_i = \frac{\partial W^{[e]}}{\partial u^{[i]}}} \tag{7.83b}$$

Example 7.10 demonstrates the use of Equations (7.83a) and Equations (7.83b).

## 7.13.6 Lagrange polynomials in three dimensions

The tetrahedron element is the simplest element that can be used for modeling regions with curved surfaces. A linear representation of generalized coordinate is shown below.

$$u(x) = C_0 + C_1 x + C_2 y + C_2 z \tag{7.84}$$

The constants $C$'s in Equation (7.84) can be found in terms of the nodal values of $u$. The functions multiplying the nodal values of $u$ would be the Lagrange polynomials in three dimensions.

Once more we will develop the Lagrange polynomials in three dimensional natural coordinates called **volume coordinates**. Figure 7.19 shows a tetrahedron with point $P$ whose coordinates are $(x, y, z)$. The volume of the four tetrahedron formed by joining point $P$ to the vertices can be found. The volume of the tetrahedron opposite to the vertex $I$ is labeled as $V_I$ and is shown in Figure 7.19. In a similar manner the other tetrahedron volumes can be labeled as $V_J$, $V_K$, and $V_L$. The volume $V_I$ will be the same irrespective of the location of point $P$ on a surface parallel to the surface $JKL$. This unique value of area $V_I$ makes the volume coordinates possible. We non-dimensionalize the volume of the four tetrahedron with the total volume of the entire tetrahedron to obtain the volume coordinates shown in Equation (7.85a). Note only three of the volume coordinates are independent.

$$L_I = \frac{V_I}{V} \qquad L_J = \frac{V_J}{V} \qquad L_K = \frac{V_K}{V} \qquad L_L = \frac{V_L}{V} \qquad V = V_I + V_J + V_K + V_L \qquad L_I + L_J + L_K + L_L = 1 \tag{7.85a}$$

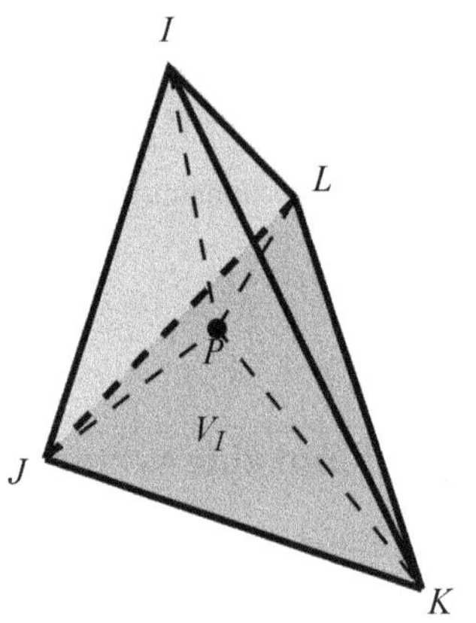


**Figure 7.19** Volume Coordinates

To relate volume of each tetrahedron to the coordinates we consider the 3 vectors from point $P$ to three vertices and use Equation (7.75). Thus $V_I$ is given by Equation (7.85b).

$$V_I = \frac{1}{6}\begin{vmatrix} (x_J-x) & (y_J-y) & (z_J-z) \\ (x_K-x) & (y_K-y) & (z_K-z) \\ (x_L-x) & (y_L-y) & (z_L-z) \end{vmatrix} \tag{7.85b}$$

Other volumes can be written in a similar manner. The Lagrange polynomials are similar to those we saw with area coordinates, with the difference that number increases due to the fourth vertex $L$. Thus the for linear Lagrange we have

$$\text{Linear} \qquad \mathcal{L}_1 = L_I \qquad \mathcal{L}_2 = L_J \qquad \mathcal{L}_3 = L_K \qquad \mathcal{L}_4 = L \tag{7.85c}$$

In a similar manner higher order polynomials can be constructed. The following identity (Zienkiewicz and Taylor [1989]) along with identities in Equation (7.80) will be useful in calculation of element stiffness matrix and the right hand side vector.

$$\iiint_T L_I^m L_J^n L_K^p L_L^q \, dx dy dz = (6V)\frac{m!n!p!q!}{(m+n+p+q+3)!} \tag{7.86}$$

There are no additional new concepts in implementation in three dimensions over those in two dimensions. However, the algebra in implementation in three dimensions becomes significantly more. Hence, we will not consider this topic any more.

## EXAMPLE 7.9

Obtain the element stiffness matrix and the element right hand side vector for an axial member using quadratic Lagrange polynomials (Figure 7.14*b*). Approximate the axial rigidly $EA$ and distributed load $p_x$ with constant values that exists at the center of the axial rod. Also assume the left and right end nodes have a concentrated axial forces $F^{[1]}$ and $F^{[3]}$, respectively.

**PLAN**

The kinematic functions are the Lagrange polynomials that can be substituted into the bilinear and linear functional to obtain the element stiffness matrix and the element right hand side vector.

**SOLUTION**

From Equations (7.7a) and (7.21a) the bi-linear functional and linear functionals are

$$B(u_1, u_2) = \int_L \left[EA\frac{du_1}{dx}\frac{du_2}{dx}\right]dx \qquad l(u_1) = \int_L p_x u_1 dx + F^{[1]}u_1(x_1) + F^{[3]}u_1(x_3) \tag{E1}$$

Transforming to natural coordinates we obtain

$$\xi = x/L \qquad dx = Ld\xi \qquad du/dx = (du/d\xi)/L \tag{E2}$$

Bilinear functional and linear functional can be written as shown below.

$$B(u_1, u_2) = \frac{EA}{L}\int_0^1 \frac{du_1}{d\xi}\frac{du_2}{d\xi}d\xi \qquad l(u) = p_x L\int_0^1 u_1(\xi)d\xi + F^{[1]}u_1(\xi = 0) + F^{[3]}u_1(\xi = 1) \tag{E3}$$

The axial displacement in terms of Lagrange polynomials can be written as

$$u_1(\xi) = \sum_{i=1}^{3} u^{[i]}\mathcal{L}_i(\xi) \qquad \mathcal{L}_1(\xi) = (1-\xi)(1-2\xi) \qquad \mathcal{L}_2(\xi) = 4(1-\xi)\xi \qquad \mathcal{L}_3(\xi) = -\xi(1-2\xi) \tag{E4}$$

The derivatives of Lagrange polynomials can be written as

$$\frac{d\mathcal{L}_1}{d\xi} = 4\xi - 3 \qquad \frac{d\mathcal{L}_2}{d\xi} = 4(1-2\xi) \qquad \frac{d\mathcal{L}_3}{d\xi} = 4\xi - 1 \tag{E5}$$

Noting that $f_i = \mathcal{L}_i$ in Equation (7.66a), we obtain the stiffness matrix and the right hand side vector as given below.

$$K_{ij} = B(\mathcal{L}_i, \mathcal{L}_j) = \frac{EA}{L}\int_0^1 \frac{d\mathcal{L}_i}{d\xi}\frac{d\mathcal{L}_j}{d\xi}d\xi \qquad R_i = l(\mathcal{L}_i) = p_x L\int_0^1 \mathcal{L}_i d\xi + F^{[1]}\mathcal{L}_i(\xi = 0) + F^{[3]}\mathcal{L}_i(\xi = 1) \tag{E6}$$

The element of the stiffness matrix can be calculated as shown below.

$$K_{11} = \frac{EA}{L}\int_0^1 \frac{d\mathcal{L}_1}{d\xi}\frac{d\mathcal{L}_1}{d\xi}d\xi = \frac{EA}{L}\int_0^1 (4\xi - 3)^2 d\xi = \frac{EA}{12L}(4\xi - 3)^3\Big|_0^1 = \frac{7}{3}\frac{EA}{L} \tag{E7}$$

$$K_{12} = \frac{EA}{L}\int_0^1 \frac{d\mathcal{L}_1}{d\xi}\frac{d\mathcal{L}_2}{d\xi}d\xi = \frac{EA}{L}\int_0^1 (10\xi - 8\xi^2 - 3)d\xi = \frac{4EA}{L}\left(5\xi^2 - \frac{8\xi^3}{3} - 3\xi\right)\Big|_0^1 = -\left(\frac{8EA}{3L}\right) \tag{E8}$$

$$K_{13} = \frac{EA}{L}\int_0^1 \frac{d\mathcal{L}_1}{d\xi}\frac{d\mathcal{L}_3}{d\xi}d\xi = \frac{EA}{L}\int_0^1 (16\xi^2 - 16\xi + 3)d\xi = \frac{EA}{L}\left(\frac{16\xi^3}{3} - 8\xi^2 + 3\xi\right)\Big|_0^1 = \left(\frac{EA}{3L}\right) \tag{E9}$$

$$K_{22} = \frac{EA}{L}\int_0^1 \frac{d\mathcal{L}_2}{d\xi}\frac{d\mathcal{L}_2}{d\xi}d\xi = \frac{16EA}{L}\int_0^1 (1 - 2\xi)^2 d\xi = -\frac{16EA}{6L}(1 - 2\xi)^3\Big|_0^1 = \frac{16EA}{3L} \tag{E10}$$

$$K_{23} = \frac{EA}{L}\int_0^1 \frac{d\mathcal{L}_2}{d\xi}\frac{d\mathcal{L}_3}{d\xi}d\xi = \frac{4EA}{L}\int_0^1 (6\xi - 8\xi^2 - 1)d\xi = \frac{4EA}{L}\left(3\xi^2 - \frac{8}{3}\xi^3 - \xi\right)\Big|_0^1 = -\left(\frac{8EA}{3L}\right) \tag{E11}$$

$$K_{33} = \frac{EA}{L}\int_0^1 \frac{d\mathcal{L}_3}{d\xi}\frac{d\mathcal{L}_3}{d\xi}d\xi = \frac{EA}{L}\int_0^1 (4\xi - 1)^2 d\xi = -\frac{EA}{12L}(4\xi - 1)^3\Big|_0^1 = \frac{7}{3}\frac{EA}{L} \tag{E12}$$

The element right hand side vector can be calculated as shown below.

$$R_1 = p_x L\int_0^1 \mathcal{L}_1 d\xi + F^{[1]}\mathcal{L}_1(\xi = 0) + F^{[3]}\mathcal{L}_1(\xi = 1) = p_x L\int_0^1 (1 - \xi)(1 - 2\xi)d\xi + F^{[1]}$$

$$R_1 = p_x L\int_0^1 (1 + 2\xi^2 - 3\xi)d\xi + F^{[1]} = p_x L\left(\xi + \frac{2}{3}\xi^3 - \frac{3}{2}\xi^2\right)\Big|_0^1 + F^{[1]} = \frac{p_x L}{6} + F^{[1]} \tag{E13}$$

$$R_2 = p_x L\int_0^1 \mathcal{L}_2 d\xi + F^{[1]}\mathcal{L}_2(\xi = 0) + F^{[3]}\mathcal{L}_2(\xi = 1) = 4p_x L\int_0^1 (1 - \xi)\xi d\xi = 4p_x L\left(\frac{\xi^2}{2} - \frac{\xi^3}{3}\right)\Big|_0^1 = \frac{2p_x L}{3} \tag{E14}$$

$$R_3 = p_x L\int_0^1 \mathcal{L}_3 d\xi + F^{[1]}\mathcal{L}_3(\xi = 0) + F^{[3]}\mathcal{L}_3(\xi = 1) = p_x L\int_0^1 -\xi(1 - 2\xi)d\xi + F^{[3]}$$

$$R_3 = p_x L\int_0^1 (2\xi^2 - \xi)d\xi + F^{[3]} = p_x L\left(\frac{2}{3}\xi^3 - \frac{\xi^2}{2}\right)\Big|_0^1 + F^{[3]} = \frac{p_x L}{6} + F^{[3]} \tag{E15}$$

The stiffness matrix and the right hand side vector can be written as

$$\textbf{ANS. } [K^{[e]}] = \frac{EA}{3L}\begin{bmatrix} 7 & -8 & 1 \\ -8 & 16 & -8 \\ 1 & -8 & 7 \end{bmatrix} \qquad \{R^{[e]}\} = \frac{p_x L}{6}\begin{Bmatrix} 1 \\ 4 \\ 1 \end{Bmatrix} + \begin{Bmatrix} F^{[1]} \\ 0 \\ F^{[3]} \end{Bmatrix} \tag{E16}$$

## COMMENTS

1. This example demonstrates the application of Rayleigh-Ritz method at element level to obtain element stiffness and right hand side vector for use in finite element method.
2. If we represent the element displacement vector as $\{u^{[e]}\}^T = \{u^{[1]}\ u^{[2]}\ u^{[3]}\}^T$, then the element potential energy is given by Equation (7.82a). Summing the potential energies from other element requires an assembly process of element stiffness and element right hand side vector discussed briefly next.
3. We observe the relationship of the displacements of element nodes to the displacement of nodes in the global mesh, thus defining the addition of each term in the matrix and the right-hand side vector.
4. The summation $F^{[1]}$ and $F^{[3]}$ at the boundary node from different element is replaced by the applicable external concentrated force acting at the node. If there is no external force acting at the node then the summation is set to zero.
5. Noting that the displacements must be kinematically admissible, that is, satisfy the zero displacements boundary conditions, we delete the rows and columns corresponding to the nodes on the boundary. The net outcome is a set of algebraic equation $[K]\{u\} = \{R\}$, where $[K]$, $\{u\}$, and $\{R\}$ are the stiffness matrix, displacement vector, and the load vector of the global mesh.

## EXAMPLE 7.10

An elastic plane of thickness $h$ is in plane stress and subjected to body forces $F_x(x,y)$ and $F_y(x,y)$. For a triangular element, using linear Lagrange polynomial, obtain the first row of the element stiffness matrix and the right hand side vector. Approximate the body forces, the material constant, and thickness with constant values that exists at the centroid of the triangle.

**PLAN**

The strain energy is given by Equation (7.17e) and the work potential by Equation (7.7e). Using Equations (7.83a) and (7.83b) we can find the terms of element stiffness matrix and the right hand side vector.

**SOLUTION**

We write $u$ and v in terms of linear Lagrange polynomials as shown below.

$$u = L_I u^{[1]} + L_J u^{[2]} + L_K u^{[3]} \qquad \mathrm{v} = L_I \mathrm{v}^{[1]} + L_J \mathrm{v}^{[2]} + L_K \mathrm{v}^{[3]} \tag{E1}$$

We define the generalized displacement vector as

$$\{u\}^T = \{u^{[1]}, \mathrm{v}^{[1]}, u^{[2]}, \mathrm{v}^{[2]}, u^{[3]}, \mathrm{v}^{[3]}\}^T \tag{E2}$$

From Equations (7.17e) and (7.7e) we can write the strain energy and work potential as

$$U^{(e)} = \frac{Eh}{2(1-\nu^2)}\iint_A \left[\left\{\left(\frac{\partial u}{\partial x}\right)^2 + 2\nu\left(\frac{\partial u}{\partial x}\right)\left(\frac{\partial \mathrm{v}}{\partial y}\right) + \left(\frac{\partial \mathrm{v}}{\partial y}\right)^2\right\} + \frac{(1-\nu)}{2}\left(\frac{\partial u}{\partial y} + \frac{\partial \mathrm{v}}{\partial x}\right)^2\right] dx\,dy \tag{E3}$$

$$W^{(e)} = h\iint_A [F_x(x,y)u(x,y) + F_y(x,y)\mathrm{v}(x,y)]dx\,dy = hF_x\iint_A u\,dx\,dy + hF_y\iint_A \mathrm{v}\,dx\,dy \tag{E4}$$

The derivatives of displacements [Equation (E1)] can be written using Equations (7.78a), (7.78b), and (7.78c) as given below.

$$\frac{\partial u}{\partial x} = \left(\frac{y_{JK}}{2A}\right)u^{[1]} + \left(\frac{y_{KI}}{2A}\right)u^{[2]} + \frac{y_{IJ}}{2A}u^{[3]} \qquad \frac{\partial u}{\partial y} = \left(\frac{-x_{JK}}{2A}\right)u^{[1]} + \left(\frac{-x_{KI}}{2A}\right)u^{[2]} + \left(\frac{-x_{IJ}}{2A}\right)u^{[3]} \tag{E5}$$

$$\frac{\partial \mathrm{v}}{\partial x} = \left(\frac{y_{JK}}{2A}\right)\mathrm{v}^{[1]} + \left(\frac{y_{KI}}{2A}\right)\mathrm{v}^{[2]} + \frac{y_{IJ}}{2A}\mathrm{v}^{[3]} \qquad \frac{\partial \mathrm{v}}{\partial y} = \left(\frac{-x_{JK}}{2A}\right)\mathrm{v}^{[1]} + \left(\frac{-x_{KI}}{2A}\right)\mathrm{v}^{[2]} + \left(\frac{-x_{IJ}}{2A}\right)\mathrm{v}^{[3]} \tag{E6}$$

The terms of element stiffness matrix can be determined using Equations (7.83a) as shown below.

$$K_{11} = \frac{\partial^2 U^{[e]}}{\partial u^{[1]}\partial u^{[1]}} = \frac{Eh}{(1-\nu^2)}\iint_A \left[\frac{\partial}{\partial u^{[1]}}\left(\frac{\partial u}{\partial x}\right)\frac{\partial}{\partial u^{[1]}}\left(\frac{\partial u}{\partial x}\right) + \frac{(1-\nu)}{2}\frac{\partial}{\partial u^{[1]}}\left(\frac{\partial u}{\partial y}\right)\frac{\partial}{\partial u^{[1]}}\left(\frac{\partial u}{\partial y}\right)\right]dx\,dy$$

$$\textbf{ANS. } K_{11} = \frac{Eh}{(1-\nu^2)}\iint_A \left[\left(\frac{y_{JK}}{2A}\right)^2 + \frac{(1-\nu)}{2}\left(\frac{-x_{JK}}{2A}\right)^2\right]dx\,dy = \frac{Eh[2y_{JK}^2 + (1-\nu)x_{JK}^2]}{8A(1-\nu^2)} \tag{E7}$$

$$K_{12} = \frac{\partial^2 U^{[e]}}{\partial u^{[1]}\partial \mathrm{v}^{[1]}} = \frac{Eh}{(1-\nu^2)}\iint_A \left[\nu\frac{\partial}{\partial u^{[1]}}\left(\frac{\partial u}{\partial x}\right)\frac{\partial}{\partial \mathrm{v}^{[1]}}\left(\frac{\partial \mathrm{v}}{\partial y}\right) + \frac{(1-\nu)}{2}\frac{\partial}{\partial u^{[1]}}\left(\frac{\partial u}{\partial y}\right)\frac{\partial}{\partial \mathrm{v}^{[1]}}\left(\frac{\partial \mathrm{v}}{\partial x}\right)\right]dx\,dy$$

$$\textbf{ANS. } K_{12} = \frac{Eh}{(1-\nu^2)}\iint_A \left[\nu\left(\frac{y_{JK}}{2A}\right)\left(\frac{-x_{JK}}{2A}\right) + \frac{(1-\nu)}{2}\left(\frac{-x_{JK}}{2A}\right)\left(\frac{y_{JK}}{2A}\right)\right]dx\,dy = \frac{-Eh(1+\nu)x_{JK}y_{JK}}{8A(1-\nu^2)} \tag{E8}$$

$$K_{13} = \frac{\partial^2 U^{[e]}}{\partial u^{[1]}\partial u^{[2]}} = \frac{Eh}{(1-\nu^2)}\iint_A \left[\frac{\partial}{\partial u^{[1]}}\left(\frac{\partial u}{\partial x}\right)\frac{\partial}{\partial u^{[2]}}\left(\frac{\partial u}{\partial x}\right) + \frac{(1-\nu)}{2}\frac{\partial}{\partial u^{[1]}}\left(\frac{\partial u}{\partial y}\right)\frac{\partial}{\partial u^{[2]}}\left(\frac{\partial u}{\partial y}\right)\right]dx\,dy$$

$$\textbf{ANS. } K_{13} = \frac{Eh}{(1-\nu^2)}\iint_A \left[\left(\frac{y_{JK}}{2A}\right)\left(\frac{y_{KI}}{2A}\right) + \frac{(1-\nu)}{2}\left(\frac{-x_{JK}}{2A}\right)\left(\frac{-x_{KI}}{2A}\right)\right]dx\,dy = \frac{Eh[2y_{JK}y_{KI} + (1-\nu)x_{JK}x_{KI}]}{8A(1-\nu^2)} \tag{E9}$$

$$K_{14} = \frac{\partial^2 U^{[e]}}{\partial u^{[1]}\partial \mathrm{v}^{[2]}} = \frac{Eh}{(1-\nu^2)}\iint_A \left[\nu\frac{\partial}{\partial u^{[1]}}\left(\frac{\partial u}{\partial x}\right)\frac{\partial}{\partial \mathrm{v}^{[2]}}\left(\frac{\partial \mathrm{v}}{\partial y}\right) + \frac{(1-\nu)}{2}\frac{\partial}{\partial u^{[1]}}\left(\frac{\partial u}{\partial y}\right)\frac{\partial}{\partial \mathrm{v}^{[2]}}\left(\frac{\partial \mathrm{v}}{\partial x}\right)\right]dx\,dy$$

$$\textbf{ANS. } K_{14} = \frac{Eh}{(1-\nu^2)}\iint_A \left[\nu\left(\frac{y_{JK}}{2A}\right)\left(\frac{-x_{KI}}{2A}\right) + \frac{(1-\nu)}{2}\left(\frac{-x_{JK}}{2A}\right)\left(\frac{y_{KI}}{2A}\right)\right]dx\,dy = \frac{-Eh[2\nu y_{JK}x_{KI} + (1-\nu)x_{JK}y_{KI}]}{8A(1-\nu^2)} \tag{E10}$$

$$K_{15} = \frac{\partial^2 U^{[e]}}{\partial u^{[1]}\partial u^{[3]}} = \frac{Eh}{(1-\nu^2)}\iint_A \left[\frac{\partial}{\partial u^{[1]}}\left(\frac{\partial u}{\partial x}\right)\frac{\partial}{\partial u^{[3]}}\left(\frac{\partial u}{\partial x}\right) + \frac{(1-\nu)}{2}\frac{\partial}{\partial u^{[1]}}\left(\frac{\partial u}{\partial y}\right)\frac{\partial}{\partial u^{[3]}}\left(\frac{\partial u}{\partial y}\right)\right]dx\,dy$$

$$\textbf{ANS. } K_{15} = \frac{Eh}{(1-\nu^2)}\iint_A \left[\left(\frac{y_{JK}}{2A}\right)\left(\frac{y_{IJ}}{2A}\right) + \frac{(1-\nu)}{2}\left(\frac{-x_{JK}}{2A}\right)\left(\frac{-x_{IJ}}{2A}\right)\right]dx\,dy = \frac{Eh[2y_{JK}y_{IJ} + (1-\nu)x_{JK}x_{IJ}]}{8A(1-\nu^2)} \tag{E11}$$

$$K_{16} = \frac{\partial^2 U^{[e]}}{\partial u^{[1]} \partial v^{[3]}} = \frac{Eh}{(1-v^2)} \iint_A \left[ v \frac{\partial}{\partial u^{[1]}}\left(\frac{\partial u}{\partial x}\right) \frac{\partial}{\partial v^{[3]}}\left(\frac{\partial v}{\partial y}\right) + \frac{(1-v)}{2} \frac{\partial}{\partial u^{[1]}}\left(\frac{\partial u}{\partial y}\right) \frac{\partial}{\partial v^{[3]}}\left(\frac{\partial v}{\partial x}\right) \right] dx dy$$

$$\textbf{ANS. } K_{16} = \frac{Eh}{(1-v^2)} \iint_A \left[ v\left(\frac{y_{JK}}{2A}\right)\left(\frac{-x_{IJ}}{2A}\right) + \frac{(1-v)}{2}\left(\frac{-x_{JK}}{2A}\right)\left(\frac{y_{IJ}}{2A}\right) \right] dx dy = \frac{-Eh[2v y_{JK} x_{IJ} + (1-v) x_{JK} y_{IJ}]}{8A(1-v^2)} \qquad (E12)$$

The first term in the right hand side vector can be determined using Equation (7.83b) and Equation (7.80) as shown below.

$$\textbf{ANS. } R_i = \frac{\partial W^{[e]}}{\partial u^{(1)}} = hF_x \iint_A \frac{\partial u}{\partial u^{(1)}} dx dy = hF_x \iint_A L_I dx dy = hF_x\left(\frac{2A}{3!}\right) = \left(\frac{AhF_x}{3}\right) \qquad (E13)$$

**COMMENTS**

1. In the above calculations, when we differentiated with two nodal values of $u$, such as $K_{11}$, $K_{13}$, and $K_{15}$, then only terms that could contribute non-zero values were $(\partial u / \partial x)^2$ and $(\partial u / \partial y + \partial v / \partial x)^2$, in other words, terms that had a square of the derivatives. Similarly, in $K_{12}$, $K_{14}$, and $K_{16}$ where we differentiated with nodal values of $u$ and v, the terms that would contribute were $(\partial u / \partial x)(\partial v / \partial y)$ and $(\partial u / \partial y + \partial v / \partial x)^2$.
2. The term $(\partial v / \partial y)^2$ did not contribute towards any term in the first row of element stiffness matrix, but will contribute in the calculations of terms in the second row.
3. The element stiffness matrix is 6 x 6, requiring calculation of only 21 terms because of matrix symmetry. Conceptually, the calculations are not difficult but the algebra can be tedious. It is for this reason that post text problems will require calculation of few terms only.
4. Note the use Equation (7.80) in calculation of the right hand side vector. For quadratic Lagrange polynomials, we would need to use Equation (7.80) in calculations of the terms in the element stiffness matrix also.

## 7.14 CLOSURE

In this chapter we established the principals of variational calculus that can be used in obtaining boundary value problems and solution by approximate method of Rayleigh-Ritz. We saw that potential energy is the functional that is minimized using variational calculus. When the conditions for minimization are satisfied at each and every point on the body, including the boundary points, then we obtain the boundary value problems. When the conditions for minimization are satisfied in an average sense over the entire body then we obtain the approximate solution by Rayleigh-Ritz. Finite element method in mechanics of materials can be seen as an application of Rayleigh-Ritz in which the kinematically admissible functions are polynomial approximations that are piecewise continuous over elements that are obtained by discretization of the entire body.

Minimum potential energy replaces the module of equilibrium in the logic depicted in Figure 2.7. We still need to obtain strains from our kinematic approximations and the corresponding stresses from our approximation of material model. Substituting the strains and stresses into strain energy density and then integrating over the cross-sectional areas for one-dimensional structural members and over the thickness for two-dimensional bodies, we obtained the strain energy for structural members and elastic body. Obtaining the work potential was relatively straight forward giving us the potential energy for the structural members, which could be the minimized to obtain either the boundary value problem or an approximate solution. Obtaining potential energy with all the complexities described in Chapter 2 and Chapter 5 is relatively straight forward. But in case of geometric nonlinearity, variational calculus approach is significantly simpler than the differential calculus approach for obtaining boundary value problems. We demonstrated this for beam bending problem in this chapter and will see additional demonstrations in the next chapter for two dimensional problems with geometric nonlinearities.

We saw that obtaining boundary value problems using variational calculus is elegant and the derivation is compact. In Chapter 8, the potential energy and the minimizing formulas will be cast using indicial notation. The indicial notation will make the derivations of boundary value problems even more compact and more elegant.

## 7.15 SYNOPSIS OF EQUATIONS

| | |
|---|---|
| | **Work Potential** |
| Axial | $W_A = \int_L p_x(x)u(x)dx + \sum_{q=1}^{m} F_q u(x_q) = l(u)$ |
| Torsion of circular shafts | $W_T = \int_L t(x)\phi(x)dx + \sum_{q=1}^{m} T_q \phi(x_q) = l(\phi)$ |
| Symmetric bending of beams | $W_B = \int_L p_y(x)\mathrm{v}(x)dx + \sum_{q=1}^{m_1} F_q \mathrm{v}(x_q) + \sum_{q=1}^{m_2} M_q \frac{d\mathrm{v}}{dx}(x_q) = l(\mathrm{v})$ |
| Bending of thin plates | $W_P = \iint_A p_z(x,y)w(x,y)dx\,dy = l(w)$ |
| Plane stress elasticity | $W_E = h\iint_A [F_x(x,y)u(x,y) + F_y(x,y)\mathrm{v}(x,y)]dx\,dy = l(u,\mathrm{v})$ |
| | **Strain Energy** |
| Axial | $U_A = \frac{1}{2}\int_L EA\left(\frac{du}{dx}\right)^2 dx$ |
| Torsion of circular shafts | $U_T = \frac{1}{2}\int_L GJ\left(\frac{d\phi}{dx}\right)^2 dx$ |
| Symmetric bending of beams | $U_B = \frac{1}{2}\int_L EI_{zz}\left(\frac{d^2\mathrm{v}}{dx^2}\right)^2 dx$ |
| Thin Plates | $U_P = \frac{D}{2}\iint_A \left\{\left(\frac{\partial^2 w}{\partial x^2}\right)^2 + \left(\frac{\partial^2 w}{\partial y^2}\right)^2 + 2\nu\left(\frac{\partial^2 w}{\partial x^2}\right)\left(\frac{\partial^2 w}{\partial y^2}\right) + 2(1-\nu)\left(\frac{\partial^2 w}{\partial x \partial y}\right)^2\right\}dx\,dy$ |
| Plane Stress Elasticity | $U_E = \frac{Eh}{2(1-\nu^2)}\iint_A \left[\left\{\left(\frac{\partial u}{\partial x}\right)^2 + 2\nu\left(\frac{\partial u}{\partial x}\right)\left(\frac{\partial \mathrm{v}}{\partial y}\right) + \left(\frac{\partial \mathrm{v}}{\partial y}\right)^2\right\} + \frac{(1-\nu)}{2}\left(\frac{\partial u}{\partial y} + \frac{\partial \mathrm{v}}{\partial x}\right)^2\right]dx\,dy$ |
| | **Bilinear functional form of strain energy** |
| Axial | $U_A = \frac{1}{2}\int_L \left[EA\frac{du_1}{dx}\frac{du_2}{dx}\right]dx = \frac{1}{2}B(u_1, u_2)$ |
| Torsion of circular shafts | $U_T = \frac{1}{2}\int_L \left[GJ\frac{d\phi_1}{dx}\frac{d\phi_2}{dx}\right]dx = \frac{1}{2}B(\phi_1, \phi_2)$ |
| Symmetric bending of beams | $U_B = \frac{1}{2}\int_L \left[EI_{zz}\frac{d^2\mathrm{v}_1}{dx^2}\frac{d^2\mathrm{v}_2}{dx^2}\right]dx = \frac{1}{2}B(\mathrm{v}_1, \mathrm{v}_2)$ |
| Thin plates | $U_P = \frac{D}{2}\iint_A \left[\frac{\partial^2 w_1}{\partial x^2}\frac{\partial^2 w_2}{\partial x^2} + \frac{\partial^2 w_1}{\partial y^2}\frac{\partial^2 w_2}{\partial y^2} + \nu\left(\frac{\partial^2 w_1}{\partial x^2}\frac{\partial^2 w_2}{\partial y^2} + \frac{\partial^2 w_2}{\partial x^2}\frac{\partial^2 w_1}{\partial y^2}\right) + 2(1-\nu)\frac{\partial^2 w_1}{\partial x \partial y}\frac{\partial^2 w_2}{\partial x \partial y}\right]dx\,dy = \frac{1}{2}B(w_1, w_2)$ |
| Plane stress elasticity | $U_E = \frac{Eh}{2(1-\nu^2)}\iint_A \left[\left\{\frac{\partial u_1}{\partial x}\frac{\partial u_2}{\partial x} + \nu\left(\frac{\partial u_2}{\partial x}\frac{\partial \mathrm{v}_1}{\partial y} + \frac{\partial u_1}{\partial x}\frac{\partial \mathrm{v}_2}{\partial y}\right) + \frac{\partial \mathrm{v}_1}{\partial y}\frac{\partial \mathrm{v}_2}{\partial y}\right\} + \frac{(1-\nu)}{2}\left(\frac{\partial u_1}{\partial y} + \frac{\partial \mathrm{v}_1}{\partial x}\right)\left(\frac{\partial u_2}{\partial y} + \frac{\partial \mathrm{v}_2}{\partial x}\right)\right]dx\,dy$ <br> $= \frac{1}{2}B(u_1, \mathrm{v}_1, u_2, \mathrm{v}_2)$ |
| | **Stationary value of line Integral** |
| Functional | $I(u) = \int_a^b H(u^{(r)}, u^{(r-1)}, u^{(r-1)}, \ldots, u^{(1)}, u^{(0)}, x)\ dx$ |
| Differential Equation | $\frac{\partial H}{\partial u^{(0)}} - \frac{d}{dx}\left(\frac{\partial H}{\partial u^{(1)}}\right) + \frac{d^2}{dx^2}\left(\frac{\partial H}{\partial u^{(2)}}\right) - \frac{d^3}{dx^3}\left(\frac{\partial H}{\partial u^{(3)}}\right) + \cdots + (-1)^r\frac{d^r}{dx^r}\left(\frac{\partial H}{\partial u^{(r)}}\right) = 0$ |
| Boundary Conditions at at $x = a$ and at $x = b$ | $-\left(\frac{\partial H}{\partial u^{(1)}}\right) + \frac{d}{dx}\left(\frac{\partial H}{\partial u^{(2)}}\right) - \frac{d^{(2)}}{dx^2}\left(\frac{\partial H}{\partial u^{(3)}}\right) + \bullet\bullet + (-1)^r\frac{d^{(r-1)}}{dx^{(r-1)}}\left(\frac{\partial H}{\partial u^{(r)}}\right) = 0$ or $\delta u = 0$ <br> $\left(\frac{\partial H}{\partial u^{(2)}}\right) - \frac{d}{dx}\left(\frac{\partial H}{\partial u^{(3)}}\right) + \bullet\bullet + (-1)^r\frac{d^{(r-2)}}{dx^{(r-2)}}\left(\frac{\partial H}{\partial u^{(r)}}\right) = 0$ or $\delta u^{(1)} = 0$ <br> $-\left(\frac{\partial H}{\partial u^{(3)}}\right) + \bullet\bullet + (-1)^r\frac{d^{(r-3)}}{dx^{(r-3)}}\left(\frac{\partial H}{\partial u^{(r)}}\right) = 0$ or $\delta u^{(2)} = 0$ <br> $\bullet = 0$ or $= 0$ <br> $\bullet = 0$ or $= 0$ <br> $\bullet = 0$ or $= 0$ <br> $\frac{\partial H}{\partial u^{(r)}} = 0$ or $\delta u^{(r-1)} = 0$ |

### Stationary value of an area integral with first order of derivatives.

Functional

$$I(u) = \iint_A H(u_{,x}, u_{,y}, u, x, y)\,dx\,dy$$

Differential Equation

$$\frac{\partial H}{\partial u} - \frac{\partial}{\partial x}\left(\frac{\partial H}{\partial u_{,x}}\right) - \frac{\partial}{\partial y}\left(\frac{\partial H}{\partial u_{,y}}\right) = 0 \qquad x, y \text{ in } A$$

Boundary Conditions

$$\left(\frac{\partial H}{\partial u_{,x}}\right)n_x + \left(\frac{\partial H}{\partial u_{,y}}\right)n_y = 0 \qquad \text{or} \qquad \delta u = 0 \qquad x, y \text{ on } \Gamma$$

### Stationary value of an area integral with second order of derivatives

Functional

$$I(u) = \iint_A H(u_{,xx}, u_{,xy}, u_{,yx}, u_{,yy}, u_{,x}, u_{,y}, u, x, y)\,dx\,dy$$

Notation

$$\boldsymbol{m}_{nn} = \left(\frac{\partial H}{\partial u_{,xx}}\right)n_x^2 + \left(\frac{\partial H}{\partial u_{,xy}} + \frac{\partial H}{\partial u_{,yx}}\right)n_x n_y + \left(\frac{\partial H}{\partial u_{,yy}}\right)n_y^2$$

$$\boldsymbol{m}_{nt} = -\left(\frac{\partial H}{\partial u_{,xx}} - \frac{\partial H}{\partial u_{,yy}}\right)n_x n_y + n_x^2\left(\frac{\partial H}{\partial u_{,yx}}\right) - n_y^2\left(\frac{\partial H}{\partial u_{,xy}}\right)$$

$$\boldsymbol{q}_x = \left(\frac{\partial H}{\partial u_{,x}}\right) - \frac{\partial}{\partial x}\left(\frac{\partial H}{\partial u_{,xx}}\right) - \frac{\partial}{\partial y}\left(\frac{\partial H}{\partial u_{,xy}}\right)$$

$$\boldsymbol{q}_y = \left(\frac{\partial H}{\partial u_{,y}}\right) - \frac{\partial}{\partial x}\left(\frac{\partial H}{\partial u_{,yx}}\right) - \frac{\partial}{\partial y}\left(\frac{\partial H}{\partial u_{,yy}}\right)$$

$$\boldsymbol{q}_n = \boldsymbol{q}_x n_x + \boldsymbol{q}_y n_y \qquad V_n = \boldsymbol{q}_n + \frac{\partial \boldsymbol{m}_{nt}}{\partial s}$$

$$V_n = \boldsymbol{q}_n - \frac{\partial}{\partial s}\left(\frac{\partial H}{\partial u_{,xx}} - \frac{\partial H}{\partial u_{,yy}}\right)n_x n_y + \frac{\partial}{\partial s}\left(\frac{\partial H}{\partial u_{,yx}}\right)n_x^2 - \frac{\partial}{\partial s}\left(\frac{\partial H}{\partial u_{,xy}}\right)n_y^2 - \left(\frac{\partial H}{\partial u_{,xx}} - \frac{\partial H}{\partial u_{,yy}}\right)\frac{n_x^2 - n_y^2}{R_{cur}} - \left(\frac{\partial H}{\partial u_{,xy}} + \frac{\partial H}{\partial u_{,yx}}\right)\frac{2n_x n_y}{R_{cur}}$$

Differential equation

$$\frac{\partial \boldsymbol{q}_x}{\partial x} + \frac{\partial \boldsymbol{q}_y}{\partial y} + \left(\frac{\partial H}{\partial u}\right) = 0 \qquad x, y \text{ in } A$$

Boundary Conditions

$$\left[\boldsymbol{m}_{nn} = 0 \qquad \text{or} \qquad \delta\left(\frac{\partial u}{\partial n}\right) = 0\right] \qquad \text{and} \qquad [V_n = 0 \qquad \text{or} \qquad \delta u = 0] \qquad x, y \text{ on } \Gamma$$

Corner condition

$$(\Delta \boldsymbol{m}_{nt})_C = 0 \qquad \text{or} \qquad \delta u_C \qquad x, y \text{ on corner } \Gamma_C$$

Potential energy

$$\Omega = U - W = \frac{1}{2}B(u, u) - l(u)$$

### Rayleigh-Ritz

Approximation

$$u = \sum_{j=1}^{n} C_j f_j$$

Algebraic equations

$$[K]\{C\} = \{R\} \qquad \text{where} \qquad K_{jk} = B(f_j, f_k) \qquad R_j = l(f_j)$$

Potential energy at equilibrium.

$$\Omega^* = -\frac{1}{2}\sum_{j=1}^{n} C_j^* R_j = -\frac{1}{2}\{C^*\}^T\{R\}$$

### Finite Element Method

Potential Energy

$$\Omega = \sum_{i=1}^{n} \Omega^{(i)} = \sum_{i=1}^{n} (U^{(i)} - W^{(i)})$$

Approximation

$$u = \sum_{i=1}^{n} u^{(i)} \mathcal{L}_i$$

Stiffness Matrix and Right hand side vector

$$K_{jk} = B(\mathcal{L}_j, \mathcal{L}_k) = \frac{\partial^2 U^{(e)}}{\partial u^{(j)} \partial u^{(k)}} \qquad R_j = l(\mathcal{L}_j) = \frac{\partial W^{(e)}}{\partial u^{(j)}}$$

Lagrange polynomials in 1-D using natural coordinates

Linear $\quad \mathcal{L}_1(\xi) = 1 - \xi \qquad \mathcal{L}_2(\xi) = \xi$

Quadratic $\quad \mathcal{L}_1(\xi) = (1 - \xi)(1 - 2\xi) \qquad \mathcal{L}_2(\xi) = 4(1 - \xi)\xi \qquad \mathcal{L}_3(\xi) = -\xi(1 - 2\xi)$

Area coordinates

$$L_I = A_I / A \qquad L_J = A_J / A \qquad L_K = A_K / A \qquad L_I + L_J + L_K = 1$$

$$A_I = [(x_J - x)(y_K - y) - (x_K - x)(y_J - y)]/2 \qquad A_J = [(x_K - x)(y_I - y) - (x_I - x)(y_K - y)]/2$$

$$A_K = [(x_I - x)(y_J - y) - (x_J - x)(y_I - y)]/2 \qquad A = [(x_J - x_I)(y_K - y_I) - (x_K - x_I)(y_J - y_I)]/2$$

Lagrange polynomials in 2-D using area coordinates

Linear $\quad \mathcal{L}_1 = L_I \qquad \mathcal{L}_2 = L_J \qquad \mathcal{L}_3 = L_K$

Quadratic

$$\mathcal{L}_1 = L_I(2L_I - 1) \qquad \mathcal{L}_3 = L_J(2L_J - 1) \qquad \mathcal{L}_5 = L_K(2L_K - 1)$$

$$\mathcal{L}_2 = 4L_I L_J \qquad \mathcal{L}_4 = 4L_J L_K \qquad \mathcal{L}_6 = 4L_K L_I$$

Integration formulas

$$\iint_A L_I^m L_J^n L_K^p\,dx\,dy = (2A)\frac{m!\,n!\,p!}{(m+n+p+2)!} \qquad \int_a^b L_I^m L_J^n L_K^p\,ds = (b-a)\frac{m!\,n!\,p!}{(m+n+p+1)!}$$

## PROBLEMS

### Sections 7.1-7.3

**7.1** Show that the strain energy for axial members is given by Equation (7.17a).

**7.2** Show that the strain energy for torsion of circular shafts is given by Equation (7.17b).

**7.3** Show that the strain energy for Timoshenko beams (see Example 2.6) is given by equation below.

$$U_B = \frac{1}{2}\int_L \left[EI_{zz}\left(\frac{d\psi}{dx}\right)^2 + GA\left(\frac{dv}{dx}-\psi\right)^2\right]dx \qquad (7.87)$$

where, v is the deflection of the beam and $\psi$ is the rotation of the beam cross section from the vertical.

**7.4** Show that the strain energy of an unsymmetric beam is given by

$$U_B = \frac{1}{2}\int_L \left[EI_{zz}\left(\frac{d^2v}{dx^2}\right)^2 + 2EI_{yz}\left(\frac{d^2v}{dx^2}\right)\left(\frac{d^2w}{dx^2}\right) + EI_{yy}\left(\frac{d^2w}{dx^2}\right)^2\right]dx \qquad (7.88)$$

**7.5** Using Equation (6.7b) show that the strain energy for plane elasticity is given by the equation below.

$$U_E = \frac{h}{2}\iint_R \left[\frac{G}{(\kappa-1)}\left\{\begin{matrix}(\kappa+1)\left(\frac{\partial u}{\partial x}\right)^2 + 2(3-\kappa)\frac{\partial u}{\partial x}\frac{\partial v}{\partial y} \\ +(\kappa+1)\left(\frac{\partial v}{\partial y}\right)^2 + G\left(\frac{\partial u}{\partial y}+\frac{\partial v}{\partial x}\right)^2\end{matrix}\right\}\right]dx\,dy \qquad (7.89)$$

**7.6** Show that the strain energy for orthotropic plates (see Example 5.2) is given by

$$U_P = \frac{1}{2}\iint_R \left[\begin{matrix}D_{11}\left(\frac{\partial^2 w}{\partial x^2}\right)^2 + 2D_{12}\left(\frac{\partial^2 w}{\partial x^2}\right)\left(\frac{\partial^2 w}{\partial y^2}\right) \\ + D_{22}\left(\frac{\partial^2 w}{\partial y^2}\right)^2 + 4D_{33}\left(\frac{\partial^2 w}{\partial x \partial y}\right)^2\end{matrix}\right]dx\,dy \qquad (7.90)$$

**7.7** Show that the strain energy for *Mindlin-Reissner* plate is given by

$$U_P = \frac{1}{2}\iint_R \left\{\begin{matrix}D\left[\left(\frac{\partial \psi_x}{\partial x}\right)^2 + 2\nu\left(\frac{\partial \psi_x}{\partial x}\right)\left(\frac{\partial \psi_y}{\partial y}\right) + \left(\frac{\partial \psi_y}{\partial y}\right)^2\right] \\ + \frac{D(1-\nu)}{2}\left(\frac{\partial \psi_x}{\partial y}+\frac{\partial \psi_y}{\partial x}\right)^2 \\ + Gh\left[\left(-\psi_x+\frac{\partial w}{\partial x}\right)^2 + \left(-\psi_y+\frac{\partial w}{\partial y}\right)^2\right]\end{matrix}\right\}dx\,dy \qquad (7.91)$$

### Sections 7.4-7.7

**7.8** A symmetric beam is subjected to transverse load $p_y$ and its temperature is increased by $\Delta T(x, y)$. The temperature distribution is such that it produces no axial force at a cross section. Obtain the differential equation from potential energy that includes temperature change. Use $m_T = \int_A Ey\alpha\Delta T dA$ as in Problem 2.13.

**7.9** A Timoshenko beam is loaded with just the transverse load $p_y(x)$. Obtain the boundary value problem from the potential energy using results of Problem 7.3.

**7.10** An unsymmetric beam is loaded with transverse loads $p_y(x)$ and $p_z(x)$. Obtain the boundary value problem from the potential energy using the results of Problem 7.4.

### Sections 7.8-7.10

**7.11** An elastic plane is loaded with body forces $F_x(x, y)$ and $F_y(x, y)$. Obtain the boundary value problem from the potential energy using the results of Problem 7.5.

**7.12** An orthotropic plate is loaded with just the transverse load $p_z(x, y)$. Obtain the boundary value problem from the potential energy using results of Problem 7.6.

**7.13** A *Mindlin-Reissner* plate is loaded with just the transverse load $p_z(x, y)$. Obtain the differential equations from the potential energy using results of Problem 7.7.

### Section 7.11

**7.14** A symmetric beam is subjected to transverse load $p_y$, an axial distributed force $p_x$, and temperature increase $\Delta T(x, y)$. The strain is given by Equation (7.60). Starting with the displacement field $u(x) = u_o - y dv/dx$, where $u_o$ and v are the axial and bending displacement and only a function of $x$. Obtain the stress formula and differential equations for the displacements $u_o$ and v that includes temperature change and large strain. Use $n_T = \int_A E\alpha\Delta T dA$ and $m_T = \int_A Ey\alpha\Delta T dA$ to represent axial force and bending moment due to temperature change.

**7.15** Fig. P7.15 shows a laminated beam cross section that is symmetric about the $y$ axis. The beam is subjected to transverse load $p_y$ and an axial distributed force $p_x$. The strain is given by Equation (7.60). Starting with displacement field $u(x) = u_o - y dv/dx$, where $u_o$ and v are the axial and bending displacement and only a function of $x$. The origin of $y$ is such that $\int_A Ey dA = 0$. Obtain the stress formula and differential equations for the displacements $u_o$ and v that includes large strain for the laminated beam.

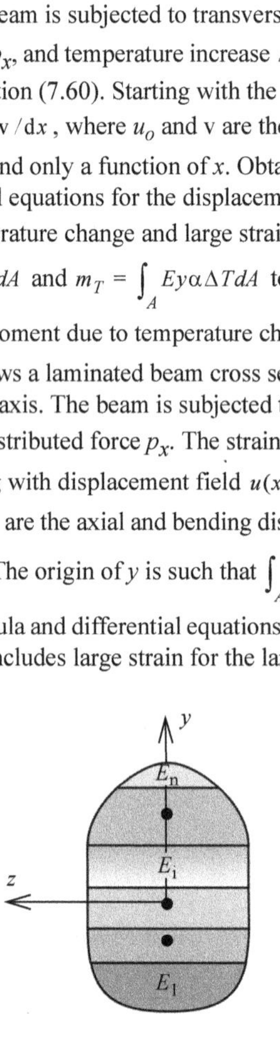


**Fig. P7.15**

**7.16** An axial member has a cross section shown in Figure 7.10 on page 192. The stress-strain equation of axial member material is given by $\sigma = E\varepsilon^{0.6}$. The member is subjected to an axial distributed force $p_x(x)$. Obtain the boundary value problem for the axial displacement $u_o$ if all assumption of classical axial member are valid except for the Hooke's law

**7.17** A circular hollow shaft has an inside radius $R_i$ and outside radius $R_o$. The stress-strain equation of shaft material is given by $\tau = G\gamma^{0.8}$ The shaft is subjected to a distributed torque $t(x)$. Obtain the boundary value problem for the shaft rotation $\phi$ if all assumptions of classical circular shaft are valid except for the Hooke's law.

**7.18** The I cross section of a symmetric beam is shown in Fig. P7.18. The stress-strain equation of the beam material is given by

$\sigma = 70\varepsilon^{0.75}$ GPa. The beam is subjected to a transverse distributed force $p_y(x)$. Obtain the boundary value problem for the beam deflection v if all assumptions of classical beam are valid except for the Hooke's law.

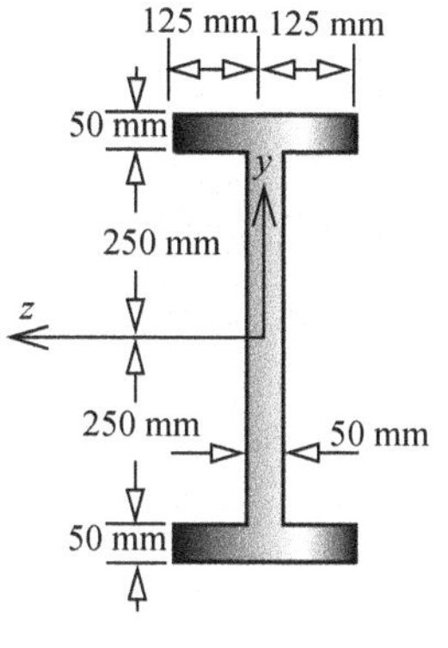


**Fig. P7.18**

## Section 7.12

**7.19** For the beam shown in Fig. P7.19, determine one- and two-parameters Rayleigh-Ritz solutions, using the approximation for the bending displacement given. Calculate the potential energy function for the one- and two-parameter problems.Assume that *EI* is constant.

$$\mathrm{v}(x) = \sum_{i=1}^{n} C_i\left(1 - \cos\frac{2i\pi x}{L}\right)$$

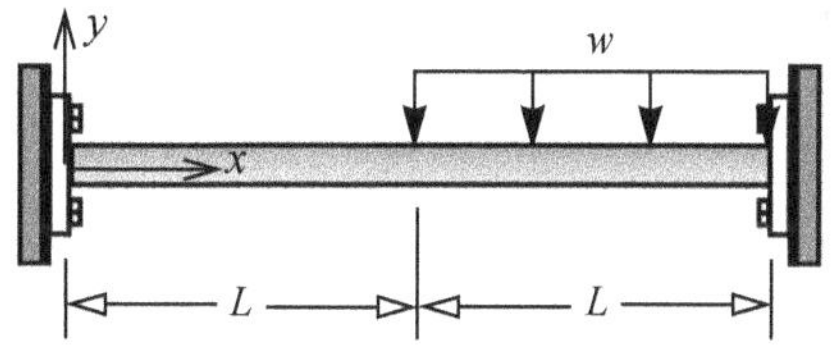


**Fig. P7.19**

**7.20** For the beam shown in Fig. P7.20, determine the one- and two-parameter Rayleigh-Ritz solutions, using the approximation for the bending displacement given. Calculate the potential energy function for the one- and two-parameter problems. Assume that *EI* is constant.

$$\mathrm{v}(x) = \sum_{i=1}^{n} C_i x^i (x - 2L)$$

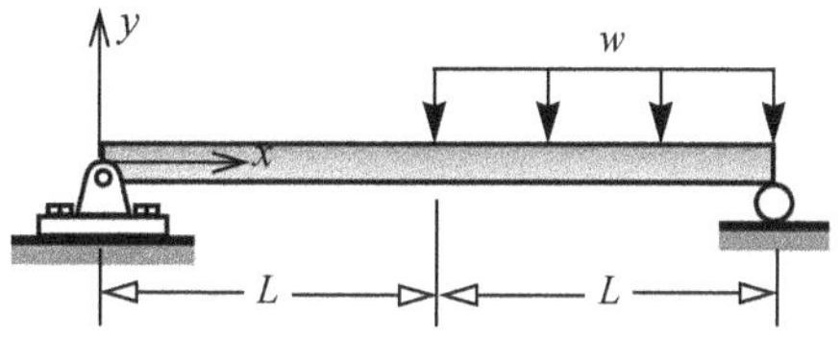


**Fig. P7.20**

**7.21** For the beam shown in Fig. P7.19, determine the one- and two-parameter Rayleigh-Ritz displacement solutions, using the approximation for the bending displacement given. Calculate the potential energy function for the one- and two-parameter problems. Assume that *EI* is constant.

$$\mathrm{v}(x) = \sum_{i=1}^{n} C_i \sin\frac{i\pi x}{2L}$$

**7.22** For the beam shown in Fig. P7.22, determine a two-parameter Rayleigh-Ritz displacement solution, using a polynomial approximation for the bending displacement.Assume that *EI* is constant.

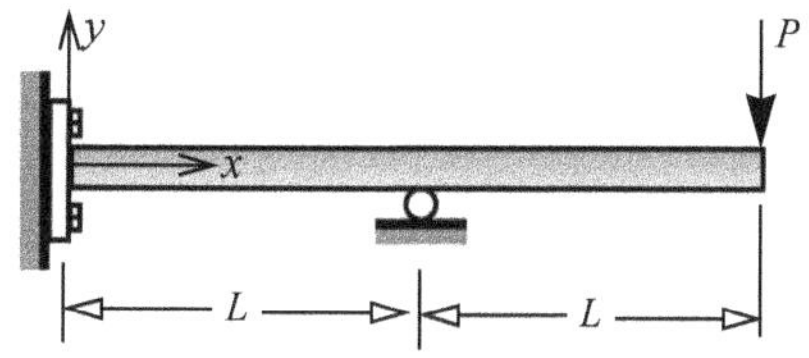


**Fig. P7.22**

**7.23** A square plate is uniformly loaded as shown in Fig. P7.23. Using Rayleigh-Ritz method determine deflection *w* at the center of the plate and the potential energy for the approximation given below.

$$w(x, y) = C_1 \sin(\pi x/a)[1 + \cos(\pi y/a)] + C_2 \sin(3\pi x/a)[1 - \cos(2\pi y/a)]$$

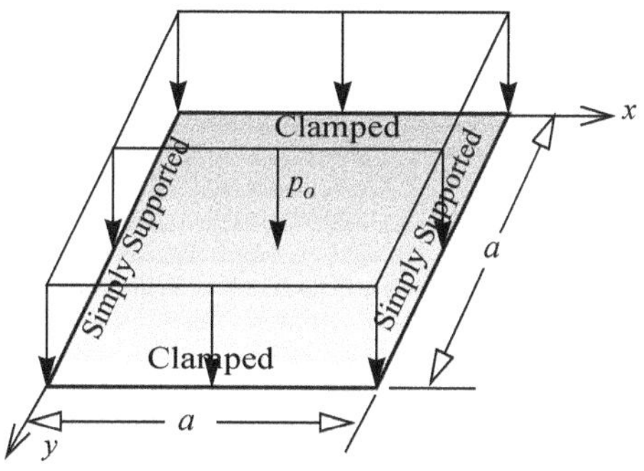


**Fig. P7.23**

**7.24** A square plate is uniformly loaded as shown in Fig. P7.24. Using Rayleigh-Ritz method determine deflection *w* at the center of the plate and the potential energy for the approximation given below.

$$w(x, y) = C_1 \sin(\pi x/a) + C_2 \sin(3\pi x/a)\cos\pi(y/a) + C_3 \sin(3\pi x/a)\cos(2\pi y/a)$$

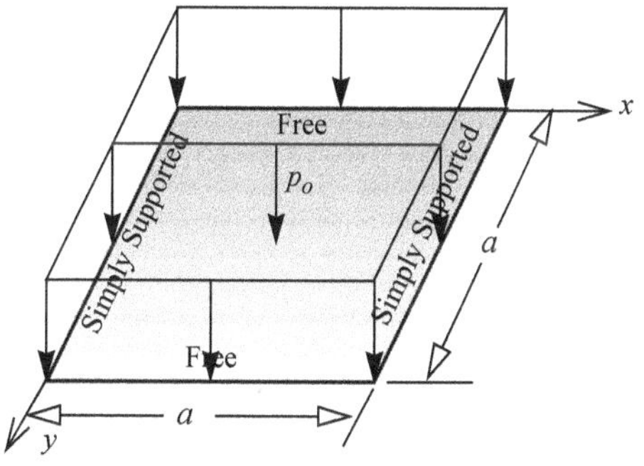


**Fig. P7.24**

**7.25** A square plate is uniformly but partially loaded at the center as shown in Fig. P7.25. Using Rayleigh-Ritz method determine deflection *w* at the center of the plate and the potential energy for the approximation given below.

$$w(x, y) = C_1 \sin(\pi x/a)\sin(\pi y/a) + C_2 \sin(3\pi x/a)\sin(\pi y/a) + C_3 \sin(\pi x/a)\sin(3\pi y/a)$$

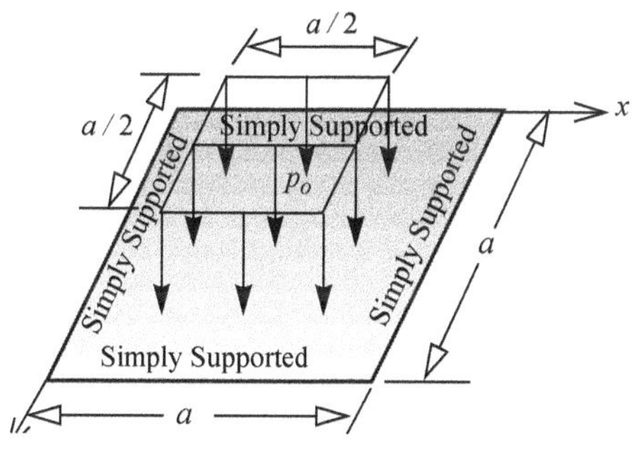


**Fig. P7.25**

## Section 7.13

**7.26** Obtain the cubic Lagrange interpolation functions in natural coordinates for the 4 nodes shown in Fig. P7.26.

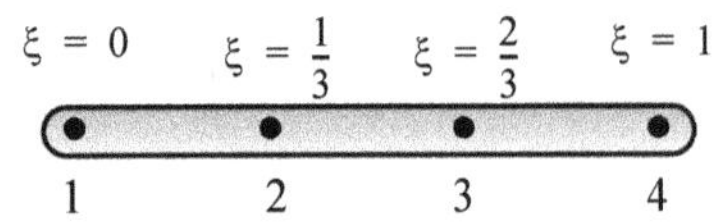


**Fig. P7.26**

**7.27** Obtain the cubic Lagrange interpolation function for triangular elements for nodes 1, 2, 3 and 10 in Figure 7.18*c*.

**7.28** Fig. P7.28 shows positive directions for displacement v, slope $\theta = \partial v / \partial x$, force *F*, and moment *M* at the end nodes of a symmetric beam element. Approximate the bending rigidity *EI* and transverse load $p_y$ with constant values that exists at the center of the beam. Obtain the element stiffness matrix and right hand side vector for the beam element using the polynomial approximation in natural coordinate ($\xi$) given below.

$$v(\xi) = [1 - 3\xi^2 + 2\xi^3]v_1 + [L(\xi - 2\xi^2 + \xi^3)]\theta_1$$
$$+ [3\xi^2 - 2\xi^3]v_2 + L(-\xi^2 + \xi^3)\theta_2$$

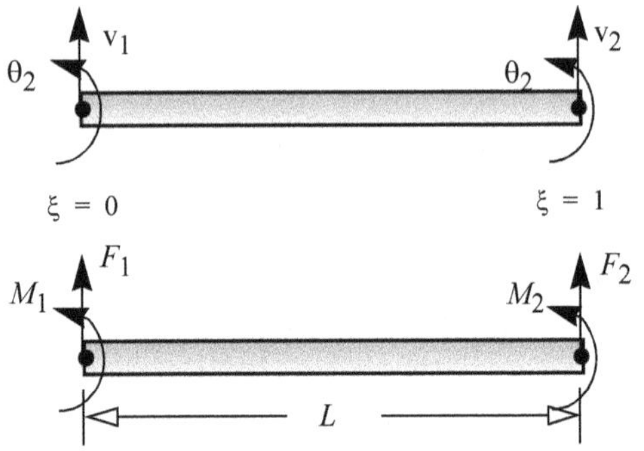


**Fig. P7.28**

**7.29** An elastic plane of thickness *h* is in plane stress and subjected to body forces $F_x(x, y)$ and $F_y(x, y)$. Approximate the body forces, the material constants, and thickness with constant values that exists at the centroid of the triangle.For a triangular element, using linear Lagrange polynomial, obtain the second row of the element stiffness matrix and the right hand side vector.

# 8 Indicial Notation

## LEARNING OBJECTIVES

1. Understand indicial notation and the rules of manipulating the indices.
2. Understand the applications of indicial notation to mechanics of materials equations.

Indicial notation is a compact way of writing and deriving equations. Writing a direction of a vector or components of a matrix using subscripts is using indicial notation in its most basic form. But manipulating these subscripts or indices to derive equations requires learning rules that have evolved over a period of time. Once learned, it becomes a challenge to work with equations in traditional form of expanded notation. Indicial notation also forms the basic language of tensor calculus.

In this chapter we first establish the notation and the rules for manipulating the indicies. We cast and derive the familiar equations of classical plate theory of Chapter 5 and equations of elasticity of Chapter 6 in indicial notation. We then cast two dimensional equations of variational calculus of Chapter 7 in indicial notation and use the equations to derive boundary value problems. We then study large strain and incorporate it in plate theory and elasticity.

## 8.1 BASIC DEFINITIONS

We start with two familiar concepts of work and moment and show the compactness of the equations that can be achieved using indicial notation.

A force vector and displacement vector in Cartesian coordinates may be written as $\vec{F} = \{F_x, F_y, F_z\}$ and $\overrightarrow{\Delta u} = \{\Delta u_x, \Delta u_y, \Delta u_z\}$. The work done ($W$) by the force $\vec{F}$ in moving a point by an amount $\overrightarrow{\Delta u}$ is given by the scalar (dot) product shown below.

$$W = \vec{F} \bullet \overrightarrow{\Delta u} = F_x \Delta u_x + F_y \Delta u_y + F_z \Delta u_z \qquad \textbf{(8.1a)}$$

The moment vector $\vec{M} = \{M_x, M_y, M_z\}$ due to the force at a point from which a position vector is drawn $\vec{r} = \{r_x, r_y, r_z\}$ is given by the vector (cross) product as

$$\vec{M} = \vec{r} \times \vec{F} = \{r_y F_z - r_z F_y, r_z F_x - r_x F_z, r_x F_y - r_y F_x\} \qquad \textbf{(8.1b)}$$

The above equations are said to be written in expanded form. In indicial notation we replace the subscripts $x, y, z$ with 1, 2, and 3, respectively. The above equations can be written as

$$W = \sum_{i=1}^{3} F_i \Delta u_i \qquad \textbf{(8.1c)}$$

$$M_1 = r_2 F_3 - r_3 F_1 \qquad M_2 = r_3 F_1 - r_1 F_3 \qquad M_3 = r_1 F_2 - r_2 F_1 \qquad \textbf{(8.1d)}$$

### 8.1.1 Summation convention

In the Equation (8.1c), $i$ could be replaced by $j$ or $k$ and it would not matter. Thus, an index used for purpose of summation is called the **dummy index**. A further compaction of notation is achieved by not writing the summation sign with the understanding that a *repeated index implies summation*. With this understanding, Equation (8.1c) is written as

$$W = F_i \Delta u_i \qquad \textbf{(8.2)}$$

Furthermore, we will write a vector just as $F_i$ or $F_j$ or $F_k$. The indicies *i, j, or k* are referred to as **free index**. One of the rules of manipulation in indicies is that the *free indicies in every term of an equation must be the same*. This is analogous to the requirement that every term in an equation must have the same dimension, and is a very useful rule in preventing mistakes in the derivations using indicial notation.

Sometimes there are exceptions made to the rule of free and dummy index usage. In such cases it is *explicitly* declared that an index is not a free index or summation is not carried over a dummy index.

A quantity with multiple indices will be referred to as component of a matrix, which is more familiar then calling them tensors. Thus, $A_{ij}$ and $A_{ijk}$ are matrices. The size of the matrix will be determined by the values each of the indices can take. If *i*, *j*, and *k* can only be 1 (*x*) and 2 (*y*) values, then $A_{ij}$ will be 2 x2 matrix and $A_{ijk}$ will be 2 x2 x2 matrix. If *i*, *j*, and *k* can be 1 (*x*) 2 (*y*) and 3 (*z*) values, then $A_{ij}$ will be 3 x 3 matrix and $A_{ijk}$ will be 3 x 3 x 3 matrix. We record the two rules for future use.

- *repeated index implies summation* unless it is explicitly stated otherwise.
- *free indices in every term of an equation must be the same* unless explicitly stated that an index is not free and is being used symbolically.

A free index can be replaced by another free index and a dummy index can be replaced by another dummy index. As we shall see this is done in the same equation from one step to next during the algebraic manipulation of equations.

### 8.1.2 Kronecker δ function

Kronecker delta function or just delta function is defined as

$$\boxed{\delta_{ij} = \begin{cases} 1 & i = j \\ 0 & i \neq j \end{cases}} \qquad \textbf{(8.3a)}$$

Thus, Kronecker delta function is an identity matrix. Consider the product of a vector with Kronecker delta function,

$$\delta_{ij}F_j = \delta_{i1}F_1 + \delta_{i2}F_2 + \delta_{i3}F_3 = \begin{cases} F_1 & i = 1 \\ F_2 & i = 2 \text{ or} \\ F_3 & i = 3 \end{cases}$$

$$\boxed{\delta_{ij}F_j = F_i} \qquad \textbf{(8.3b)}$$

The multiplication changed the index from *j* to *i*. This substitution of index is equally true for a product of Kronecker delta function with any matrix component. Thus, $\delta_{ij}A_{jk} = A_{ik}$ and $\delta_{ij}A_{jkm} = A_{ikm}$. Another useful observation is recorded below

$$\delta_{ii} = \delta_{11} + \delta_{22} = 2 \quad \text{in two dimension} \quad \text{and} \quad \delta_{ii} = \delta_{11} + \delta_{22} + \delta_{33} = 3 \quad \text{in three dimension} \qquad \textbf{(8.3c)}$$

The product of the delta function with a function that is anti-symmetric in its subscript will result in a zero value as shown below.

$$\boxed{\delta_{jk}A_{jk} = 0 \qquad \text{IF } A_{jk} = -A_{kj}} \qquad \textbf{(8.3d)}$$

### 8.1.3 Permutation function

The permutation function is defined as

$$\boxed{e_{ijk} = \begin{cases} 0 & \text{if any } i, j, \text{ or } k \text{ are same,} \\ 1 & \text{if } ijk \text{ are in cyclic order,} \\ -1 & \text{if } ijk \text{ are noncyclic order.} \end{cases}} \qquad \textbf{(8.4)}$$

By cyclic order we have $e_{123} = e_{231} = e_{321} = 1$. By non-cyclic order we have $e_{213} = e_{321} = e_{132} = -1$. The moment components in Equation (8.1d) can now be written in the compact form as

$$M_i = e_{ijk}r_jF_k \qquad \textbf{(8.5)}$$

In the above equation *i* is the free index and *j* and *k* are the dummy indices over which we have to sum. Consider now $M_1$. With *i* = 1, the non-zero values of permutation function will be for *j* and *k* with values of 2 and 3. We thus obtain

$$M_1 = e_{123}r_2F_3 + e_{132}r_3F_2 = r_2F_3 - r_3F_1$$

In a similar manner we can show other components of moment from Equation (6.6) are same as those given in Equation (8.1d). We define a two-dimensional permutation function as

$$\boxed{e_{ij} = \begin{cases} 0 & \text{if any } i = j \\ 1 & \text{if } ij \text{ are in cyclic order,} \\ -1 & \text{if } ij \text{ are noncyclic order.} \end{cases}} \qquad \textbf{(8.6)}$$

The following identities between permutation symbol and the delta function can be proven by substituting the values of 1, 2, and 3.

$$\boxed{e_{ijk}e_{imn} = \delta_{jm}\delta_{kn} - \delta_{jn}\delta_{km} \qquad e_{km}e_{nm} = \delta_{km}} \qquad \textbf{(8.7)}$$

The product of the permutation function with a function that is symmetric in its subscript will result in a zero value as shown below.

$$e_{ijk}A_{jk} = 0 \qquad \text{and} \qquad e_{jk}A_{jk} = 0 \qquad \text{IF } A_{jk} = A_{kj} \tag{8.8}$$

### 8.1.4 Derivative notation

The partial derivative of a function with respect to a coordinate is shown by a comma followed by the index of the direction as shown below.

$$\frac{\partial u_i}{\partial x_j} = u_{i,j} \tag{8.9}$$

We have the basic tools for use of indicial notation in our familiar equations.

## 8.2 EQUATIONS OF ELASTICITY AND THIN PLATE IN INDICIAL NOTATION

In this section we cast the equations of classical plate theory of Chapter 5 and equations of elasticity of Chapter 6 in indicial notation. The variables used in the equations were defined in Chapter 5 and Chapter 6.

In Chapter 1, we observed that

tensor normal strains = engineering normal strains        tensor shear strains = (engineering shear strains) / 2

In this chapter we will use *tensor strain definition* as is the general convention in indicial notation. The strain-displacement relationships of Equation (6.1) for tensor strains can be written as

$$\varepsilon_{ij} = (u_{i,j} + u_{j,i})/2 \tag{8.10a}$$

The Generalized Hooke's law of Equation (1.42) relating stresses ($\sigma_{ij}$) to tensor strain can be written as

$$\varepsilon_{ij} = [(1+\nu)\sigma_{ij} - \nu\delta_{ij}\sigma_{kk}]/E \tag{8.10b}$$

The equilibrium equations relating stresses to body forces ($F_i$) of Equations (6.8d) through (6.8d) can be written as

$$\sigma_{ij,j} + F_i = 0 \tag{8.10c}$$

The stress vector, that is traction of Equation (1.10), can be written as

$$S_i = \sigma_{ij}n_j \tag{8.10d}$$

The strain energy density of Equation (7.13), using tensor strain can be written as

$$U_0 = \sigma_{ij}\varepsilon_{ij}/2 \tag{8.10e}$$

**EXAMPLE 8.1**

Starting with Generalized Hooke's law, obtain stresses in terms of strain in indicial notation in three dimensions.

**PLAN**

We find $\sigma_{kk}$ and then obtain $\sigma_{ij}$ from Equation (8.10b).

**SOLUTION**

In Equation (8.10b) we substitute $i = j$ to obtain

$$\varepsilon_{jj} = [(1+\nu)\sigma_{jj} - \nu\delta_{jj}\sigma_{kk}]/E \tag{E1}$$

Noting that in three dimension $\delta_{jj} = 3$ and $j$ and $k$ are both dummy index, we obtain

$$\varepsilon_{jj} = [(1+\nu)\sigma_{jj} - 3\nu\sigma_{jj}]/E = (1-2\nu)\sigma_{jj}/E \qquad \text{or} \qquad \sigma_{kk} = E\varepsilon_{kk}/(1-2\nu) \tag{E2}$$

Substituting $\sigma_{kk}$ in Equation (E1) we obtain the desired result as shown below.

$$E\varepsilon_{ij} = (1+\nu)\sigma_{ij} - \nu\delta_{ij}\left(\frac{E\varepsilon_{kk}}{1-2\nu}\right) \qquad \text{or} \qquad (1+\nu)\sigma_{ij} = E\varepsilon_{ij} + \frac{E\nu}{(1-2\nu)}\delta_{ij}\varepsilon_{kk} \tag{E3}$$

$$\textbf{ANS. } \sigma_{ij} = \frac{E}{(1+\nu)}\varepsilon_{ij} + \frac{E\nu}{(1-2\nu)(1+\nu)}\delta_{ij}\varepsilon_{kk} = 2G\left[\varepsilon_{ij} + \frac{\nu}{(1-2\nu)}\delta_{ij}\varepsilon_{kk}\right] \tag{8.11a}$$

**COMMENTS**

1. Lame's constant is defined as

$$\lambda = \frac{E\nu}{(1-2\nu)(1+\nu)} \tag{E4}$$

and noting the definition of shear modulus *G*, the preferred form of the solution is

$$\sigma_{ij} = 2G\varepsilon_{ij} + \lambda\delta_{ij}\varepsilon_{kk} \tag{8.11b}$$

2. Equation (8.10b) represents three equations. Equation (E2) represents the sum of the three equations. We would have to substitute the sum in the three equations represented by Equation (8.10b) to obtain the three equations of Equation (8.11a).

## EXAMPLE 8.2

Starting with the displacement equation and assuming all assumptions are valid in classical plate theory, obtain equations in indicial notation for strain, stresses, stress resultants, equilibrium equations, differential equation, and strain energy.

**PLAN**

We start with the displacement expression in Equation (5.2) and follow the logic outlined in Section 5.2 to obtain the required quantities.

**SOLUTION**

1. *Displacements*: From Equation (5.2) the displacements in indicial notation can be written as

$$u_i \approx -zw_{,i} \qquad w \approx w(x_1, x_2) \tag{8.12a}$$

2. *Strains*: From Equation (8.10a), the tensor strains are

$$\varepsilon_{ij} = \frac{1}{2}(u_{i,j} + u_{j,i}) = -zw_{,ij} \tag{8.12b}$$

3. *Stresses*: In Equation (8.10b), the indicies $i$ and $j$ can only take the values of 1 and 2 for plane stress state. Letting $i = j$ in Equation (8.10b) and noting that in plane stress $\delta_{jj} = 2$ we obtain

$$\varepsilon_{jj} = [(1+\nu)\sigma_{jj} - 2\nu\sigma_{jj}]/E \qquad \text{or} \qquad E\varepsilon_{kk} = (1-\nu)\sigma_{kk} \qquad \text{or} \qquad \sigma_{kk} = E\varepsilon_{kk}/(1-\nu)$$

$$E\varepsilon_{ij} = (1+\nu)\sigma_{ij} - \nu\delta_{ij}\left(\frac{E\varepsilon_{kk}}{1-\nu}\right) \qquad \text{or} \qquad (1+\nu)\sigma_{ij} = E\varepsilon_{ij} + \frac{E\nu}{(1-\nu)}\delta_{ij}\varepsilon_{kk}$$

$$\sigma_{ij} = \frac{E}{(1-\nu^2)}[(1-\nu)\varepsilon_{ij} + \nu\delta_{ij}\varepsilon_{kk}] \tag{8.12c}$$

Substituting Equation (8.12b) into Equation (8.12c) we obtain

$$\sigma_{ij} = -\frac{Ez}{(1-\nu^2)}[(1-\nu)w_{,ij} + \nu\delta_{ij}w_{,kk}] \tag{8.12d}$$

4. *Stress resultants*: The moment and shear force resultants of Equations (5.8) and (5.9) can be written as

$$\boldsymbol{m}_{ij} = \int_h z\sigma_{ij}dz \qquad \boldsymbol{q}_i = \int_h \tau_{iz}dz \tag{8.12e}$$

Substituting Equation (8.12d) into Equation (8.12e) we obtain

$$\boldsymbol{m}_{ij} = -\int_h \frac{Ez^2}{(1-\nu^2)}[(1-\nu)w_{,ij} + \nu\delta_{ij}w_{,kk}]dz = -\frac{E}{(1-\nu^2)}[(1-\nu)w_{,ij} + \nu\delta_{ij}w_{,kk}]\int_h z^2dz$$

$$\boldsymbol{m}_{ij} = -D[(1-\nu)w_{,ij} + \nu\delta_{ij}w_{,kk}] \tag{8.12f}$$

5. *Equilibrium equations*: The equilibrium equations (5.17a), (5.17b), and (5.17c) can be written as

$$\boldsymbol{q}_{i,i} = -p_z(x_1, x_2) \qquad \boldsymbol{q}_i = \boldsymbol{m}_{ji,j} \tag{8.12g}$$

6. *Differential equation*: Substituting Equation (8.12f) into Equation (8.12g) we obtain

$$\boldsymbol{q}_i = -D[(1-\nu)w_{,ij} + \nu\delta_{ij}w_{,kk}]_{,j} = -D[(1-\nu)w_{,ikk} + \nu w_{,kki}] = -Dw_{,kki} = -D(\nabla^2 w)_{,i} \tag{8.12h}$$

$$[-Dw_{,kki}]_{,i} = -p_z(x_1, x_2) \qquad \text{or} \qquad Dw_{,kkii} = D(\nabla^4 w) = p_z(x_1, x_2) \tag{8.12i}$$

7. *Strain Energy*: Substituting Equation (8.12a) and Equation (8.12c) into Equation (8.10e) and integrating over the volume of the plate we obtain

$$U = \frac{1}{2}\iint_A\left[\int_h \sigma_{ij}\varepsilon_{ij}dz\right]dx_1dx_2 = \frac{1}{2}\iint_R\left[\int_h \frac{Ez}{(1-\nu^2)}[(1-\nu)w_{,ij} + \nu\delta_{ij}w_{,kk}]zw_{,ij}dz\right]dx_1dx_2$$

$$U = \frac{1}{2}\iint_A\left[\frac{E}{(1-\nu^2)}[(1-\nu)w_{,ij} + \nu\delta_{ij}w_{,kk}]w_{,ij}\int_h z^2dz\right]dx_1dx_2 = \frac{1}{2}\iint_A D[(1-\nu)w_{,ij}w_{,ij} + \nu\delta_{ij}w_{,ij}w_{,kk}]dx_1dx_2$$

$$U = \frac{1}{2}\iint_A D[(1-\nu)w_{,ij}w_{,ij} + \nu w_{,ii}w_{,kk}]dx_1dx_2 \tag{8.12j}$$

**COMMENTS**

1. This example highlights the compactness in deriving plate equations using indicial notation.
2. Note $\nabla^2 w = w_{,kk}$ and $\nabla^4 w = w_{,kkii}$ in Equations (8.12h) and (8.12i).
3. In this example we obtained the differential equation by first casting the equilibrium equations (5.17a), (5.17b), and (5.17c) into indicial notation. In Example 8.5, we obtain the boundary value problem using variational calculus.

**EXAMPLE 8.3**

Obtain the differential equation governing displacements in a three dimensional elastic body subjected to body forces $F_i$.

**PLAN**

We substitute Equation (8.10a) into Equation (8.11b) and the result then into Equation (8.10c) to obtain the differential equations.

**SOLUTION**

Substituting Equation (8.10a) into Equation (8.11b), we obtain

$$\sigma_{ij} = G(u_{i,j} + u_{j,i}) + \lambda\delta_{ij}(u_{k,k} + u_{kk})/2 = G(u_{i,j} + u_{j,i}) + \lambda\delta_{ij}u_{k,k} \quad \text{(E1)}$$

Substituting the above equation into Equation (8.10c) we obtain

$$G(u_{i,jj} + u_{j,ij}) + \lambda\delta_{ij}u_{k,kj} + F_i = 0 \quad \text{(E2)}$$

Noting the substituting property of Kronecker delta we obtain

$$G(u_{i,jj} + u_{j,ij}) + \lambda u_{k,ki} + F_i = 0 \quad \text{(E3)}$$

Replacing the dummy index $j$ with $k$ we obtain

$$G(u_{i,kk} + u_{k,ki}) + \lambda u_{k,ki} + F_i = 0 \text{ or} \quad \text{(E4)}$$

$$\textbf{ANS. } Gu_{i,kk} + (G+\lambda)u_{k,ki} + F_i = 0 \quad \text{(E5)}$$

**COMMENTS**

1. Note that each term in Equation (E1) had the free subscripts $i$ and $j$. In Equations (E3) through (E5), each term had free subscripts $i$.
2. Were we to do the derivation in long hand, each of the above equations would be three equations, with each term representing 3 terms. This is the power and simplicity of indicial notation.

## 8.3 FUNDAMENTAL SOLUTIONS

In Chapter 3, we introduced the concept of fundamental solutions associated with point singularities. In elasticity many fundamental solutions are available for two and three dimensional problems. These solutions provide us with a rich set of problems for understanding the use of indicial notations in derivations of mechanics problems—the sole purpose of introducing the topic in this chapter.

In Sections 3.1.2 and 3.1.3, we saw that point force was a source singularity and point moment was a doublet singularity. In a similar manner there is a source singularity called dislocation and a doublet singularity called displacement discontinuity. Figure 8.1 shows a dislocation. Two adjoining points along a line displace relative to one another in a **dislocation**. Since displacement is a vector, the intensity of dislocation has components which we will label as $D_x$ and $D_y$ or in indicial notation as $D_i$. We can have point strain singularity called **inclusion singularity** and point stress singularity called **inhomogeneity singularity**. In each case the displacement field due to the singularity will be given and we will calculate the strain and stress from it using indicial notation.

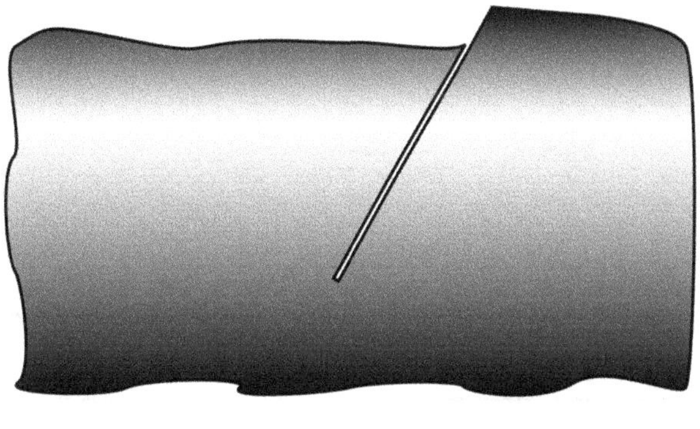

**Figure 8.1** Dislocation singularity.

For sake of simplicity, we shall assume that the singularity is located at the origin. The field point for evaluating displacements, strains, or stresses is located at a radial distance $r$ with components $r_i$ as shown in Figure 8.2. The following identities can be easily established.

$$r^2 = r_i r_i \qquad r_{i,j} = \delta_{ij} \qquad r_{,j} = r_j/r \qquad [ln(r)]_{,j} = r_j/r^2 \quad \textbf{(8.13a)}$$

From Figure 8.2, we can write the angle $\theta$ and its derivatives with respect to each coordinate and then write the derivatives using indicial notation as shown below.

$$\theta = atan\left(\frac{r_2}{r_1}\right) \qquad \frac{\partial\theta}{\partial r_1} = -\frac{r_2}{r^2} \qquad \frac{\partial\theta}{\partial r_2} = \frac{r_1}{r^2} \qquad \theta_{,j} = e_{mj}\frac{r_m}{r^2} \quad \textbf{(8.13b)}$$

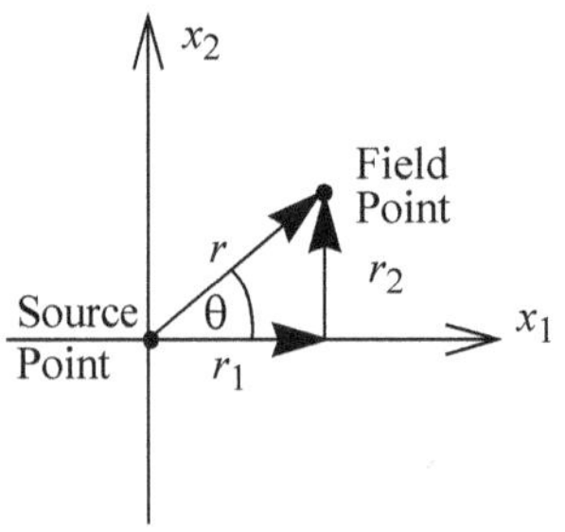


**Figure 8.2** Notation for use in fundamental solutions.

## EXAMPLE 8.4

The displacement field due to a dislocation in an elastic body in plane strain is given by the equation below. Obtain the strain invariant $\varepsilon_{ii}$.

$$u_i = \frac{1}{8\pi(1-\nu)}\left[4(1-\nu)\delta_{ik}\theta - 2(1-2\nu)e_{ki}ln(r) + 2e_{kn}\frac{r_i r_n}{r^2}\right]D_k \quad \text{(E1)}$$

where $D_k$ is the intensity of dislocation.

**PLAN**

We can obtain the $u_{i,j}$ from the given displacements. The strain invariant $\varepsilon_{ii}$ is equal to $u_{i,i}$ and can be obtained from solution of part *a*.

**SOLUTION**

We write the derivative of displacements using Equations (8.13a) and (8.13b) as shown below.

$$u_{i,j} = \frac{1}{8\pi(1-\nu)}\left[4(1-\nu)\delta_{ik}\left(e_{mj}\frac{r_m}{r^2}\right) - 2(1-2\nu)e_{ki}\left(\frac{r_j}{r^2}\right) + 2e_{kn}\left\{\frac{\delta_{ij}r_n + \delta_{nj}r_i}{r^2} - \frac{2r_i r_n r_j}{r^4}\right\}\right]D_k \quad \text{(E2)}$$

Substituting $i = j$ into the above equation we obtain

$$\varepsilon_{ii} = u_{i,i} = \frac{1}{8\pi(1-\nu)}\left[4(1-\nu)\delta_{ik}\left(e_{mi}\frac{r_m}{r^2}\right) - 2(1-2\nu)e_{ki}\left(\frac{r_i}{r^2}\right) + 2e_{kn}\left\{\frac{\delta_{ii}r_n + \delta_{ni}r_i}{r^2} - \frac{2r_i r_n r_i}{r^4}\right\}\right]D_k \quad \text{(E3)}$$

We use the substituting property of $\delta_{ik}$, the fact $\delta_{ii} = 2$ in two dimension, $r_i r_i = r^2$ and one dummy index can be replaced by another dummy index to obtain the following

$$\varepsilon_{ii} = \frac{1}{8\pi(1-\nu)}\left[4(1-\nu)\left(e_{mk}\frac{r_m}{r^2}\right) - 2(1-2\nu)e_{km}\left(\frac{r_m}{r^2}\right) + 2e_{km}\left\{\frac{2r_m + r_m}{r^2} - \frac{2r_m}{r^2}\right\}\right]D_k \quad \text{(E4)}$$

Noting that $e_{km} = -e_{mk}$ we obtain

$$\varepsilon_{ii} = \frac{1}{8\pi(1-\nu)}[4(1-\nu) + 2(1-2\nu) - 2]\left(e_{mk}\frac{r_m}{r^2}\right)D_k \quad \text{(E5)}$$

$$\textbf{ANS. } \varepsilon_{ii} = \frac{(1-2\nu)}{2\pi(1-\nu)}\left(e_{mk}\frac{r_m}{r^2}\right)D_k \quad \text{(E6)}$$

**COMMENTS**

1. The new features in taking derivatives of fundamental solutions are Equations (8.13a) and (8.13b).
2. From Equation (E2) we could find the strain $\varepsilon_{ij}$ and using Equation (8.11a) for plane strain we can obtain stresses $\sigma_{ij}$.

## 8.4 VARIATIONAL CALCULUS

In Section 7.8, we derived the boundary value problem by considering the stationary value of area integrals. In this section we write the results using indicial notation. The variables used in the equations below are defined in Section 7.8. We saw that we will have to take derivatives of the functional $H$ with respect to derivatives of $u$. The rule below must be followed.

- If $H$ is defined in terms of $u_{,i}$ then take the derivative $\partial H/\partial u_{,k}$ and use the fact that $\partial u_{,i}/\partial u_{,k} = \delta_{ik}$.

### 8.4.1 Stationary value of a functional with first order derivatives

We define the functional of Equation (7.43a) using indicial notation as shown below.

$$I(u) = \iint_A H(u_{,x_1}, u_{,x_2}, u, x_1, x_2)dx_1 dx_2 \quad \textbf{(8.14a)}$$

The associated boundary value problem given by Equations (7.44a) and (7.44b) can be written as

$$\text{Differential Equation:} \quad \frac{\partial H}{\partial u} - \left(\frac{\partial H}{\partial u_{,i}}\right)_{,i} = 0 \qquad x_k \text{ in } A \tag{8.15a}$$

$$\text{Boundary Conditions:} \quad \frac{\partial H}{\partial u_{,i}} n_i = 0 \qquad \text{or} \qquad \delta u_j = 0 \qquad x_k \text{ on } \Gamma \tag{8.15b}$$

## 8.4.2 Stationary value of a functional with second order derivatives

We define the functional of Equation (7.45) using indicial notation as shown below.

$$I(u) = \iint_A H(u_{,x_1x_1}, u_{,x_1x_2}, u_{,x_2x_2}, u_{,x_1}, u_{,x_2}, u, x_1, x_2)dx_1dx_2 \tag{8.16a}$$

We write the terms defined in Equations (7.52) and (7.54) in indicial notation as shown below.

$$\boldsymbol{m}_{nn} = \left(\frac{\partial H}{\partial u_{,ij}}\right) n_i n_j \qquad \boldsymbol{m}_{nt} = \left(\frac{\partial H}{\partial u_{,ij}}\right) e_{mi} n_m n_j \qquad -\boldsymbol{q}_i = \left(\frac{\partial H}{\partial u_{,i}}\right) - \left(\frac{\partial H}{\partial u_{,ij}}\right)_{,j} \tag{8.16b}$$

where, $e_{mi}$ is the permutation function defined in Equation (8.6).

The associated boundary value problem given by Equations (7.56a), (7.56b), and (7.56c) can be written as

$$\text{Differential Equation:} \quad \boldsymbol{q}_{i,\,i} + \frac{\partial H}{\partial u} = 0 \qquad x_k \text{ in } A \tag{8.17a}$$

$$\text{Boundary Conditions:} \quad \left.\begin{matrix} \boldsymbol{m}_{nn} = 0 & \text{or} & \delta\left(\frac{\partial u}{\partial n}\right) = 0 \\ & \text{and} & \\ \frac{\partial \boldsymbol{m}_{nt}}{\partial s} + \boldsymbol{q}_i n_i = 0 & \text{or} & \delta u = 0 \end{matrix}\right\} \quad x_k \text{ on } \Gamma \tag{8.17b}$$

$$\text{Corner Condition:} \quad (\Delta \boldsymbol{m}_{nt})_C = 0 \qquad \text{or} \qquad \delta u_C \qquad x_k \text{ on corner } \Gamma_C \tag{8.17c}$$

### EXAMPLE 8.5

Obtain the boundary value problem for thin plate bending with only transverse distributed forces using indicial notation and variational calculus.

**PLAN**

We can write the potential energy expression in indicial notation using Equation (8.12j). We can then use Equations (8.17a), (8.17b), and (8.17c) to write the boundary value problem.

**SOLUTION**

The potential energy for the plate can be written using Equation (8.12j) as shown below.

$$\Omega(w) = \frac{1}{2}\iint_A D[(1-\nu)w_{,ij}w_{,ij} + \nu w_{,ii}w_{,kk}]dx_1dx_2 - \iint_A p_z(x_1, x_2)w(x_1, x_2)dx_1dx_2 \tag{E1}$$

The functional in Equation (8.16a) can be written as

$$H(w_{,x_1x_1}, w_{,x_1x_2}, w_{,x_2x_2}, w_{,x_1}, w_{,x_2}, w, x_1, x_2) = \frac{D}{2}[(1-\nu)w_{,ij}w_{,ij} + \nu w_{,nn}w_{,kk}] - p_z w \tag{E2}$$

The derivatives of the functional $H$ can be written as shown below.

$$\frac{\partial H}{\partial w} = -p_z \qquad \frac{\partial H}{\partial w_{,l}} = 0 \qquad \frac{\partial H}{\partial w_{,lm}} = \frac{D}{2}\left[2(1-\nu)w_{,ij}\frac{\partial w_{,ij}}{\partial w_{,lm}} + 2\nu\frac{\partial w_{,nn}}{\partial w_{,lm}}w_{,kk}\right] = D[(1-\nu)w_{,lm} + \nu\delta_{lm}w_{,kk}] = -\boldsymbol{m}_{lm} \tag{E3}$$

In the above equation we made use of the fact $\partial w_{,ij}/\partial w_{,lm} = \delta_{il}\delta_{jm}$, $w_{,ij}\delta_{il}\delta_{jm} = w_{,lm}$ and $\partial w_{,nn}/\partial w_{,lm} = \delta_{nl}\delta_{nm} = \delta_{lm}$. From Equation (8.16b) and changing the free indicies from $l$ and $m$ to $i$ and $j$, we obtain the following equations.

$$\boldsymbol{m}_{nn} = \left(\frac{\partial H}{\partial w_{,ij}}\right) n_i n_j = -\boldsymbol{m}_{ij} n_i n_j \qquad \boldsymbol{m}_{nt} = \left(\frac{\partial H}{\partial w_{,ij}}\right) e_{mi} n_m n_j = -\boldsymbol{m}_{ij} e_{mi} n_m n_j \tag{E4}$$

$$-\boldsymbol{q}_i = \left(\frac{\partial H}{\partial w_{,i}}\right) - \left(\frac{\partial H}{\partial w_{,ij}}\right)_{,j} = -D[(1-\nu)w_{,ij} + \nu\delta_{ij}w_{,kk}]_{,j} = -D[(1-\nu)w_{,ijj} + \nu\delta_{ij}w_{,kkj}] = -D[(1-\nu)w_{,ijj} + \nu w_{,kki}] \text{ or}$$

$$\boldsymbol{q}_i = D w_{,kki} \tag{E5}$$

From Equation (8.17a) we obtain

$$q_{i,i} + \left(\frac{\partial H}{\partial w}\right) = 0 \text{ or} \tag{E6}$$

$$\textbf{Differential equation: } Dw_{,kkii} - p_z = 0 \quad \text{or} \quad Dw_{,kkii} = p_z \tag{E7}$$

From Equations (8.17b) and (8.17c) we obtain the boundary conditions as

$$\textbf{Boundary Conditions:} \quad \begin{matrix} m_{nn} = 0 & \text{or} & \delta\left(\frac{\partial w}{\partial n}\right) = 0 \\ & \text{and} & & x_k \text{ on } \Gamma \\ \frac{\partial m_{nt}}{\partial s} + q_i n_i = 0 & \text{or} & \delta w = 0 \end{matrix} \tag{E8}$$

$$\textbf{Corner Condition: } (\Delta m_{nt})_C = 0 \quad \text{or} \quad \delta w_C \quad x_k \text{ on corner } \Gamma_C \tag{E9}$$

**COMMENTS**

1. The use of indicial notation and variational calculus reduces the derivation of plate boundary value problem to half a page as shown in this example.
2. Though the derivation has become compact, it has also become very mathematical. To understand the meaning of various terms it may require expanding the terms in long notation and drawing figures.

## 8.5 NONLINEAR STRAINS

Our objective is to determine the change in the distance between two *infinitely* close points of a body as it deforms. In Section 1.5 we saw that there are two definitions of large strains. Lagrangian strain is computed from deformation by using the original *undeformed geometry* as the reference geometry. Eulerian strain is computed from deformation by using the *final deformed* geometry as the reference geometry. The Lagrangian description is usually used in solid mechanics. The Eulerian description is usually used in fluid mechanics. For small deformations it does not matter if we define strain using the initial undeformed geometry or the deformed geometry and the two definitions yield the same results. However, if deformations are large then the two definitions yield different analysis.

In this section we develop nonlinear measures of strains. Initially we use the expanded form to develop the concepts in two dimensions, then generalize it to three dimensions using indicial notation.

### 8.5.1 A perspective of reference frame

In this section we develop a perspective on the origin of geometric nonlinearity. We define the following:

$x$ and $y$ are the coordinates of a point in Cartesian system $(X, Y)$ on the *undeformed* body.

$\xi$ and $\eta$ are the coordinates of a point in Cartesian system $(X, Y)$ on the *deformed* body.

$u$ and v are the displacement components in the $x$ and $y$ direction.

The point $P$ with coordinates $x, y$ becomes point $P^*$ with coordinates $\xi$, $\eta$ due to deformation. We can write:

$$\xi = x + u(x, y) \qquad \eta = y + \mathrm{v}(x, y) \tag{8.18}$$

Consider three points $P_1$, $P_2$ and $P_3$ on a straight line $x = x_o$ as shown in Figure 8.3*a*. Equation (8.18) can be written as

$$\xi = x + u(x_o, y) \qquad \eta = y + \mathrm{v}(x_o, y) \tag{8.19}$$

Equation (8.19) represents a curve in the $\xi$, $\eta$ coordinate system if $u$ and v are not linear in $y$.

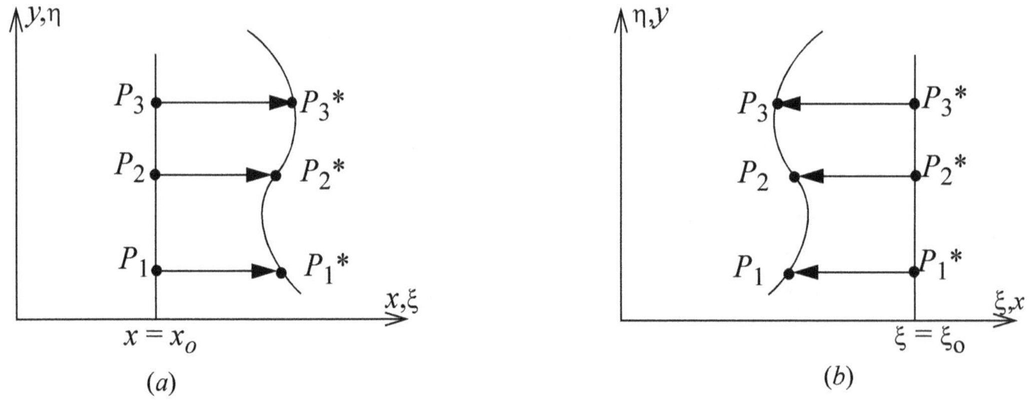


**Figure 8.3** Transformation of straight lines into curves due to deformation.

In a similar manner consider three points $P_1^*$, $P_2^*$ and $P_3^*$ on a straight line $\xi = \xi_o$ shown in Figure 8.3*b*. Equation (8.18) can be written as

$$\xi_o = x + u(x, y) \qquad \eta = y + v(x, y) \tag{8.20}$$

Equation (8.18) can be solved to yield

$$x = f_1(\xi_o, \eta) \qquad y = f_2(\xi_o, \eta) \tag{8.21}$$

Equation (8.21) represents a curve in the $x$, $y$ coordinates provided $f_1$ and $f_2$ are not a linear function of $\eta$.

We see that a straight line becomes a curve in the process of deformation. Thus our coordinate system of straight edges will become curvilinear coordinate system (shown with $\frown$) when we move from deformed to undeformed reference frame as shown in Figure 8.4. The derivatives in strain will thus have different forms in the deformed and undeformed states. In a similar manner the differential areas $dxdy$ [or $d\xi d\eta$ ], which is a rectangle will become a quadrilateral with curved sides $d\widehat{x}\,d\widehat{y}$ [or $d\widehat{\xi}\,d\widehat{\eta}$ ], impacting the form in which we describe our stress. The various measures (definitions) of stress and strain are an outcome of the reference frame we use to describe them.

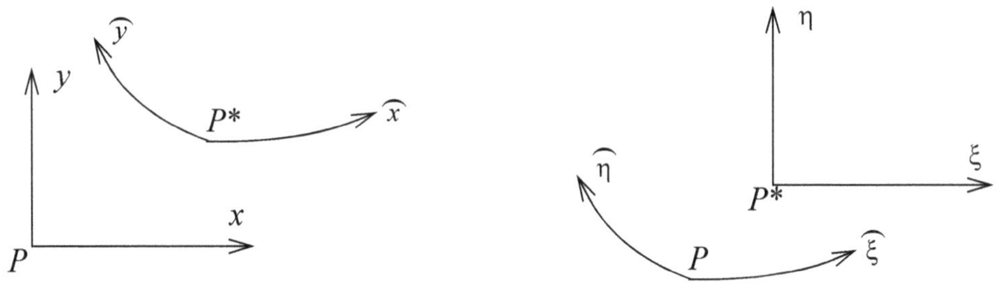


**Figure 8.4** Transformation of Cartesian coordinates to curvilinear coordinate system.

## 8.5.2 Lagrangian strain in two dimension

The point $P(x, y)$ deforms to point $P^*(\xi, \eta)$, where as the point $Q(x+dx, y+dy)$ deforms to $Q^*(\xi+d\xi, \eta+d\eta)$. Equation (8.18) can be written as

$$d\xi = dx + \frac{\partial u}{\partial x}dx + \frac{\partial u}{\partial y}dy = \left(1 + \frac{\partial u}{\partial x}\right)dx + \frac{\partial u}{\partial y}dy \qquad d\eta = dy + \frac{\partial v}{\partial x}dx + \frac{\partial v}{\partial y}dy = \frac{\partial v}{\partial x}dx + \left(1 + \frac{\partial v}{\partial y}\right)dy \tag{8.22}$$

The square of infinitesimal distance $PQ$ and $P^*Q^*$ can be written as:

$$ds^2 = dx^2 + dy^2 \qquad ds^{*2} = d\xi^2 + d\eta^2 \tag{8.23}$$

Substituting Equation (8.22) into Equation (8.23), we obtain

$$ds^{*2} - ds^2 = \left[\left(1 + \frac{\partial u}{\partial x}\right)^2 - 1 + \left(\frac{\partial v}{\partial x}\right)^2\right]dx^2 + \left[\left(1 + \frac{\partial v}{\partial y}\right)^2 - 1 + \left(\frac{\partial u}{\partial y}\right)^2\right]dy^2 + 2\left[\left(1 + \frac{\partial u}{\partial x}\right)\frac{\partial u}{\partial y} + \left(1 + \frac{\partial v}{\partial y}\right)\frac{\partial v}{\partial x}\right]dxdy$$

$$ds^{*2} - ds^2 = \left[2\frac{\partial u}{\partial x} + \left(\frac{\partial u}{\partial x}\right)^2 + \left(\frac{\partial v}{\partial x}\right)^2\right]dx^2 + \left[2\frac{\partial v}{\partial y} + \left(\frac{\partial v}{\partial y}\right)^2 + \left(\frac{\partial u}{\partial y}\right)^2\right]dy^2 + 4\left[\frac{\partial u}{\partial y} + \frac{\partial v}{\partial x} + \frac{\partial u}{\partial x}\frac{\partial u}{\partial y} + \frac{\partial v}{\partial y}\frac{\partial v}{\partial x}\right]dxdy$$

$$ds^{*2} - ds^2 = 2\varepsilon_{xx}dx^2 + 2\varepsilon_{yy}dy^2 + 2(\varepsilon_{xy} + \varepsilon_{yx})dxdy \tag{8.24}$$

$$\boxed{\varepsilon_{xx} = \frac{\partial u}{\partial x} + \frac{1}{2}\left[\left(\frac{\partial u}{\partial x}\right)^2 + \left(\frac{\partial v}{\partial x}\right)^2\right] \qquad \varepsilon_{yy} = \frac{\partial v}{\partial y} + \frac{1}{2}\left[\left(\frac{\partial v}{\partial y}\right)^2 + \left(\frac{\partial u}{\partial y}\right)^2\right] \qquad \varepsilon_{xy} = \varepsilon_{yx} = \frac{1}{2}\left[\frac{\partial u}{\partial y} + \frac{\partial v}{\partial x} + \frac{\partial u}{\partial x}\frac{\partial u}{\partial y} + \frac{\partial v}{\partial y}\frac{\partial v}{\partial x}\right]} \tag{8.25}$$

Consider the conventional definition of strain in the direction of $PQ$.

$$E_{PQ} = \frac{ds^* - ds}{ds} \qquad ds^* = (1 + E_{PQ})ds \tag{8.26}$$

Equation (8.24) can now be written as:

$$ds^{*2} - ds^2 = (ds^* - ds)(ds^* + ds) = E_{PQ}(2 + E_{PQ})ds = 2E_{PQ}(1 + E_{PQ}/2)ds \text{ or}$$

$$E_{PQ}(1 + E_{PQ}/2)ds^2 = \varepsilon_{xx}dx^2 + \varepsilon_{yy}dy^2 + \varepsilon_{xy}dxdy \tag{8.27}$$

Suppose we now consider a line element $PQ$ is in the $x$ direction ($E_{PQ}$=$E_{xx}$). In such a case d$s$ = d$x$, d$y$ = 0, and d$z$ = 0. From Equation (8.27) we obtain:

$$E_{xx}(1 + E_{xx}/2) = \varepsilon_{xx} \qquad or \qquad E_{xx} = \sqrt{1 + 2\varepsilon_{xx}} - 1 \tag{8.28}$$

If strains are small then by binomial expansion $E_{xx} \approx (1 + (2\varepsilon_{xx})/2 + \text{ higher order terms}) - 1 \approx \varepsilon_{xx}$. Thus, we see that $E_{xx}$ is analogous to our classical definition of strain. In a similar manner if we neglect the quadratic terms in Equations (8.25) through (8.25), we obtain the classical definition of small strain. Equations (8.25) through (8.25) are called the Lagrangian strain as we used the original geometry as the reference geometry.

### 8.5.3 Eulerian strain in two dimension

Once more the point $P(x, y)$ deforms to point $P^*(\xi, \eta)$, where as the point $Q(x+dx, y+dy)$ deforms to $Q^*(\xi+d\xi, \eta+d\eta)$. But now we use the deformed geometry as our reference geometry to describe the problem, that is, $u$ and v in Equation (8.18) are now functions of $\xi, \eta$. We rewrite Equation (8.18) as

$$\xi = x + u(\xi,\eta) \qquad \eta = y + \mathrm{v}(\xi,\eta) \qquad \text{or} \qquad x = \xi - u(\xi,\eta) \qquad y = \eta - \mathrm{v}(\xi,\eta) \tag{8.29}$$

We now consider the infinitesimal lengths as:

$$dx = d\xi - \left(\frac{\partial u}{\partial \xi}d\xi + \frac{\partial u}{\partial \eta}d\eta\right) = \left(1 - \frac{\partial u}{\partial \xi}\right)d\xi - \frac{\partial u}{\partial \eta}d\eta \qquad dy = d\eta - \left(\frac{\partial \mathrm{v}}{\partial \xi}d\xi + \frac{\partial \mathrm{v}}{\partial \eta}d\eta\right) = -\frac{\partial \mathrm{v}}{\partial \xi}d\xi + \left(1 - \frac{\partial \mathrm{v}}{\partial \eta}\right)d\eta \tag{8.30}$$

The square of infinitesimal distance $PQ$ and $P^*Q^*$ can be written as in Equation (8.24). We now write:

$$ds^{*2} - ds^2 = \left[1 - \left(1 - \frac{\partial u}{\partial \xi}\right)^2 - \left(\frac{\partial \mathrm{v}}{\partial \xi}\right)^2\right]d\xi^2 + \left[1 - \left(1 - \frac{\partial \mathrm{v}}{\partial \eta}\right)^2 - \left(\frac{\partial u}{\partial \eta}\right)^2\right]dy^2 + 2\left[\left(1 - \frac{\partial u}{\partial \xi}\right)\frac{\partial u}{\partial \eta} + \left(1 - \frac{\partial \mathrm{v}}{\partial \eta}\right)\frac{\partial \mathrm{v}}{\partial \xi}\right]d\xi d\eta$$

$$ds^{*2} - ds^2 = \left[\frac{\partial u}{\partial \xi} - \left(\frac{\partial u}{\partial \xi}\right)^2 - \left(\frac{\partial \mathrm{v}}{\partial \xi}\right)^2\right]d\xi^2 + \left[\frac{\partial \mathrm{v}}{\partial \eta} - \left(\frac{\partial \mathrm{v}}{\partial \eta}\right)^2 - \left(\frac{\partial u}{\partial \eta}\right)^2\right]dy^2 + 2\left[\frac{\partial u}{\partial \eta} + \frac{\partial \mathrm{v}}{\partial \xi} - \frac{\partial u}{\partial \xi}\frac{\partial u}{\partial \eta} - \frac{\partial \mathrm{v}}{\partial \eta}\frac{\partial \mathrm{v}}{\partial \xi}\right]d\xi d\eta$$

$$ds^{*2} - ds^2 = 2E_{\xi\xi}d\xi^2 + 2E_{\eta\eta}d\eta^2 + 2(E_{\xi\eta} + E_{\eta\xi})d\xi d\eta \tag{8.31}$$

$$\boxed{E_{\xi\xi} = \frac{\partial u}{\partial \xi} - \frac{1}{2}\left[\left(\frac{\partial u}{\partial \xi}\right)^2 + \left(\frac{\partial \mathrm{v}}{\partial \xi}\right)^2\right] \qquad E_{\eta\eta} = \frac{\partial \mathrm{v}}{\partial \eta} - \frac{1}{2}\left[\left(\frac{\partial \mathrm{v}}{\partial \eta}\right)^2 + \left(\frac{\partial u}{\partial \eta}\right)^2\right] \qquad E_{\xi\eta} = E_{\eta\xi} = \frac{1}{2}\left[\left(\frac{\partial u}{\partial \eta} + \frac{\partial \mathrm{v}}{\partial \eta}\right) - \frac{\partial u}{\partial \xi}\frac{\partial u}{\partial \eta} - \frac{\partial \mathrm{v}}{\partial \eta}\frac{\partial \mathrm{v}}{\partial \xi}\right]} \tag{8.32}$$

### 8.5.4 Indicial notation and non-linear strain in three dimension

We represent $x_i$ and $\xi_i$ as the coordinates in the undeformed and deformed state and $u_i$ the displacement vector. Here,

$x_1$, $x_2$, and $x_3$ represent $x$, $y$, and $z$, respectively.

$\xi_1$, $\xi_2$, and $\xi_3$ represent $\xi$, $\eta$, and $\zeta$, respectively.

$u_1$, $u_2$, and $u_3$ represent $u$, v, and $w$, respectively.

Equations (8.25) and (8.32) can be written as

$$\boxed{\varepsilon_{ij} = \frac{1}{2}\left[\frac{\partial u_i}{\partial x_j} + \frac{\partial u_j}{\partial x_i} + \frac{\partial u_k \partial u_k}{\partial x_i \partial x_j}\right] = \frac{1}{2}[u_{i,j} + u_{j,i} + u_{k,i}u_{k,j}] \qquad i,j = x, y, z} \tag{8.33}$$

$$\boxed{E_{ij} = \frac{1}{2}\left[\frac{\partial u_i}{\partial \xi_j} + \frac{\partial u_j}{\partial \xi_i} - \frac{\partial u_k \partial u_k}{\partial \xi_i \partial \xi_j}\right] = \frac{1}{2}[u_{i,j} + u_{j,i} - u_{k,i}u_{k,j}] \qquad i,j = \xi, \eta, \zeta} \tag{8.34}$$

The strain tensor $\varepsilon_{ij}$ is called the Green's strain tensor. The strain tensor $E_{ij}$ is called the Almansi's strain tensor. In other words, the Lagrangian description of finite strain in tensor form is the **Green's strain tensor** and the Eulerian description of finite strain in tensor form is the **Almansi's strain tensor.** For small strain the two tensor forms are the same and are called **Cauchy's strain tensors**. Note both strain tensors are symmetric:

$$\varepsilon_{ij} = \varepsilon_{ji} \qquad E_{ij} = E_{ji} \tag{8.35}$$

#### EXAMPLE 8.6

Lagrangian strain with moderately large rotation for plate bending is given as

$$\varepsilon_{ij} = \frac{1}{2}(u_{i,j} + u_{j,i}) + \frac{1}{2}w_{,i}w_{,j} \qquad i,j = x, y \tag{8.36}$$

Assume the displacement field is $u_i = u_{oi} - zw_{,i}$, where $u_{oi}$ and $w$ are the inplane and bending displacements, respectively. All assumptions of classical plate theory except small strain approximation are valid. $p_i$ and $p_z$ are the inplane and bending distributed loads per unit area. Obtain the stress formulas and the statement of boundary value problems for displacements $u_{oi}$ and $w$.

**PLAN**

We will follow the logic described in Chapter 5 to develop the stress formulas and use potential energy and variational equations of Section 8.4.2 to develop the boundary value problem.

**SOLUTION**

1. *Strains*: Substituting the displacement field into Equation (8.36) we obtain

$$\varepsilon_{ij} = (u_{oi,j} + u_{oj,i})/2 + (w_{,i}w_{,j})/2 - zw_{,ij} \tag{E1}$$

We note that the first two terms are not functions of $z$. For ease of manipulating algebraic equations we introduce the following function and write the strain as shown below.

$$f_{ij} = (u_{oi,j} + u_{oj,i})/2 + (w_{,i}w_{,j})/2 \qquad \varepsilon_{ij} = f_{ij} - zw_{,ij} \tag{E2}$$

2. *Stresses*: Substituting the strain into Equation (8.12c) we obtain stresses as

$$\sigma_{ij} = \frac{E}{(1-\nu^2)}[(1-\nu)\varepsilon_{ij} + \nu\delta_{ij}\varepsilon_{kk}] = \frac{E}{(1-\nu^2)}[(1-\nu)f_{ij} + \nu\delta_{ij}f_{kk}] - \frac{Ez}{(1-\nu^2)}[(1-\nu)w_{,ij} + \nu\delta_{ij}w_{,kk}] \tag{E3}$$

3. *Stress resultants*: The stress resultants are given by Equations (5.7), (5.8) and (5.9). They can be written in indicial notation as shown below.

$$\boldsymbol{n}_{ij} = \int_h \sigma_{ij}dz \qquad \boldsymbol{m}_{ij} = \int_h z\sigma_{ij}dz \qquad \boldsymbol{q}_i = \int_h \tau_{iz}dz \tag{E4}$$

We make use of the fact that material is homogenous and $z$ is measured from midplane. Thus the following identities can be used in the evaluation of integrals over the thickness.

$$\int_h dz = h \qquad \int_h zdz = 0 \qquad \int_h z^2dz = h^3/12 \tag{E5}$$

Substituting Equation (E3) into Equation (E4) and using the above equation we obtain the stress resultants as shown below.

$$\boldsymbol{n}_{ij} = \boldsymbol{n}_{ji} = \int_h \frac{E}{(1-\nu^2)}[(1-\nu)f_{ij} + \nu\delta_{ij}f_{kk}]dz = \frac{Eh}{(1-\nu^2)}[(1-\nu)f_{ij} + \nu\delta_{ij}f_{kk}] \tag{E6}$$

$$\boldsymbol{m}_{ij} = -\int_h \frac{Ez^2}{(1-\nu^2)}[(1-\nu)w_{,ij} + \nu\delta_{ij}w_{,kk}]dz = -\left[\frac{Eh^3}{12(1-\nu^2)}\right][(1-\nu)w_{,ij} + \nu\delta_{ij}w_{,kk}] = -D[(1-\nu)w_{,ij} + \nu\delta_{ij}w_{,kk}] \tag{E7}$$

4. *Stress formula*: Substituting Equations (E6) and (E7) into Equation (E4) we obtain our result for stresses.

**ANS.** $\sigma_{ij} = \dfrac{\boldsymbol{n}_{ij}}{h} + \dfrac{12\boldsymbol{m}_{ij}z}{h^3}$

5. *Potential Energy*: The work and strain energy can be found and potential energy written as shown below.

$$W = \int_A (p_i u_{oi} + p_z w)dA \tag{E8}$$

$$U = \frac{1}{2}\int_A\left[\int_h \sigma_{ij}\varepsilon_{ij}dz\right]dA = \frac{E}{2(1-\nu^2)}\int_A\left[\int_h [(1-\nu)\varepsilon_{ij}\varepsilon_{ij} + \nu\varepsilon_{ii}\varepsilon_{jj}]dz\right]dA$$

$$U = \frac{E}{2(1-\nu^2)}\int_A\left[\int_h [(1-\nu)(f_{ij} - zw_{,ij})(f_{ij} - zw_{,ij}) + \nu(f_{ii} - zw_{,ii})(f_{jj} - zw_{,jj})]dz\right]dA$$

$$U = \frac{E}{2(1-\nu^2)}\int_A\left[[(1-\nu)f_{ij}f_{ij} + \nu f_{ii}f_{jj}]\int_h dz - [2(1-\nu)f_{ij}w_{,ij} + 2\nu f_{ii}w_{,jj}]\int_h zdz + [(1-\nu)w_{,ij}w_{,ij} + \nu w_{,ii}w_{,jj}]\int_h z^2dz\right]dA$$

$$U = \frac{Eh}{2(1-\nu^2)}\int_A [(1-\nu)f_{ij}f_{ij} + \nu f_{ii}f_{jj}]dA + \frac{D}{2}\int_A [(1-\nu)w_{,ij}w_{,ij} + \nu w_{,ii}w_{,jj}]dA \tag{E9}$$

$$V = U - W = \frac{Eh}{2(1-\nu^2)}\int_A [(1-\nu)f_{ij}f_{ij} + \nu f_{ii}f_{jj}]dA + \frac{D}{2}\int_A [(1-\nu)w_{,ij}w_{,ij} + \nu w_{,ii}w_{,jj}]dA - \int_A (p_i u_{oi} + p_z w)dA \tag{E10}$$

6. *Functional and its derivatives*: The highest order of derivative is two in potential energy. We write the potential energy as a functional described in Section 8.4.2.

$$V = \int_A H(u_{oi,j}, u_{oi}, w_{,ij}, w_{,i}, w, x) \quad dA \text{ where} \tag{E11}$$

$$H(u_{oi,j}, u_{oi}, w_{,ij}, w_{,i}, w, x) = \frac{Eh}{2(1-\nu^2)}[(1-\nu)f_{ij}f_{ij} + \nu f_{ii}f_{jj}] + \frac{D}{2}[(1-\nu)w_{,ij}w_{,ij} + \nu w_{,ii}w_{,jj}] - (p_i u_{oi} + p_z w) \tag{E12}$$

The derivatives of $H$ with respect to $u_{ol}$ and its derivatives are as shown below.

$$\frac{\partial H}{\partial u_{ol,m}} = \frac{Eh}{2(1-\nu^2)}\left[2(1-\nu)f_{ij}\frac{\partial f_{ij}}{\partial u_{ol,m}} + 2\nu f_{ii}\frac{\partial f_{jj}}{\partial u_{ol,m}}\right] \qquad \frac{\partial H}{\partial u_{ol}} = p_i\frac{\partial u_{oi}}{\partial u_{ol}} = p_i\delta_{il} = p_l \tag{E13}$$

$$\frac{\partial f_{ij}}{\partial u_{ol,m}} = \frac{1}{2}\left(\frac{\partial u_{oi,j}}{\partial u_{ol,m}} + \frac{\partial u_{oj,i}}{\partial u_{ol,m}}\right) = \frac{1}{2}(\delta_{il}\delta_{jm} + \delta_{jl}\delta_{im}) \qquad \frac{\partial f_{jj}}{\partial u_{ol,m}} = \frac{1}{2}(\delta_{jl}\delta_{jm} + \delta_{jl}\delta_{jm}) = \delta_{lm} \tag{E14}$$

$$\frac{\partial H}{\partial u_{ol,m}} = \frac{Eh}{(1-\nu^2)}\left[\frac{(1-\nu)}{2}f_{ij}(\delta_{il}\delta_{jm} + \delta_{jl}\delta_{im}) + \nu f_{ii}\delta_{lm}\right] = \frac{Eh}{(1-\nu^2)}\left[\frac{(1-\nu)}{2}(f_{lm} + f_{ml}) + \nu f_{ii}\delta_{lm}\right]$$

$$\frac{\partial H}{\partial u_{ol,m}} = \frac{Eh}{(1-\nu^2)}[(1-\nu)f_{lm} + \nu f_{ii}\delta_{lm}] = \boldsymbol{n}_{lm} = \frac{Eh}{2(1-\nu^2)}[(1-\nu)(u_{ol,m} + u_{om,l} + w_{,l}w_{,m}) + \nu(u_{oi,i} + w_{,i}w_{,i})\delta_{lm}] \tag{E15}$$

The derivatives of $H$ with respect to $w$ and its derivatives are as shown below.

$$\frac{\partial H}{\partial w_{,lm}} = \frac{D}{2}\left[2(1-\nu)w_{,ij}\frac{\partial w_{,ij}}{\partial w_{,lm}} + 2\nu w_{,ii}\frac{\partial w_{,jj}}{\partial w_{,lm}}\right] = D[(1-\nu)w_{,ij}(\delta_{il}\delta_{jm}) + \nu w_{,ii}\delta_{lm}]$$

$$\frac{\partial H}{\partial w_{,lm}} = D[(1-\nu)w_{,lm} + \nu w_{,ii}\delta_{lm}] = -\boldsymbol{m}_{lm} \tag{E16}$$

$$\frac{\partial H}{\partial w_{,l}} = \frac{Eh}{2(1-\nu^2)}\left[2(1-\nu)f_{ij}\frac{\partial f_{ij}}{\partial w_{,l}} + 2\nu f_{ii}\frac{\partial f_{jj}}{\partial w_{,l}}\right] = \frac{Eh}{(1-\nu^2)}\left[(1-\nu)f_{ij}\left(\frac{\delta_{il}w_{,j}+\delta_{jl}w_{,i}}{2}\right) + \nu f_{ii}\frac{\delta_{jl}w_{,j}+\delta_{jl}w_{,j}}{2}\right]$$

$$\frac{\partial H}{\partial w_{,l}} = \frac{Eh}{(1-\nu^2)}\left[(1-\nu)\left(\frac{f_{jl}w_{,j}+f_{il}w_{,i}}{2}\right) + \nu f_{ii}w_{,l}\right] = \frac{Eh}{(1-\nu^2)}[(1-\nu)f_{jl}w_{,j} + \nu f_{ii}w_{,l}]$$

$$\frac{\partial H}{\partial w_{,l}} = \frac{Eh}{(1-\nu^2)}[(1-\nu)f_{jl} + \nu f_{ii}\delta_{jl}]w_{,j} = \boldsymbol{n}_{jl}w_{,j} \tag{E17}$$

$$\frac{\partial H}{\partial w} = -p_z \tag{E18}$$

7. *Boundary value problem*: From Equation (8.15a) and changing the free indicies from $l$ to $i$ and $m$ to $j$ we have.

$$\frac{\partial H}{\partial u_{oi}} - \left(\frac{\partial H}{\partial u_{oi,j}}\right)_{,j} = p_i - \boldsymbol{n}_{ij,j} = 0 \quad \text{or} \quad \boldsymbol{n}_{ij,j} = p_i \tag{E19}$$

**Differential Equation 1:** $\frac{Eh}{2(1-\nu^2)}[(1-\nu)(u_{oi,j} + u_{oj,i} + w_{,i}w_{,j}) + \nu(u_{ok,k} + w_{,k}w_{,k})\delta_{ij}]_{,j} = p_i \quad x_k \text{ in } A$ (E20)

From Equation (8.16b) and changing the free indicies from $l$ to $i$ and $m$ to $j$ we have

$$\boldsymbol{m}_{nn} = \left(\frac{\partial H}{\partial w_{,ij}}\right)n_in_j = -D[(1-\nu)w_{,ij} + \nu w_{,kk}\delta_{ij}]n_in_j \tag{E21}$$

$$\boldsymbol{m}_{nt} = \left(\frac{\partial H}{\partial w_{,ij}}\right)e_{mi}n_mn_j = -D[(1-\nu)w_{,ij} + \nu w_{,kk}\delta_{ij}]e_{mi}n_mn_j \tag{E22}$$

$$-\boldsymbol{q}_i = \left(\frac{\partial H}{\partial w_{,i}}\right) - \left(\frac{\partial H}{\partial w_{,ij}}\right)_{,j} = \boldsymbol{n}_{ji}w_{,j} - D[(1-\nu)w_{,ij} + \nu w_{,kk}\delta_{ij}]_{,j} = \boldsymbol{n}_{ji}w_{,j} - D[(1-\nu)w_{,ijj} + \nu w_{,ijj}]$$

$$\boldsymbol{q}_i = Dw_{,ijj} - \boldsymbol{n}_{ji}w_{,j} \tag{E23}$$

From Equation (8.17a) we have the second differential equation as shown below.

$$\boldsymbol{q}_{i,i} + \frac{\partial H}{\partial w} = [Dw_{,ijj} - \boldsymbol{n}_{ji}w_{,j}]_{,i} - p_z = Dw_{,iijj} - (\boldsymbol{n}_{ji}w_{,j})_{,i} - p_z = Dw_{,iijj} - (\boldsymbol{n}_{ji,i}w_{,j} + \boldsymbol{n}_{ji}w_{,ji}) - p_z = 0 \tag{E24}$$

Substituting Equation (E19) into the above equation we obtain the second differential equation

**Differential Equation 2:** $Dw_{,iijj} = p_z + p_jw_{,j} + \boldsymbol{n}_{ji}w_{,ji} \quad x_k \text{ in } A$ (E25)

From Equation (8.15b) we have the first set of boundary conditions

$$\frac{\partial H}{\partial u_{oj,i}}n_i = \boldsymbol{n}_{ji}n_i = 0 \quad \text{or} \quad \delta u_{oj} = 0 \quad x_k \text{ on } \Gamma \tag{E26}$$

**Boundary Condition 1:** $(1-\nu)(u_{oj,i} + u_{oi,j} + w_{,j}w_{,i}) + \nu(u_{om,m} + w_{,m}w_{,m})\delta_{ji}]n_i \quad \text{or} \quad \delta u_{oj} = 0 \quad x_k \text{ on } \Gamma$ (E27)

From Equations (8.17b) and (8.17c) we obtain the boundary conditions as

**Boundary Conditions 2:**

$$\left[\boldsymbol{m}_{nn} = 0 \quad \text{or} \quad \delta\left(\frac{\partial w}{\partial n}\right) = 0\right] \quad \text{and} \quad \left[\frac{\partial \boldsymbol{m}_{nt}}{\partial s} + \boldsymbol{q}_in_i = 0 \quad \text{or} \quad \delta w = 0\right] \quad x_k \text{ on } \Gamma \tag{E28}$$

**Corner Condition:** $(\Delta \boldsymbol{m}_{nt})_C = 0 \quad \text{or} \quad \delta w_C \quad x_k \text{ on corner } \Gamma_C$ (E29)

The above differential equations and boundary conditions describe the boundary value problem.

## COMMENTS

1. Derivation of the boundary value problem without the indicial notation and variational calculus would be algebraically very tedious.
2. The inplane and bending problem cannot be decoupled for large strain. We see from Equation (E23) the inplane forces have a component in the plate transverse direction in the same way we saw for beams in Figure 7.9.
3. Clearly the differential equations are nonlinear. However, consider the case when $p_i = 0$ and $\boldsymbol{n}_{ji}$ are constants. Then the differential equation 1 in the form of Equation (E19) is satisfied and differential equation 2 reduces to a linear differential equation, which written in expanded form is as given below.

$$D\nabla^4 w = p_z + \boldsymbol{n}_{xx}(\partial^2 w/\partial x^2) + 2\boldsymbol{n}_{xy}(\partial^2 w/\partial x\partial y) + \boldsymbol{n}_{yy}(\partial^2 w/\partial y^2) \tag{E30}$$

The above equation is plate buckling equation when $\boldsymbol{n}_{ji}$ are compressive.

## 8.6 CLOSURE

In this chapter we studied the language of indicial notation and the rules of manipulating induces. An important check in the derivation is that every term in an equation must have the same free index. We re-derived equations of plate, elasticity, and variational calculus. We saw that lengthy derivations and tedious algebra are dramatically reduced by the use of indicial notation. We modified the classical plate theory to obtain stress resultants, stresses, and boundary value problem for plate bending with large strain.

This chapter brings this book to a close.

## 8.7 SYNOPSIS OF EQUATIONS

| | |
|---|---|
| | **Elasticity** |
| Tensor strain | $\varepsilon_{ij} = (u_{i,j} + u_{j,i})/2$ |
| Isotropic constitute equations | $\varepsilon_{ij} = [(1+\nu)\sigma_{ij} - \nu\delta_{ij}\sigma_{kk}]/E$ ; $\sigma_{ij} = 2G\varepsilon_{ij} + \lambda\delta_{ij}\varepsilon_{kk}$ ; $\lambda = (E\nu)/[(1-2\nu)(1+\nu)]$ ; $G = E/[2(1+\nu)]$ |
| Plane stress | $\sigma_{ij} = E[(1-\nu)\varepsilon_{ij} + \nu\delta_{ij}\varepsilon_{kk}]/(1-\nu^2)$ |
| Equilibrium | $\sigma_{ij,j} + F_i = 0$ |
| Traction (stress vector) | $S_i = \sigma_{ij}n_j$ |
| Strain energy density | $U_0 = \sigma_{ij}\varepsilon_{ij}/2$ |
| | **Classical plate** |
| Displacements | $u_i \approx -zw_{,i}$ $\quad w \approx w(x_1, x_2)$ |
| Strains | $\varepsilon_{ij} = -zw_{,ij}$ |
| Stresses | $\sigma_{ij} = -Ez[(1-\nu)w_{,ij} + \nu\delta_{ij}w_{,kk}]/(1-\nu^2)$ |
| Stress resultants | $\boldsymbol{m}_{ij} = \int_h z\sigma_{ij}dz = -D[(1-\nu)w_{,ij} + \nu\delta_{ij}w_{,kk}]$ ; $\boldsymbol{q}_i = \int_h \tau_{iz}dz = \boldsymbol{m}_{ji,j} = -Dw_{,kki} = -D(\nabla^2 w)_{,i}$ |
| Equilibrium | $\boldsymbol{q}_{i,i} = \boldsymbol{m}_{ji,j} = -p_z(x_1, x_2)$ |
| Differential equation | $Dw_{,kkii} = D(\nabla^4 w) = p_z(x_1, x_2)$ |
| Boundary variables | $\boldsymbol{m}_{nn} = -\boldsymbol{m}_{ij}n_in_j$ $\quad \boldsymbol{m}_{nt} = -\boldsymbol{m}_{ij}e_{mi}n_mn_j$ |
| Boundary conditions | $\left[\boldsymbol{m}_{nn} = 0 \quad \text{or} \quad \delta\left(\frac{\partial w}{\partial n}\right) = 0\right]$ and $\left[\frac{\partial \boldsymbol{m}_{nt}}{\partial s} + \boldsymbol{q}_in_i = 0 \quad \text{or} \quad \delta w = 0\right]$<br>$(\Delta \boldsymbol{m}_{nt})_C = 0$ or $\delta w_C$ on corners. |
| Strain energy | $U = (D/2)\iint_A [(1-\nu)w_{,ij}w_{,ij} + \nu w_{,ii}w_{,kk}]dx_1dx_2$ |
| | **Variational calculus** |
| First order derivative | $\frac{\partial H}{\partial u} - \left(\frac{\partial H}{\partial u_{,i}}\right)_{,i} = 0 \quad x_k$ in $A$ ; $\frac{\partial H}{\partial u_{,i}}n_i = 0$ or $\delta u_j = 0 \quad x_k$ on $\Gamma$ |
| Second order derivatives | $\boldsymbol{m}_{nn} = \left(\frac{\partial H}{\partial u_{,ij}}\right)n_in_j$ ; $\boldsymbol{m}_{nt} = \left(\frac{\partial H}{\partial u_{,ij}}\right)e_{mi}n_mn_j$ ; $-\boldsymbol{q}_i = \left(\frac{\partial H}{\partial u_{,i}}\right) - \left(\frac{\partial H}{\partial u_{,ij}}\right)_{,j}$ ; $\boldsymbol{q}_{i,i} + \frac{\partial H}{\partial u} = 0$ ;<br>$\left[\boldsymbol{m}_{nn} = 0 \quad \text{or} \quad \delta\left(\frac{\partial u}{\partial n}\right) = 0\right]$ and $\frac{\partial \boldsymbol{m}_{nt}}{\partial s} + \boldsymbol{q}_in_i = 0$ or $\delta u = 0$<br>$(\Delta \boldsymbol{m}_{nt})_C = 0$ or $\delta u_C$ on corners. |
| | **Large strain** |
| Lagrangian strain | $\varepsilon_{ij} = \frac{1}{2}\left[\frac{\partial u_i}{\partial x_j} + \frac{\partial u_j}{\partial x_i} + \frac{\partial u_k \partial u_k}{\partial x_i \partial x_j}\right] = \frac{1}{2}[u_{i,j} + u_{j,i} + u_{k,i}u_{k,j}] \quad i,j = x, y, z$ |
| Eulerian strain | $E_{ij} = \frac{1}{2}\left[\frac{\partial u_i}{\partial \xi_j} + \frac{\partial u_j}{\partial \xi_i} - \frac{\partial u_k \partial u_k}{\partial \xi_i \partial \xi_j}\right] = \frac{1}{2}[u_{i,j} + u_{j,i} - u_{k,i}u_{k,j}] \quad i,j = \xi, \eta, \zeta$ |

# PROBLEMS

## Section 8.1

**8.1** Starting with Equations (8.34) and (8.35) write the strains $\varepsilon_{xx}$, $E_{\xi\xi}$, $\varepsilon_{xy}$, and $E_{\xi\eta}$ in three dimension using the expanded notation ($x$, $y$, $z$, $u$, v, $w$, $\xi$, $\eta$, $\zeta$).

**8.2** Starting with Equations (8.34) and (8.35) write the strains $\varepsilon_{yy}$, $E_{\eta\eta}$, $\varepsilon_{yz}$, and $E_{\eta\zeta}$ in three dimension using the expanded notation ($x$, $y$, $z$, $u$, v, $w$, $\xi$, $\eta$, $\zeta$).

**8.3** Starting with Equations (8.34) and (8.35) write the strains $\varepsilon_{zz}$, $E_{\zeta\zeta}$, $\varepsilon_{zx}$, and $E_{\zeta\xi}$ in three dimension using the expanded notation ($x$, $y$, $z$, $u$, v, $w$, $\xi$, $\eta$, $\zeta$).

**8.4** In two dimensions, the stresses are given by Equation (6.7b). Write a single stress equation using indicial notation.

**8.5** Using the results of Problem 8.4 write in indicial notation the strain energy density for two dimensions in terms of derivatives of displacements.

## Section 8.2

**8.6** Starting with Equation (8.10b), assuming plane stress, obtain the stresses in terms of strain.

**8.7** Starting with the displacement equation of *Mindlin-Reissner plate theory* given below, obtain equations in indicial notation for strain, stresses, stress resultants, equilibrium equations, and differential equations [see Example 5.3].

$$u_i \approx -z\psi_i(x_1, x_2) \qquad w \approx w(x_1, x_2)$$

**8.8** Show the strain energy for *Mindlin-Reissner plate theory* in indicial notation is given by the equation below.[see Equation (7.91)].

$$U_P = \frac{1}{2}\iint_A \left\{ D\begin{bmatrix} \frac{(1-\nu)}{4}(\psi_{i,j}+\psi_{j,i})(\psi_{i,j}+\psi_{j,i}) + \nu\psi_{j,j}\psi_{k,k} \\ + Gh(w_{,i}-\psi_i) \end{bmatrix} \right\} dx_1 dx_2$$

## Section 8.3

**8.9** The solution for displacements due to a point force $F_k$ in an infinite elastic body in plane strain is given by the equation below. Obtain the stresses $\sigma_{ij}$.

$$u_i = \frac{1}{8G\pi(1-\nu)}\left[-(3-4\nu)\delta_{ik}ln(r) + \frac{r_i r_k}{r^2}\right]F_k$$

**8.10** The solution for displacements due to a point moment $M_z$ in an infinite elastic body in plane strain is given by the equation below. Obtain the stresses $\sigma_{ij}$.

$$u_i = \frac{1}{2\pi}\left[\frac{e_{ik}r_k}{r^2}\right]M_z$$

**8.11** The solution for displacements due to a point force in an semi-infinite elastic body in plane strain is given by the equation below. Obtain the strain invariant $\varepsilon_{ii}$.

$$u_i = \frac{1}{2\pi G}\left[-2(1-\nu)\delta_{ik}ln(r) + (1-2\nu)e_{ik}\theta + \frac{r_i r_k}{r^2}\right]F_k$$

## Section 8.4

**8.12** Starting from potential energy, obtain the differential equation governing displacements in an elastic body in plane stress subjected to body forces $F_i$.

**8.13** Starting from potential energy, obtain the differential equation governing displacements in an elastic body in plane strain subjected to body forces $F_i$.

**8.14** A *Mindlin-Reissner* plate is loaded with just the transverse load $p_z(x_1, x_2)$. Obtain the differential equations from the potential energy using results of Problem 8.8.

# A Appendix A Bibliography


1. Barber, J. R. (1992) Elasticity, Kluwer Academic Publishers.
2. Boresi, A. P., and Chong, K. P. (1987) *Elasticity in Engineering Mechanics,* Elsevier.
3. Fung, Y. C. (1965) Foundations of Solid Mechanics, Prentice-Hall.
4. Kreyszig, E. (1979) *Advanced Engineering Mathematics,* John Wiley & Sons.
5. Lanczos, C. (1986) *The Variational Principles of Mechanics,* Dover.
6. Mendelson, A. (1968) *Plasticity: Theory and Application,* Macmillan.
7. Muskhelishvili, N. I. (1963) *Some Basic Problems of the Mathematical Theory of Elasticity,* P. Noordhoff.
8. Reddy, J. N. (1993) An Introduction to the Finite Element Method, 2nd ed., McGraw-Hill.
9. Sneddon, I. N. (1980) *Fourier Transform,* McGraw-Hill.
10. Synge, J. L., and Schild, A. (1978) *Tensor Calculus,* Dover.
11. Szilard, R. (1974) *Theory and Analysis of Plates: Classical and Numerical Methods,* Prentice-Hall.
12. Timoshenko, S. and Woinowsky-Krieger, S. (1959) *Theory of Plates and Shells*, McGraw-Hill.
13. Vable, M (2013) Intermediate Mechanics of Materials, 2nd Ed., Expanding Educational Horizons, ISBN: 978-0-9912446-0-7
14. Wahl A.M. and Lobo G. (1930) *Transactions of ASME*, v52
15. Zienkiewicz, O. C., and Taylor R. L. (1989) *The Finite Element Method,* 4th ed., McGraw-Hill.

# B Appendix B
# Answers To Selected Problems

## CHAPTER 1

**1.9** $\sigma_{nn} = 57.99$ MPa (C) ; $\tau_{nt} = 91.56$ MPa; $\sigma_1 = 278.4$ MPa (T);
$\sigma_2 = 49.9$ MPa (C) ; $\sigma_3 = 116.5$ MPa (C); $\tau_{oct} = 172.6$ MPa;
$\tau_{max} = 197.5$ MPa ; $\theta_x = 48.9^o$ $\theta_y = 108.8^o$ $\theta_z = 47.1^o$ ;
$J_1 = 0$ ; $J_2 = -44706.3$ MPa$^2$ ; $J_3 = 3237002.7$ MPa$^3$

**1.13** $\delta_{AP} = 0.0647$ mm extension ; $\delta_{BP} = 0.2165$ mm extension

**1.15** $u(L) - u(0) = 2KL^2$

**1.19** $\varepsilon_{xx} = 8500\mu$ ; $\varepsilon_{yy} = -8500\mu$; $\gamma_{xy} = -9000\mu$

**1.24** $\varepsilon_{nn} = 315.84\mu$ ; $\varepsilon_1 = 759.6\mu$ ; $\varepsilon_2 = 296.1\mu$ ; $\varepsilon_3 = -15.7\mu$ ;
$\gamma_{max} = 775.3\mu$ ; $\theta_x = 15.4°$ ; $\theta_y = 76.7°$ ; $\theta_z = 82.3°$

**1.25 (a)** $\varepsilon_{xx} = 260\mu$ ; $\varepsilon_{yy} = 590\mu$ ; $\gamma_{xy} = -1650\mu$ ; $\sigma_{zz} = 0$ ;
$\varepsilon_{zz} = -400\mu$ **(b)** $\varepsilon_{xx} = 132\mu$ ; $\varepsilon_{yy} = 462\mu$ ; $\gamma_{xy} = -1650\mu$ ;
$\sigma_{zz} = 80\text{MPa(T)}$ ; $\varepsilon_{zz} = 0$

**1.31** $\sigma_1 = 80.8$ MPa(T) ; $\sigma_2 = 55.8$ MPa(T) ; $\sigma_3 = 17.4$ MPa(T) ;
$\tau_{max} = 31.7$ MPa ; $\theta_x = 22.9^o$ , $\theta_y = 69.5^o$ , $\theta_z = 80.3^o$

**1.35 (a)** $\sigma_{zz} = 0$ ; $\varepsilon_{xx} = -283\mu$ ; $\varepsilon_{yy} = -0.353\mu$ ; $\gamma_{xy} = -1300\mu$ ;
$\varepsilon_{zz} = 730\mu$ **(b)** $\varepsilon_{zz} = 0$ ; $\varepsilon_{xx} = -4.3\mu$ ; $\varepsilon_{yy} = -134\mu$ ;
$\sigma_{zz} = 21.9 ksi(C)$ ; $\gamma_{xy} = -1300\mu$

**1.39** $k = 1.29$

## CHAPTER 2

**2.1** $V_y = 0$ ; $V_z = Kta[\alpha - sin\alpha\, cos\alpha]$ ;
$T = 2a^2tK[sin\alpha - \alpha\, cos\alpha]$ CCW; $e_y = 2a\dfrac{[sin\alpha - \alpha\, cos\alpha]}{\alpha - sin\alpha\, cos\alpha}$ ; $e_z = 0$

**2.4** $\tau_{max} = \dfrac{0.1188 T_{ext}}{r^3}$; $\phi_B = 0.0523\left(\dfrac{T_{ext}L}{Gr^4}\right)$ $CW$

**2.15** $(\sigma_{xx})_1 = -2E\alpha T_0$; $(\sigma_{xx})_2 = -2E\alpha T_0$

**2.17** $T_{ext} = 599.6$ $in-kips$; $\phi_C = 0.05$ $rad$ $ccw$

**2.29** $a = 5.804$ in ; $\sigma_{xx} = \begin{cases} 0.0254\boldsymbol{M}_z(-y)^{0.4} \text{ ksi} & y < 0 \\ -0.0254\boldsymbol{M}_z(y)^{0.4} \text{ ksi} & y > 0 \end{cases}$

**2.33** $F = (f_{max}L)/3$ ; $u_L - u_0 = (f_{max}L^2)/(4EA)$

**2.36** $\dfrac{dv}{dx}(L) = \dfrac{wL^3}{80EI}$ ; $M_A = \dfrac{11wL^2}{120}$ $CW$ ; $R_A = \dfrac{61wL}{120}$ $up$

**2.44** $v(x) = (w/24)[-\langle x - L\rangle^4 + 4wLx^3 - 12wL^2x^2 + 4L^3x]$ ;
$v(L) = -[wL^4/(6EI)]$

**2.48** $v(3L) = -\left(\dfrac{2wL^4}{9EI}\right)$

## CHAPTER 3

**3.5** $G = \dfrac{1}{6EI}[\langle x - \xi\rangle^3 - x^3 + 3\xi x^2]$ ; $H(x, \xi) = \dfrac{1}{2EI}[-\langle x - \xi\rangle^2 + x^2]$

**3.6** $v(x) = -\left(\dfrac{wx^2}{24EI}\right)[x^2 - 4Lx + 6L^2]$

**3.13** $v(x = 5) = -0687$ in. ; $\boldsymbol{M}_z(x = 5) = 65,248$ in.-lb ;
$v_{max} = -0687$ in. ; $(\sigma_{xx})_{max} = 1957.4$ psi (C) or (T)

**3.14** $v(x = 5) = -0.152$ in. ; $\boldsymbol{M}_z(x = 5) = -21,209$ in.-lb ;
$v_{max} = -0.397$ in. ; $(\sigma_{xx})_{max} = 5.19$ ksi (T) or (C)

**3.18** $v_{max} = 0.3810$ in. ; $V_y(0^-) = 32$ kips ;
$V_y(0^+) = -32$ kips ; $V_y(4) = 1.36$ kips

**3.19** $R_s = (-4\beta M)\dfrac{(1 - e^{-\beta L}cos\beta L - e^{-\beta L}sin\beta L)}{1 - e^{-2\beta L} + 2e^{-\beta L}sin\beta L}$ ;
$M_s = 4M\dfrac{1 - e^{-\beta L}cos\beta L}{(1 - e^{-2\beta L} + 2e^{-\beta L}sin\beta L)}$ ; $v(0) = 0.3889(10^{-3})$ ft ;
$\boldsymbol{M}_z(0) = -1.28$ ft-kips

## CHAPTER 4

**4.3** $K = 1.18$

**4.8** $tan\lambda L = \lambda EI/K$

**4.13** $L = 42 in$

**4.20** $P_{cr} = \pi^2a^4E/(24L^2)$

**4.22** $v_{max} = 0.0458 in$ ; $\sigma_{max} = 2.68 ksi(C)$

**4.26** $L_{max} = 2.09 m$

**4.28** $P_{cr} = 986$ kN

## CHAPTER 5

**5.4** $z_o = \dfrac{3h}{8}\left[\dfrac{E_3 - E_1}{E_1 + 2E_2 + E3}\right]$

**5.10** $w(0,y)=0;\ \frac{\partial w}{\partial x}(0,y)=0;\ w(x,0)=0;\ \frac{\partial^2 w}{\partial y^2}(x,0)=0;$

$$\frac{\partial^2 w}{\partial x^2}(a,y)+\nu\frac{\partial^2 w}{\partial y^2}(a,y)=0;\quad \frac{\partial^2 w}{\partial y^2}(x,b)+\nu\frac{\partial^2 w}{\partial x^2}(x,b)=0;$$

$$\frac{\partial^3 w}{\partial x^3}(a,y)+(2-\nu)\frac{\partial^3 w}{\partial x\partial y^2}(a,y)=0;$$

$$\begin{bmatrix} -(2-x/a)\left\{\frac{\partial^3 w}{\partial y^3}(x,b)+(2-\nu)\frac{\partial^3 w}{\partial x^2\partial y}(x,b)\right\} \\ +\frac{6}{a}(1-\nu)\frac{\partial^2 w}{\partial x\partial y}(x,b) \end{bmatrix}=0$$

**5.14** $w(r)=\frac{p_o}{64D(1+\nu)}[(1+\nu)r^4-2(3+\nu)r^2R_o^{\,2}+(5+\nu)R_o^{\,4}];$

$$\boldsymbol{m}_{rr}=-\left[\frac{p_o(3+\nu)}{16}(R_o^{\,2}-r^2)\right];\quad \boldsymbol{m}_{\theta\theta}=\frac{p_o}{16}[(3+\nu)R_o^{\,2}-(1+3\nu)r^2]$$

**5.17** $w_{max}=1.3522\frac{p_oR_i^4}{D};\ (\boldsymbol{m}_{rr})_{max}=0.05927p_oR_i^2;$

$$(\boldsymbol{m}_{\theta\theta})_{max}=-1.382p_oR_i^2$$

# CHAPTER 6

**6.7** $\sigma_{rr}=\frac{1}{R_o^2-R_i^2}\left[-(p_oR_o^2-p_iR_i^2)-\frac{R_i^2R_o^2(p_i-p_o)}{r^2}\right];$

$$\sigma_{\theta\theta}=\frac{1}{R_o^2-R_i^2}\left[-(p_oR_o^2-p_iR_i^2)+\frac{R_i^2R_o^2(p_i-p_o)}{r^2}\right];$$

$$u_r=\left[-(1-2\nu)(p_oR_o^2-p_iR_i^2)r+\frac{R_i^2R_o^2(p_i-p_o)}{r}\right]/[2G(R_o^2-R_i^2)]$$

**6.16** $u(x,y)=\sigma(2xy-\nu x^2-y^2)/(2E)+\alpha_x+\beta y;$

$$v(x,y)=\sigma(2xy-\nu y^2-x^2)/(2E)+\alpha_y-\beta x$$

**6.17** $u(x,y)=\alpha_x+\beta y\qquad v(x,y)=\frac{\tau}{G}x+\alpha_y-\beta x$

**6.18** $\sigma_{xx}=2E(y-x);\ \sigma_{yy}=0;\ \tau_{xy}=2Ey;$

$$u(x,y)=2xy-x^2+2y^2+\nu y^2-ay;$$

$$v(x,y)=-\nu y^2+2\nu xy-x^2+ax$$

**6.20** $\phi_B=0.0701$ rads $CW$; $\tau_{max}=54.71$ MPa

**6.21** $\frac{d\phi}{dx}=\frac{15\sqrt{3}\boldsymbol{T}}{Gh^4}$; $\tau_{max}=(15\sqrt{3}/2)(\boldsymbol{T}/h^3)$

**6.22** $\tau_{max}=50$ MPa ; $\phi_D-\phi_A=0.019$ rads $CCW$

**6.23** $GK=937.5$ kips in$^2$ ; $\tau=16$ ksi

**6.24** $GK=G(8at^3/3)$ ; $\tau=3\boldsymbol{T}/(8at^2)$

**6.25** $GK=G(\pi+2)Rt^3/3$ ; $\tau=\frac{3\boldsymbol{T}}{(\pi+2)Rt^2}$

# CHAPTER 7

**7.14** $\frac{d}{dx}\left\{EA\left[\frac{du_o}{dx}+\frac{1}{2}\left(\frac{dv}{dx}\right)^2\right]-N_T\right\}=p_x;$

$$\frac{d^2}{dx^2}\left(EI_{zz}\frac{d^2v}{dx^2}\right)-EA\left[\frac{du_o}{dx}+\frac{1}{2}\left(\frac{dv}{dx}\right)^2-N_T\right]\frac{d^2v}{dx^2}=p_y+p_x\frac{dv}{dx}-\frac{d^2M_z}{dx^2}$$

**7.18** $0.2393(10^9)\frac{d^2}{dx^2}\left[\left(\frac{d^2v}{dx^2}\right)^{0.75}\right]=p_y;$

$$\left(\frac{d^2v}{dx^2}\right)^{0.75}=0\quad\text{or}\quad\delta\left(\frac{dv}{dx}\right)=0\quad\text{at } x=0 \text{ and } x=L;$$

$$\frac{d}{dx}\left[\left(\frac{d^2v}{dx^2}\right)^{0.75}\right]=0\quad\text{or}\quad\delta v=0\quad\text{at } x=0 \text{ and } x=L$$

**7.20** **(a)** $C_1=\left(\frac{wL^2}{12EI}\right);\ \Omega_1=-0.0278\frac{w^2L^5}{EI}$; (b) $C_1=\left(\frac{7wL^2}{96EI}\right);$

$$C_2=\left(\frac{wL}{96EI}\right);\ \Omega_2=-\left(\frac{0.0291wL^5}{EI}\right)$$

**7.23** $w\left(\frac{a}{2},\frac{a}{2}\right)=4.304(10^{-3})\frac{p_oa^4}{D};\ \Omega^*=-1.39(10^{-3})\frac{p_o^2a^6}{D}$

**7.28** $K]=\frac{2EI_{zz}}{L}\begin{bmatrix}6 & 3L & -6 & 3L\\ 3L & 2L^2 & -3L & L^2\\ -6 & -3L & 6 & -3L\\ 3L & L^2 & -3L & 2L^2\end{bmatrix}\qquad \{R\}=\frac{p_yL}{12}\begin{Bmatrix}6\\ L\\ 6\\ -L\end{Bmatrix}+\begin{Bmatrix}F_1\\ M_1\\ F_2\\ M_2\end{Bmatrix}$

# CHAPTER 8

**8.1** $\varepsilon_{xx}=\frac{\partial u}{\partial x}+\frac{1}{2}\left[\left(\frac{\partial u}{\partial x}\right)^2+\left(\frac{\partial v}{\partial x}\right)^2+\left(\frac{\partial w}{\partial x}\right)^2\right];$

$$\varepsilon_{xy}=\frac{1}{2}\left[\frac{\partial u}{\partial y}+\frac{\partial v}{\partial x}\right]+\frac{1}{2}\left[\frac{\partial u\partial u}{\partial x\partial y}+\frac{\partial v\partial v}{\partial x\partial y}+\frac{\partial w\partial w}{\partial x\,\partial y}\right];$$

$$E_{\xi\xi}=\frac{\partial u}{\partial \xi}-\frac{1}{2}\left[\left(\frac{\partial u}{\partial \xi}\right)^2+\left(\frac{\partial v}{\partial \xi}\right)^2+\left(\frac{\partial w}{\partial \xi}\right)^2\right];$$

$$E_{\xi\eta}=\frac{1}{2}\left[\frac{\partial u}{\partial \eta}+\frac{\partial v}{\partial x}\right]-\frac{1}{2}\left[\frac{\partial u\partial u}{\partial\xi\partial\eta}+\frac{\partial v\partial v}{\partial\xi\partial\eta}+\frac{\partial w\partial w}{\partial\xi\,\partial\eta}\right]$$

**8.4** $\sigma_{ij}=2G\left[\varepsilon_{ij}+\frac{(3-\kappa)}{2(\kappa-1)}\delta_{ij}\varepsilon_{kk}\right]$

**8.6** $\sigma_{ij}=\frac{E}{(1-\nu^2)}[(1-\nu)\varepsilon_{ij}+\delta_{ij}\nu\varepsilon_{kk}]$

**8.9** $\sigma_{ij}=\frac{1}{4\pi(1-\nu)}\left[(1-2\nu)\frac{\delta_{ij}r_k-\delta_{ik}r_j-\delta_{jk}r_i}{r^2}-\frac{2r_ir_kr_j}{r^4}\right]F_k$

# Index

Airy stress function 148
Almansi's strain tensor 222
Axisymmetric plates 134
Axisymmetric problems 146

Beams on elastic foundations 80
Bilinear functional 169
Bilinear material model 23
Boundary value problem 51
Buckling 97
Buckling modes 98

Cauchy's strain tensors 222
Characteristic equation 7, 9, 17
Circular plates 132
Classical thin plate theory 111
Column 97
Compatibility equations 143
Compressive stress 2
Couple stress 2
Critical buckling load 98

Deformation 13
Deviatoric stress matrix 11
Dislocation 217
Displacement 13
Doublet singularity 71
Dummy index 213

Eccentric loading 104
Elastic–perfectly plastic material model 23
Engineering strains 14
Euler load 98
Eulerian strain 13, 220, 222

Failure envelopes 28
Failure Theories 26
Field point 71
Finite beams 86
Finite element method 196
First variation 168
Force singularity 71
Free index 213
Free surfaces 5
Functional 169
Fundamental solutions 71

Geometric nonlinearity 186
Green's strain tensor 222

Inclusion singularity 217
Indicial notation 213
Influence function 71
inhomogeneity singularity 217

Kinematic condition 168
Kinematically admissible functions 168

Lagrange polynomials 197, 201, 203
Lagrangian strain 13, 220, 221
Laminated plate 119
Linear functional 169
Linear Strain-Hardening Material Mode 23
Local buckling 97

Material nonlinearity 188
Maximum normal stress theory 28
Maximum octahedral shear stress theory 27
Maximum shear stress 9
Maximum shear stress theory 27
Membrane analogy 158
Mindlin-Reissner plate theory 122
Minimum potential energy 173
Modified Mohr's theory 28
Modulus of foundation 80
Moment singularity 72

Nadai-Levy solution 128
Navier's solution 124
Nonlinear material models 23
Nonlinear strains 220
Normal stress 2

Octahedral plane 10
Octahedral stresses 10
Orthotropic plate 120

Plane stress 5
Plane stress and plane strain 144
Power law material model 24
Prandtl's method 154
Principal angles 7
Principal axis 7
Principal coordinate directions 17
Principal direction 7
Principal planes 7
Principal strains. 17
Principal stress 7

Rayleigh-Ritz method 190
Rectangular plates 124, 128
Rotating disks 147

Saint-Venant's method 154
Secant formula 105
Second variation 168
Shear stress 2
Singularity 71
Slenderness ratio 99
Source point 71
Source singularity 71
Stability 97

Stationary value 168
Strain energy 170
Stress element 4
Stress invariants 9
Stress on a surface 1
Stress resultants 35
Stress transformation 5
Stress vector 6
Structural buckling 97
Symmetric bilinear functional 169

Tensile stress 2
Torsion of non-circular shafts 153
Torsion of thin-walled open section 160
Traction 6
Tresca's yield criterion 27

Virtual displacement 167
Virtual work 173
Von-Mises stress 28

Work 169

www.ingramcontent.com/pod-product-compliance
Lightning Source LLC
Chambersburg PA
CBHW061407210326
41598CB00035B/6125
* 9 7 8 0 9 9 1 2 4 4 6 4 5 *